Recent Advances in Polymer Nanocomposites:
Synthesis and Characterisation

# Recent Advances in Polymer Nanocomposites: Synthesis and Characterisation

*By*

S. Thomas
G.E. Zaikov
S.V. Valsaraj
A.P. Meera

CRC Press
Taylor & Francis Group
Boca Raton  London  New York

CRC Press is an imprint of the
Taylor & Francis Group, an **informa** business

First published 2010 by VSP

Published 2019 by CRC Press
Taylor & Francis Group
6000 Broken Sound Parkway NW, Suite 300
Boca Raton, FL 33487-2742

© 2010 by Taylor & Francis Group, LLC
CRC Press is an imprint of Taylor & Francis Group, an Informa business

First issued in paperback 2019

No claim to original U.S. Government works

ISBN 13: 978-0-367-44598-0 (pbk)
ISBN 13: 978-90-04-17297-5 (hbk)

**Visit the Taylor & Francis Web site at**
**http://www.taylorandfrancis.com**

**and the CRC Press Web site at**
**http://www.crcpress.com**

# Contents

# 1 Nanoparticles: Synthesis, Characterization, and Applications

*Anu Tresa Sunny and Sabu Thomas*

School of Chemical Sciences, Mahatma Gandhi University, Kerala, India 686 560

sabut552001@yahoo.co.in, anutresa@gmail.com

**Abstract**

This chapter will present a general overview of the state-of-the-art for the variety of different synthesis routes to prepare various nanomaterials having different dimensions with a focus on wet chemical synthesis and emerging new approaches. The chapter is organized into the three parts: the first part will give an introduction to the area of nanoscience and technology; the second part will discuss the main methodologies and emerging promising approaches for the synthesis of nanostructured materials with strong application potential in the context of quantum dots; and finally a brief discussion of the characterization techniques and applications of nanomaterials followed by the importance of health and safety practices in the nanomaterials' work place.

**Keywords:** nanoparticles, nanostructures, mesoporous materials

## 1.1. Introduction to Nanoscience and Nanotechnology

### 1.1.1. A Brief Historical Overview

Before trying to understand and discuss about synthesis, characterization and application of nanomaterials, it is perhaps best to start with the broad area that is known as nanoscience and nanotechnology. Nanoscience and nanotechnology is an emerging area of science that concerns itself with the study of materials that have very small dimensions as well as the development of reliable processes and techniques for the synthesis and characterization of nanosized materials over a wide range of dimensions. It also refers to the fundamental understanding and resulting technological advances arising from the exploitation of new physical, chemical, and biological properties of systems that are intermediate in size, between isolated atoms and molecules and bulk materials, where the transitional properties between the two limits can be controlled. It cannot really be called chemistry, physics, or biology; researchers from all domains are studying very small things in order to better understand our world.

The word "nano" is derived from the Greek word "nanos," which means dwarfs. Richard Feynman (1918–1988), the Nobel laureate physicist, first mentioned the concept of "nanosized materials" (not yet using that name) in his speech (APS meeting Dec 29, 1959) titled *There's Plenty of Room at the Bottom* [1]. Later in 1974, Professor Norio Taniguchi from Tokyo Science University, Japan, coined the term *nanotechnology* to describe the arts and science of manipulating atoms and molecules to create new systems, materials, and devices. The term nanotechnology was then reintroduced and popularized by California scientist and author Eric Drexler. In 1981, the advent of the scanning tunneling microscope enabled atom clusters to be seen, while in 1991 IBM demonstrated the ability to arrange individual xenon atoms using an atomic force instrument [2, 3].

## 1.1.2. Nanomaterials

"Nanomaterials" are the materials (crystalline or amorphous) that have one or more dimensions in the range of 1 to 100 nm. The prefix "nano" means one billionth. In general, these particles (nanopowders, nanofibers, tubes, and thin films) could exist in powder form, dispersed in some medium, or as solid films [4]. Figure 1.1 shows the size of the nanoscale relative to some things we are more familiar with. We can see that the difference between a nanometer and a person is roughly the same as the difference between a person and celestial orbits. From the point of view of a chemist, the basic building blocks of matter are atomic nuclei and electrons in which the electrons orbit around the single nucleus and the number of electrons depends on the element. Anything smaller than a nanometer, in size is just a loose atom or small molecule. To get a sense of nanoscale, DNA strands are around 10 nm long, a bacterial cell measures a few hundreds of nanometers across, while a man is billions of nanometer tall (Figure 1.1).

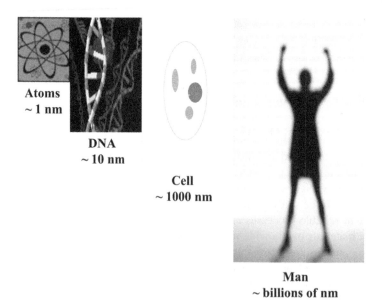

**Atoms**
**~ 1 nm**

**DNA**
**~ 10 nm**

**Cell**
**~ 1000 nm**

**Man**
**~ billions of nm**

**Figure 1.1**   Scheme showing the size of nanoscale relative to some things we are more familiar with.

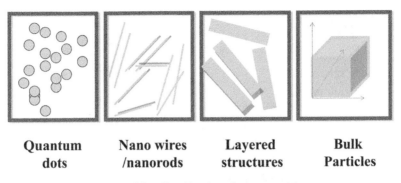

| Quantum<br>dots | Nano wires<br>/nanorods | Layered<br>structures | Bulk<br>Particles |

**Figure 1.2**    Classification of nanomaterials.

On the basis of the dimensionality, the materials can be generally classified [4] as:

- *Zero-dimensional structures*:    Quantum dots or simply *"nanoparticles."*
- *One-dimensional structures*:    Nanowires and nanorods.
- *Two-dimensional structures*:    Thin films, layered structures.
- *Special nanomaterials*:    Carbon fullerenes, carbon nanotubes, Micro- and mesoporous materials, core shell structures, and nanocomposites.

Figure 1.2 shows a schematic representation of some examples of nanomaterials, which use the nomenclature adopting the number of growing dimensions [4]. Nanoparticles or zero-dimensional structures, such as metal oxides and their clusters, are typically defined as being less than 100 nm in all three dimensions. One-dimensional materials like nanotubes, nanofiber, etc., will grow preferentially in one dimension. Two-dimensional materials (grown preferentially in two dimensions), such as thin films, layered materials including layered silicate, so-called nanoclay, form part of a class of material with interesting properties, especially by its "sui-generis" structure, which is built by the stacking of "two-dimensional" units known as layers, along the basal direction. Already the early theoretical predictions have shown that nanomaterials when compared to their bulk counterpart's exhibit dramatically improved mechanical properties in addition to their interesting functional properties [4, 5].

Nanomaterials have been important in the materials field for quite a long time. An early example was the incorporation of gold nanoparticles in stained glass in 10 AD at sizes in the nanorange as gold can exhibit a range of colors [3, 6]. The famous "Purple of Cassius" of 17th century consisted of colloidal gold particles and tin dioxide. In 1818, Jermias Benjamine Richter suggested that purple color of drinkable gold results from gold particles of finest degree of subdivision, whereas yellow color originates from the aggregation of fine particles. In 1857, Faraday had reported the formation of deep red solutions of colloidal gold by the phosphorus reduction of $HAuCl_4$ in carbon disulfide. He has also observed a reversible color change for a thin film of so prepared nanoparticles from bluish purple to green upon mechanical compression [3]. Also nanosized carbon black particles have been used to reinforce tires for 100 years [7]. At length-scales comparable to atoms and molecules, quantum effects strongly modify properties of matter, like color, reactivity, magnetic, or dipolar moment, etc.

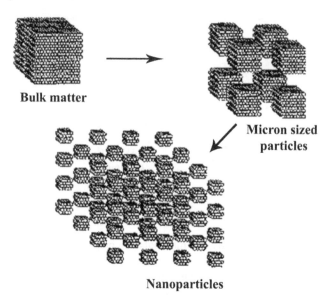

**Bulk matter**

**Micron sized
particles**

**Nanoparticles**

**Figure 1.3**    A typical illustration of surface area of bulk matter and nanoparticles.

In the last decade, low-dimensional materials have exhibited a wide range of electronic and optical properties that depend sensitively on both size and shape and are of both fundamental and technological interest. The reasons for the change in behavior when solids are in the nano regime may be an increased relative surface area (producing increased chemical reactivity) and the increasing dominance of quantum effects (with effects on materials optical, magnetic, or electrical properties). It has been widely pointed out that while a cube measuring 1 cm on a side possesses 6 cm$^2$ of surface area, the same 1 cm cube, subdivided into many 1 nm cubes, possesses a surface area equivalent to that of a football field! (Figure 1.3).

A tiny piece of any solid would have many of its atoms at its edges, which are unstable. The nanosized particles when compared to their bulk counterparts possess more edges, which makes large fraction of its atoms to locate in the grain or interphase boundaries. Two divergent views exist on the nature of grain boundaries in nanomaterials. The classic interpretation suggests that the nanosized materials have more fractions of its atoms in the grain boundaries and the nature of these boundaries in nanomaterials is identical to that in the coarse-grained materials. But new sophisticated computer simulation techniques reveal the absence of a long-range periodicity in the nanomaterials. It has also been suggested that the grain boundary energy and the width distributions are narrower while grain boundaries are wider when compared to that of coarse-grained materials. This results in increased solubility, diffusion, and also considerable changes in the thermodynamic behavior (e.g., lower melting temperature of nanophase) [4, 8, 9].

Fast electronic systems, extremely sensitive sensor devices to probe confined environments, and multiplexed techniques for high-throughput analysis represent some of today's most prominent nanotechnological needs. The ultimate realization of these technological advancements will be based on our ability to synthesize and organize matter into controlled geometries on the nanoscale. During the past few years, the exploration of synthetic techniques for the fabrication of nanostructured materials having controllable morphologies has emerged as a fast-growing subfield of nanotechnology research.

Advanced functional materials incorporating well-defined nanoarchitectures have shown great potential for nanotechnological applications, such as miniaturized nanoelectronics, ultrafast quantum computing, high-density memory/data storage media, ultrasensitive chemical sensing/biosensing, generation of high-efficiency catalytic substrates, and high-throughput templating for the growth/attachment of other types of bio- or inorganic nanomaterials. Of particular interest are zero, one- and two-dimensional arrays of patterned nanostructures (nanoparticles, nanowires, nanotubes, layered nanostructures, etc.), which have been shown to display unique optoelectronic, magnetic, or catalytic properties that can be tuned by varying their size and/or interparticle separation distance [10]. For example, patterned gold nanoparticles display plasmon optical properties that can be applied in surface-enhanced Raman scattering detection systems with high sensitivities [8], whereas nanowires with high surface ratios and diameters in the 10–200 nm range display interesting optical and electrical properties that are highly desirable in electrochemical sensing technologies, field-emission systems, and lasers [10]. Nanoporous materials displaying molecular sieve properties can act as chemical sensor elements, wherein the plasmon properties of the pores can be used in optical detection systems (Raman, optical waveguides). One of the most important technological challenges that remain to be addressed, however, is the development of effective patterning methods to control materials assembly on a nanometer scale. At present, there is a wide variety of top-down and bottom-up fabrication techniques that are capable of creating nanostructured arrays with varying degrees of speed, cost, and structural quality. We shall provide a brief overview of some of these techniques below, but for more detailed descriptions, the reader is referred to several excellent recent reviews [11–14] and books [15].

## 1.2. Synthesis of Nanosized Materials

In recent years, a wide range of nanostructured materials in various forms such as nanosized powders, nanosized fibers, nanotubes, thin films, etc., of metals, semiconductors, dielectrics, magnetic materials, polymers, and other organic compounds are introduced by means of a number of processing routes. It is seen that the properties of these particles are quite sensitive to their sizes. The intensive research and the contemporary technological advances make it difficult to keep up to date with all the achievements in the field of nanotechnology. In this section, we are trying our best to briefly summarize the various synthetic routes used for the production of low-dimensional nanoparticles with strong application potentials [10–16].

The drive for finding novel routes for the synthesis of nanomaterials has gained considerable momentum in recent years, owing to the ever-increasing demand for smaller particle sizes. Since nanomaterials have little mass and are dominated by surface area and size effects, the processes and equipment for their manufacturing are expected to differ significantly from conventional approaches.

Conventional methods, called top-down techniques (Figure 1.4), include chopping down the bulk material by mechanical means and the resulting colloidal particles are subsequently stabilized by colloidal protecting agents. The simplest example is ball milling. Generally top-down strategies involve either (1) using macroscopic tools to first transfer a computer-generated pattern onto a larger piece of bulk material, and then "sculpting"

**Top-down Method**

Creates nano structures out of
macro structures by breaking
down matter into more building
blocks.

**Bottom-up Method**

Building complex systems in the
nano regime by combining
simple stomic level components
through self-assembly of atoms
or molecules

**Figure 1.4**    Nanofabrication approaches.

a nanostructure by physically removing material (e.g., through wet/dry etching); or (2) using macroscopic tools to directly add/rearrange ("write") materials on a substrate. In the first category, the most common techniques are based on photolithography, which is cost-effective and relatively fast, but its resolution is ultimately limited by optical diffraction effects to typically 0.2–0.5 μm. Electron and ion-based lithographic methods, on the other hand, permit the creation of ordered nanostructured arrays with high resolution (i.e., 50 nm features and/or spacing) and allow very good control over particle shape and spacing; however, their throughput is limited; line-by-line pattern generation (a serial technique) is considered very slow when compared with a parallel technique (such as photolithography) in which the entire surface is simultaneously patterned all at once. The second category of top-down approaches includes scanning probe lithographic (SPL) techniques (e.g., dip-pen nanolithography (DPN) and scanning tunneling microscopy (STM)), micro contact printing (μCP), and nano imprint lithography (NIL). Currently, patterns generated using DPN can be as small as 15 nm, whereas STM offers the unparallel capability to position individual atoms to pattern structures with ultrahigh, subnanometer precision. However, DPN and STM are also serial techniques and are, therefore, not suitable for high-volume manufacturing technologies, although this is a drawback that may eventually be overcome by the introduction of massively parallel microfabricated probe tip arrays. In summary, top-down approaches offer a wide range of structures of high quality/yield, but are generally neither cost nor time-effective, and for some methods, resolution below the 100 nm range is not easily achievable. The other drawbacks of this method are relatively broad size distribution, varied particle shape or geometry, and presence of significant amount of impurities from the milling medium [10–16].

While in bottom-up manufacture, nanoparticles are synthesized by atom–atom or molecule–molecule assembling of particles (e.g., homogeneous nucleation from liquid or vapor, or by heterogeneous nucleation on substrates); that is, bottom-up approach takes the advantage of the physicochemical interactions for the hierarchical synthesis of ordered nanoscale structures through the self-assembly of the basic building blocks. When such processes are properly controlled, they lend themselves extremely well to the synthesis of nanomaterials, for the obvious reason that they involve molecules of the appropriate size and mobility in the liquid state. Therefore, the process offers the chemist a unique opportunity to design and engineer original and novel nanomaterials. Some of them take advantage of the versatility and complexity of the hydrolysis and condensation reactions of metal alkoxides, while others exploit the nanoporosity of gels as a host for various molecules or inorganic clusters. In all cases, the chemistry of the systems is the main controlling factor for the resulting nanostructure. Examples discussed are semiconducting materials, ferroelectric, metallic, and oxide nanocrystallites, as well as organic–inorganic nanohybrids. In this approach, the challenge is to make structures large enough, and of sufficient quality, for use as materials. Currently, the most common types of bottom-up fabrication procedures are those based on (1) wet chemistry routes, such as precipitation, reduction, surfactant assisted methods, sol–gel processes, and (2) the use of a templating substrate, such as chemically or topologically patterned surfaces, inorganic mesoporous structures, and organic supramolecular complexes (mainly block copolymer (BCP) systems). In the following sections, we will be focusing more on the usefulness of wet chemistry routes in the synthesis of materials comprising nanoscale architectures.

## 1.2.1. Simple Precipitation

One of the oldest techniques for the synthesis of quantum dots or simply nanoparticles is the precipitation of products from bulk solutions. In this method, the precursors are dissolved in a common solvent followed by the addition of a precipitating agent. In most cases, a further heat treatment is needed on the collected precipitate. These reactions can generate a wide range of materials like ferrites, tungstates, perovskites, etc. The main attracting features of this method are its cost effectiveness, production in large quantities, phase purity; however, it is rather difficult to control the size as only kinetic factors are available to control growth [16]. Uniformity in terms of particle size and morphology can be achieved by tuning the synthesis parameters like precursor concentration, reaction time, pH of the reaction medium, use of surfactants, type of the precipitating agents used, etc. [10–20]. This method is suitable when nanoparticles are required in large quantities.

## 1.2.2. Template-Assisted Synthesis

The method is mainly targeted on one-dimensional nanomaterials (e.g., nanorods or nanowires) and their superstructures with high crystallinity. As the name indicates, it requires either a carefully and costly predesigned template or rather strict/drastic reaction conditions (e.g., channels, controlled growth on a surface, vapor phase synthesis, etc.) to tailor the size and morphology. The template-assisted synthesis of nanoparticles is well studied [22–29]. Recently preparation of a wide range of compounds, especially magnetic nanoparticles (e.g., ferrites [24]), metal oxides (e.g., ZnO [25], GeO$_2$ [26], sulfides [22]), etc., was reported which makes use of carbon nanotubes (CNTs) as templates. A schematic illustration is shown in Figure 1.5. Other examples include solution phase electrodeposition for the formation of various nanoparticles with anode aluminum oxide (AAO) as template. This methodology was popularized by Martin and co-workers [27, 28] and Routkevitch et al. [29]. The nanorods/nanowires will nucleate from the bottom of the channels of the template and grow continuously along the channels, achieving continuous structures. Also the pore diameters of the templates can be easily adjusted in the order of ∼10 nm sizes enabling quantum confinement [22]. Moreover, higher pore densities and aspect ratio of

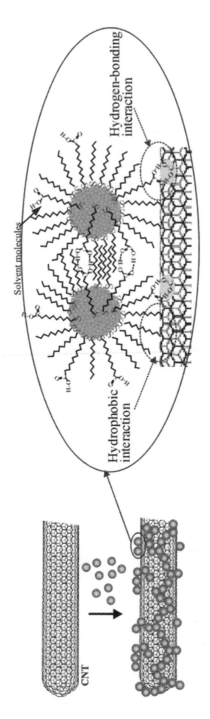

**Figure 1.5**    Schematic illustration of template-assisted synthesis of nanoparticles using CNTs.

the templates extend the applicability of the method. The templates have to be removed by heating [22] (e.g., CNTs)/by treatment with suitable solvents (e.g., AAO) [27, 28] in order to get the pure nanosolids, provided such post-treatments may or may not affect the phase stability, purity, etc., of the as-prepared nanosolids.

### 1.2.3. Reduction

These include the reduction of metal precursors using suitable reducing agents. Glavee et al. reported the reduction of various transition metals like iron, cobalt, nickel [30–32] using sodium borohydride, resulting in the formation of spherical nanoparticles. While a report from Wang et al. [33] shows that copper oxide nanowires can be fabricated in the presence of a surfactant using hydrazine hydrate as the reducing agent. Also mixing of metal salt precursors at suitable ratios before reduction can form alloys [34, 35]. Unfortunately in this method, side reactions can affect the phase purity of the products. By carefully controlling the reaction conditions viz. of suitable selection of solvent, optimizing the reaction temperature, avoiding moisture content, one can bring the side reactions under control.

### 1.2.4. Sol–Gel Processing

Classical solid-state reactions require a high calcination temperature and hence induce the sintering and aggregation of particles. Compared to this, sol–gel method is a useful and attractive technique for the preparation of nanosized materials including glass, ceramic powders, thin films, fibers, monoliths [36] because of its advantages: good stoichiometric control and the production of ultrafine particles with a size distribution in a relatively short processing time at lower temperatures. Traditionally, the sol–gel process involves hydrolysis followed by condensation. A schematic illustration is given in Figure 1.6. Metal alkoxides are good precursors since they readily undergo hydrolysis giving rise to "sol," once the hydrolysis occurred, the sol can react further and condensation/polymerization occurs, leading to a gel. The gel on calcination gives the ultra-fine nanosolids. The particle size ranges from 10 to 500 nm (650–1250°C) and is mainly dependent on the calcination temperature.

The main influencing parameters of the process are nature of the solvent, reaction temperature, nature and the initial concentration of the precursors, presence of any catalysts, pH, and mechanical agitation. The use of a chelating agent, for example, polyacrylic acid (PAA) [37], has got the advantage of reducing the calcination temperature since the heat of combustion of chelating agent is utilized for the crystallization of the particles. PAA has more carboxylic groups to form chelates with mixed cations and results in a sol. It also greatly aids in the formation of a cross-linked gel, which may provide more homogenous mixing of cations and less tendency for the aggregation during calcinations [35–37]. Nanocrystalline powders such as ferrites [36–38], $TiO_2$ [39], $SiO_2$ [39], yttrium iron garnets [40], manganites [41], metal oxide–organic nanocomposites [39], metal–ceramic nanocomposite powders [42] were also synthesized using different metal salt precursors by this method.

In the design of complex nanoarchitectures, the main advantage of the sol–gel process is the versatility in the control of size, distribution, and arrangement of nanopores, for example, through self-assembly [43]. From the above dealt references, it is clear that this nanoporosity is functionalized using organic molecules or polymers as the gel developed or

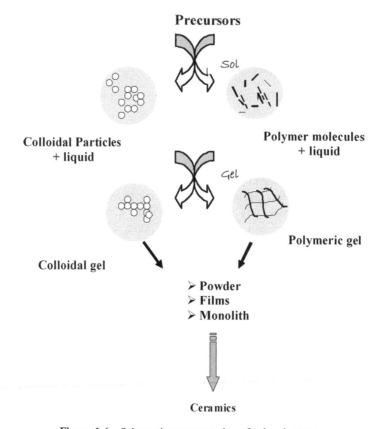

**Figure 1.6**    Schematic representation of sol–gel process.

even after as the gel has dried. Furthermore, through careful heat treatment and dependent upon the chemistry of the starting solution and the thermal treatment, many kinds of nanoparticles (oxides, sulfides, metals, and semiconductors) can be grown within those nanopores, resulting in families of nanomaterials heretofore very difficult to fabricate. The versatility of the process is in large part due to the rich and varied chemistry of organometallic precursors, combined with the low processing temperature. As a result, the sol–gel process is likely to continue attracting the attention of chemists interested in designing advanced nanomaterials.

## 1.2.5. Surfactant-Assisted Synthesis

Agglomeration can occur during any of the stages: synthesis, drying, handling, and/or post-drying. To produce monodispersed particles, a variety of stabilizers in the form of donor ligands, polymers, and/or surfactants can be used. Surface active agents or simply "surfactants" can be used to control the dispersion during chemical synthesis and also to disperse the as-synthesized particles. A surfactant is any substance that affects the surface or interfacial tension of the medium in which it is dissolved. A surfactant is composed of a hydrophilic head and a hydrophobic tail. These are used during the nanoparticle synthesis in order to reduce interparticle interaction through an increase in repulsive forces leading to the formation of stable colloids. The narrow size distribution of the particles can be maintained as long as agglomeration, Ostwald ripening, and secondary nucleation do not occur [10–16].

Surfactant molecule may spontaneously aggregate together to form micelles or microemulsions, with a diameter ranging from 10 to 100 nm. Surfactant micelles may enclose precursor molecules to form amphiphilic nanoreactors. Micelles impose kinetic and thermodynamic constraints on particle formation, tailoring the size, shape, and polydispersity of the nanoparticles. Thus surfactants play a crucial role in the compartmentalization of nanoparticles.

Micelle formation can happen in any medium but the size of micelles is dependent on the solvent to surfactant ratio [16]. Micelles can be "direct" or "inverse," depending on the orientation of "heads and tails" of surfactants molecules in the aggregates. In the case of direct micelles [44], the hydrophilic portion of the surfactant molecules will form the outer surface of the aggregate while for inverse or reverse micelles will have the hydrophobic portion outside [45].

Surfactants can be anionic, cationic, or nonionic. Commonly used and most characterized surfactants systems are polyethylene monolaurate (a nonionic surfactant), carboxybetaine 12 (a zwitterionic surfactant), Aerosol-OT (sodium dioctyl sulfosuccinate, anionic surfactant), CTAB (cetyl trimethyl ammonium bromide), SDS (sodium dodecylsulfate), CETAC ($C_{16}H_{13}N\text{-}(CH_3)_3^+Cl^-$, a cationic surfactant) [16].

## 1.2.6. Polyol Method

In polyol method, a polyol (e.g., any alcohol containing two or more hydroxyl groups acts as a solvent, as well as the reducing agent and at the same it also performs the function of a surfactant. The metal salt precursors when dissolved in a polyol form an intermediate in which when subjected to further chemical reaction gives the nanoparticles. Upon increasing the reaction temperature, the particle size increases. A wide range of nanocrystalline powders of metals [42–50], metal oxides [18], ferrites [51], alloys [52], bimetallic clusters [53], etc., can be prepared by this method.

## 1.2.7. Hydrothermal Methods

This involves aqueous reactions carried out under high pressure at temperatures over 200°C using autoclaves or high-pressure reactors. Size and morphological control can be achieved by controlling the conditions of time and temperature. The nature and concentration of the precursor and pH plays a crucial role in determining the phase purity of the nanoparticles. Traditionally, it proceeds through hydrolysis followed by oxidation in the presence of water as the medium accelerating the kinetics of the reaction [16, 54]

## 1.2.8. Simple Decomposition Methods

One of the simplest methods to prepare nanoparticles is the decomposition of precursors either by heat (thermolysis), light (photolysis), or by sound (sonolysis). This method requires the use of surfactants or capping agents to prevent the agglomeration of nanoparticles. In many cases, low-valent organometallic compounds and several organic derivatives of transition metals were used as the precursors in the preparation of colloidal nanoparticles with a variation in the size and shape, since they can be decomposed at relatively low temperatures to form the final product. By varying the concentration of the organic species as well as the stabilizing agents and decomposition temperature, the size and shape of the particles can be controlled. But the main drawback of this method is that most of the reactions involve air-sensitive reactants and also the presence of a surfactant is essential to tailor the particle size [10–16].

Thermolysis is the decomposition of precursors into desired products under suitable thermal treatment. Thermolysis of organometallic precursor nickel bis-cycloocta-1,5-diene $(Ni(COD)_2)$ in the presence of hexa decyl amine (HAD) or TOPO results in the formation of nanorods, spheres, or tear-shaped nanorods depending on the concentration of the organic species [55].

In photolysis, short pulses of high-energy photons are applied to the metal salt solution, which in turn starts the reduction process in the whole solution instantaneously [55, 56]. The limitation of photo, gamma, and laser irradiation methods lies in the restriction of low precursor concentration in solution. Consequently, these methods are suitable for the large-scale preparation of nanomaterials.

In sonolysis, ultrasound or acoustic waves are used to decompose the precursors. Generally sonolysis is carried out on volatile organometallic precursors in low vapor pressure solvents to optimize the yield of nanostructured particles. But the powders formed are usually amorphous, agglomerated, and porous, which on further annealing gives the crystalline phase [58, 59]. If the organic solvents have high boiling points, highly porous amorphous powders were obtained [16].

## 1.3. Characterization, Application, and Environmental Risk Assessment of Nanomaterials

Particle characterization has become an indispensable tool in various fields of research and development and in the manufacture and quality control of many materials and products in our day-to-day life. In material characterization, there are two classes of properties that need to be addressed, compositional and crystal structure elucidation and determination of geometrical properties like size, shape, polydispersity, and surface characteristics of particulate matter. There are many technologies that have been developed and successfully employed for particle characterization; a brief description of some of the techniques is presented in this section. Computer-controlled scanning probe microscopy enables a real-time, hands on nanostructure manipulation. Optical tweezers provide another approach to holding and moving nanometer structures, a capability especially useful in investigating the dynamics of molecules and particles.

### 1.3.1. Size and Morphology Determinations

The distinguishing characteristics of nanoparticles are, of course, their size and morphology. A definite volume of nanoparticles generally possess a distribution of sizes. Also the nanoparticles do not exist as single crystals; they always tend to remain as aggregates. Different techniques will give different results depending on whether the technique is sensitive to the median size or mean size and whether that median is number weighted or volume weighted. Some techniques will yield a crystallite size, while some others give the size of aggregates.

Direct inspections of nanometer scale structures can be carried out using transmission electron microscopy (TEM) and scanning electron microscopy (SEM). TEM will generally report the total particle size along with details of size distribution. If the material consists of more than one phase and phases possess enough contrast, then the individual phases may also be visible. A review article by Thomas and Hutten [60] shows that TEM is a versatile tool in investigating the physical and magnetic structure of nanostructured materials.

In many cases, a detailed knowledge of the physical nature and chemical composition of the surfaces on solids on a submicrometer scale is obtained using SEM. In order to obtain an electron microscopic image, the surface of a solid sample is swept in a raster pattern with a finely focused beam of electrons. A raster is a scanning pattern similar to that used in a cathode ray tube in which an electron beam is first swept across a surface in a straight line, and then returned to its starting position and shifted downward by a standard increment. This process is repeated until an area of the surface has been scanned. Several types of signals are produced from the surface when it is scanned with an energetic beam of electrons backscattered and secondary electrons serve as the basis of SEM. Review article by Willard et al. [16] points out that dynamic light scattering (DLS), also called photon correlation spectroscopy, is an important supplemental technique for determining the sizes of particles in solution or for *in situ* studies, particularly when the size distribution is approximately log normal.

## 1.3.2. Phase Identification

X-ray powder diffraction (XRD) is a technique routinely used for identifying the crystalline phase as well as for determining the particle size of crystalline materials. XRD pattern is a set of lines or peaks, each of different intensity and position ($d$-spacing or Bragg angle, $\theta$), on either a strip of photographic film or on a length of chart paper. For a given substance, the line positions are essentially fixed and are characteristic of that substance. The intensities may vary somewhat from sample to sample, depending on method of sample preparation and instrumental conditions. For the phase identification purposes, principal note is taken of line positions together with a semiquantitative consideration of intensities [60, 61].

XRD can also be used to measure the average crystal size in a powdered sample, provided the average diameter is less than about 2000 Å. Crystallite size can be estimated using Scherrer formula, which is given as

$$D = K\lambda/(\beta \cos \theta), \tag{1.1}$$

where $D$ is crystallite size, $\beta$ FWHM, full width half maximum characteristic of the peak, $\theta$ glancing angle, $\lambda$ wavelength of copper $K_\alpha$ X-ray used ($\lambda = 0.1542$ nm), and $K$ is the shape factor; its value found to be 0.89 [38].

The lines in a powder diffraction pattern are of finite breadth but if the particles are very small, the lines are broader than usual. The broadening increases with decreasing particle size. The limit is reached with particle diameters in the range roughly 20 to 100 Å; then lines are so broad that they effectively "disappear" into the background radiation. For particles that are markedly nonspherical, it may be possible to estimate the shape since different lines in the powder pattern are broadened to differing degrees. Crystalline solids give diffraction patterns that have a number of sharp lines. Noncrystalline solids (e.g., glasses, gels) give diffraction patterns that have a number of very broad humps [38, 61]. But the line broadening can also occur due to some other factors like strain; Scherrer analysis generally provides a lower limit on mean crystallite size [16].

In some cases, discrimination of phases can be achieved with the help of selected area electron diffraction (SAED), but at this scale double diffraction, calibrating issues, and the close proximity of diffraction spots can make variations in the observations [16]. Also diffraction can provide only very limited information about amorphous phases, quantitative information about multiple phases, point defects, etc. Synchroton-based techniques are well suited for these purposes.

### 1.3.3. Spectral Characterization

Various spectroscopic methods like ultraviolet-visible spectroscopy (UV-vis), infrared (IR) and Raman spectroscopy, nuclear magnetic resonance spectroscopy (NMR), X-ray absorption spectroscopy (XAS), and Mössbauer spectroscopy have been adequately used in literature for the characterization purposes.

UV-vis spectroscopy is very useful in characterizing metallic or semiconductor type nanoparticles whose plasmon resonance lies in the UV-visible range of electromagnetic spectrum. The absorption maxima ($1$) are dependent on the size and shape of the particles, as well as how close the particles are relative to each other [62].

IR spectroscopy has been used as a surface probe, commonly done to investigate the surface adsorption on the nanomaterials [62]. To a lesser extent, surface enhanced Raman spectroscopy (SERS) has been used to probe the surfaces usually systems containing colloidal metal particles [64].

NMR is a classic method for characterizing the nanoparticulates associated with ligands and for probing intermetallic interactions [65, 66]. There is a great deal of information on the use of XAS to investigate geometric and electronic structures of metallic nanomaterials. A detailed review of the spectral characterization of nanomaterials has been published by Bonneman et al. in 2004 [17].

### 1.3.4. Application of Nanomaterials

Nanotechnology clearly opens a new era in the field of science and engineering. It is estimated that sales of nanomaterials were $1.5 million in 1999 and grew to $430 by the end of 2003. Recent studies by the experts at National Science Foundation show that with this annual growth rate, the market size would be more than $700 billion by the end of 2008 and will exceed $1,000 billion by 2015. To describe the numerous potential and actual applications of nanotechnology in the automotive and aeronautical fields, in electronics, communications, biology, medicine, energy, and more is beyond the scope of this chapter since this list is growing rapidly.

The volume of an object decreases as the third power of its linear dimensions, and only by the second power for its surface area; therefore, nanomaterials show a massive increase in the surface area to volume ratio and also in a given volume when the size of the constituent particles decreases, the portion of atoms at or near the surface increases exponentially, creating innumerable sites for bonding with surrounding materials. This explains why huge improvements are possible in the mechanical, chemical, and electrical properties of solids when they were in the nano regime.

The stability of nanoparticles is important since these tiny particles tend to coalesce during preparation or applications. So it is very difficult to prepare and store large quantities of nanoparticles because of this inherent tendency to agglomerate. A detailed under-standing of the coalescence process is necessary before the use of nanoparticles in many applications. In nanosolids, the quantum size effects and the large number of surface atoms influences its behavior. The high-resolution transmission electron microscopy (HRTEM) studies [65–67] of nanoparticles under ordinary conditions and during heating reveals that the nanoparticles reorient themselves to the least lattice mismatch between each other before merging and also the surface atomic migration as well as the temperature has impact on the coalescence process. However, a detailed mechanism of nanoparticle coalescence has not been fully understood yet.

The subject of nanomaterials is of great vitality and offers immense opportunities. It is truly interdisciplinary and encompasses chemistry, physics, biology, materials, and engineering. Some of the important applications and technologies based on nanomaterials are the following:

- production of nanomaterials of ceramics and other materials;
- preparation of high-performance nanocomposites;
- development of nanoelectrochemical systems (NEMS);
- applications of nanotubes for various purposes;
- DMA chips for chemical/biochemical assays;
- gene targeting/drug targeting;
- defect-free electronics for future applications.

Presently, there are no regulations specifically governing nanoproducts. Absolutely nothing is known about the effect of nanoparticles on the environment. In free form, they can be released in the air or water during the lifecycle of the finished product. Even the largest particles of nanodust are impossible for the naked eye to see. How nanoparticles behave inside an organism is still not well understood. Since nanoparticles are 100% nonbiodegradable, since recently used nanostructured materials constitute a wide range of metals and hazardous metal oxides. Nobody knows what will happen in the future, so how to remove or recycle them from the environment is a problem requiring immediate attention.

## 1.3.5. Risk Assessment of Nanomaterials

Nanomaterials can be defined as engineered materials with one or more dimensions below 100 nm, having large surface areas per unit of volume, and novel electrical and magnetic properties that differ from conventional materials. Nanomaterials can be produced in bulk amounts, typically produced in the chemical or the polymer industry in metric tons (e.g., titanium dioxide and carbon black), and novel nanomaterials with targeted properties fulfilling specific functions (e.g., CNTs and quantum dots). The diversity of nanomaterials challenges us to consider how we prioritize them for environmental risk assessment.

There are already many products containing nanomaterials in the market today, and their unique properties have raised expectations for more applications ranging from lightweight materials, drug-delivery systems, and catalytic converters to usage in food, cosmetics, and leisure products.

Nanomaterials pose many new questions on risk assessment that are not yet completely answered. Thus, voluntary industrial risk assessment initiatives can be considered vital to the environmental health and safety issues associated with nanomaterials. It was found that the nanomaterials exhibited such a diversity of properties that a categorization according to risk and material issues could not be made. Fate of nanomaterials in the use and disposal stage received little attention by industry and the majority of institutions did not foresee unintentional release of nanomaterials throughout the lifecycle. The development of risk and safety decision frameworks in industry seems, therefore, necessary to ensure that the potential risks of engineered nanomaterials are taken into consideration. The introduction of nanoparticulate materials into more and more applications will inevitably also result in their introduction into environmental compartments and ecosystems. It is, therefore, likely that also an exposure of the human body to nanomaterials will take place. One of the recent surveys conducted among various companies working with nanomaterials in Europe showed that 65% of these companies did not perform any risk assessment of their nanomaterials, while 32.5% performed risk assessments sometimes or always [69].

Furthermore, in this survey no factors were identified that could provide any explanation of why some companies conducted risk assessment and why others did not. Also the survey detected a lack of any systemic approach among industry players in regard to assessing risks associated with nanomaterials. Although most nanomaterial applications may indeed be quite safe, there is still the issue of concern that consumers may be exposed to unassessed risks. Developing proactive risk management strategies appears to be an urgent task for minimizing the risk of harm to the environment and the public health. How much responsibility the individual firm should take in a globalized market is an issue of considerable debate in policy. Nevertheless, it may be necessary for the regulators to take measures to ensure that risks associated with nanomaterials are properly assessed by the industry. A first step could be to initiate a database with information on the properties of the different nanomaterials produced and handled in industry. Such a database would assist in categorizing nanomaterials with respect to, for example, chemical properties, toxicity, and consumer use. The voluntary reporting scheme in place in the UK has received very few contributions from the industry. A legally enforced information duty of industrial producers seems, therefore, to be the most effective solution to ensure quality and coverage. Actively initiating risk management strategies may also help industry address any public concern related to the possible risks of nanomaterials [69, 70].

Industry, scientists, governmental bodies, and environmental advocacy groups find regulatory interventions useful, but they are of different opinions as to whether regulations should be evidence- or precaution-oriented, voluntary, or top-down controlled. Voluntary initiatives have been under consultation in the United States and the UK. At the moment, regulatory bodies do not know to which extent they should regulate this area. Improved scientific knowledge on the potential hazards and risks associated with nanomaterials is needed to determine the type and extent of regulations. Regulations often demand that certain risk assessment activities or precautionary measures are conducted by industry, such as the Toxic Substances Control Act for regulating chemicals in the United States [69, 70]. However, given that nanomaterials may cause harm and that there are currently no regulations that take the specific properties of NPM into account, the responsibility for safe production and products is mostly left with industry. Risk assessment procedures and precautionary measures initiated by the industry are, therefore, vital to manage the environmental health and safety [69, 70]. It is, therefore, of utmost importance to investigate industrial initiatives in this area. In an article [69] published in the journal of *Environmental Science and Technology* in 2008, it was given that no such investigations were publicly available by this time.

## 1.4. Conclusion

The advances in the area of nanotechnology provide a rich set of materials useful for probing the fundamental nature of matter. As we have learnt, matter behaves differently below 100 nm than it does at macroscale. Nanosized materials have unique structures and tunable properties, making them suitable for many real-world applications. The nanoparticles have a wide range of applications from energy, catalytic, magnetic, and electronic to biomedical, pharmaceutical, and cosmetic industries. The stability of nanoparticles is important for most of these applications. Most nanosolids are metastable, if given an opportunity they would combine to become larger. Therefore one of the current hottest issues in the area of nanoscience and nanotechnology is the retention of "nanosize" at higher temperatures either during processing or in service. It can be said that nowadays most of the synthetic methodologies are based on the soft chemical routes. A new generation of solids and synthetic nanomaterials is emerging; it now remains to be seen how practical application of these materials will grow from the seeds over the coming years. Within a few decades, we will use these nanomachines to manufacture consumer goods at the molecular level.

# References

1. Feynman, F.P. 1960. There is Plenty of Room at the Bottom. www.zyvex.com/nanotech/Feynman.html.
2. Hunt,W.H. JOM. October, 2004. http://www.foley.com/files/tbl_s31Publications/FileUpload137/2623/Nanomaterials.pdf.
3. www.en.wikipedia.org.
4. Hannula S.P.; Koskinen J.; Haimi, E. Nowak R. ENN. 2004, 5, 131.
5. Shipway, A.N.; Katz, E.; Wilner, I. Chem. Phys. Chem. 2000, 1, 18.
6. Nanoscience and nanotechnologies: Opportunities and Uncertainties, July 29, 2004. www.royalsoc.ac.uk/policy/.
7. Stix,G. 2001. Little Big Science. www.sciam.com/article.
8. Seigel, W. Mater. Sci. Eng. A. 1993, 168, 189.
9. Phillopot, S.R.; Wolf, D.; Gleiter, H. J. Appl. Phys. 1995, 78, 847.
10. Sotiropoulou, S; Sierra-Sastre, Y; Mark, S.S.; Batt, C.A. Chem. Mater. 2008, 20, 821.
11. Tseng, A.A.; Notargiacomo, A.; Chen, T.P. J. Vac. Sci. Technol., B 2005, 23(3), 877.
12. Huck, W.T.S. Angew. Chem., Int. Ed. 2007, 46(16), 2754.
13. Li, X.-M.; Huskens, J.; Reinhoudt, D.N. J. Mater. Chem. 2004, 14(20), 2954.
14. Gates, B.D.; Xu, Q.; Stewart, M.; Ryan, D.; Willson, C.G.; Whitesides, G.M. Chem. Rev. 2005, 105(4), 1171.
15. Madou, M.J. Fundamentals of Microfabrication: The Science of Miniaturization, 2nd edn.; CRC Press, Boca Raton, FL, 2002.
16. Willard, M.A.; Kurihara, L.K.; Carpenter, E.E.; Calvin, S.; Harris, V.G. ENN. 2004, 1, 815.
17. Bonnemann, H.; Nagabhushana, K.S. ENN, 2004, 1, 777.
18. George T.; Joseph S.; Sunny A.T.; Mathew, S. J. Nanopart Res. 2008, 10, 567.
19. Massart, R.; Cabuil, V.J. Chim. Phys. Chim. Biol, 1987, 84, 967.
20. Lee, W.-J.; Fang, T.-T. J. Mater. Sci. 1995, 30, 4349.
21. Sankaranaryanan, V.K.; Khan, D.C. J. Magn. Magn. Mater. 1996, 153, 337.
22. Yu, S-H.; Yang, J.; Qian, Y. ENN. 2004, 4, 607.
23. Frosch, C. J.; Thurmond, C.D. J. Phys. Chem.1962, 66, 877.
24. Pham-Huu, C.; Keller, N.; Estournes, C.; Ehret, G.; Ledoux, M.J. Chem. Commun. 2002, 1882.
25. Kim, H.; Sigmund, W. Appl. Phys. Lett. 2002, 81, 2025.
26. Zhang, Y.; Zhum J.; Zhang, Q.; Yan, Y.; Wang, N.; Zhang, X. Chem. Phys. Lett. 2000, 317, 504.
27. Klein, J.D.; Herrick, R.D.; Palmer, D.; Saitor, M.J.; Brumlik, C.J.; Martin, C.R. Chem. Mater. 1993, 5, 902.
28. Martin, C.R. Science. 1994, 266, 1961.
29. Routkevitch, D.; Bigioni, T.; Mosksovits, M.; Xu, J. M. J. Phys. Chem. 1996, 100, 14037.
30. Glavee, G.N.; Klabunde, K.J.; Sorenson, C.M.; Hadjipanayis, G.C. Inorg. Chem. 1995, 34, 28.
31. Glavee, G.N.; Klabunde, K.J.; Sorenson, C.M.; Hadjipanayis, G.C. Langmuir. 1993, 9, 162.
32. Glavee, G.N.; Klabunde, K.J.; Sorenson, C.M.; Hadjipanayis, G.C. Langmuir 1994, 10, 4726.
33. Wang, W.; Wang, G.; Wang, X.; Zhan, Y.; Liu, Y.; Zheng, C. Adv. Mater. 2002, 14(1), 67.
34. Yedra, A.; Barquin, L.F.; Calderon, R.G.; Pankhurst, Q.A.; Sal, J.C.G. J. Non-Cryst. Solids. 2001, 20, 287.
35. Morup, S.; Sethi, S.A.; Linderoth, S.; Bender Koch, C.; Bentzon, M.D. J. Mater. Sci. 1992, 27, 3010.
36. Pierre A.C: Introduction to Sol–Gel Processing, Kluwer Academic Publishers, Boston, 1998.

37. Chen D.H.; He X.R. Mater. Res. Bull. 2001, 36, 1369.
38. Suryanarayana C. Bull. Mater. Sci. 1994, 17, 307.
39. Birnker, C.J.; Sherrer, G.W. Sol–Gel Science, Academic Press, New York, 1990.
40. Sanchez, R.D.; Rivaz, J.; Vaqueiro, P.; Lopez-Quintela, M.A.; Caerio, D. J. Magn. Magn. Mater. 2002, 92, 247.
41. Mathur, S.; Shen. J. Sol–Gel Sci. Tech. 2002, 25, 147.
42. Wang, J.P.; Luo, H. L. J. Appl. Phys. 1994, 75, 7425.
43. Mackenzie J.D.; Bescher E. P. Acc. Chem. Res. 2007, 40, 810.
44. Liu, C.; Rondinone, A.J.; Zhang, Z.J. Pure. Appl. Chem. 2000, 72, 37.
45. Agnoli, F.; Zhou, W.L.; O'Connor, C.J. Adv. Mater. 2001, 13, 1697.
46. Viau, G.; Ravel, F.; Acher, O.; Fievet-Vincent, F.; Fievet , F. J. Appl. Phys. 1994, 76, 6570.
47. Viau, G.; Ravel, F.; Acher, O.; Fievet-Vincent, F.; Fievet , F. J. Magn. Magn. Mater. 1995, 377, 140.
48. Viau, G.; Fievet-Vincent, F.; Fievet , F. J. Mater. Chem. 1996, 6, 1047.
49. Viau, G.; Fievet-Vincent, F.; Fievet, F. Solid State Ionics. 1996, 84, 259.
50. Viau, G.; Fievet-Vincent, F.; Fievet , F.;  Toneguzzo, P.; Ravel, F.; Acher, O. J. Appl. Phys. 1997, 81, 2749.
51. Jungk, H.O.; Feldmann, C. J. Mater. Res. 2000, 15, 2244.
52. Toneguzzo, P.; Acher, O.; Viau, G.; Fievet-Vincent, F.; Fievet , F. Adv. Mater. 1998, 10, 1032.
53. Chow, G.M.; Kurihara, L.K.; Schoen, P.E. U.S. Patent, June 2, 1998, 5, 759, 230.
54. Wang, M.L.; Shih, Z.W.; Lin, C.H. J. Cryst. Growth. 1994, 134, 47.
55. Cordente, N.; Respaud, M.; Senocq, F.; Casanove, M.J.; Amiens. C.; Chaudret, B. Nano. Lett. 2001, 1, 565.
56. Yonezawa, T.; Mol. Catal. 1993, 83, 167.
57. Toshima, N.; Takahashi, T.; Hirai, H. Chem. Lett. 1985, 1245.
58. Suslick, K.S.; Hyeon, T.; Fang, M. Chem. Mater. 1996, 8, 2172.
59. Suslick, K,S.; Fang, M.; Hyeon, T.J. Am. Chem. Soc. 1996, 118, 11960.
60. Thomas, G.; Hutten, A. Nanostruct. Mater. 1997, 9, 271.
61. Rozman, M.; Drofenik, M. J. Am. Ceram. Soc.1995, 78, 2449.
62. Schimid, G. Ed. Clusters and Colloids, VCH, Weinheim, 1994, 506.
63. Sheppard, N.; Nguyen, T.T. Advances in Infrared & Raman Spectroscopy, (Clark, R.J.H.; Hester, R. Eds.) Heyden, London, 1978, 5, 67.
64. Cao, Y.C.; Jin, R.; Mirkin, C.A. Science, 2002, 297, 1536.
65. Tong, Y.Y.; Wieckowski, A.; Oldfield, E. J. Pjys.Chem.B. 2002, 106, 2434.
66. Long, N.J.; Marzke, R.F.; Mckelvy, M.; Glaunsinger, W.S. Ultramicroscopy, 1986, 20, 15.
67. Yeadon, M,; Yang, J.C.; Averback, R.S.; Bullard, J.W.; Olynick, D.L.; Gibsonm J.M. Appl. Phys. Lett. 1997, 71, 1631.
68. Zhang, W.J.; Miser, D.E. J. Nanopart Res. 2006, 8, 1027.
69. Helland, A.; Scheringer, M.; Siegrist, M.; Kastenholz, H.G.; Wiek, A.; Scholz, A.W. Environ. Sci. Technol. 2008, 42, 640.
70. Conti, J.A.; Killpack, K.; Gerritzen, G.; Huang, L.; Mircheva, M.; Delmas, M.; Harthorn, B.H.; Appelbaum, R.P.; Holden, P.A. Environ. Sci. Technol. 2008, 42, 3155.

# 2 Processing of Polymer Nanocomposites: New Developments and Challenges

*Sophie Peeterbroeck[1], Michaël Alexandre[1], and Philippe Dubois[2]*

[1]Materia Nova ASBL, Avenue N .Copernic 1, B-7000 MONS, Belgium
[2]Service des Matériaux Polymères et Composites, Université de Mons-Hainaut,
Place du Parc 20, B-7000 MONS, Belgium
philippe.dubois@umh.ac.be

**Abstract**

Organomodified nanoclays and carbon nanotubes can be used to prepare polymer nanocomposites with enhanced properties. Several methods have been developed to easily obtain well-dispersed nanofillers in polymer matrices, leading to materials with satisfactory properties. Some processes are characteristic for thermoplastics and some others are only applicable to thermosets and cross-linkable elastomers. However, some processes can also, after adequate modifications, be applied to both thermoplastics and thermosets. Actually, industry is searching for easy and universal method to prepare polymer nanocomposites, so researchers have found some alternative or some "combined classical ways" such as compatibilization ways or polymerization-filling technique to obtain nanocomposite materials with finely dispersed nanofillers throughout the polymer matrix.

## 2.1. Introduction: Overview on Polymer Nanocomposites

Nature uses composites for all its hard materials. These are complex structures consisting of continuous and discontinuous fibrous or particulate materials embedded in an organic matrix acting as glue. For instance, wood is a composite of fibrous cellulose embedded within lignin and bone is a composite of collagen and other proteins gluing calcium phosphate crystals. The shells of mollusks are made of layers of hard aragonite separated by a protein binder.

Modification of organic polymers through the incorporation of additives results in mixtures characterized by unique micro- and macrostructures responsible for their properties. The primary reasons for using additives are as follows:

- modification and enhancement of properties;
- cost reduction; and
- improvement and control of processing characteristics.

Inorganic additives for polymer composites have been variously classified as reinforcements, fillers, or reinforcing fillers, which are characterized by typical morphologies (fibers, flakes, spheres, particulates). The additives may be continuous, for example, long fibers or ribbons, or discontinuous, for example, short fibers or platelets. Familiar examples are the well-known long fiber based thermoset laminates used in the automotive and aeronautic industries that are usually classified as high-performance polymer composites. On the other hand, discontinuous additives are usually dispersed in thermoplastics and are classified as lower performance polymer composites.

Reinforcing fillers are characterized by relatively high aspect ratio, $\alpha$, defined as the ratio of length to diameter for fibers and tubes, or the ratio of diameter to thickness for platelets and flakes. A useful parameter for characterizing the effectiveness of a filler is the ratio of its surface area, $A$, to its volume, $V$, which needs to be as high as possible for an effective reinforcement.

In general, parameters affecting the properties of polymer composites, whether filled with continuous or discontinuous reinforcements, include:

- the properties of the additives (inherent properties, size, shape);
- the composition;
- the interaction between the components; and
- the method of fabrication.

The concentration and inherent properties of the additive, as well as its interactions with the matrix, are important parameters controlling the processability of the polymer composites. In developing reinforcing fillers, process or material modifications aim at increasing the aspect ratio of the dispersed particles and improving their compatibility and interfacial adhesion with the chemically dissimilar polymer matrix. Such modifications may enhance and optimize not only the primary function of the reinforcing filler (for instance, to modify the mechanical properties) but also may introduce or enhance additional functions (for instance, barrier, or fire properties).

Global demand for fillers/reinforcing fillers, including calcium carbonate, aluminum trihydrate, talc, kaolin, mica, wollastonite, glass fiber, aramid fiber, carbon fiber, and carbon black for the plastics industry was estimated to be about 15 million tons a year in 2003. Use-markets are building/construction, transportation, furniture, industrial/machinery, electrical/electronics, packaging, automotive. Among polymers, PVC is still the plastic with the highest filler usage, followed by polyolefins, nylons, and polyesters.

Some exciting new application areas for composites containing nanofillers such as nanoclays, nanosilicates, carbon nanotubes (CNTs), ultrafine $TiO_2$, talc, and hydroxyapatite are:

- as structural materials with improved mechanical properties, barrier properties, electrical conductivity and flame retardancy;
- as high-performance materials with improved UV absorption;
- as barrier packaging for reduced oxygen degradation;
- as bioactive components for tissue engineering applications.

Recently, polymer/layered silicate (PLS) nanocomposites have attracted great interest due to their remarkable improvement in materials properties, when compared to virgin polymer or conventional micro- and macrocomposites, such as high moduli, increased strength and heat resistance, decreased gas permeability and flammability, and increased biodegradability of biodegradable polymers, for a low amount of introduced filler (usually less than 5 wt%) [1–18].

Although the intercalation chemistry of polymers when mixed with appropriately modified layered silicates and synthetic layered silicates is known for a long time, two major findings have stimulated the revival of interest in these materials: first, the report from the Toyota research group of a nylon-6/montmorillonite (MMT) nanocomposite [1], for which very small quantities of layered silicate resulted in pronounced improvements of thermal and mechanical properties of the polymer matrix; and second, the possible melt mixing without the use of organic solvents reported by Vaia et al. [19]. Today, efforts have been conducted globally, using almost all types of polymer matrices.

There is also a considerable interest in fabricating nanocomposites containing carbon nanotubes both from the point of view of fundamental theory and the development of applications. The first polymer nanocomposite using carbon nanotubes as a filler was reported in 1994 by Ajayan et al. [20]. The reported exceptional mechanical and physical properties combined with the low density of carbon nanotubes have motivated several researchers to investigate experimentally the mechanics of nanotube-based composite materials. However, Salvetat and collaborators [21] have demonstrated that the aggregates of nanotubes can affect the elastic properties to a large extent. So, uniform dispersion within the polymer matrix and improved nanotube/matrix wetting and adhesion represent critical issues in the processing of these nanocomposites.

## 2.2. Nanocomposites Based on Nanoclays (Layered Silicates)

### 2.2.1. The Most Commonly Used Silicates

The commonly used layered silicates for the preparation of PLS nanocomposites belong to the same general family of 2:1 phyllosilicates. Their crystal structure consists of layers made up of two tetrahedrally coordinated silicon atoms fused to an edge-shared octahedral sheet of either aluminum or magnesium hydroxide (Figure 2.1). The layer thickness is around 1 nm, and the lateral dimensions of these layers may vary from 30 nm to several microns or larger, depending on the particular layered silicate. Stacking of the layers leads to a regular van der Waals gap between the layers called the interlayer spacing.

Montmorillonite, hectorite, and saponite are the most commonly used layered silicates.

Two particular characteristics of layered silicates are generally considered for PLS nanocomposites. The first is the ability of the silicate particles to disperse into individual layers. The second characteristic is the ability to fine-tune their surface chemistry through ion-exchange reactions with organic and inorganic cations. These two characteristics are interrelated since the degree of dispersion of layered silicate in a particular polymer matrix depends on the interlayer environment.

### 2.2.2. Organomodification by Organic Cations

Natural layered silicates usually contain hydrated $Na^+$ or $K^+$ ions [22]. Obviously, in this pristine state, layered silicates are only miscible with hydrophilic polymers, such as poly(ethylene oxide) (PEO) [23] or poly(vinyl alcohol) (PVA) [24]. To render layered silicates miscible with other polymer matrices, one must convert the normally hydrophilic silicate surface to an organophilic one, favoring, therefore, the intercalation of a larger range of synthetic polymers. Generally, this can be performed by ion-exchange reactions

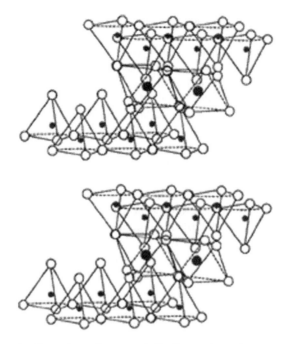

**Figure 2.1**   Schematic illustration of a 2:1 phyllosilicate (from http://www.mycoad.com/eng_5_15.htm).

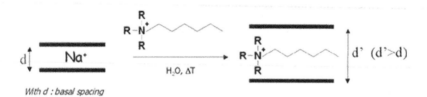

**Figure 2.2**   Sketch for the organomodification of montmorillonite.

(Figure 2.2) with cationic surfactants including primary, secondary, tertiary and quaternary alkylammonium or alkylphosphonium cations. Alkylammonium or alkylphosphonium cations lower the surface energy of the inorganic host and improve the wetting characteristics of the polymer matrix, and result in a larger interlayer spacing. Additionally, the alkylammonium or alkylphosphonium cations can provide functional groups that can react with the polymer matrix, or in some cases initiate the polymerization of monomers to improve the strength of the interface between the inorganic and the polymer matrix.

## 2.2.3.  The Various Processing Techniques

The intercalation methods are divided into three main groups: intercalation from solution, *in situ* intercalative polymerization, and melt intercalation. Intercalation of polymer or prepolymer from solution is based on a solvent system in which the polymer or prepolymer is soluble and the silicate layers are swellable. The layered silicate is first swollen in a solvent, and when the polymer and layered silicate solutions are mixed, the polymer chains intercalate and displace the solvent within the interlayer of the silicate. After that, the solvent is removed. In the *in situ* intercalative polymerization method, the layered silicate is swollen within the liquid monomer or a monomer solution so the polymer formation can occur between the intercalated sheets. Polymerization can be initiated either by heat or radiation, by the diffusion of a suitable initiator or by an (organic) initiator or catalyst located beforehand through cation exchange inside the interlayer before the swelling step.

The melt intercalation method involves heating a mixture of polymer and organomodified layered silicate above the softening point of the polymer. This last method has great advantages over either *in situ* intercalative polymerization or polymer solution intercalation, because of the absence of organic solvents and it is compatible with current industrial process, such as extrusion and injection molding.

The resulting morphology of the composite and its related properties are directly linked to the nature of the components, but also strongly depend on the method employed in the preparation of the so-called nanomaterials.

### 2.2.3.1. *Intercalation in Solution*

This technique, partially devoted to water-soluble polymers, and called exfoliation–adsorption, consists to form highly swollen stacks of clays or even to exfoliate the clay into single layers using a solvent in which the polymer, or the prepolymer, is soluble. Indeed, layered silicates can be easily swollen and sometimes dispersed in an adequate solvent because of the weak forces that stack the layers together. The clay exfoliation process can be assisted further through mechanical stirring or by exposure to ultrasound. The polymer can, therefore, crawl into the interlayer spacing of the swollen clay more easily. In the final step, the solvent is evaporated, usually under vacuum. The mixture can also be precipitated into a large excess of a nonsolvent of the polymer. During these processes, the sheets tend to stack again, sandwiching the polymer chains. This process leads at best to a well-ordered multilayer structure. During polymer intercalation from solution, a relatively large number of solvent molecules need to be desorbed from the host to accommodate the incoming polymer chains. The desorbed solvent molecules gain one translational degree of freedom, and the resulting entropic gain compensates for the decrease in conformational entropy of the confined polymer.

Several nanocomposites can be prepared by this technique, such as those based on PEO, poly(imide) by using poly(amic acid) as precursor, nitrile-based copolymer, polyethylene-based polymer, poly(methyl methacrylate) (PMMA) through emulsion polymerization but also polylactide and poly($\varepsilon$-caprolactone) [25–30].

Intercalation in solution remains an interesting process for the preparation of intercalated nanocomposites, especially for water-soluble polymers. It is also the only possible process for the production of nanocomposite based on polymer that cannot be molten such as poly(imide). However, due to the large amount of solvents and the relatively low degree of exfoliation that can be achieved through this process, it remains only scarcely used for the preparation of delaminated nanocomposites.

### 2.2.3.2. *In situ Polymerization Method*

The second used technique to prepare nanocomposite is the *in situ* polymerization of the monomer directly in the presence of the clay. In comparison to melt intercalation, where it is the polymer chains that have to crawl into the interlayer space of the filler, the intercalation of the monomers, small organic molecules compared to the corresponding macromolecules, is thermodynamically easier. In some cases, the progressive growth of the macromolecule during the propagation step allows for the exfoliation of the clay platelets, whereas the melt blending leads to intercalated or exfoliated species, depending on thermodynamic factors. However, the major inconvenient of this *in situ* intercalative polymerization technique consists in the extrapolation of the process to large industrial scale, or the use of organic solvent when the monomer is a solid with a high melting temperature.

Practically, the layered silicate is first swollen within the monomer, liquid, molten, or in solution, so as the polymer formation can occur in-between the intercalated sheets. This step requires a certain period of time, which depends on the polarity of the monomer molecules, the surface nature of the clay, and the swelling temperature. The nature of the organically modified layered silicate employed is thus critical in order to maximize the interaction between the layers surface and the monomer and promote its intercalation into the interlayer spacing, preliminarily to the polymerization initiation. In the case where a solid monomer is put in solution, the swelling phenomena of clays results from a balance between the interlayer cohesive forces and the attractive forces between the solvent and the interlayer cations [31]. The role of the solvent is to improve the wetting and penetration of the monomer in the clay interlayers.

Polymerization is then initiated by heat, irradiation, or the diffusion of a suitable initiator/catalyst. Alternatively, an (organic) initiator or catalyst can be anchored to the silicate layer surface by cationic exchange before the swelling step by the monomer. In this case, the clay organomodifier plays the role of both a swelling and compatibilizing agent as well as the polymerization promoter.

Such a technique gives the possibility to obtain intercalated as well as exfoliated nanocomposites depending on both monomer and clay used. It was largely employed in order to prepare thermoplastic-based nanocomposites, with several types of polymer matrices, that is, polyacetylene, polystyrene (PS), polyethylene, PMMA, polyurethane, aliphatic polyesters (poly($\varepsilon$-caprolactone), polylactide,...), or polyamides [32–42].

As an example, *in situ* intercalative polymerization was studied as an alternative solution to mechanical blending for the preparation of nylon-6 clay nanocomposites. Indeed, the elaboration of nylon-6/clay nanocomposites by melt kneading commercial nylon-6 and sodium MMT in a twin screw extruder gave a phase separated conventional (micro)composite. So, polymerization of $\varepsilon$-caprolactam (the nylon-6 monomer) was envisaged directly in the interlayer space of montmorillonite to disperse each platelet of silicate into a nylon-6 matrix at the molecular level. The choice of the montmorillonite ion exchanged with 12-aminolauric acid was driven by two crucial parameters: the good swelling of the clay into the liquid monomer and the possibility to initiate the ring-opening polymerization of the $\varepsilon$-caprolactam by the carboxylic acid groups left available by the organomodifier. The advantages of single silicate layers uniformly dispersed at the nanometer scale in nylon-based nanocomposites were first demonstrated by the Toyota's Research Centre in Japan [43]. Even at filler content as low as 1wt%, these materials exhibit largely and extraordinarily improved mechanical resistance (tensile and impact stress), a better heat resistance, and thermal stability compared to conventional composites or to the virgin matrix.

Besides thermoplastic matrices, *in situ* intercalative polymerization has also been explored to prepare thermoset-based nanocomposites or to prepare nanostructure with elastomeric matrix [44, 45]. In addition to mechanical properties, the barrier properties could also be improved as the permeability in the nanocomposite is dramatically reduced in respect with those characterizing the polymer, as one can consider the silicate layers impermeable to a variety of gases and solvents. Finally, depending on the type of polymer matrix, the nanocomposites can display interesting properties in the frame of ionic conductivity or in optical applications.

*In situ* polymerization processes provide a very powerful technique to prepare exfoliated nanocomposites, especially when initiation is performed from species anchored onto the layered silicate sheets. In this case, the continuous growth (especially when controlled polymerization processes are used) of the polymer chains from the clay surface promotes exfoliation, even if the thermodynamic interactions between the organomodified clay and the polymer matrix would lead to intercalation and/or phase separation. However, the molecular parameters of the *in situ* produced polymer matrix are dependent on the amount of initiator, which is also dependent on the amount of clay in the final materials, meaning that the properties of the matrix is intimately linked to the quantity of layered silicate, that might reduce the scope of this technique in directly producing industrially interesting materials.

### 2.2.3.3. *Melt Intercalation*
*The Simple Melt Blending*   In melt intercalation, the layered silicates are mixed with the polymer matrix in the molten state. Depending on the mixing conditions, as the temperature or the application of shear (e.g., extruder, two-roll mixer, or counter-rotating chamber), but also if the clay layer surface is sufficiently compatible with the polymer, the polymer chains can crawl into the interlayer space of the filler in order to form either intercalated or exfoliated morphologies.

The advantages of direct melt intercalation are considerable. The absence of solvent makes direct intercalation an environmentally and economically advantageous method.

The thermodynamics driving the intercalation of polymer chains in the molten state into the interlayer spacing of a layered aluminosilicate was studied at first by Vaia and Giannelis [46]. They found that, assuming that the configurations and interactions of the various constituents of a nanocomposite are independent, the free energy change of nanocomposite formation is separable into independent entropic and enthalpic terms. The entropic term is then the sum of the configurational changes associated with the polymer and the silicate (including the alkyl ammonium chains in organomodified clays). In fact, although the confinement of the polymer chains inside the silicate galleries results in a decrease in the overall entropy of the macromolecular chains, this entropic penalty of polymer confinement may be compensated, for gallery heights up to the length of the fully extended tethered chains, by the increased conformational freedom of the tethered chains as the layers separate.

When the total entropy change is small due to a low increase in the gallery spacing, limited changes in the internal energy of the system will determine if intercalation is thermodynamically possible. Complete layer separation depends on the formation of favorable polymer-organomodified layered silicate interactions that overcome the penalty of polymer confinement. The enthalpy of mixing has been, therefore, classified in two components: unfavorable apolar interactions arising from interactions between polymer and surfactant (apolar) chains, and polar interactions that originate from the Lewis acid/Lewis base character of the polar layered silicates interacting with the polymer chains.

Since in most conventional organomodified silicates the surfactant chains are apolar, dispersion forces dominate polymer–surfactant interactions. On the other hand, a favorable energy decrease is associated with the many favorable polymer–surface polar interactions. For alkylammonium-modified layered silicates, a favorable energy change is accentuated by maximizing the magnitude and number of favorable polymer–surface interactions while minimizing the magnitude of unfavorable apolar interactions between the polymer and the tethered surfactant chains. If complete layer separation is achieved, the total entropy change is near zero again, and the potential stability of the nanocomposite may be thought of in terms of a blend of two macromolecules. Similar works using self-consistent field theory were carried out to study the intercalation of macromolecules in unmodified MMT [47, 48].

It was shown that it is theoretically possible to promote the exfoliation of an unmodified MMT by melt blending it with a polymer mixture comprising the desired polymer matrix and a small amount of end-functionalized polymer whose terminal function could strongly interact with the silicate layers. This end-functionalized polymer could also be replaced by a diblock copolymer, one sequence of which intercalate while the other one is either of the same nature than the polymer matrix or at least compatible with it. Under these conditions, but without taking into account any kinetic factors, exfoliation would occur at a fraction of end-functionalized polymer or copolymer as low as 5 mol% [47].

Beside the intercalation thermodynamics, Vaia and co-workers have studied in details the kinetics of polymer chains intercalation into layered clays using XRD and TEM [49]. In this way, it was observed that increasing the blending temperature or lowering the molecular weight of the polymer chains lead both to an increase in the intercalation rate.

While the importance of chemistry used to modify the surface of the clay (usu-ally montmorillonite) was well known [50, 51], the role of the process conditions has been for a long time forgotten. It was necessary to await works by Paul's research team on mixtures of either polyamide 6 (PA6) or poly(propylene) with two different organically modified montmorillonites (one modified by dimethyl di(hydrogenated tallow) ammonium (Cloisite®15A) and the other by bis(2-hydroxyethyl) methyl tallow ammo-nium (Cloisite®30B)) to really appreciate the influence of the process on the resulting materials. They employed, for the mechanical kneading, various extruder configurations, with single or twin screws, allowing them to apply low, medium, or high shear in function of the geometry of the screws.

Increasing the mean residence time in the extruder using back-mixing generally im-proves the exfoliation and dispersion. However, there appears to be an optimum extent of back-mixing, and an optimum shear intensity; excessive shear intensity or back-mixing apparently causes poorer delamination and dispersion. From XRD and TEM analyses, one can clearly observe the influence of the configuration of the extruder, leading to intercalated or exfoliated nanocomposites.

It is also important to note that for polymers that require high melt-processing tempera-tures, the thermal stability of the organic components of the modified clay becomes a sig-nificant issue for several reasons [52]. For example, degradation of alkyammonium cations through Hoffman degradation mechanism upon high-temperature polymer processing can promote some degradation in polyamide or aliphatic polyesters. In addition, a variety of metal ions exist in natural clays and the degradation products may trigger various reactions and color formation [53].

On the basis of these results, the authors propose a mechanism for delamination and dispersion of the filler in the polymer matrix (Figure 2.3). The proposed mechanism, based on the relationship between the compatibility of the chemistry of the clay treatment/resin matrix and the process conditions used to prepare a nanocomposite, is described in three cases. The first case (Figure 2.3, point (1)) is chemistry dependent. When the clay chemical

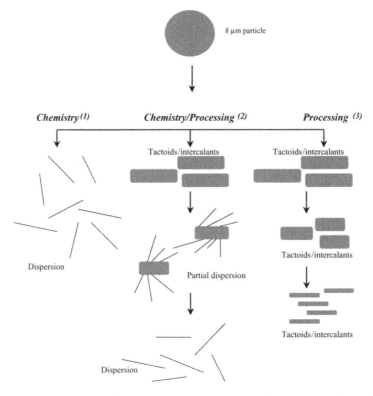

**Figure 2.3** Proposed mechanism of how the organoclay particles disperse into polymer during melt processing – influence of both chemistry and extrusion conditions on the resulting morphology, adapted from [20].

treatment and the resin are compatible, almost any set of processing conditions (except for the single screw extruder, but including a Brabender counter-rotating chamber or a two-roll mixer) can be used to form an exfoliated nanocomposite. In the second case (Figure 2.3, point (2)), clay chemical treatment and polymer are marginally compatible. In this situation, the process conditions can be optimized to give a good exfoliated nanocomposite. The organoclay chemical treatment and the resin matrix are compatible enough so that extrusion conditions can be varied to optimize delamination and dispersion. When there is no apparent compatibility of the clay chemical modification and the polymer matrix (Figure 2.3, point (3)), processing can be optimized to give intercalants or tactoids that are minimized in size at best. However, even under optimized processing conditions, no trace of exfoliation is observed. Moreover, two delamination pathways can be envisaged (Figure 2.4). In pathway 1, stacks of platelets are decreased in height by sliding platelets apart from each other, a pathway that requires high shear intensity. Pathway 2 shows polymer chains entering the clay galleries pushing the end of the platelets apart. This pathway does not require high shear intensity but involves diffusion of polymer into the clay galleries (driven by either physical or chemical affinity of the polymer for the organoclay surface) and is thus facilitated by increased residence time in the extruder. As more polymer enters and goes further in between clay platelets, especially near the edge of the clay galleries, the platelets peel apart. It is important to note that the platelet in the

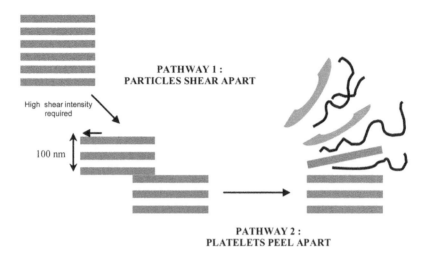

**Figure 2.4**   Delamination pathways. adapted from [20].

stack may be able to curl away as polymer enters from the edge since the platelets are quite flexible and able to bend. From the TEM observations, Paul and his co-workers have demonstrated that the two pathways are involved.

*The Masterbatch Technique*   Undoubtedly, the key-challenge remains reaching a high level of nanoparticle dissociation ultimately leading to their fine individual dispersion upon melt blending within the selected polymer matrix. In that context, it is worth pointing out the exceptional efficiency of the *in situ* intercalative polymerization process as catalyzed directly from an initiator or a catalyst anchored at the nanofiller surface that allows for the complete destructuration of the native filler aggregates. Dissociated nanoparticles are accordingly coated, their surface being homogeneously covered/grafted by the *in situ* grown polymer chains. Interestingly enough, such surface-treated nanoclays can be added as "masterbatch" in commercial polymeric matrices. As a result of the predestructuration of the nanofillers by this "grafting-from" technique, it comes out that the resulting polymer nanocomposites display much higher thermomechanical properties even at very low nanofiller loading. The key-role of *in situ* intercalative polymerization thus promoted from the nanofiller surface has been recently highlighted [54].

A two-step route to polymer/layered silicate nanocomposites characterized by a large extent of nanoplatelet delamination is presented. It consists first in the preparation of poly($\varepsilon$-caprolactone) (PCL)-grafted clay masterbatches ($\sim$30 wt% clay) by *in situ* intercalative polymerization of $\varepsilon$-caprolactone (CL) in bulk. The CL polymerization is promoted at the surface of organomodified clays bearing targeted amounts of hydroxyl groups used as initiating species. The influence of hydroxyl concentration on clay platelet coverage by PCL has been studied by atomic force microscopy. In a second step, these masterbatches are all readily dispersed in commercial matrices (PCL, PVC, chlorinated polyethylene) via rather conventional melt blending processes. The morphology of the resulting nanocomposites has been characterized as well as their thermal and mechanical properties which are compatible with a high level of exfoliation.

In the same way, a two-step masterbatch process for preparing nylon-6 nanocomposites that provides good exfoliation and low melt viscosities (for shorter cycle times in injection molding) has been investigated by Shah et al. [55]. In the first step, masterbatches of high molecular weight (HMW) nylon-6 with different clay contents were prepared by melt processing to get good exfoliation. In the second step, the masterbatch was diluted with low molecular weight (LMW) nylon-6 to the desired MMT content to reduce the melt viscosity. It was difficult to produce masterbatches containing more than 20 wt% MMT (or 28.5 wt% organoclay) owing to problems of stranding the extrudate arising from its lower melt strength and of pelletizing the solidified strand because of its hardness. Masterbatches containing 4 and 8.25 wt% MMT were quite well exfoliated, and nanocomposites prepared by diluting them with LMW nylon-6 exhibited properties close to those seen with composites based on HMW nylon-6 alone. On the other hand, masterbatches containing 14 and 20 wt% MMT were not so well exfoliated; however, mechanical property, TEM, and WAXD analysis of nanocomposites prepared by diluting these masterbatches revealed better exfoliation than corresponding nanocomposites prepared directly from LMW nylon-6. A distinct trade-off between the tensile modulus of these nanocomposites and the reduction of melt viscosity was observed. Nanocomposites prepared from HMW masterbatches with a lower MMT concentration (8.25 wt%) offer a significant decrease in melt viscosity over those prepared directly from HMW nylon-6, for a limited reduction in modulus. On further increasing the MMT content of the masterbatch, the trade-off becomes less favorable. However, if it is absolutely necessary to have throughout rates similar to LMW nylon-6, the use of nanocomposites prepared from a more concentrated masterbatch (8.25 wt%) could offer up to a 10% improvement in modulus over nanocomposites prepared from LMW nylon-6 only. While two extrusion steps were used in this work, the concept illustrated could be implemented in a single extrusion through the use of larger twin-screw extruders that have downstream feed ports. In this case, the organoclay and HMW nylon-6 would be fed to the hopper, while LMW nylon-6 could be injected in a downstream feed port. However, one drawback of this method may be found in the large shear stress needed during extrusion masterbatch preparation that could significantly degrade sensitive polymers and induce a decrease in the properties of the final nanocomposite.

*Processes Using Supercritical Carbon Dioxide*   The first strategy involves subjecting the nanoscale filler, clay (or graphite), to supercritical carbon dioxide ($scCO_2$) for a certain period of time and then rapidly depressurizing the medium. High diffusivity and low viscosity of $scCO_2$ enable its diffusion between the layers. During depressurization, it expands and pushes the layers apart resulting in significant delaminated structures. A second strategy involves introducing a LMW organic compound soluble in $CO_2$ (i.e., a coating agent) in the $scCO_2$ processing step. Again, favorable transport properties of the solution facilitate diffusion of the organic compound along with $CO_2$ between the layers. During depressurization, the coating agent precipitates and coats the layers while $CO_2$ expands and pushes the layers apart. A third strategy exploits $scCO_2$ ability to swell and plasticize a polymer and facilitate its mixing with the fillers. The polymer and the nanofiller are mixed thoroughly and processed in $scCO_2$ at a temperature and pressure suitable for the given polymer/filler system followed by depressurization to remove $CO_2$. The degree of filler dispersion is a function of the processing temperature, pressure, and depressurization rate. As examples, polydimethylsiloxane (PDMS)/clay, PS/clay, and polyvinylmethylether (PVME)/clay nanocomposites have been investigated [56–58]. Some benefits of $scCO_2$ processing include the following:

(1) use of natural clay instead of chemically modified organoclays;
(2) lower processing temperatures avoiding thermal degradation;
(3) elimination of organic solvents in processing.

The studies indicate that scCO$_2$ processing produces significant dispersion of nanoscale fillers resulting in improved properties of the polymer nanocomposites. More precisely, the authors found that the presence of scCO$_2$ promotes significant increase in the basal spacing of the clay and thereby may enhance the ease of the polymer intercalation into the galleries of the clay.

## 2.3. Nanocomposites Based on CNTs

### 2.3.1. The Carbon Nanotubes

Carbon nanotubes possess high flexibility, low mass density, and large aspect ratio (typically ca. 300–1000) [59]. CNTs have a unique combination of mechanical, electrical, and thermal properties that make nanotubes excellent candidates to substitute or complement the conventional nanofillers in the fabrication of multifunctional polymer nanocomposites and great enthusiasm exists among researchers around the world to explore the huge potential of these specific nanoparticles. Some nanotubes are stronger than steel, lighter than aluminum, and more conductive than copper. For example, theoretical and experimental results on individual single-wall carbon nanotubes (SWNT) show extremely high tensile modulus (640 GPa to 1 TPa) and tensile strength (150–180 GPa) [60, 61]. Depending on their structural parameters, SWNT can be metallic or semiconducting, which further expands their range of applications. The first TEM evidence for the tubular nature of some nanosized carbon filaments is believed to have appeared in 1952 in the *Journal of Physical Chemistry of Russia* [62]. In 1991, a Japanese microscopist, Sumio Iijima, also observed graphitic carbon needles, ranging from 4 to 30 nm in diameter and up to 1 mm in length, as by-products of the arc-discharge evaporation of carbon in an argon environment and his paper reached a tremendous impact [63].

Nanotubes can be synthesized in two structural forms, SWNTs, and multiwall carbon nanotubes (MWNTs). SWNTs consist of a single sheet of carbon rolled up into a cylinder and typically have diameters from 1 to 5 nm, whereas MWNTs have an arrangement of coaxial tubes of carbon sheets forming a tube-like structure (Figure 2.5). Each MWNT consists of 2–50 of these tubes. Usually MWNTs have inner diameters of 1.5–15 nm and outer diameters of 2.5–30 nm [64]. The nanotubes observed by Iijima were MWNTs.

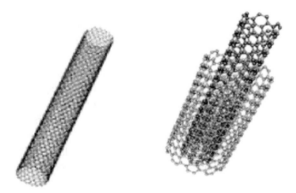

**Figure 2.5**  Schematic representation of a single-wall (A) and a multiwall carbon nanotube (B) (from www.ewels.info).

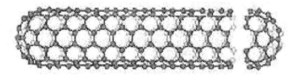

**Figure 2.6**  Schematic representation of SWNT with its end cap (from www.nanomedicine.com).

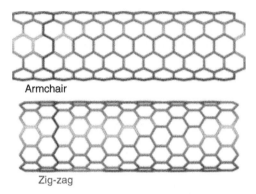

Armchair

Zig-zag

**Figure 2.7**  Schematic representation of an armchair tube and a zigzag tube (from www. chemsoc.org).

The terminations of the nanotubes are often called caps or end caps, and consist of a "hemisphere" of fullerene. Each cap contains six pentagons and an appropriate number and placement of hexagons that are organized to fit perfectly to the long cylindrical section (Figure 2.6).

The primary symmetry classification of a CNT defines either achiral or chiral CNTs. More precisely, the CNTs are described using one of three morphologies: armchair, zigzag (Figure 2.7), and chiral [65]. The electrical conductivity of nanotubes depends on their chirality [66].

An achiral CNT is defined by a CNT whose mirror image has an identical structure to the original one. There are only two cases of achiral nanotubes: armchair and zigzag nanotubes. The names of armchair and zigzag arise from the shape of the cross-sectional ring. Chiral nanotubes exhibit a spiral symmetry whose mirror image cannot be superposed on to the original one. This tube is called a chiral nanotube since such structures are called axially chiral in the chemical nomenclature.

Another structural aspect of CNTs is their ability to self-organize themselves into "ropes." CNT ropes are much longer than any individual nanotube. These ropes consist of many tubes (typically 10–100) running together along their length, in van der Waals contact with each other.

CNTs are typically 100–1000 nm in length, whereas CNT ropes are essentially endless, branching off from another one, and then joining others. These ropes are useful for offering very long conductive pathways well below what would normally be the loading required to reach a "percolation threshold" for conductivity [67].

Due to the size scale of nanotubes, a direct measurement of properties of interest such as mechanical properties and thermal/electrical conductivity is difficult using conventional experimental methods. Therefore, mathematical modeling has been primarily used to evaluate those properties. These models are often derived from models used for similar structures, such as graphite, with modifications to take into account the peculiar tubular shape. For example, molecular dynamics, empirical potentials and first-principles total-energy, continuum shell, and empirical lattice model were used to describe the elastic properties of nanotubes [68].

The empirical lattice models, as used to calculate the elastic properties of graphite, led to tensile modulus values in the range of 1 TPa for single- and multiwall nanotubes [69]. These values compared well with diamond and out-performed conventional carbon fibers. These exceptionally high theoretical values have led to increasing interest in nanotubes as both structural and conductive materials and have, therefore, led to the development of techniques to measure nanotube properties experimentally. The methods to measure the elastic properties of individual nanotubes include transmission electron microscopy [70, 71] and atomic force microscopy [72, 73]. The elastic properties of bundles of nanotubes have also been measured [74, 75] and the experimental values measured ranged from significantly below theoretical values to values in agreement with the models. These methods have highlighted tensile modulus and strength values for single- and multiwall nanotubes ranging from 270 GPa to 1 TPa and 11 GPa to 200 GPa, respectively [76].

In addition to the mechanical properties, research has been performed to determine the electrical and thermal conductivity of nanotubes.

The thermal conductivity $k$ of CVD-grown MWNTs measured from 4 to 300 K was found to vary as $T^2$, similar to that of graphite. As the diameter of the MWNT increases, the thermal conductivity versus $T$ becomes similar to the bulk measurements [77].

In a similar fashion, the static electrical conductive and superconductive nature of nanotubes was modeled based on the conductivity of the graphite sheet structure. The dimensional scale of the individual nanotubes eliminated many of the conventional methods used to measure conductivity. The transport experiments involved both two- and four-probe measurements on individual MWNTs [78, 79], SWNT bundles [80, 81], and isolated SWNTs [82].

Experimental four-probe measurements of individual nanotubes showed resistivities in the range of $5.1 \times 10^{-6}$–$5.8$ $\Omega$ cm (or conductivities ranging from 0.17 to 196 S/cm). As with the mechanical properties, the variation in the conductivities lies in the procedures used and the differences in the nanotube structure.

After quantifying the remarkable properties of CNTs, the challenge has consisted in translating these properties from nanoscale to macroscale. Currently many methods for producing macroscale nanotube materials have been investigated.

The most common method has been to incorporate them in a polymer matrix as a reinforcing material.

### 2.3.1.1. *The Different Processing Ways*
There is a considerable interest in fabricating polymeric composite materials containing CNTs from the point of view of both fundamental theory and application development.

The exceptional mechanical and physical properties combined with the low density of nanotubes have motivated researchers to experimentally investigate the properties of nanotube-based composite materials. In this respect, CNTs have been well recognized as nanostructural materials that can be used to improve mechanical, thermal, and electrical properties of polymer-based composites.

Various techniques are used to disperse CNTs in polymer matrices. The most straight-forward method is the direct melt blending with thermoplastics or the *in situ* formation of CNT-based composites by cross-linking reactions in the case of thermoset resins.

Previously, due to their limited availability and high cost, most CNT-reinforced composites were prepared in laboratories using the so-called solution-evaporation method [83]. This general procedure involved the mixing of the polymer dissolved in a solvent and the CNTs dispersed in the same solvent, with the aid of ultrasonication and finally casting films from the mixed solution. Haggenmueller et al. [84] developed an alternative melt-mixing method consisting of a combined solution-evaporation method to prepare thin films followed by repeated compression molding to obtain uniform films. Composites based on polypropylene (PP), acrylonitrile butadiene styrene copolymer (ABS), and PS have been prepared with adequately dispersed CNTs [85].

The dry powder mixing method has also been employed to produce PMMA composites [86]. This method is a combination of several protocols including solution-evaporation, sonication, kneading, and extrusion.

Until now, no simple method has been highlighted to easily prepare nanocomposites with well-dispersed CNTs. It has been proposed that the nanotubes should provide load transfer in the same way chopped glass fibers do in conventional composite systems. The property response of polymer/nanotube composites has, however, varied from no change to moderate increase in mechanical properties (25–50% stiffness increase; 80–150% toughness increase), electrical properties (with a percolation threshold below 0.1 wt%; conductivity 2 S/m), and thermal properties (125% increase at room temperature). However, optimal property improvements have rarely been achieved due to deficiencies in nanotube dispersion and alignment as presented below.

*Dispersion by Direct Mixing*     Undoubtedly, the direct mixing is a useful process from an industrial standpoint.

Accordingly, most of the researchers have largely studied the effects of CNTs on the properties of the polymer host and they highlighted some interesting results.

Theoretical work has also been carried out with the aim of modeling the mechanical properties of fiber-reinforced composites. While some of these models [87] are quite sophisticated, the simplest and most common is generally considered: the Halpin–Tsaï model. Accordingly, it was demonstrated that a good dispersion state of the fillers is necessary to get the best enhancement of the properties.

Unfortunately, whatever the polymer matrix used, nonhomogeneous dispersion of CNTs is generally observed. This is due to the strong $\pi-\pi$ interactions stabilizing the nanotube bundles.

Hereafter, some representative examples found in the scientific literature showing the typical effects of CNTs on the properties and the morphology of materials obtained by direct mixing will be developed.

*Melt Blending with Pristine CNTs*     Melt processing is a common alternative, which is particularly useful when dealing with thermoplastic polymers. Amorphous polymers, such as PS, can be processed above their glass transition temperature, while semicrystalline polymers, such as polyethylene (PE), need to be heated above their melting temperature to induce sufficient softening. Advantages of this technique are its speed and simplicity, not to mention its compatibility with standard industrial techniques [88, 89].

However, it is important for the processing conditions to be optimized, not just for different nanotube types, but also for the whole range of polymer–nanotube combinations. This is because nanotubes can affect melt properties such as viscosity, resulting in unexpected polymer degradation under conditions of high shear rates [90].

As mentioned above, in spite of nonhomogeneous dispersion, some interesting properties have been highlighted. For example, in PMMA/MWNTs composites readily fabricated by simple melt-blending, the storage modulus of the polymer was significantly increased, particularly at high temperature, by the incorporation of nanotubes. Moreover, MWNTs also promoted a stabilizing effect to the PMMA matrix since its onset of thermal degradation occurs at higher temperature [86, 91].

In PP/MWNTs, PS/MWNTs, and ABS/MWNTs composites prepared by shear mixing, it was demonstrated that the nanotube concentration at which electrical conductivity appeared varied with the host polymer [92]. In PP, this was at an MWNTs content as low as 0.05 vol%, while higher concentrations were required for PS and particularly for ABS. The removal of defects and impurities from the nanotubes by graphitization at high temperature improved the electrical properties of the composites.

In PP/CNTs nanocomposites, the crystallization behavior, the thermal degradation, the flammability properties, and the strength properties have also been studied. It has been shown that even with poor dispersion, SWNTs act as nucleating agent for PP crystallization and that the nanotube additives greatly reduced the heat release rate of PP [93–96]. The strength properties of polypropylene were enhanced with SWNTs. For example, for 1 wt% loading of nanotubes in PP, the tensile strength was increased by 40%. At the same time, the modulus increased by 55%.

In PE/MWNTs nanocomposites in which both individual and agglomerations of MWNTs were observed, a percolation threshold at about 7.5 wt% was obtained and the electrical conductivity of PE was increased significantly by 16 orders of magnitude, from $10^{-20}$ to $10^{-4}$ S/cm. The temperature of crystallization and fraction of crystalline PE were also modified by incorporating MWNTs [97].

In polycarbonate (PC)/CNTs composites, the rheological and dielectrical properties have been determined [98, 99]. In this case, the rheological percolation threshold coincides with the electrical conductivity percolation threshold, which was found to be between 1 and 2 wt% nanotubes and it was also noticed that the percolation threshold is strongly dependent on the measurement temperature. It changes from about 5 to 0.5 wt% MWNTs by increasing the measurement temperature from 170 to 280°C. This temperature dependence cannot be explained by a classical liquid–solid transition but has been related to the existence of a combined nanotube–polymer network.

The effects of addition of CNTs in polyamides have also been studied. CNTs reinforced PA6 composites prepared by simple melt compounding showed significant improvement in mechanical properties compared with neat PA6 [100]. The tensile modulus, tensile strength, and hardness of the composites are greatly improved by about 115%, 120%, and 67%, respectively, when incorporating only 1 wt% MWNTs. SEM observations of the fracture surfaces of the composite indicate that a satisfactory dispersion of MWNTs throughout PA6 matrix and a strong interfacial adhesion between the matrix and MWNTs have been successfully achieved, which are responsible for the significant enhancements in mechanical properties.

The crystallization and melting behaviors of neat PA6 and PA6/MWNTs composites studied by DSC show two crystallization exotherms for PA6/MWNTs composites instead of a single exotherm for the neat matrix [101]. XRD results indicate that the nucleation sites provided by CNTs seem to be favorable to the formation of thermodynamically stable $\alpha$-phase crystals of PA6. The dominant $\alpha$-phase crystals in PA6/MWNTs composites may play an important role in the remarkable enhancement of mechanical properties.

All these examples show some interesting effects in terms of mechanical reinforcement, crystallization behavior, and percolation thresholds but these effects are always dependent on the studied matrix. It is clear that the direct melt preparation method does not generally give very good dispersion of the nanofillers in the different polymer matrices because the aggregates of nanotubes are highly stabilized and the compatibility between the polymer chains and CNTs is most often low. Moreover, too large shear stresses during the melt blending process can also promote CNTs breakage that can limit the expected increase in mechanical and electrical properties.

*Direct Dispersion of CNTs in Epoxy Resin*    Another easy direct mixing procedure consists in dispersing the MWNTs into a thermoset such as an epoxy resin with relatively low viscosity and using intensive shear mechanical stirring. Some representative examples with interesting properties are exposed hereafter.

Recently, an original "Velcro" effect at the nanoscale has been highlighted in microcracks (induced failure) of epoxy composites. It is due to the attractive van der Waals forces between the nanotubes present on each of the surfaces of epoxy material [102]. This could be interesting for application by creating nanotube-filled adhesive polymer films where adhesion between the nanotubes-filled polymer surfaces could be greatly enhanced by some kind of reversible "Velcro" effect.

The electrical properties of composites, produced by dispersing the high-conductivity CNTs in the insulating matrix of epoxy resin, have been investigated in the vitreous state. The experimental results showed a low percolation threshold $\rho c = 0.3$ wt% [103].

The effect of the aspect ratios of SWNTs and MWNTs to obtain an efficient stress transfer as well as the effect of the dispersion state of CNTs on rheological, mechanical, electrical, and thermal properties of epoxy-based nanocomposites have also been studied [104, 105]. From the morphological observation, it was found that the use of a solvent during the CNTs dispersion process leads to the better deaggregation of CNTs. It was also found that the nanocomposites containing poorly dispersed CNTs exhibited, from rheological point of view, a more solid-like behavior than ones with well-dispersed CNTs. Tensile strength and elongation at break of the nanocomposites with various dispersion states of CNTs were measured. Both well and poorly dispersed CNTs composites showed an electrical percolation threshold lower than 0.5 wt%.

Again, this direct preparation method does not give highly dispersed CNTs because of the strong interactions between the nanotubes and the lack of compatibility between the epoxy resin and CNTs.

Thus, there is a real need to enhance the compatibility between the components of the nanocomposites.

*Compatibilization Techniques*    Several compatibilization methods have already been developed to efficiently deaggregate the nanotube bundles and to disperse the individualized CNTs by direct mixing in polymer matrices in order to optimize the properties of the resulting materials.

To improve polymer/CNTs interactions, it is possible to either chemically modify the polymer chains in order to enhance their interactions with the $\pi$-system of CNTs or to functionalize the walls of the CNTs in order to improve the wetting of the nanotubes, as

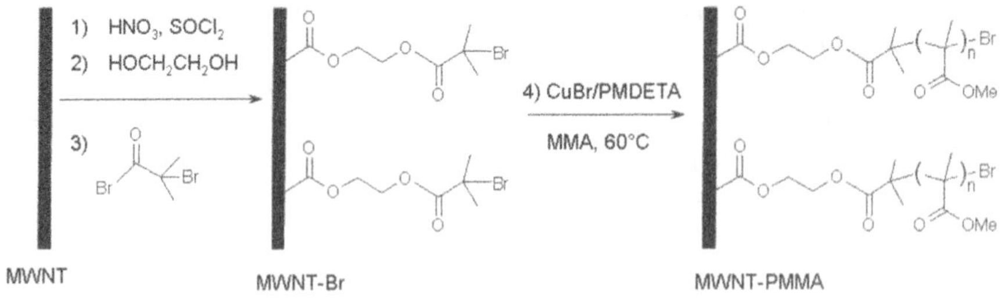

**Figure 2.8** ATRP "grafting from" approach to functionalize MWNTs by PMMA chains (adapted from [109]).

well as their dispersion in the polymer matrix. One can also use a third component such as a surfactant, in order to assist the exfoliation of the CNTs bundles and the subsequent incorporation of the individualized CNTs into the polymer matrix. All these methods are briefly discussed hereafter and illustrated with representative examples. The chemical functionalization and more precisely the sidewall carboxylic acid functionalization of SWNTs have been successfully used as a first step to prepare nanocomposites with enhanced properties [106]. The reaction of SWNTs with succinic or glutaric acid acyl peroxides in o-dichlorobenzene at 80–90°C resulted in the addition of 2-carboxyethyl or 3-carboxypropyl groups, respectively, to the sidewalls of the SWNTs. These acid-functionalized SWNTs were converted to acid chlorides by derivatization with $SOCl_2$ and then to amides with $\alpha, \omega$-diamines. The degree of SWNTs sidewall functionalization with the acid-terminated groups was estimated as 1 in 24 carbons on the basis of TG-MS data. The efficiency of the functionalization has been showed by an improved solubility in polar solvents (e.g., in alcohols and water) of the acid-functionalized SWNTs in comparison with the pristine SWNTs. Their incorporation into polymers is thus made easier.

Hydroxyl-functionalized SWNTs have been used to reinforce PVA [107]. The idea was that the surface–OH groups would hydrogen bond with the –OH of the PVA. Reasonably good modulus enhancement was observed with an increase from 2.4 to 4.3 GPa on addition of 0.8 wt% nanotubes, and a strength increase from 74 to 107 MPa was also observed.

The modification of CNTs with polymer chains leads also to a successful processing way to enhance the chemical compatibility and dispersion properties of CNTs, which enable both a more extensive characterization and subsequent chemical reactivity [108]. The modifications of CNTs with polymers could also improve the interfacial interaction with polymers by noncovalent attachment (polymer wrapping and adsorption) and by covalent attachment (via "grafting to" and "grafting from" techniques).

As a typical example, we find in literature, for example, an *in situ* ATRP "grafting from" approach to functionalize MWNTs by PMMA chains. Four steps are included in this strategy (Figure 2.8) [109]:

1. Carbonyl chloride groups functionalized MWNTs were prepared via the reaction of thionyl chloride with carboxylic acid functionalized MWNTs previously produced by oxidation of the native MWNTs with $HNO_3$;
2. Hydroxyl groups were introduced on the surface of MWNTs by reaction of MWNT-COCl with ethylene glycol generating MWNT-OH;
3. MWNT-Br were formed by reacting MWNT-OH with 2-bromo-2-methylpropionyl bromide;
4. MMA was polymerized from MWNT-Br, catalyzed by a soluble copper-based catalyst.

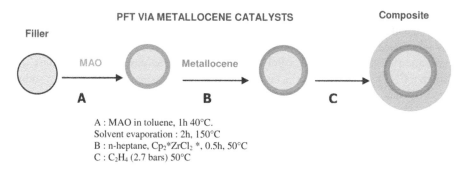

PFT VIA METALLOCENE CATALYSTS

Composite

Filler

MAO

Metallocene

A

B

C

A : MAO in toluene, 1h 40°C.
Solvent evaporation : 2h, 150°C
B : n-heptane, Cp₂*ZrCl₂ *, 0.5h, 50°C
C : C₂H₄ (2.7 bars) 50°C

**Figure 2.9**  Scheme of the polymerization-filling technique (PFT) applied to CNTs. MAO stands for methylaluminoxane and $Cp_2^*ZrCl_2$ for bis(pentamethyl-$\eta$-5-cyclopentadienyl) zirconium(IV) dichloride.

The MWNT-PMMA shows a relatively good solubility in non- to weakly polar solvents such as THF and $CHCl_3$, and a poor solubility in strong polar solvents. Interestingly enough, the approach can be extended to the copolymerization reaction, which may open an avenue for exploring and preparing novel CNT-based nanocomposites and molecular devices with tailor-made structures and properties.

The functionalization of SWNTs with $\varepsilon$-caprolactam via the "grafting from" process to prepare polyamide-functionalized CNTs by anionic ring-opening polymerization promoted from the nanotube surface has also been reported [110]. The SWNTs surface was activated by sodium that initiates the polymerization of the lactam monomer. This technique allowed to graft PA6 from about 70% of the treated SWNTs.

The surface of SWNTs was also successfully modified with polyethylene via *in situ* Ziegler–Natta polymerization [111]. In this case, the SWNTs were treated with triisobutylaluminum in hexane in order to anchor an active Ziegler–Natta catalyst ($MgCl_2/TiCl_4$) at the surface of the nanotubes. Upon ethylene addition, polyethylene was formed and precipitated onto the SWNTs. Scanning electron microscopy and solubility measurements showed that the surface of the SWNTs was covered with a PE. It is reported that after tentative solvent extraction, most of the PE chains remained at the nanotube surface more likely due to the strong interaction between SWNTs and PE.

In a similar way, Bonduel and colleagues obtained homogeneous surface coating of long MWNTs by *in situ* polymerization of ethylene as catalyzed directly from the nanotube surface-treated by a highly active metallocene-based complex (Figure 2.9) [112, 113]. This treatment allows for the break-up of the native nanotube bundles leading upon further melt blending with HDPE to performant PE nanocomposites.

In the same area of noncovalent immobilization of polymers at the surface of CNTs, surprising periodic pattern of polymeric materials on individual CNTs using a controlled polymer crystallization method has also been reported [114]. CNTs were designed with polymer lamellar crystals, standing perpendicular to the CNT axis, resulting in so-called nanohybrid shish-kebab (NHSK) structures.

The periodicity of the polymer lamellae varies from 20 to 150 nm. The kebabs are approximately 5–10 nm thick (along CNT direction) with a lateral size of ~20 nm to micrometers, which can be readily controlled by varying crystallization conditions. Both polyethylene and nylon-6,6 were successfully designed on CNTs. The formation mechanism was attributed to size-dependent soft epitaxy. The reported method opens a gateway to periodically patterning polymers and different functional groups on individual CNT in an ordered and controlled manner, an attractive research field that is yet to be explored.

As mentioned above, it is also possible to use surfactant molecules, to assist the incorporation of the CNTs into a polymer matrix. In most cases, the surfactants used to disperse CNTs are anionic surfactants such as sodium dodecyl sulfate (SDS), sodium dodecyl benzene sulfate (SDBS) but also some type of polysaccharide such Gum Arabic (GA). Generally, bundles of surfactant-dispersed CNTs are sonicated in an aqueous medium. During sonication, the provided mechanical energy overcomes the van der Waals interactions in the CNT bundles and leads to CNT "exfoliation," whereas at the same time surfactant molecules adsorb onto the surface of the CNTs walls [115]. The colloidal stability of the dispersion of CNTs with adsorbed surfactant molecules on their surface is guaranteed by electrostatic and/or steric repulsion [116–119].

The organization of surfactant molecules on the CNT surface has already been extensively investigated. Three main arrangements have been considered: structureless random adsorption on the CNT walls without any preferential arrangement of the head and tail, hemi-micellar adsorption on the CNT surface, and encapsulation of the CNTs in a cylindrical surfactant micelle. The majority of experimental (AFM and adsorption measurements) and theoretical studies dealing with the organization of surfactant molecules on CNT walls support the hemi-micelle configuration [120–125].

The incorporation of the filler into the polymer matrix is achieved by obtaining a colloidal system, that is, a mixture of CNTs and polymer particles, both stabilized by surfactant molecules. This colloidal suspension is then dried and processed to get a composite in the solid state; in most cases the dispersion and exfoliation of the CNTs are preserved in the polymer matrix.

The colloidal system could be stabilized by *in situ* polymerization. Alternatively, the colloidal system can be prepared by direct mixing of the CNTs and the polymer host particles, after the polymerization has already been carried out [126–128].

A relatively new approach to incorporate CNTs into a polymer matrix is based on the use of latex technology [129]. First, the surfactant is incorporated to the CNT aqueous mixture.

The resulting CNTs suspension is then centrifuged to remove catalyst particles and large, nonexfoliated CNTs bundles. The aqueous supernatant contains exfoliated CNTs or very small CNT bundles consisting mostly of two or three tubes. Depending on the quality of the CNTs, 15–50% of the material is lost during the centrifugation step. The supernatant is mixed with latex particles. After freeze-drying and subsequent melt-processing, a composite consisting of homogeneously dispersed SWNTs in a polymer matrix of choice can be finally obtained. Authors claim that the advantages of this technique are obvious: it is easy, versatile, reproducible, reliable, and allows a good incorporation of predominantly individual CNTs into a highly viscous polymer matrix. It does not require the use of toxic and inflammable solvents, which is safe and environmentally friendly. Besides, no difficult synthesis of special polymers has to be carried out. Basically, it also seems to exhibit a great degree of flexibility with respect to the choice of the matrix: a homogeneously dispersed CNTs network can be obtained for any kind of polymer, which can be produced by emulsion polymerization or which can be brought into a latex form. Authors claim that amorphous polymer latex such as PS, PMMA, or even semicrystalline polymer (e.g., polyethylene) can be used [130].

The investigations by Grunlan and co-workers have led to an alternative method, which consists of using a polymer brought into a latex formed in an artificial way, like poly(vinyl acetate) (PVAc), instead of a polymer latex directly synthesized by emulsion polymerization, for example [129]. Untreated SWNTs produced via the HiPCO process (29 wt% of metal catalyst impurities in the batch used) were exfoliated by sonication. Authors used GA as a stabilizing agent. Once stabilized, the SWNT dispersion was mixed with a PVAc emulsion to create a stable colloidal system leading to conductive composites after drying. The percolation threshold is about 0.04 wt % with maximum conductivity of 101 S/m at 4 wt% of SWNT loading. Grunlan claims that water-based CNT–polymer composites should have a lower percolation threshold than similar composites of which the preparation method is based on polymer solutions or melts. Due to the ability of the polymer solution or melt to surround added filler, CNTs can freely organize during drying or cooling, whereas in suspension or in polymer latex, solid particles create excluded volume. Consequently, the free volume available for the CNTs to form a conductive network is reduced and CNTs are pushed into the interstitial space between the polymer particles during the film formation process. This effect should help to significantly reduce the value of the percolation threshold. It has to be mentioned that this assumption is true as long as the drying process of the CNT–dispersion/polymer–emulsion system respects the polymer latex particle shape and does not result in flow of the latex particles.

Researchers have used these compatibilizing methods to obtain better dispersion of the CNTs in polymer matrices by direct mixing (melt blending and dispersion in solution). Some illustrating examples are given below.

*Direct Mixing with Compatibilized CNTs*    The effect of sidewall functional group on dispersion state of CNTs in polymer matrices has been studied.

For example, fluorinated single-wall CNTs (F-SWNTs) were dispersed in a PE matrix and the mechanical properties of these F-SWNT composites have been determined [131]. The study demonstrates that in comparison with polyethylene composites filled with pristine (purified) nanotubes, improved dispersion and interfacial and mechanical properties are achieved for F-SWNT-loaded matrices due to chemical functionalization. In addition, the observed partial removal of functional groups from the F-SWNTs during melt processing with polyethylene by shear mixing suggests a possibility of *in situ* covalent bonding between the nanotubes and the matrix, which ultimately results in mechanical reinforcement of the composite.

To optimize the dispersion of the nanofillers in epoxy resin and enhance the resulting properties, the processing-aids frequently applied are stirring in a solvent, using functionalized CNTs, or ultrasonication. Some pertinent examples are exposed hereafter.

The influence of the amino-functionalized MWNTs (obtained by ball-milling process) on the thermal properties of epoxy composites as prepared by dispersing the nanotubes in the polyetheramine curing agent, which seems to stabilize the nanotube suspension, has been reported [132]. There was a clear dependence of the nanotube content on the matrix $T_g$; an increase in the amount of nanotubes led to an increase in $T_g$, but the samples containing functionalized nanotubes showed a stronger influence on $T_g$ in comparison to composites containing the same amount of nonfunctionalized nanotubes. These results show the influence of functionalization of CNT surface on the interfacial interactions between the polymer and CNTs by reducing the mobility of the matrix.

SWNTs modified by a $H_2SO_4/HNO_3$ acid treatment and subsequent fluorination show considerable improvement in the dispersion of SWNTs in an epoxy composite [133]. The functionalized nanotubes were observed to be highly dispersed and well integrated in the epoxy composites. The enhancement of mechanical properties was indicated by a 30%

increase in modulus and 18% increase in tensile strength. The functionalized SWNTs can be integrated into epoxy composites through the formation of strong covalent bonds in the epoxy ring opening esterification and curing chemical reactions.

A very good dispersion of amino functionalized DWNTs in an epoxy resin by a standard calandering technique have also been observed and compared to theoretical predictions [134]. The investigation of the mechanical properties resulted in an increase in strength, Young's modulus, and strain to failure at a nanotube content as low as 0.1 wt%. The correlation of the experimentally obtained Young's modulus showed a good agreement with a modified Halpin–Tsaï theory.

A promising preliminary study has highlighted the very simple, rapid, and straight amino-functionalization of CNTs via Ar+$N_2$ microwave plasma in the postdischarge chamber avoiding any structural alteration and polluting substances. Parameters such as plasma power, gas flow, pressure, mechanical stirring, and treatment time need to be tuned up for improving the selectivity of this surface treatment [135].

Homogeneously HDPE-coated CNTs (cMWNTs coated by ∼40 wt% of HDPE) have been used as "a predispersed masterbatch" for the preparation of EVA/MWNTs nanocomposites [54]. cMWNTs have been prepared by PFT and dispersed (3 wt%) in a commercially available polymer matrix, that is, ethylene-*co*-vinyl acetate copolymer (EVA with 27 wt% VA unit). Even though EVA (at least with VA content higher than 10 wt%) and PE are known to be nonmiscible polymers [136, 137], TEM analysis clearly indicates that the cMWNTs are individually and homogeneously dispersed in the studied EVA matrix. Such morphology appears in sharp contrast with the simple melt blend of EVA and pristine (i.e., uncoated) MWNTs (pMWNTs) where large CNTs aggregates can be observed. As assessed by DSC, two melting endotherms at 72°C and 133°C, typical for EVA and PE, respectively, are measured, indicating that the two polymers in the blend are effectively immiscible and preserve their own crystallinity. The addition of 3 wt% of pMWNTs to EVA allows increasing the Young's modulus from 12 to 19 MPa, but the very high dispersion of the nanotubes in the nanocomposite based on the dispersion of the cMWNTs in the same EVA matrix allows reaching a very high Young's modulus of 32 MPa. Since the crystallinity of the matrix is not influenced by the addition of the MWNTs, the main cause for this very high modulus enhancement has to find its origin in the tremendous improvement of the cMWNTs dispersion throughout the polymeric matrix.

A similar process was described by Kaminsky et al. to produce CNT/polypropylene nanocomposites [138].

All these examples show that perturbation of the $\pi$–$\pi$ interactions stabilizing the nanotube bundles is really needed to optimally enhance the dispersion state, and thus the composites properties.

### 2.3.1.2. Dispersion in Solution

The solution mixing process is often used to prepare thin films of polymer/CNT nanocomposites. For example, Shaffer and Windle [139] have been able to process PVA/CNT composites for mechanical and electrical characterization. Careful mixing was required in order to prevent aggregation. They report a successful route for the fabrication of large composite films containing nanotubes based on the formation of a stable colloidal intermediate, which should be broadly applicable to a range of nanotubes and polymers. The presence of nanotubes stiffens the so-obtained compositions and the electrical conductivities show typical percolation behavior. However rather low conductivity was recorded as regards to other composites based on very low loading of nanotubes in epoxy resins.

It rapidly appears that a compatibilization process (activation via sonication for several hours, functionalization of the outer surface of CNTs, etc.) is also necessary to achieve a satisfactory dispersion. Some representative examples of nanocomposites prepared by this process, and the resulting effects, are exposed hereafter.

In poly(styrene-*co*-butyl acrylate)/MWNTs nanocomposites, a clear percolation threshold has been observed for a quite low value (below 3 wt%) with a good load transfer to the matrix and the thermal stability of the resin has been also strongly improved [127].

Cadek et al. studied, theoretically and experimentally, the effects of the addition of CNTs on mechanical properties of PVA and poly(9-vinyl carbazole) (PVK). Both Young's modulus and hardness increased in reasonable agreement with Halpin–Tsaï theory [140]. Cadek et al. also studied the role of nanotube surface area and interfacial adhesion in the reinforcement of PA6 with six different types of CNTs and for different nanotube loading levels. The reinforcement scales linearly with the total nanotube surface area in the films indicating that low-diameter MWNTs represent the best type of tube for reinforcement [141–143].

The effect of CNTs on the glass transition temperature ($T_g$) of PS filled with SWNTs, functionalized or not, has also been studied. The increase in $T_g$ is due to the reduced dynamics of the polymer chains in the composites [144]. The dispersion state and the resulting crystallization behavior of poly($\varepsilon$-caprolactone)/MWNTs (PCL/MWNTs) composites prepared via the mixing of a PCL polymer solution with MWNTs bearing carboxylic groups have been studied [145]. TEM micrographs showed that MWNTs were well separated and uniformly dispersed in PCL matrix. DSC isothermal results revealed that introducing MWNTs into the PCL triggered a heterogeneous nucleation induced by a change in the crystallite growth process as observed by optical microscopy.

Systems consisting of CNTs and conjugated polymers may form the basis for nanocomposite materials for electronic device applications, as well as for tailoring the optoelectronic properties of conjugated polymers. For example, composites made with aligned CNTs in a polypyrrole (PPy) matrix have shown exceptional charge storage capacities, which may in the future lead to potential applications in supercapacitors and secondary batteries [146].

It has also been shown that the conductivity of several nanocomposite systems such as those based on poly(*m*-phenylenevinylene-*co*-2,5-dioctyloxy-*p*-phenylenevinylene) (PmPV)/CNTs and polypyrrole (PPy)-CNTs is dominated by percolation [147]. For nanocomposites made of CNTs-conjugated polymers, a mixed conduction process exists, since both the filler and the polymer matrix are conductive. It is worth noting that the conductivity of most conjugated polymers is strongly dependent on molecular weight distribution, defect concentration, chain conformation, and degree of purity. Consequently, it is not possible to describe the conductivity behavior of such systems by a model simply based on the percolation theory. For example, Coleman prepared PmPV/SWNT composites by mixing SWNTs with PmPV in toluene. The calculated percolation threshold of the resulted composites was located between 8 and 9 wt% of CNTs. The incorporation of SWNTs increased the conductivity by 10 orders of magnitude, from $2 \times 10^{-10}$ S/m for unfilled PmPV polymer to 3 S/m at 36 wt% of SWNTs [148].

The coagulation method [149] has been used to prepare different composites with higher properties. This consists in adding CNTs to an appropriate solvent (e.g., DMF in case of PMMA nanocomposites) up to a determined concentration. Based on the desired weight fraction of CNTs in the final composite, an appropriate quantity of polymer (PMMA) is dissolved into the CNTs/solvent mixture. The suspension is then dripped into a large amount of nonsolvent. The polymer precipitates immediately due to its insolubility in the solvent/nonsolvent mixture. The precipitating polymer chains entrapped the CNTs. After filtration and drying under vacuum, the raw polymer/CNTs composites are isolated.

For example, in PMMA/SWNTs composites, prepared via coagulation method, a uniform dispersion of the nanotubes in the polymer matrix has been obtained [150]. Interestingly, a rheological threshold ($\sim$0.12 wt%) smaller than the percolation threshold of electrical conductivity ($\sim$0.39 wt%) was observed. This difference is understood in terms of the smaller nanotube–nanotube distance required for electrical conductivity as compared to that required to improve polymer chain mobility.

The flammability properties of PMMA/SWNTs nanocomposites prepared by this coagulation method have been enhanced [151]. The difference in the dispersion of the nanotubes affects the heat release rate of the burning nanocomposites. The observed effects of the concentration of SWNTs on the flammability properties are based on physical processes in the condensed phase (thermal stability of the nanocomposites shows little effect due to the concentration of the tubes). A structured layer containing a network of nanotube without any major cracks was formed during the burning tests and covered the entire sample surface of the nanocomposite. The peak of heat release rate of this type of composite is reduced by about 50%. The covering layer acts as a heat shield that slows down the thermal degradation of PMMA.

The effect of CNTs on electrical conductivity and rheological properties of poly(ethylene terephthalate) (PET) prepared by coagulation method have been determined [152]. It has been demonstrated that when the MWNT loading increases, the nanocomposites undergo transition from electrically insulative to conductive at room temperature, while the melts show transition from liquid-like to solid-like viscoelasticity. The percolation threshold at 0.6 wt% for rheological property and 0.9 wt% for electrical conductivity has been found out. The low percolation thresholds result from homogeneous dispersion of MWNTs in PET matrix and high aspect ratio of MWNTs. The lower rheological percolation threshold has been mainly attributed to the fact that a denser MWNT network is required for electrical conductivity, while a less dense MWNT network sufficiently impedes PET chain mobility (related to the rheological percolation threshold).

The alignment of SWNTs in thermoplastic polyurethane (TPU) films obtained by solution casting process has also been observed when a surfactant was used [153]. Polarized Raman spectroscopy and SEM analysis on TPU/SWNTs composites show an efficient macroscopic alignment of the CNTs. The solvent–polymer interaction seems to be responsible for this phenomenon. TPU is a linear block copolymer consisting of alternating hard and soft segments. The segments aggregate into macrodomains yielding a structure that consists of glassy hard domains and rubbery soft domains. When TPU samples are exposed to THF and the SWNTs are dispersed under ultrasonication, the chains segments became relaxed and aligned, and the SWNTs are driven by the tendency of the soft chain segments toward orientation during the polar solvent swelling and moisture curing stage. The authors postulated that mechanical and electrical properties can be improved but did not measure them.

A model showing that the tensile strength increases with the thickness of the interface region has been proposed by Coleman et al. [154]. He suggested that composite strength can be optimized by maximizing the thickness of the crystalline coating or the thickness of the interfacial space partially occupied by functional groups.

More practically, purified MWNTs have been treated via amidation reaction with octadecylamine and then solution mixed at various loading with a P(MMA-*co*-EMA) copolymer [155]. A large improvement in Young's modulus and tensile strength with the increase in modified MWNTs content has been shown and the $T_g$ increases from 89°C to 106°C for 10 wt% MWNTs. Good dispersion and great interfacial bonding between the filler and the polymer may be the key again for the improvement of the mechanical properties.

Other specific treatments of MWNTs, consisting in polymerization reactions on MWNTs before dispersion in a polymer matrix, have shown interesting effects [156, 157]. For example, PMMA chains have been grafted onto MWNTs by atom transfer radical polymerization (ATRP) and composites of such PMMA-grafted MWNTs and poly(styrene-*co*-acrylonitrile) (SAN) prepared by solution casting from THF. Since PMMA is miscible to SAN, the two polymers mix intimately to facilitate the dispersion of PMMA-grafted MWNTs in the SAN matrix. The incorporation of 1 wt% PMMA-grafted CNTs into SAN leads to large increases in storage modulus at 40°C, Young's modulus, tensile strength, ultimate strain, and toughness. Morphological studies showed a finer dispersion of PMMA-grafted MWNTs in the SAN matrix.

Another example, CNTs have been functionalized with PVA by esterification reactions [158]. More precisely, SWNTs and MWNTs can be functionalized by PVA in carbodiimide-activated esterification reactions. Similar to the parent PVA, the so-functionalized CNTs samples are soluble in highly polar solvents such as DMSO and water. Their common solubilities have allowed the intimate mixing of the PVA-functionalized nanotubes with the matrix polymer for the wet casting of nanocomposite thin films. PVA is an excellent host polymer matrix for composite films with high optical quality. A typical procedure for the wet casting of a PVA–nanotube composite thin film is to mix an aqueous solution of PVA-functionalized CNTs with a highly viscous aqueous solution of PVA. The mixture, first sonicated for a short time, is then stirred for 4–12 h. The homogeneous solution was dropped onto a glass substrate for film casting; the recovered thin film was dried at room temperature and peeled off. The films were of high optical quality, transparent, and the dispersion of the nanotubes proves homogeneous up to 3 wt% of MWNTs. This work demonstrates that the functionalization of CNTs by the matrix polymer itself is an effective way for homogeneous nanotube dispersion yielding high-quality polymer/CNT nanocomposite materials.

All these examples found in the scientific literature demonstrate that the homogeneous dispersion of CNTs in polymeric matrices by direct mixing is not so-easy, but it can lead to interesting thermomechanical properties. However, the effects of dispersing such nanofillers have not been fully explained and understood yet. The native aggregates of CNTs, stabilized by numerous $\pi$-interactions, are not easily disrupted, and thus they affect the properties of the resulting nanocomposites. Uniform dispersion within a polymer matrix and improved nanotube/matrix wetting and adhesion are critical issues to achieve optimal properties for these (nano)composites.

To overcome this main drawback, different compatibilization methods have been developed and used with satisfactory results depending on the matrix used.

In spite of the numerous studies reported in the scientific literature, no universal preparation method combining direct mixing and compatibilization approach both leading to polymer nanocomposites with well-dispersed CNTs has been reported so far. But one very promising process, the so-called masterbatch process, involving coating of CNTs by a thin layer of HDPE *in situ* obtained by PFT proved to perturb the $\pi$–$\pi$ interactions between the tubes in such a way that native bundles are broken down. This leads to de-aggregated fillers and facilitates the dispersion of the fillers in a large variety of polymer matrices, for example, HDPE, EVA, polycarbonate, etc.

## 2.4. Conclusion

New processes have been developed successfully to readily produce nanocomposites with well-dispersed nanofillers. The melt blending certainly represents the most common way to prepare polymeric nanocomposites based on organoclays or carbon nanotubes. But it appears that a compatibilization step is needed to favor the delamination/dispersion of the nanofillers throughout the matrix. The "polymerization filling technique" (PFT) seems to be one of the most effective processing-aid developed during the last few years.

## References

1. Okada A., Kawasumi M., Usuki A., Kojima Y., Kurauchi T., Kamigaito O. In: Schaefer DW, Mark JE, editors. *Polymer Based Molecular Composites. MRS Symposium Proceedings,* Pittsburgh, USA, **1990**, vol. 171, 45.
2. Giannelis E.P., *Adv. Mater.,* **1996**, 8, 2.
3. Giannelis E.P., Krishnamoorti R., Manias E., *Adv. Polym. Sci.,* **1999,** 138, 107.
4. LeBaron P.C., Wang Z., Pinnavaia T.J., *Appl. Clay Sci.,* **1999**, 15, 11
5. Vaia R.A., Price G., Ruth P.N., Nguyen H.T., Lichtenhan J., *Appl. Clay Sci.,* **1999**, 15, 67.
6. Biswas M., Sinha Ray S., *Adv. Polym. Sci.,* **2001**, 155, 167.
7. Giannelis E.P., *Appl. Organomet. Chem.,* **1998**, 12, 675.
8. Xu R., Manias E., Snyder A.J., Runt J., *Macromolecules,* **2001**, 34, 337.
9. Bharadwaj R.K., *Macromolecules,* **2001**, 34, 1989.
10. Messersmith P.B., Giannelis E.P., *J. Polym. Sci., Part A: Polym. Chem.,* **1995**, 33:1047–57.
11. Yano K., Usuki A., Okada A., Kurauchi T., Kamigaito O., *J. Polym. Sci. Part A: Polym. Chem.,* **1993**, 31, 2493–8.
12. Kojima Y., Usuki A., Kawasumi M., Fukushima Y., Okada A., Kurauchi T., Kamigaito O., *J. Mater. Res.,* **1993**, 8, 1179.
13. Gilman J.W., Kashiwagi T., Lichtenhan J.D., *SAMPE J.,* **1997**, 33, 40.
14. Gilman J.W., *Appl. Clay Sci.,* **1999**, 15, 31.
15. Dabrowski F., Le Bras M., Bourbigot S., Gilman J.W., Kashiwagi T., *Proceedings of the Euro-Fillers'99,* Lyon-Villeurbanne, France; 6–9 September **1999**.
16. Bourbigot S., LeBras M., Dabrowski F., Gilman J.W., Kashiwagi T., *Fire Mater.,* **2000**, 24, 201.
17. Gilman J.W., Jackson C.L., Morgan A.B., Harris Jr R., Manias E., Giannelis E.P., Wuthenow M., Hilton D., Phillips S.H., *Chem. Mater.,* **2000**, 12, 1866.
18. Sinha R.S., Yamada K., Okamoto M., Ueda K., *Nano Lett.,* **2002**, 2, 1093.
19. Vaia R.A., Ishii H., Giannelis E.P., *Chem. Mater.,* **1993**, 5, 1694
20. Ajayan P.M., Stephan O., Colliex C.,Trauth D., *Science,* **1994**, 265, 1212.
21. Salvetat J.P., Briggs G.A.D., Bonard J.M., Bacsa R.R., Kulik A.J., Stöckli T., Burnham N.A., Forro L., *Phys. Rev. Lett.,* **1999**, 82, 944.
22. Brindly S.W., Brown G., editors. *Crystal Structure of Clay Minerals and their X-ray Diffraction,* Mineralogical Society, London, **1980**.
23. Aranda P., Ruiz-Hitzky E., *Chem. Mater.* **1992**, 4, 1395.
24. Greenland D.J., *J. Colloid. Sci.* **1963**, 18, 647.
25. Wu J., Lerner M.N., *Chem. Mater.,* **1993**, 5, 83.
26. Yano K., Usuki A., Okada A., Kurauchi T., Kamigaito O., *J. Polym. Sci.: Part A: Polym. Chem.,* **1993**, 31, 2493.
27. Jeon H.G., Jung H.-T., Lee S., Hudson S.D., *Polym. Bull.,* **1998**, 41, 107.
28. Lee D.C., Jang L.W., *J. Appl. Polym. Sci.* **1996**, 61, 1117.
29. Ogata N., Jimenez G., Kawai H., Ogihara T., *J. Polym. Sci.: Part B: Polym. Phys.* **1997**, 35, 389.

30. Jimenez G., Ogata N., Kawai H., Ogihara T., *J. Appl. Polym. Sci.* **1997**, 64, 2211.
31. Biswas M., Sinha Ray S., *Adv. Polym. Sci.*, **2001**, 155, 170.
32. Sahoo S.K., Kim D.W., Kumar J., Blumstein A., Cholli A.C., *Macromolecules*, 2003, **36**, 2777.
33. Akelah A., Moet A., *J. Mater. Sci.* **1996**, 31, 3589.
34. Zeng C., Lee L.J., *Macromolecules*, **2001**, 34, 4098
35. Weimer M.W., Chen H., Giannelis E.P., Sogah D.Y., *JACS.*, **1999**, 121, 1615.
36. Alexandre M., Dubois Ph., Jérôme R., Garcia-Marti M., Sun T., Garces J.M., Millar D.M., Kuperman A. WO Patent WO 9947598A1, **1999.**
37. Dubois Ph., Alexandre M., Jérôme R., *Macromol. Symp.*, **2003**, 194, 13.
38. Okamoto M., Morita S., Kim Y.H., Kotaka T., Tateyama H., *Polymer*, **2001**, 42, 1801.
39. Yao K. J., Song M., Houston D.J., Luo D.Z., *Polymer*, **2002**, 43, 1017.
40. Lepoittevin B., Pantoustier N., Devalckenaere M., Alexandre M., Kubies D., Calberg C., Jérôme R., Dubois Ph., *Macromolecules*, **2002**, 35, 8385.
41. Paul M.-A., Alexandre M., Degée Ph., Calberg C., Jérôme R., Dubois Ph., *Macromol. Rapid Comm.*, **2003**, 24, 561.
42. Okada A., Usuki A., *Mater. Sci. Eng.*, **1995**, C3, 109.
43. Messersmith P.B., Giannelis E.P., *Chem. Mater.*, **1994**, 6, 1719.
44. Zilg C., Mülhaupt R., Finter J., *Macromol. Chem. Phys.*, **1999**, 200, 661.
45. Giannelis E.P., Krishnamoorti R., Manias E., *Adv. Polym. Sci.,* **1999**, 138, 108.
46. Vaia R.A., Giannelis E.P., *Macromolecules*, **1997**, 30, 7990.
47. Balazs A.C., Singh C., Zhulina E., *Macromolecules*, **1998**, 31, 8370.
48. Fisher H.R., Gielgens L.H., Koster T.M.P., *Acta Polym.*, **1999**, 50, 122.
49. Vaia R.A., Jandt K.D., Kramer E.J., Giannelis E.P., *Macromolecules*, **1995**, 28, 8081.
50. Chavarria F., Paul D.R., *Polymer*, **2006**, 47, 7760.
51. Cui L., Ma X., Paul D.R., *Polymer*, **2007**, 48, 6325.
52. Fornes T.D., Yoon P.J., Paul D.R., *Polymer*, **2003**, 44, 7545.
53. Yoon P.J., Hunter D.L., Paul D.R., *Polymer*, **2003**, 44, 5341.
54. Peeterbroeck S., Lepoittevin B., Pollet E., Benali S., Broekaert C., Alexandre M., Bonduel D., Viville P., Lazzaroni R., Dubois Ph., *Polym. Eng. Sci.*, **2006,** 46, 1022.
55. Shah R. K , Paul D.R., *Polymer*, **2006**, 47, 4075.
56. Horsch, S.E., Kannan R.M. Gulari E., Wayne State University, US2006 NSTI Nanotechnology Conference and Trade Show – Nanotech **2006** – 9th Annual.
57. Horsh S., Serhatkulu G., Gulari E., Rangaramanujam M. K., *Polymer*, **2006**, 47, 7485.
58. Nguyen Q.T., Baird D.G., *Adv. Polym. Tech.*, **2006**, 25, 270.
59. Thostenson E., *Compos. Sci. Tech.,* **2001**, 61, 1899.
60. Lu J.P., *J. Phys. Chem. Solids,* **1997**, 58(11), 1649.
61. Treacy M.M.J., Ebbesen T.W., Gibson J.M., *Nature,* **1996**, 381, 678.
62. Radushkevich L.V., Lukyanovich V.M., *Zurn. Fisic. Chim.,* **1952,** 26, 88.
63. Iijima S., *Nature*, **1991**, 354, 56.
64. Thostenson E.T., Chou T-W., *J. Phys. D : Appl. Phys.,* **2003**, 36, 571.
65. Belin T., Epron F., *Mater. Sci. Eng.,* **2005**, B119, 105.
66. Reich S., Thomsen C., Maultzsch J., *Carbon Nanotubes: Basic Concepts and Physical Properties*, Wiley, New York, **2004.**
67. Kilbride B.E., Coleman J.N., Fraysse J., Fournet P., Cadek M., Drury A., et al., *J. Appl. Phys.,* **2002**, 92, 4024.
68. Overney G., Zhong W, Tomanek D., *Z Phys. D-At. Mol. Clusters,* **1993,** 27(1), 93.
69. Lu J.P., *J. Phys. Chem. Solids,* **1997**, 58(11), 1649.
70. Treacy M.M.J., Ebbesen T.W., Gibson J.M., *Nature,* **1996**, 381, 678
71. Yu M., Lourie O., Dyer M.J., Kelly T.F., Ruoff R.S., *Science,* **2000,** 287, 637.
72. Wong E.W., Sheehan P.E., Lieber C.M., *Science,* **1997**, 277, 1971.
73. Salvetat J.P., Kulik A.J., Bonard J.M., Briggs G.A.D., Stockli T., Metenier K., et al., *Adv. Mater.,* **1999,**11(2), 161.
74. Salvetat J.P., Briggs G.A.D., Bonard J.M., Bacsa R.R., Kulik A.J., Stöckli T., Burnham N.A., Forro L., *Phys. Rev. Lett.,* **1999**, 82, 944.
75. Yu M.F., Files B.F., Arepalli S., Ruoff R.S., *Phys. Rev. Lett.,* **2000**, 84, 5552.
76. Popov V.N., *Mater. Sci. Eng.,* **2004**,R 43, 61.
77. Yi W., Lu L., Zhang D.L., Pan Z.W., Xie S.S., *Phys. Rev.,* **1999,** B 59, R9015.

78. Ebbesen T.W., Lezec H.J., Hiura H., Bennett J.W., Ghaemi H.F., Thio T., *Nature,* **1996**, 382, 54.
79. Dai H., Wong E.W., Lieber C.M., *Science,* **1996**, 272, 523.
80. Bockrath M., Cobden D.H., McEuen P.L., Chopra N.G., Zettle A., Thess A., Smalley R.E., *Science,* **1997**, 275, 1922.
81. Fischer J.E., Dai H., Thess A., Lee R., Hanjani N.M., Dehaas D.L., Smalley R.E., *Phys. Rev. B,* **1997**, 55, R4921.
82. Tans S.J., Devoret M.H., Dai H., Thess A., Smalley R., Geerlings L.J., Dekker C., *Nature,* **1997** , 386, 474.
83. Mitchell C.A., *Macromolecules,* **2002**, 35, 8825–8830.
84. Haeggenmuller R., Gommans H.H., Rinzler A.G., Fischer J.E., Winey K.I., *Chem. Phys. Lett.,* **2000**, 330, 219.
85. Qian D., Dickey E.C., Andrews R., Rantell T., *Appl. Phys. Lett.,* **2000**, 76, 2868.
86. Cooper C.A., Ravich D, Lips D, Mayer J, Wagner HD, *Compos. Sci. Technol.,* **2002**, 62, 1105.
87. Tucker C.L., Liang E., *Compos. Sci. Technol.,* **1999**, 59, 655–671.
88. Andrews R., Jacques D., Minot M., Rantell T., *Macromol. Mater. Eng.,* **2002**, 287, 395.
89. Breuer O., Sundararaj U., *Polym. Compos.,* **2004**, 25, 630.
90. Potschke P., Bhattacharyya A.R., Janke A., Goering H., *Comp. Interf.,* **2003**, 10, 389.
91. Jin Z., Pramoda K.P., Xu G., Goh S.H., *Chem. Phys. Lett.,* **2001**, 337, 43.
92. Bhattacharyya A.R., Sreekumar T.V., Liu T., Ericson L.M., Hauge R.H., Smalley R.E., *Polymer,* **2003**, 44, 2373.
93. Lopez-Manchado M.A., Valentini L., Biagiotti J., Kenny J.M., *Carbon,* **2005**, 43, 1499.
94. Kashiwagi T., Grulke E., Hilding J., Awad W., Douglas J., *Macromol. Rapid Comm.,* **2002**, 23, 761.
95. Kashiwagi T., Grulke E., Hilding J., Groth K., Harris R., Butler K., Shields J., Kharchenko S., Douglas J., *Polymer,* **2004**, 45, 4227.
96. Kearns J.C., Shambaugh R.L., *J. Appl. Polym. Sci.,* **2002**, 86, 2079
97. Mc Nally T., Potschke P., Halley P., Murphy M., Martin D., Bell S.E.J., Brennan G.P., Bein D., Lemoine P., Quinn J.P., *Polymer,* **2005**, 46, 8222.
98. Pötschke P., Fornes R.D., Paul D.R., *Polymer,* **2002**, 43, 3247.
99. Pötschke P., Abdel-Goad M., Alig I., Dudkin S., Lellinger D., *Polymer,* **2004**, 45, 8863.
100. Zhang W.D., Shen L., Phang I.Y., Liu T., *Macromolecules,* **2004**, 37, 256.
101. Phang I.Y., Ma J., Shen L., Liu R., Zhang W-D., *Polym. Int.,* **2006**, 55, 71.
102. Ajayan P.M., Schadler L.S., Giannaris C., Rubio A., *Adv. Mater.,* **2000**, 12 (10), 750.
103. Barrau S., Demont P., Peigney A., Laurent C., Lacabanne C., *Macromolecules,* **2003**, 36 (14), 5187.
104. Fidelus J.D., Wiesel E., Gojny F.H., Schulte K., Wagner H.D., *Compos. A,* **2005**, 36, 1555.
105. Song Y.S., Youn J.R., *Carbon,* **2005**, 43, 1378.
106. Peng H., Alemany L.B., Margrave J.L., Khabashesku V.N., *JACS,* **2003**, 125, 15174.
107. Liu L., Barber A.H., Nuriel S., Wagner H.D., *Adv. Funct. Mater.,* **2005**, 15(6), 975.
108. Liu P., *Eur. Polym. J.,* **2005**, 41, 2693.
109. Kong H., Gao C., Yan D., *JACS Communications,* **2004**, 126, 412
110. Qu L., Veca L.M., Lin Y., Kitaygorodskiy A., Chen B., McCall A.M., Connell J.W., Sun Y-P., *Macromolecules,* **2005**, 38, 10328.
111. Tong X., Liu C., Cheng H-M., Zhao H., Yang F., Zhang X., *J. Appl. Polym. Sci.,* **2004,** 92, 3697.
112. Bonduel D., Mainil M., Alexandre M., Monteverde F., Dubois Ph., *Chem. Comm.,* **2005**, 781.
113. Bonduel D., Bredeau S., Alexandre M., Monteverde F., Dubois Ph., *J. Mater. Chem.* **2007**, 17, 2359.
114. Li L., Li C.L., Ni C., *JACS,***2006**, 128 (5), 1692.
115. Strano M. S., Moore C. M., Miller M. K., Allen M. J., Haroz E. H., Kittrel, C., Hauge R. H., Smalley R.E.J., *Nanosci. Nanotechnol.,* **2003**, 3, 81.
116. Islam M. F., Rojas E., Bergey D. M., Johnson A. T., Yodh A. G. *Nano Lett.,* **2003**, 3, 269.
117. Bandhyopadhyaya R., Nativ-Roth E., Regev O., Yerushalmi-Rozen R., *Nano Lett.,* **2002**, 2, 25.
118. Shvartzman-Cohen R., Levi-Kalisman Y., Nativ-Roth E., Yerushalmi- Rozen R., *Langmuir,* **2004**, 20, 6085.
119. Szleifer I., Yerushalmi-Rozen R., *Polymer,* **2005**, 46, 7803.
120. Yurekli K., Mitchell C.A., Krishnamoorti R., *JACS,* **2004**, 126, 9902.
121. Richard C., Balavoine F., Schultz P., Ebbesen T. W., Mioskowski C., *Science,* **2003**, 300, 77.
122. Matarredona O., Rhods H., Li Z., Harwell J.H., Balzano L., Resasco D.E., *J. Phys. Chem. B,* **2003**, 107, 13357.
123. Manne S., Cleveland J. P., Gaub H. E., Stucky G. D., Hansma P. K., *Langmuir,* **1994**, 10, 4409.
124. Wanless E. J., Ducker W. A., *J. Phys. Chem.,* **1996**, 100, 3207.
125. Kiraly Z., Findenegg G. H., Klumpp E., Schlimper H., Dekany I., *Langmuir,* **2001**, 17, 2420.

126. Regev O., El Kati P.N.B., Loos J., Koning C.E., *Adv. Mater.,* **2004**, 16, 248.
127. Dufresne A., Paillet M., Putaux J.L., Canet R., Carmona F., Delhaes P., Cui S., *J. Mater Sci.,* **2002**, 37, 3915.
128. Grunlan J.C., Mehrabi A. R., Bannon M.V., Bahr J. L., *Adv. Mater.,* **2004**, 16, 150.
129. Grossiord N., Loos J., Koning C. E., *J. Mater. Chem.,* **2005**, 15, 2349.
130. Bauers F. M., Mecking S., *Macromolecules,* **2001**, 34, 1165.
131. Schofner M.L., Khabashesku V.N., Barrera E.V., *Chem. Mater.,* **2006**, 18, 906.
132. Gojny F., Schulte K., *Compos. Sci. Technol.,* **2004**, 2303.
133. Zhu J., Kim J.D., Peng H., Margrave J.L., Khabashesku V.N., Barrera E.V., *Nano Lett.,* **2003**, 3 (8), 1107.
134. Gojny F.H., Wichmann M.H.G., Köpke U., Fiedler B., Schulte K., *Compos. Sci. Technol.,* **2004**, 64, 2363.
135. Ruelle B., Peeterbroeck S., Gouttebaron R., Godfroid T., Monteverde F., Dauchot J.P., Alexandre M., Hecq M., Dubois P., *J. Mater. Chem.,* **2007**, 17, 157.
136. John B., Varughese K. T., Oommen Z., Pötschke P., Thomas S., *J. Appl. Polym. Sci.,* **2003**, 87, 2083.
137. Khonakdar H.A., Wagenknecht U., Jafari S.H., Hassler R., Eslami H., *Adv. Polym. Technol.,* **2004**, 23, 307.
138. Funck A., Kaminsky W., *Comp. Sci. Technol.,* **2007**, 67, 906.
139. Shaffer M.S.P., Windle A.H., *Adv. Mater,.* **1999**, 11 (11), 937.
140. Cadek M., Coleman J.N., Barron V., Hedicke K., Blau W.J., *Appl. Phys. Lett.,* **2002**, 81 (27), 5123.
141. Cadek M., Coleman J.N., Ryan K.P., Nicolosi V., Bister G., Fonseca A., Nagy J.B., Szostak K., Beguin F., Bau W.J., *Nano Lett.,* **2004**, 4 (2), 353.
142. Pham J. Q., Mitchell C.A., Bahr J.L., Tour J.M., Krishnamoorti R., Green P.F., *J. Polym. Sci. Part B: Polym. Phys.,* **2003**, 41, 3339.
143. Wu T-M., Chen E-C., *J. Polym. Sci. Part B: Polym. Phys.,* **2006**, 44, 598.
144. Frackowiak E., *Dekker Encyclopedia of Nanoscience and Nanotechnology,* Marcel Dekker, New York, **2004**, 537.
145. Coleman J.N., Curran S., Dalton A.B., Davey A.P., McCarthy B., Blau W.J., Barklie R.C., *Phys. Rev. B,* **1998**, 58, 7492.
146. Long Y., Chen Z., Zhang X., Zhang J., Liu Z., *J. Phys. D,* **2004**, 37, 1965.
147. Chen L., Pang X-J., Qu M-Z., Zhang Q-T, Wang B., Zhang B-L., Yu Z-L., *Compos. A,* **2006**, 37, 1485.
148. Coleman J.N., Curran S., Dalton A.B., Davey A.P., McCarthy B., Blau W., Barklie R.C., *Phys. Rev.,* **1998,** B 58, 7492.
149. Du F., Scogna R.C., Zhou W., Brand S., Fisher J.E., Winey K.I., *Macromolecules,* **2004**, 37, 9048.
150. Kashiwagi T., Du F., Winey K.I., Groth K.M., Shields J.R., Bellayer S.P., Kim H., Douglas J.F., *Polymer,* **2005**, 46, 471.
151. Du F., Fisher J.E.,Winey K.I., *J. Polym. Sci. Part B: Polym. Phys.,* **2003,** 41, 3333.
152. Hu G., Zhao C., Zhang S., Yang M., Wang Z., *Polymer,* **2006**, 47, 480
153. Chen W., Tao X., *Macromol. Rapid. Comm.,* **2005**, 26, 1763 .
154. Coleman J. N., Cadek M., Blake R., Nocolosi V., Tyan K.P., Belton C., Fonseca A., Nagy J.B., Gun'ko Y.K., Blau W.J., *Adv. Funct. Mater.,* **2004**, 14 (8), 791.
155. Yang J., Hu J., Wang C., Qin Y., Guo Z., *Macromol. Mater. Eng.,* **2004**, 289, 828.
156. Wang M., Pramoda K.P., Goh S.H., *Polymer,* **2005**, 46, 11510.
157. Hill D.E., Lin Y., Rao A.M., Allard L.F., Sun Y-P., *Macromolecules,* **2002**, 35, 9466.
158. Lin Y., Zhou B., Shiral Fernando K.A., Liu P., Allard L.F., Sun Y-P., *Macromolecules,* **2003**, 36, 7199.

# 3 *In-Situ* Generation of Polyolefin Nanocomposites

*Walter Kaminsky, Andreas Funck, and Katrin Scharlach*

Institut für Technische und Makromolekulare Chemie, Universität Hamburg, Germany

kaminsky@chemie.uni-hamburg.de

## Abstract

Polyolefine nanocomposites were generated *in situ* in a two-step process. First, the metallocene or another single-site catalyst is absorbed on the surface of the nanofiller; then by addition of ethene or propene, a polyolefin film is formed, covering the nanoparticles layered silicates and oxides, or fibers. Polyolefin nanocomposites produced by *in-situ* generation have better mechanical properties than material produced by mechanical blending. Metallocene/methylaluminoxane catalysts are soluble in hydrocarbons and, therefore, they can cover perfectly the surface of particles or permeate in the layers of layered silicates and oxides. Metallocenes and other single-site catalysts allow the synthesis of polymers with a precisely defined microstructure, tacticity, and stereoregularity as well as new copolymers with superior properties such as film clarity, tensile strength, and lower extractables. These polymer properties can be enlarged by the incorporation of nanofillers. The resulting polyethylene or polypropylene nanocomposites cause a tremendous boost to the physical and chemical properties such as a dramatically improved stiffness, high gas barrier properties, significant flame retardancy, and high crystallization rates.

## Abbreviations

| | |
|---|---|
| $\eta^*$ | complex viscosity |
| CB | carbon black |
| CNF | carbon nanofiber |
| CNT | carbon nanotube |
| DSC | differential scanning calorimetry |
| EVA | ethylene vinyl acetate |
| FTIR | Fourier transform infrared spectroscopy |
| $G'$ | melt elasticity |
| $G''$ | loss modulus |
| GPC | gel permeation chromatography |
| HDPE | high-density polyethylene |
| HDT | heat distortion temperature |
| iPP | isotactic polypropylene |

| | |
|---|---|
| LDPE | low-density polyethylene |
| LLDPE | linear low-density polyethylene |
| M 250 | monosphers |
| MAO | methylaluminoxane |
| MMT | montmorillonite |
| MWCNT | multiwalled carbon nanotube |
| PE | polyethylene |
| PFT | polymer filling technique |
| POM | polarizing optical microscopy |
| rt | room temperature |
| SEM | scanning electron microscopy |
| sPP | syndiotactic polypropylene |
| SWCNT | single-walled carbon nanotube |
| $\tau^{1/2}$ | half-time of crystallization |
| $T_c$ | melting temperature |
| TEM | transmission electron microscopy |
| TIBA | triisobutylaluminium |
| TOF | turnover frequency |
| UHMWiPP | ultrahigh molecular weight isotactic polypropylene |
| UHMWPE | ultrahigh molecular weight polyethylene |
| WAXD | wide angle X-ray diffraction |
| wt% | weight percent |
| $X_c$ | crystallinity |
| XRD | X-ray diffraction |

## 3.1. Introduction

Polyolefins such as polyethylene (LDPE, HDPE, LLDPE) and polypropylene are the most used thermoplastic polymers. In 2006, the production of these polyolefins reached 105 million tons worldwide [1]. They are often filled with organic or inorganic components to increase their strength, impact resistance, or conductivity and to reduce permeability of gases [2]. Within the last years, much research in academic and industrial laboratories has focused on the field of polyolefin nanocomposites because of their high potential as materials with novel properties such as improved mechanical properties, increased heat distortion temperature (HDT), reduced permeability and flammability [3–5]. Exceptionally strong materials could be synthesized by the soft polyolefin matrix with nanosized, rigid filler particles. The properties of the nanocomposites are not only influenced by the kind of fillers, but also by the microstructure of the polyolefins and the preparation process. A lot of work has been carried out to use layered silica and metal oxides [3], clay, self-assembled nanoboehmites [6], or montmorillonite (MMT) [7–9] as fillers to obtain composite materials, which show a lower permeability for gases such as oxygen, nitrogen, steam. In the past, most composites are commonly prepared by mechanical blending of the particles or fibers above the melting. Blending or melt compounding of polyolefins with nanoparticles is hard to achieve, especially at high filler content and lead to aggregation, intercalation which decreases the mechanical properties [10]. Above 60 wt%, filler results in highly particle aggregation. For mixing in solution, the solubility of polyethylene (HDPE) and polypropylene is too low. Another problem is the hydrophilic nature of most inorganic fillers and the hydrophobic nature of the polyolefins. The differences result in weak interfacial adhesion between filler and polymer matrix and low mechanical properties. Therefore fillers must be modified by surface active agents. Both disadvantages can be solved by *in-situ* generation where the catalyst is absorbed on the surface of the nanofillers, changing the surface to a hydrophobic one. In a second step, the activated fillers are used as catalysts for olefin polymerization. Each particle or fiber is covered by a polyolefin film. For bigger particles also heterogeneous nanosized Ziegler–Natta catalysts can be used [11]. For nanoparticles, homogeneous catalysts are preferred to cover the surface with active sites.

Metallocene/methylaluminoxane (MAO) and other single-site catalysts are soluble in hydrocarbons and, therefore, they can cover perfectly the surface of particles or permeate in the layers of layered silicates and oxides. Aluminum flake filled polyethylene (PE) was first prepared by this procedure showing high thermal but low electrical conductivity [12, 13].

Metallocene/MAO catalysts are highly active for the production of precisely designed polyolefins and engineering plastics [14–18]. Especially zirconocene complexes and half sandwich titanium complexes have opened a frontier in the area of new polymer synthesis and processing. The transition metal complexes can be activated by MAO but also by other bulky cocatalysts such as perfluorophenylborate. By changing the ligand structure, these catalysts allow the synthesis of polymers with a tailored microstructure tacticity and stereoregularity as well as new copolymers with superior properties such as film clarity, tensile strength, and lower extractables. With these single-site catalysts and by *in-situ* polymerization introduced, nanofillers with a large aspect ratio, layered silicates, fibers into a polyolefin matrix cause a tremendous boost to the physical and chemical properties of polymers such as a dramatically improved stiffness with a negligible loss of impact strength, high gas barrier properties, significant flame retardancy, better clarity, and gloss and high crystallization rates. Even low nanoparticle contents are already sufficient to obtain new or modified material characteristics.

Especially PE, isotactic, or syndiotactic polypropylene nanocomposites are investigated and expected as materials for electronic, magnetic, domestic devices, and automotive applications with outstanding properties [19].

## 3.2. Moochers in Monospheres

The *in-situ* polymerization has been successfully used to incorporate a wide number of nanofiller types, e.g., moochers (silica balls) [20], clay, layered metal oxides, silicates, MMT, carbon black (CB), carbon nanofibers (CNF), and different types of carbon nanotubes (CNT). All incorporated nanofillers, neither the inorganic nor the organic, showed any deactivating effects to the catalyst system or a certain pretreatment is able to prevent it from deactivating the catalysts. A further advantage of this method is the versatility of metallocene/MAO catalysts, which allow the tailoring of the polymer microstructure, and therefore of the composite matrix, by the utilization of a metallocene with a suitable structure [21, 22]. Metallocene/MAO catalysts are also excellent tools in the production of certain molecular masses and, in case of polypropylenes, tacticities. PP with an isotactic, syndiotactic, or atactic configuration can be obtained by using $C_2$-, $C_{2V}$-, or $C_S$-symmetric zirconocenes [23, 24].

For example, *rac*-dimethylsilyl-bis(2-methyl-indenyl)zirconium dichloride (**1**) leads to a high isotactic polypropylene matrix with, for the catalyst, typical isotactic pentads of 95 $\pm$ 2% [24].

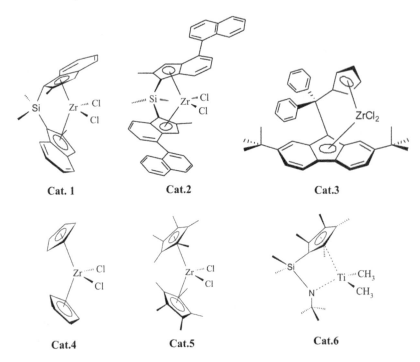

**Figure 3.1** Catalysts for the synthesis of isotactic (**1**, **2**) and syndiotactic polypropylene (**3**) and PE (**4–6**) used in the references cited.

The molecular weights were not influenced by the nanoparticles and were in the range of 320,000 to 340,000 g/mol. For the production of an (ultra) high molecular mass matrix (UHMWiPP) *rac*-dimethylsilyl-bis(2-methyl-4-(1-naphyl)indenyl)zirconium dichloride (**2**) was the catalyst used, but necessarily requires the pretreatment described in Figure 3.1. The isotactic pentads were $97 \pm 2\%$ and the molecular mass was about 1,500,000 g/mol. $C_s$-symmetric metallocenes, such as di(paramethylphenyl)methyl(cyclopentadienyl(2,7-bis-*tert*-butylfluorenyl)) zirconium dichloride (**3**) are used for sPP matrices. Some authors have reported on nanocomposites based on a PE matrix, which typically were synthesized by dicyclopentadienylzirconium dichloride (**4**) [4], or its pentamethyl-cyclopentadienyl derivate (**5**) [25]. But also half sandwich catalysts, such as *tert*-butylamidodimethyl(tetramethyl-$\eta^5$-cyclopentadienyl)silanedimethyl titanium (**6**), had successfully been used in the production of UHMWPE, co-, and terpolymers [26].

In all cases, the polymerizations were performed in dry toluene or dry heptane at 30–70°C. The first step of preparation is the absorbance of MAO onto the filler surface. If layered silica were used, the MAO diffuses into the layers and widened up the space between the layers. The second step is the addition of a transition metal compound and the formation of catalytically active sites on the surface and the addition of ethene or propene [27, 28]. The monomer pressures were in the range of 2–10 bars.

For example, 4.1 g dried nanoparticles (monosphers, M250) with a narrow particle size distribution of 250 nm were dried and dispersed in 20 ml toluene and then the dispersion was improved by sonication for 15 min; 600 mg of MAO solution was added and the mixture was stirred for 24 h at room temperature. It was then filtered using a D4 glass filter and washed 10 times with 5 ml of toluene each. After drying for 4 h under vacuum at room temperature, the resulting MAO/SiO$_2$ was stored in the glove box [20]. These catalysts were stable over weeks and loose only 20% of activity in 11 weeks. For polymerizations, a 1 L Glass reactor (Büchi AG, Ulster, Switzerland) was used that had been heated to 90°C for 1 h and then flushed with argon. The reactor was charged with 200 ml of toluene, heated to the desired polymerization temperature, and 2 ml of TIBA solution was added. The solution was saturated with propene at the desired pressure using a mass-flow controller. And 0.55 g of the MAO/SiO$_2$ nanoparticles were dispersed in 10 ml toluene added by 1.3 × 10$^{-6}$ mol zirconocene (**3**) and then introduced into the reactor using a pressure lock. The polymerization activity was 3400 kg Pol/mol Zr h cp by 30°C (cp = concentration of propene in mol/l). Polymerizations were typically quenched after 0.5–2 h by addition of 5 ml ethanol. All polymers were stirred with a quench solution (water, ethanol, hydrochloride acid) over night, filtered, washed, and dried under vacuum at 60°C.

M250 contents were determined by treatment of the M250-containing polymers by a Bunsen burner and subsequent heating at 800°C for 2 h. The resulting inorganic residue was weighted.

The viscosimetric molar weights were determined by viscosimetry at 408 K using an Ubbelohde viscometer (capillary Oa, $K = 0.005$ mm$^2$/s$^2$) as a 1 mg/ml solution in 50 ml decahydronaphthalin. Melting temperatures $T_m$ were determined by differential scanning microscopy (DSC) with a DSC 821e (Mettler Toledo) from the heating curve of the second heat at a heating rate of 20°C/min. Electron microscopy was performed on a Leo 1530 FE-REM.

The predicted potential of nanocomposites is frequently not accomplished in practice which can be attributed to an insufficient load transfer between the matrix and the filler.

As said before, a homogeneous distribution and a good interfacial adhesion are crucial for the successful preparation of nanocomposites but often difficult to achieve, especially fillers with high aspect ratio like CNTs tend to stay aggregated during the polymerization process because of their high surface energy and numerous $\pi$–$\pi$ electron interactions between the tubes. The filler had to be pretreated in a way to separate it, e.g., by ultrasound, and to prevent it from deactivating the catalysts. While spherical particles with low aspect ratio could be separated quite well by simply sonication with ultrasound with low energy input in a low-viscosity medium like toluene, the separation of high aspect filler particles is more difficult. The road to success may be the functionalization of the fillers in combination with the polymerization-filling technique (PFT), which is known to be an efficient way to improve the mechanical properties of (bulk) composites too [29]. It was initially investigated in Ziegler–Natta polymerization in the late 70s by Howard [30, 31] and Enikolopian et al. [32].

PFT originally consisted of attaching a Ziegler–Natta type catalyst onto the acidic surface of inorganic fillers; this led to a polymer growth directly on the filler surface and results in a good filler coverage with polymer and separation of the individual particles during the polymerization process. Nowadays, there are several derivatives of the PFT in context with nanocomposites and metallocene catalysts in use and the method is no longer limited to acidic filler types, but also with basic and other, e.g., metallic- or graphite-like surfaces [26]. Dubois et al. already used a derivate of the PFT in combination with metallocene/MAO catalysts to coat native nanotubes with PE. Their work based on MAO physicochemically anchored onto the nanotube surface, accomplished by a reaction of MAO and CNT firstly at 40°C (1 h), then 150°C (2 h), and solvent evaporation. Under these conditions, most of the MAO (<98 mol%) is immobilized onto the nanotube surface [25].

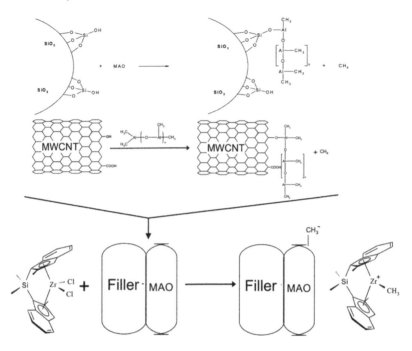

**Figure 3.2** Hydroxyl or carboxyl groups present on the filler surface can react with added MAO to form a heterogeneous cocatalyst. The MAO is now anchored, but still able to form an active complex with the metallocene. Therefore the polymer is growing directly from the surface.

Kaminsky et al. immobilized MAO by establishing a covalent bonding between MAO and hydroxyl- and carboxyl-functionalized filler surfaces, such as silica balls (monospheres®) or oxidized MWCNT, though impregnating the tubes for 24 h with MAO at rt. Purified and oxidized nanotubes are bearing functional groups such as hydroxyl or carboxyl on their surfaces. These groups can react with MAO by the formation of covalent oxygen aluminum bond, without deactivating effects for the catalysts. The MAO is now anchored, but still able to form a catalytically active complex with the metallocene (Figure 3.2).

Thus a heterogeneous cocatalyst is formed. This method causes a lower activity during the coating process, because unreacted and dissolved MAO was removed by filtration, which in turn caused low [Al]/[Zr] ratio, but it even allows an ultrasonic treatment of the impregnated nanotubes! After completion of the coating phase, the activity can be raised to a normal level by the addition of MAO. Especially for ultrahigh molecular mass matrices, this method provided outstanding results.

In both PFT derivates described, a heterogeneous cocatalyst is formed and the CNTs are covered with a thin polymer layer after the entire coating process (Figure 3.3).

In order to generate the active complex from the zirconocene and the bonded MAO, the metallocene has necessarily diffused to the heterogeneous cocatalyst. Thereby the polymerization takes place nearby the filler surface (Figure 3.4). This leads to a polymer growth directly on the nanoparticle and results in excellent filler coverage with polymer, and promotes the separation of the individual particles during the polymerization. Furthermore, the hydrophobic character of some filler materials such as CNTs supports the drawing on the fiber.

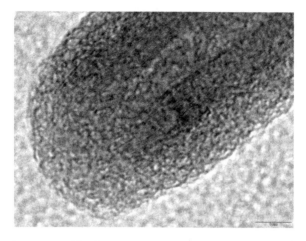

**Figure 3.3**  TEM image (magnification ×400,000) of ox. MWCNT, coated by a thin polymer film.

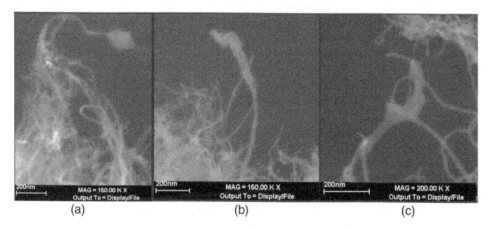

**Figure 3.4**  SEM micrograph (a, magnification × 150,000; b, magnification × 150,000; c, magnification × 200,000) of ox. MWCNT after 20 min of coating. Polymer accumulated at the edges of the nanotube, where MAO was anchored the most.

A third approach based on the PFT was reported by Dong et al. [4]. They prepared nanowires by ethylene *in-situ* polymerization with CNTs supported by a metallocene catalyst. Therefore, a toluene solution of MAO containing CNT was added to a $Cp_2ZrCl_2$ solution and the toluene was evaporated. The external surface of the CNT was covered with some particles and the resulting composites had fibrous microstructure, with a fiber diameter of 50–200 nm.

## 3.3.  Polyethylene Nanocomposites

### 3.3.1.  Clay and Layered Metal Oxides

For the fabrication of PE nanocomposites, different fillers were used. Clay and layered metal oxide nanocomposites were preferred.

Coates et al. produce PE/silicate nanocomposites by *in-situ* polymerization with a palladium catalyst [33]. The palladium catalyst $\{[2,6-Pr_2^i C_6 H_3 N-C(Me)-C(Me)-NC_6 H_3 Pr_2^i-2,6]\ Pd(CH_2)_3 CO_2 Me\}\{B[C_6 H_3-(CF_3)_2]_4\}$ was intercalated to synthetic fluorohectorite [34]. The polymerization activity was investigated. The turnover frequency (TOF/mol ethylene/(mol Pd h)) of the catalyst is 162 $h^{-1}$, which is based on the elementary analysis and it was also supposed that all intercalated palladium species produced PE. The investigation of the molecular weight mass, which was measured by a gel permeation chromatography (GPC) calibrated with polystyrene standards, showed high masses of the polymer ($M_n$ 159,000, $M_w$ 262,000). Of the exfoliated nanocomposites, XRD measurements were done. The peak of the nanoparticles disappeared during the polymerization.

PE/palgorskite nanocomposites were produced by Rong et al. by *in-situ* coordination polymerization [35]. The palgorskite is a fibrous silicate. The distribution of the nanoparticles was good in the polymer, and there are strong interaction between the filler and the PE.

More research is carried out of PE nanocomposites with MMT. The MMT could be modified in different ways.

Jin et al. used MMT for PE nanocomposites as fillers [36]. The catalyst was the Ziegler-Natta $TiCl_4/Et_3Al$ system. The catalyst activity (5–80 kg PE/(mol Ti h)) is comparatively low. The polymerization rate decreased with the polymerization time. Likely the catalyst was deactivated by the OH groups. The catalyst activity was also affected by the Al/Ti molar ratio. The activity became higher with a higher Al/Ti ratio. The melting temperatures, degrees of crystallinity, and molecular weights were also analyzed. The molecular weight of the natural sodium MMT was higher than that of the organophilically modified MMT. A reason could be the deactivation of the catalyst by the hydroxyl groups of the modified MMT. The WAXD investigations showed that the basal peak of the nanocomposites with the modified MMT disappeared, the peak of the composites with sodium MMT were shifted a little bit to a lower scattering angle, but the peak did not completely disappear. By microscopic investigations, the difference between the nanocomposites with the two fillers could be pointed out. The particles of the PE/MMT-OH were better distributed in the polymer matrix.

Wu et al. also used Ziegler–Natta catalysts, but unlike Jin et al. they supported it on organoclay/$MgCl_2$ and used $AlEt_3$ as the cocatalyst [37].

Nanocomposites with PE and MMT (cloisite $Na^+$) were also prepared by *in-situ* intercalative polymerization using an MMT/$MgCl_2$/$TiCl_4$ catalyst system [38]. The nanocomposites had superior tensile properties. The exfoliation was observed by WAXD and TEM. The melting temperatures and the crystallization temperatures did not change by the amount of filler, but the crystallinity got rapidly lower with higher amount of filler.

Antipov et al. synthesized clay/PE-nanocomposites with a Ziegler–Natta catalyst [39]. The nanoparticles were modified with dimethyldioctadecylammonium. Nanocomposites until 43 mass% were prepared. X-ray diffraction (XRD) analyses of the pure PE, the nanocomposites, and the pure MMT were taken. The modified clay without polymer determined a clearly defined basal reflex at $2\Theta = 3.3°$. This reflex could not be obtained in a sample with a filler content of less than 24 mass%. A reason could be that the filler is widely spaced or completely encased with polymer. The melting temperatures of the nanocomposites did not change with different amounts of filler, but the melting enthalpy got higher with an increase in the filler. The physicomechanical properties were researched. The properties of the nanocomposites with an amount of 4–8 mass% were the best.

PE/MMT nanocomposites were also prepared by *in-situ* polymerization with a metallocene catalyst, which were fixed at the MMT [40]. The activity of the produced nanocomposite ($2–4 \times 10^6$ g PE/mol Zr * H) was lower than the activity of the PE without filler ($10^7$

g PE/mol Zr * H). XRD analysis showed that the peak of the nanoparticles disappeared and a peak of the PE appeared with the polymerization time. The molecular weight after 15-min polymerization was higher than that after 45 min.

Two species of particles, clay and silica, were simultaneously inserted in the polymer by *in-situ* polymerization [41]. The mechanical properties were analyzed, which were good with an amount of lower 10% filler. The morphology of the nanocomposites was different to the homogeneous PE, which had a spongy structure.

The produced PE/MMT-nanocomposites were analyzed by the emphasis of the isothermal crystallization kinetic behavior [42]. It was observed that a high amount of fillers decreased the mobility and arrangement ability of the PE chains. A reason was the change of the relative rates of secondary nucleation and surface spreading, leading to regime III crystallization in the studied temperature range.

Mülhaupt et al. prepared PE nanocomposites with different silicate types and silicate content by *in-situ* polymerization [43]. The distribution of the nanoparticles in the polymers, which were produced by *in-situ* polymerization, was better than in the nanocomposites prepared by melt compounding. The particles were much smaller and had different morphologies. If the catalyst activities were low, the catalyst residues would form smaller particles. A reason could be the decomposition of the MAO activator accumulated at layered interface. The properties of the nanocomposites prepared by *in-situ* polymerization were better than the properties produced by melt compounding, which was shown by TEM and WAXD.

PE nanocomposites with MMT were also synthesized by Zhang et al. [44]. In the PE/MMT, nanocomposites were intercalated and structures delaminated, which were approved by FTIR, XRD, and SEM. The melting and the crystallization temperatures of the nanocomposites were higher than that of the pure PE and the crystallinity decreased rapidly.

Clay/PE-nanocomposites were prepared by Ray et al. with MAO and iron(II) catalysts, 2,6-bis[1-(2,6-diisopropylphenylimino)ethyl] pyridine iron(II) dichloride [45]. The degree of exfoliation depended on the amount of clay. With a higher amount of filler, the distribution of the particles in the polymer was worse. The distribution got better with a higher Al/Fe ratio. The crystallite size and the crystallinity were lower by complete exfoliation. The degree of exfoliation decreased with the use of a preformed clay-supported catalyst. The reason was the blockade of the active centers by the clay. Samples with similar molecular weights were analyzed. The melt elasticity $G'$ got higher with an increase in the extent of the clay dispersion, which was supported by the picture of hydrodynamically percolated clay platelets in the melt. The density of the nanocomposites got higher. If the extent of the exfoliation was greater, the elasticity would increase, and so the loss modulus $G''$ and the complex viscosity $\eta^*$. If the extent of the exfoliation was similar, $G'$, $G''$, and $\eta^*$ would increase with higher molecular weight. A comparison of a heterogeneously catalyzed nanocomposite with a homogenously catalyzed one showed that the melt elasticity was higher of the heterogenous one. The viscoelastic properties of the melt got higher with the amount of the clay, which was analogous with the percolation picture.

PE/MMT nanocomposites were produced by using a nickel complex [46]. The PE had bimodal molecular weight distribution and bimodal chain branching distribution. Polymerizations in the presence of boehmite fillers were realized by Mülhaupt et al. with two different metallocene catalysts, $Cp_2ZrCl_2$ and *rac*-$Me_2Si(2$-Me-benz[e]-Ind$)_2ZrCl_2$ [47]. The boehmite was used unmodified and modified with organic acids as carboxylic and sulfonic acids. The catalyst activity increased in the polymerizations with unmodified boehmite and boehmite with 2 wt% organic modifiers, but the distribution of the filler in the PE got worse. However, nanocomposites with 20 wt% boehmite had a lower catalyst activity and the fillers are homogenously dispersed. The sulfonate modified boehmite activated the heterogenous production of the PE/boehmite nanocomposites, because the sulfonate-modified boehmite supported the MAO-activated metallocene catalysts.

Another filler for nanocomposites was (Pb,Sr)TiO$_3$ (PST), which were used to prepare PE nanocomposites by *in-situ* polymerization and Ziegler–Natta catalyst [6]. The PST was homogenously distributed which was shown by TEM, XRD, and SEM. The dielectric constant of (Pb,Sr)TiO$_3$/PE-nanocomposites depended highly on the concentration.

New magnetic nanocomposites were produced by using Fe(II)-nanoparticle and Ziegler–Natta catalyst [48].

Alexandre et al. produced a lot of nanocomposites with different fillers, different catalysts, and with ethylene and octene, and a few nanocomposites additional with decadiene [3, 49]. Polymerizations were realized with kaolin particles, in which pretreatment was different, and with a Ziegler–Natta catalyst. The activities varied depending on the particles treatment. The influence of hydrogen during the polymerization of kaolin nanocomposites with a metallocene catalyst was improved. The activity of the nanocomposites produced with hydrogen is higher than the nanocomposites prepared without hydrogen. Ethylene/1-octene and ethylene/1-octene/1,9-decadiene nanocomposites were fabricated with kaolin, graphite, wollastonite, and magnesium hydroxide. All the fabricated nanocomposites had lower melting temperatures than the pure PE, which is an indication that the octene was integrated in the PE chains. The use of hydrogen during the polymerization increased the catalyst activity and controlled the molecular weight, but the hydrogen did not influence the melting temperature, which was an indication that the incorporation of the octene did not change. The incorporation of the octene changed the thermal behavior of the polymer. The different fillers had an influence on the catalyst activity. The activity of the polymerization with kaolin particles was the highest. Different bulks were produced with different fillers and the same polymerization conditions. The incorporation of the octene was dependent on the filler type. The incorporation of the octene was higher in the presence of wollastonite than in the presence of magnesium hydroxide. This led to different mechanical properties of prepared composites. The modulus of magnesium hydroxide composites was very high compared with the wollastonite composites, which depended on the decreased incorporation of the octene and a change in the filler morphology.

Wang et al. prepared PE/MT-composites with ethene oligomers and ethene [50]. The ethene oligomers were prepared with an MT-supported Zr(acac)2Cl$_2$ catalyst and AlEt$_2$Cl cocatalyst. *Rac*-Et(Ind)$_2$ZrCl$_2$ and MAO were added to the solution of the oligomer. The ratio of the two catalysts and the cocatalyst had an influence on the properties. With a higher amount of oligomer catalyst, the melting point, the crystallinity, and the molecular weight of the PE decreased, because the branched chain content of the PE increased. The X-ray analysis showed that the H-MT peak disappeared at XRD picture of the nanocomposite. This is an indication that the MT particles are very well distributed. Thermogravimetric analysis was performed. The thermal decomposition temperature was between 287 and 427°C and the incorporation of the particles between 2.1% and 12.1%.

## 3.3.2. Silicate and Spheric Particles

Schumann et al. produced nanocomposites with highly active heterogenous cocatalysts, using supports consisting of interlinked and surface-modified spherical silica nanoparticles with aminoorgano-trialkoxysilanes as spacers and bis(alkoxysilyl)organyls as linkers [51].

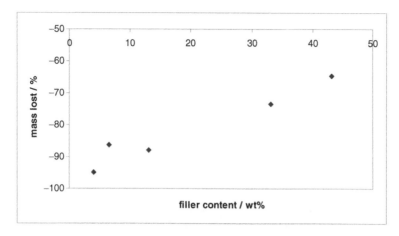

**Figure 3.5** Incorporation of the silica particles in the polymer. Thermo gravimetric measured mass loss in dependence of the filler content of PE silica nanocomposites.

They used $Cp_2ZrCl_2$ as the catalyst for the polymerizations. The treatment only with MAO was done in a one-step reaction and the treatments with TMA/MAO, TEA/MAO, or MAO/MAO in two-step reactions. The activities of these systems were comparable to MAO in homogeneous solutions and higher to the commercially available heterogenous reference system Sylopol/MAO from Witco. Fragmentation of the catalyst particles was observed by the new cocatalyst during the polymerization, which was supported by SEM and EDX and the polymerization activity profiles. Leaching tests were also done.

Alexandre et al. produced $SiO_2$/PE-nanocomposites by the PFT, which is similar to the *in-situ* polymerization [3, 49].

PE-nanocomposites either with $SiO_2$ or $CaCO_3$ were prepared in our group by *in-situ* polymerization with different catalysts, such as $[Me_2Si(Me_4Cp)(N^{tert}Bu)]TiCl_2$, $Me_2Si(Ind)_2ZrCl_2$, and $Me_2C(Flu)(Cp)ZrCl_2$ [52]. The activity depends on the pretreatment of the particles and on the type of the catalyst; $Me_2Si(Ind)_2ZrCl_2$ shows the highest activity. With more amounts of the filler, the activity decreased, not increased. The nanocomposites produced with the addition of triisobutylaluminium had higher activities than the nanocomposites produced without it. The molecular weights and the melting points decreased with more amounts of the filler. The crystallization temperatures got lower with more amounts of the filler. The PE/$SiO_2$-nanocomposites could not be used as flame retardant, because the thermal decomposition temperatures increased with lower amounts of the filler. The thermogravimetric analysis gives a relation between the mass loss and the incorporation (filler content) of the silica particles (Figure 3.5).

The used amounts of the particles were incorporated into the polymer. The variations were the remains of aluminum of the methylaluminoxane in the nanocomposites and impurities of the particles.

Figures 3.6 and 3.7 were taken from electron microscopy analysis. It can be seen that the nanoparticles are encased with PE. The nanoparticles are homogenously distributed and are connected by the PE filaments. No agglomeration of the nanoparticles was observed.

LLDPE/alumina nanocomposites were also successfully prepared by *in-situ* polymerization, using d-MMAO impregnated $Al_2O_3$ [53]. It was found that in this case, the melting temperature and the molecular weight depended on the [Al]dMMAO/[Zr] ratios.

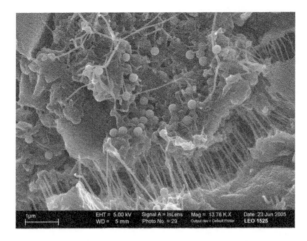

**Figure 3.6**    Silica nanocomposites with 13.1 wt% filler content.

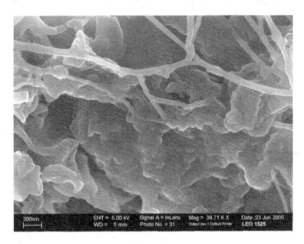

**Figure 3.7**    Calcium carbonate nanocomposites with 8.6 wt% filler content.

As another spherical nanoparticle $CaCO_3$ was homogeneously incorporated into a HDPE matrix [54].

## 3.4. PE/CB Nanocomposites

The *in-situ* polymerization of PE/CB nanocomposites were performed without pretreatment, with PFT [55] and via a catalyst-supported approach [56]. In cases where the polymer was synthesized in the presence of pristine carbon black, the filler distribution in the resulting composite was quite heterogeneous and had a morphology that changes with the graphite content, which in turn had a direct effect on the melting behavior and crystallinity. In contrast, the catalyst-supported approach provided a much better control of the PE production and led to well-dispersed and encapsulated CB particles. It is pretty much the same for the PFT, which allowed very defined and homogenous materials, even at high graphite loadings. The close interaction of the matrix with the CB surface perturbed the melting behavior in regard to the filler loading. Melting peak and crystallinity decreased as the filler level increased, which was explained due to differences in the thermal conductivity for composites of different graphite contents [55] and because of the strong interactions between PE and CB describable by means of a two-layer PE model [56]. These results were in contrast to what was found for PP/CB composites, where the $X_c$ and $T_m$ of the PP were unaffected by the presence of CB (see Section 4.3).

The activity showed a drastic loss in regard of the polymerization time in case of PE-based composites, while in the production of PP nanocomposites no filler effect was observed, neither for CB nor for other carbon-based particles like CNF or CNT.

However, PE matrices and their morphology seemed to be more influenceable by the graphite nanoparticles, which was also observable in thermal degradation analyses by TGA. The CB was responsible for a thermal degradation at lower temperatures ($\sim 300°C$) than for pure PE, which mainly degrades between 400 and 500°C.

## 3.5. PE/CNT Nanocomposites

CNTs are especially attractive class of fillers for polymers because of their intriguing mechanical and thermal properties. They represent one of the strongest and toughest materials known [57]. An entire separation, homogeneous distribution, and a good adhesion of the polymer matrix and the single- or multiwalled carbon nanotubes (SWCNT, MWCNT) are important to achieve the full potential of the resulting PE/CNT nanocomposite. The pretreatments and PFT discussed in Section 3.2 represent promising routes to improve the mechanical properties of nanotube-based composites. It was reported that because of the attractive van der Waals forces, CNTs tend to stay aggregated or in case of SWCNT bundled, if not or only treated by ultrasound. The PFT derivates, or at least an impregnation of the CNT with MAO, are necessary to achieve a satisfactory distribution of nanotubes in a polyolefin matrix [58]. Bonded or adsorbed (co-)catalyst, usually MAO, stabilized the suspension of the dispersed nanotubes due to the repulsive forces between the MAO units, which in turn prevent the CNT from reagglomerating.

The catalyst activity was not directly influenced by the presence of (dried) nanotubes. The preparation of CNT containing nanocomposites depended on the same factors like the *in-situ* polymerization of neat polyolefins. The activity is influenced by the [Al]/[Zr] ratio, which in some cases depended on the pretreatment. Of course, certain functional groups and impurities on the tube surface especially adsorbed water are able to hydrolyze MAO, or SDS (sodium dodecyl sulfat), which is often used during CNT modification to stabilize the suspension, can cause an inactivity of the metallocene, due to etherification of the zirconium center. Besides that, when using the pretreatment method developed by Kaminsky et al., where a covalent bonding is established between MAO and the filler, of course the number of functional groups on the CNT surface controls the [Al]/[Zr] ratio, at least for the time of coating, after which extra MAO can be added to ascertain the activity desired.

It is known that CNTs act as nucleating agents; therefore improvements on the crystallization behavior were expected. The crystallization temperature ($T_c$), crystallinity ($X_c$), and the half-time of crystallization ($\tau 1/2$) play an important role during the processing of these materials. By reducing the cooling time required for part solidification, cycle times can be shortened. This can be accomplished by the incorporation of nucleating agents into the neat polymer, which accelerates the crystallization. The half-time of crystallization

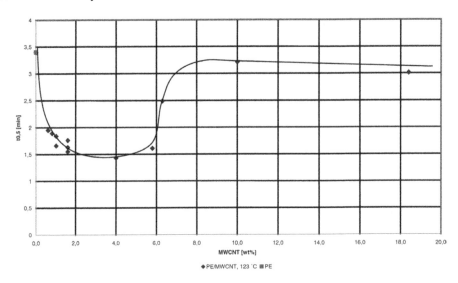

**Figure 3.8**    Half-time of crystallization in regard of the filler content (PE/MWCNT).

is the length of time needed for 50% of the crystallizable material to solidify from the melt. It depended on the (isothermal) crystallization temperature and on the filler content. The crystallization kinetic influences the morphology, which in turn affects the mechanical properties of a semicrystalline polymer [59, 60].

All incorporated amounts of CNT increased $T_c$ of the resulted nanocomposites in comparison to neat PE [61]. The reference PE had a $T_c$ of about 114°C; 0.6 wt% MWCNT led to an increase of about 1.5 K and a maximum was found at 5 wt% filler content, which caused an increase of 3.7 K. Even higher filler loadings led to lower $T_c$s, but still leveled above as what was found for pure PE. For example, a nanocomposite containing 18.0 wt% of CNT had a $T_c$ of 115.4°C. In comparison to PP/MWCNT composites, the effect on the PE matrix was not a quarter as good distinctive. Also the half-time of crystallization, representing the CNT nucleating effect, was not as pronounced as for comparable PP-based nanocomposites.

However in Figure 3.8, it can be seen that the half-time of crystallization was reduced significantly with low amounts of nanotubes at a temperature of 123°C [62]. There was a reduction of $\tau^{1/2}$ in the range of 0–6 wt% with a minimum at 4 wt% filler content. At the minimum, the solidification of 50% of the crystallizable material was about 2.5 times faster than for neat PE. Larger amount of MWCNT seemed to decelerate the crystallization until (MWCNT content > 10 wt%) it was as slow as for pure PE. It can be said that PE, which inherently crystallizes faster than PP, is much more influenced by the presence of the nanotubes and their amount.

The morphology of the PE/CNT nanocomposites prepared by *in-situ* polymerization was, in comparison to melt compounding produced ones, generally indicated by a good CNT separation, homogenous distribution in the matrix, and a good polymer wetting, which in turn indicates a tight adhesion, especially when a pretreatment, which led to an encapsulating of the tubes with polymer, was performed. *In-situ* polymerization in combination with an effective pretreatment offers the possibility to synthesize the whole composite with a tailored polyolefin matrix and any desired molecular weight or if quenched after the CNT coating/encapsulation process to produce a "masterbatch," with typically 40 wt% of polyolefin. In this masterbatch, the nanotubes are predispersed in PE and can be used for the preparation of matrices normally not accessible by *in-situ* polymerization. PE-coated CNTs, synthesized by *in-situ* polymerization, were successfully used as a predispersed masterbatch for the production of a homogenously ethylene-*co*-vinyl acetate (EVA/MWCNT) nanocomposite, using a co-rotating twin-screw mini-extruder. The MWCNT amount in the final material has been determined to be 3 wt% [63].

Not only MAO can be anchored on the tube surface, there was also a report on CNT-supported metallocenes. Nanotubes enwrapped in nano PE wires could be obtained by using supported $Cp_2ZrCl_2$ [4].

## 3.6. Polypropylene Nanocomposites

The world consumption of PP in 2005 was roughly about $37.5 \times 10^6$ tons, exhibiting an average annual growth rate of 14.9% [64]. The most common commercial form of PP is isotactic PP (iPP), which features good stiffness, high melting temperature and yield strength, good chemical resistance, and excellent moisture barrier properties [65]. Isotactic and syndiotactic PP (sPP) exhibit a high crystallinity and melting temperature. In addition, sPP shows good elastic properties in a wide deformation range [66]. It is softer and has a higher clarity than iPP. On the other hand, the low crystallization rate hinders the commercial application [67]. In summary, it can be ascertained that PP is one of the most versatile and common polyolefins, which in turn makes it a main focus, especially in combination with CNTs, of today's nanocomposite research.

### 3.6.1. Layered Silicate, Montmorillonite

PP/silicate nanocomposites were produced by Coates et al. by *in-situ* polymerization with a palladium catalyst, but the TOF was low (5.6 $h^{-1}$) [68].

Sun prepared PP/clay nanocomposites by *in-situ* polymerization without any other cocatalysts, for example, methylaluminoxan or perfluoroborates, and high pressure or high temperature [69]. The clay fillers were modified intercalation of amine complexes instead of the clay ions. In optical images and TEM, pictures were shown as agglomerates. The polypropylene in the nanocomposites had a molecular weight between 100,000 and 300,000 g/mol and the molecular weight distribution is narrow (2.3) and similar to polypropylene prepared with metallocene catalysts. The isotacticity of the polypropylene is very high (95–98%), which was measured by $^{13}C$ NMR and the crystallinity is about 65%. The Yong's tensile modulus of the nanocomposite was 600 kpsi, which is higher than that of crystalline PE. Nanocomposite with 0.1 wt% filler had a lower Yong modulus between 220 and 260 kpsi. The Yong modulus was at the similar range to pure PE. The reduced thermal expansion coefficient, improved melt flow strength, increased heat-distortion temperatures, and lowered gas diffusion were few more advantages of the nanocomposites.

Ma et al. produced PP nanocomposites with clay filler by intercalative polymerization [70]. The clay was distributed very well in the polymer and dispersed into nanoparticles, which was approved by WAXD and TEM. It was analyzed by the storage modulus, which increased with a higher amount of filler. If especially the temperatures were higher than $T_g$, the storage modulus would be three times higher than that of the pure PE. The HDT has risen up with the amount of filler. PE without filler had the lowest heat distortion temperature (416°C).

MMT was successfully used as a filler in iPP-based nanocomposites, which were synthesized by *rac*-Me$_2$Si(2-Me-4-Ph-Ind)$_2$ZrCl$_2$/MAO [71] iPP/monoalkylimidazolium-modified MMT prepared by *in-situ* polymerization with Ziegler–Natta catalyst TiCl$_4$/MgCl$_2$, which had improved the thermal stability [72, 73]. The catalytic efficiency of the catalyst under optimum polymerization condition achieved 1300 kg/(molTi h). The catalytic properties did not vary by the use of MMT. The molecular weight of the polypropylene is high and the good distribution of the silicate layers of MMT in the polypropylene matrix is shown by TEM images. The main crystallite, which was created by the PP/MMT nanocomposites, was the $\alpha$-phase crystallite.

PP/MMT-nanocomposites were prepared by Ma et al. with Ziegler–Natta catalysts [74]. The crystallization behaviors of the nanocomposites were analyzed by using differential scanning calorimeter (DSC) and polarizing optical microscope (POM). The nanoparticles in the polymer led to a decrease in crystallinity. The MMT worked as heterogenous nuclei in the nucleation of crystallization. This took to a higher crystallization rate and lower spherulite size with the increase in the amount of the filler. The nucleus density increased with the amount of the filler, which took a positive effect on the crystallization, supported by the result that the interfacial free energy per unit area perpendicular to PP chains in PP/MMT nanocomposites got higher with lower amount of the MMT filler.

PP/sodium-type MMT nanocomposites were produced with the *rac*-Et(Ind)$_2$ZrCl$_2$ catalysts by Hwu [75]. The distribution of the silicate layers of MMT observed by XRD and TEM were good. The isotactic crystallinity, bulk density, Vicker hardness, melt temperature, and the enthalpy of the nanocomposites were higher than polypropylene without amounts of filler.

Wang et al. synthesized new magnetic nanocomposites with propylene and ethylene/propylene by using Fe(II)-nanoparticle and Ziegler–Natta catalyst [76].

## 3.6.2. Silica, Spheric Particles

PP/SiO$_2$ nanocomposites with a Ziegler–Natta catalyst were produced in an SPP reactor by Garcia et al. [77]. A homogenous distribution with 3 wt% filler was observed by SEM. The crystallinity was not influenced by the amount of the silica in the polymer.

PP/SiO$_2$ nanocomposites could also be prepared with metallocene catalysts by slurry and gas phase reaction by Kaminsky et al. [78].

For the experiments in our group, solid silica balls (Monosphers M250) with an average diameter of 250 nm were used (see preparation).

To find the minimum amount of MAO needed to activate the metallocene and in order to determine whether the applied amount of MAO had any influence on the envelopment of the silica spheres, different amounts of MAO were applied to the silica spheres for heterogenization. All other conditions (solvent volume, polymerization temperature, amount of M250/MAO, polymerization time, catalyst amount) were kept constant. Several polymerizations were performed with each batch of M250/MAO to determine the optimum amount of MAO for the impregnation. The results are shown in Table 3.1.

The average activity of the M250/MAO/1 catalyst varied according to the amount of MAO applied for the preparation of the cocatalyst. The maximum activity of 3400 kg$_{Pol}$/(mol$_{Zr}$ h mol$_{Mon}$/l) was reached when an MAO amount of 420 mg/g M250 was used for the impregnation. For higher amounts of 830 mg MAO/g M250, the activity of

**Table 3.1**  Activity related to the amount of MAO applied to M250$^a$.

| mg MAO in<br>200 ml toluene | Activity<br>kg Pol/mol Zr h cp |
|:---:|:---:|
| 100 | 2700 |
| 210 | 2700 |
| 420 | 3400 |
| 600 | 2600 |
| 830 | 2400 |

$^a$ Polymerization conditions: Polymerization temperature $T_p$ 30°C, polymerization time ($t_p$) 30 min, propylene concentration ($cp$) 1.38 mol/l (2 bar), solvent 200 ml of toluene, amount of $[(p\text{-MePh})''C(Cp)_2(2,7\text{-bis-}t\text{-BuFlu})]ZrCl_2$ $1.3 \times 10^{-6}$ mol, amount of M250/MAO 0.55 g, amount of triisobutylaluminium (TIBA) 2 mmol.

the corresponding catalysts decreased to 2400 kg$_{Pol}$/(mol$_{Zr}$ h mol$_{Mon}$/l), respectively. All detected melting temperatures lay in the region of 140°C with no distinct trend visible regarding the applied amount of MAO. The same is true for the molar masses ranging from 500,000 to 600,000 g/mol.

The coverage of the silica spheres with syndiotactic polypropylene can be seen in Figure 3.9. The figure shows that every monospher is covered with a thin film of 20–50 nm of PP. There are some polypropylene fibers that contain less silica balls due to different active sites. Here 420 mg MAO were used for 1 g of M250. The content of fillers in the sample is around 40 wt% and the rrrr pentade around 91%.

To reach the goal of a better dispersion and envelopment of the silica spheres with polypropylene, the influence of the polymerization conditions was investigated. To ensure comparability among the results, one batch of M250/MAO was synthesized as described above and used as cocatalyst for all subsequent slurry polymerizations. The first condition to be varied was the polymerization temperature. All polymerizations were carried out in

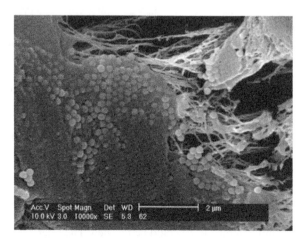

**Figure 3.9**  sPP/M250 nanocomposite (magnification ×10,000). Almost every monospher particle is covered with a thin polymer film.

**Table 3.2**  Average activity and polymer properties in relation to polymerization temperature, propylene concentration, and catalyst amount[a].

| Polymerization temperature (°C) | Propylene concentration (mol/l) | Catalyst amount (mol) | Activity $(kg_{Pol}/(mol_{Zr}$ h $mol_{Mon}/l))$ | Melting point (°C) | Molecular weight (g/mol) |
|---|---|---|---|---|---|
| 0  | 1.4 | $1.30 \times 10^{-6}$ | 600  | 149 | 790,000 |
| 30 | 0.6 | $1.40 \times 10^{-6}$ | 1800 | 137 | 370,000 |
| 30 | 1.4 | $1.30 \times 10^{-6}$ | 2300 | 141 | 560,000 |
| 30 | 3.5 | $6.00 \times 10^{-7}$ | 2500 | 142 | 640,000 |
| 60 | 1.4 | $7.00 \times 10^{-7}$ | 3000 | 121 | 220,000 |

[a] Polymerization conditions: polymerization time (30–180 min), solvent 200 ml of toluene, amount of M250/MAO 0.55 g, amount of triisobutylaluminium (TIBA) 2 mmol.

200 ml toluene mixed with 2 mmol of TIBA as scavenger. The amount of M250/MAO was kept constant at 0.55 g, and the polymerization time was between 30 an 180 min, depending on the temperature. The corresponding results for 0°C, 30°C, and 60°C are listed in Table 3.2.

As can be seen from the table, a higher polymerization temperature expectedly included a higher activity. The average activity of the polymerizations with M250/MAO/1 as cocatalyst was much lower than that of comparable homogeneous polymerizations with MAO/1. For 0°C, it was only 600 $kg_{Pol}/(mol_{Zr}$ h $mol_{Mon}/l)$ with the heterogeneous system M250/MAO/3 as compared to 2000 $kg_{Pol}/(mol_{Zr}$ h $mol_{Mon}/l)$ for the homogeneous polymerization. At 30°C, it was 2300 as compared to 5200 $kg_{Pol}/(mol_{Zr}$ h $mol_{Mon}/l)$, and at 60°C it amounted to 2400 as compared to 9500 $kg_{Pol}/(mol_{Zr}$ h $mol_{Mon}/l)$. As expected, the highest melting temperatures of 149°C were reached for polypropylenes synthesized at 0°C, then decreased to 141°C at 30°C polymerization temperature, and to 121°C at a polymerization temperature of 60°C, corresponding to a sinking syndiotacticity.

The next parameter to be investigated was the concentration of propylene in the reaction mixture. It was varied in the range from 0.6 to 3.5 mol/l. According to Table 3.2, hardly any dependence of the activity, lying between 1800 $kg_{Pol}/(mol_{Zr}$ h $mol_{Mon}/l)$ and 2500 $kg_{Pol}/(mol_{Zr}$ h $mol_{Mon}/l)$, on the concentration was found. Only at a propylene concentration as low as 0.6 mol/l, the average activity sank to 1800 $kg_{Pol}/(mol_{Zr}$ h $mol_{Mon}/l)$ and was, therefore, inferior to activities in polymerizations with a higher propylene concentration. Likewise, the melting point of approximately 137°C lay about 4°C below that of the polypropylene from the slurry polymerizations at 1.4 mol/l propylene at otherwise same conditions. For the highest propylene concentration of 3.5 mol/l, the melting temperature lies in the same range as for the polymers produced at a concentration of 1.4 mol propylene per liter. The molecular weight was naturally influenced by the monomer concentration and increased from 370,000 to 640,000 g/mol for propylene concentrations of 0.6 to 3.5 mol/l, respectively.

In a gas phase, experiments were carried out using $rac$-[Et-$(IndH_4)_2$]$ZrCl_2$, which produces isotactic polypropylene. The polymerization was carried out at 30°C for 60 min with a metallocene amount of $6.25 \times 10^{-6}$ mol. The propylene concentration was 0.2 mol/l, and 0.55 g M250/MAO prepared as described above was used as the cocatalyst. An activity of 400 $kg_{Pol}/(mol_{Zr}$ h $mol_{Mon}/l)$ could be reached. The filler content was 50% and the melting temperature of the isotactic polypropylene was 138°C. It can be seen from Figure 3.10 that the monosphers are also very well dispersed in the polymer matrix and

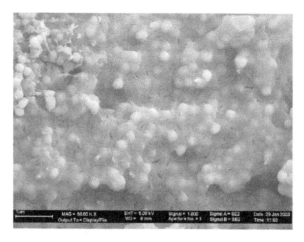

**Figure 3.10**    iPP/M250 nanocomposite (magnification ×50,000). The filler is well dispersed in the polymer matrix and almost every ionosphere particle is covered with a thin (20–100 nm) polymer layer.

the coverage is even better than in the case of the syndiotactic polypropylene produced with M250/MAO/**1**. The thickness of the polypropylene layer covering the monosphers lies approximately between 20 and 100 nm [79].

### 3.6.3. PP/CB Nanocomposites

In comparison to filler types with high aspect ratios, nanosized CB (e.g., pyrograph III, average diameter about 30 nm) was relatively easy to separate due to its spherical shape. The melting temperatures of an sPP/CB nanocomposite was in the same region as for the unfilled polymer. Also the crystallinity (about 28%) was largely unaffected by the presence of CB and so was the activity. The average activity of the system metallocene (**3**)/MAO was between 4000 and 5000 $kg_{Pol}/(mol_{Zr}$ h $mol_{Mon}/l)$, which was in the same range as the activity of 4300 $kg_{Pol}/(mol_{Zr}$ h $mol_{Mon}/l)$ for the polymerizations without filler.

On the other hand, the filler led to a slight increase in the crystallization temperature of the respective nanocomposites in comparison to the pure polymer. $T_c$ was raised by 4 to 6°C depending on the filler loading. The crystallization of sPP/CB nanocomposites proceeded faster than that of the pure PP. For example at 118°C, the half-time of crystallization of a nanocomposite containing 0.6 wt% carbon black was 3.5 min, while pure sPP's $\tau_{1/2}$ was about 8 min. The half-times of crystallization of carbon nanowires and carbon black composites were comparable, with MWCNT showing a superiority as nucleating agents, as discussed in Section 3.4.5.

### 3.6.4. PP/CNF Nanocomposites

Carbon nanowires are interesting as fillers for polymers, because they represent an intermediate between traditional bulk carbon fibers of a large diameter and CNTs with a diameter in the range of a very few nanometers. They are available in large quantities at low costs [80, 81], which makes them even more attractive. CNF can be synthesized catalytically. Gaseous hydrocarbons are decomposed in the gas phase in the presence of a transition metal catalyst to form the CNF [82].

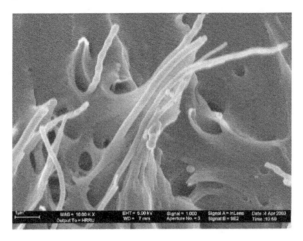

**Figure 3.11**  CNF wetted with sPP. Pullout effects can also be seen.

In contrast to CNT, the nanosized fibers are stiff and inflexible, their agglomerates are not so knotted, and so they are comparably easy to separate by ultrasound.

Incorporated CNF led only to a slight increase (about 2°C in comparison to neat sPP or iPP) in the composite melting point and largely unaffected the crystallinity, independently of their weight percentage. In contrast to that, the crystallization temperature and the half-time of crystallization were much more influenced by the presence of CNF and depended strongly on the quality of the filler distribution. For example, a nanocomposite prepared by *in-situ* polymerization, which contains only 1.0 wt% CNF, increased the crystallization temperature by 5°C, while for the same improvement 7 wt% must be added by using melt compounding [59]. Possibly the *in-situ* polymerization, or the opportunity to accomplish the prereaction in a low-viscosity solvent, was leading to a more homogenous CNF distribution. The values of the half-time of crystallization taken from isothermal DSC measurements showed a decrease in $\tau_{1/2}$ when CNFs were incorporated. Contrary to CNTs, higher contents of nanofibers (range 0.1–1.1 wt%) seemed not to modify $\tau_{1/2}$ anymore. It was decreased to roughly one-third of the value for pure iPP and one-half for pure sPP in all [83].

The morphology and the quality of the adhesion of the PP matrix to the CNF can be estimated from SEM micrographs. In Figure 3.11, it can be seen that the wetting of the fibers by the polymer is good and the fibers were separated from each other, although they were loosely agglomerated. It can also be seen from this picture that fiber pull-out has happened to a certain degree, which is obvious from the holes in the polymer matrix.

### 3.6.5.  PP/CNT Nanocomposites

There were many reports on PP/CNT nanocomposites prepared with the pretreatments discussed in Section 3.2. Thus prepared nanocomposites showed improvements especially on the crystallization behavior. Due to the fact that CNTs act as nucleating agents, they led to crystallization at higher temperatures, which was common for neat iPP.

The effect on the crystallization temperature was most emphasized. Pure iPP had a $T_c$ of about 111°C. The addition of 0.1 wt% MWCNT led to an increased $T_c$ by 4 to 5 K. At higher filler contents, the enhancement of the crystallization temperature raised rapidly

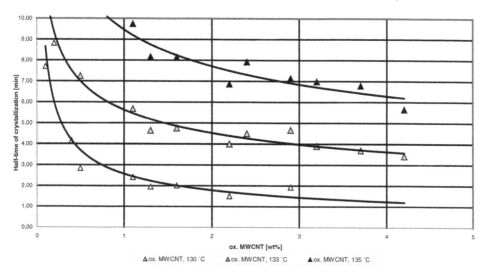

**Figure 3.12**   Half-time of crystallization of iPP/ox. MWCNT nanocomposites in dependence on the filler content at different isothermal crystallization temperatures.

to a certain extent of 9 K above pure iPP, which was reached at an MWCNT content of 1.5–2 wt%. In case of sPP/MWCNT composites, the maximum enhancement of $T_c$ was even up to 15 K higher in comparison to neat sPP, which has a crystallization temperature of about 95°C.

Figure 3.12 shows the dependence of $\tau_{1/2}$ on oxidized MWCNT content at different isothermal crystallization temperatures. The trend shown was similar for all types of MWCNT (crude, purified, oxidized, modified) and depended preliminary on the fillers distribution quality in the matrix. At the temperatures shown, pure iPP did not crystallize or it took too long to provide reliable results; on the other hand, the crystallization of the obtained nanocomposites was too fast in the range of temperature where iPP gave adequate results. However in Figure 3.12, it can be seen that the half-time of crystallization was reduced significantly with low amounts of oxidized nanotubes at all temperatures.

At 130°C, for instance, $\tau_{1/2}$was 8 min for an iPP/ox. MWCNT composite with 0.1 wt% filler content, while 2.5 wt% reduced $\tau_{1/2}$ to only 1.5 min. When the percentage of MWCNT was raised to more than 3 wt%, no appreciable reduction of $\tau_{1/2}$ was obtained anymore. For this system, that seemed to be the lower limit. It should be noted that this is still a controversy discussed in literature. The half-time of crystallization of CNT nanocomposites was found to decrease in all cited references with regard to the pure polymers, independently of which polymerization technique was used. Valentini et al. also reported a reduction for PP/EPDM/SWCNT nanocomposites prepared by melt compounding. But in this case, $\tau_{1/2}$ showed a minimum at 0.5 wt% SWCNT [84]. The authors attributed this to the fact that the SWCNTs nucleate crystallite growth on one hand but hinder the spherulite growth on the other hand.

The morphology of the PP/MWCNT nanocomposites prepared by *in-situ* polymerization was indicated by a good separation, homogenous distribution, and a good wetting with polymer, which in turn indicated a tight adhesion, especially when a pretreatment, which led to an encapsulating of the tubes with polymer, was performed. SEM micrographs taken from cryofractures showed no pull-out effect (Figure 3.13).

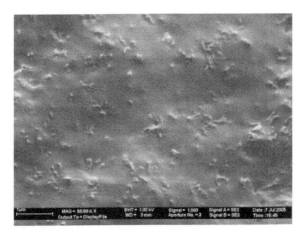

**Figure 3.13** SEM micrograph (cryofracture, magnification ×50,000) of an iPP/ox. MWCNT nanocomposite containing 1.3 wt% nanotubes. Experiment contained coating process. The nanotubes are well separated, wetted with iPP, and seem to be located under a polymer layer.

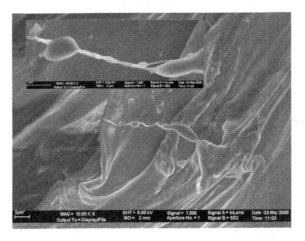

**Figure 3.14** SEM micrograph (cryofracture, magnification ×10,000, insertion ×50,000) of an iPP/ox. MWCNT nanocomposite containing 3.7 wt% nanotubes. Experiment contained coating process. The nanotube shown is still wetted and layered by polymer. The position marked shows the pristine nanotube.

There was evidence for the existence of a good load transfer between the matrix and the MWCNT (Figure 3.14). The pretreatments described in Section 3.2 also reduced the number and the extent of CNT agglomerates in comparison to the synthesis performed without pretreatments.

TEM micrographs prove the coating ability of the pretreatments discussed here. In Figure 3.15, it is obvious that almost every nanotube is coated with a 10-nm *in-situ* grown polypropylene film and that there is a good drawing on the tubes. With the PFT, it was also possible to anchor the polymerization active species, consisting of MAO and a metallocene, on the MWCNT surface. Thus impregnated nanotubes were dried and then

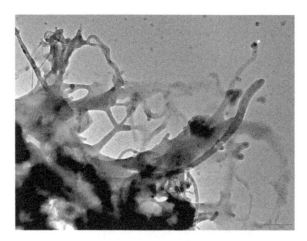

**Figure 3.15**   TEM image of polymer encapsulated ox. MWCNT. Almost every nanotube is covered by a thin and homogenous in-situ grown PP film (about 10 nm).

used to synthesize a high molecular weight PP film ($M_w \approx 1,500,000$ g/mol) on a MWCNT fleece from the gas phase.

## 3.7. Conclusions

New materials of polyolefin nanocomposites can be produced by *in-situ* polymerization. Most research groups are using single-site catalysts and derivates of the PFT, but some are working with traditional Ziegler–Natta catalysts. PE filled with layered silicates and oxides, such as MMT or boehmites, show good gas barrier properties and a significant flame retardancy. These composites are near to an industrial application. Other polyolefin-based nanocomposites, such as PE and polypropylene filled with silica, calcium carbonate, and CNFs or tubes, are still in the state of experimentation.

Especially when metallocenes were used for *in-situ* polymerization, the separation and distribution of the nanosized particles or fibers in the polyolefin matrix is very good. These nanocomposite materials show some excellent property advances, such as improved stiffness, better clarity, and gloss and high crystallization rates. It can be expected that they are at the cusp of an exponential growth in a new wave of technology-driven innovation of polymer products and processes.

## References

1. VKE, Verband Kunststofferzeugende Industrie **2006**, http/www.vke.de.
2. Mülhaupt, R. *Kunststoffe* **2004**, 94(8), 76–88.
3. Alexandre, M.; Martin, E.; Dubois, P.; Garcia-Marti, M.; Jerome, R. *Macromol. Rapid Commun.* **2000**, 21, 931–936.
4. Dong, X.; Wang, L.; Jiang, G.; Sun, T.; Zhao, Z.; Yu, H.; Chen, T. *J. Appl. Polym. Sci.* **2006**, 101, 1291–1294.
5. Heinemann, J.; Reichert, P.; Thomann, R.; Mülhaupt, R.; *Macromol. Rapid Commun.* **1999**, 20, 423–430.
6. Xalter, R.; Halbach, T.S.; Mülhaupt, R. *Macromol. Symp.* **2006**, 236, 145–150.
7. He, A.; Wang, L.; Li, J.; Dong, J.; Han, C. *Polymer* **2006**, 47, 1767–1771.
8. Vaia, R.A.; Giannelis, E.P. *Macromolecules* **1997**, 30(25), 7990–7999.

9. Benedikt, G.M.; Goodall, B.L.; Marchant, N.S.; Rhodes, L.F. *New J. Chem.* **1994**, 18(1), 105.

10. Yano, K.; Usuki, A.; Okada, A.; Kurauchi, T.; Kamigaito, O. *J Polym. Sci. Polym. Chem.* **1993**, 31(10), 2493–2498.

11. Ewangelidis, S.; Hanke, A.; Reichert, K.H. in *Transition Metals and Organometallics as Catalysts for Olefin Polymerization*; Kaminsky, W.; Sinn, H. Eds.; Springer, Berlin, **1988**, 137–148.

12. Kaminsky, W. *Macromol. Chem. Phys.* **1996,** 197, 3907–3945.

13. Dutschke, J.; Kaminsky, W.; Lüker, H. in *Polymer Reaction Engineering*; Reichert, K.H.; Geiseler, W. Eds; Hanser Publisher, Munich **1983**, 207–220.

14. Kaminsky, W. *J. Polym. Sci. Part A. Polym. Chem.* **2004**, 42, 3911–3921.

15. Coates, G.W. *Chem. Rev.* **2000**, 100, 1223–1252.

16. Scheirs, J.; Kaminsky, W. *Metallocene-Based Polyolefins*. Wiley & Sons, Chichester, UK, **2000**, Vols. I and II.

17. Razavi, A; Thewalt, U. *Coordination Chem. Rev.* **2006**, 250, 155–169.

18. Matsui, S.; Fujita, T. *Catalysis Today* **2001**, 66, 63–73.

19. Sinclair, K.B. *Macromol. Symp.* **2001**, 173, 237–261.

20. Kaminsky W.; Wiemann K., *Expected Materials for the Future* **2003,** 3, 6–9.

21. Kaminsky W., *J. Polym. Sci., Part A: Polym. Chem.* **2004**, 42, 3911.

22. Brintzinger H.H.; Fischer D.; Mülhaupt R.; Rieger B.; Waymouth R., *Angew. Chem.*, **1995**, 107, 1255.

23. Ewen J.A.; Jones R.L.; Razavi A.; Ferrara J.D., *J. Am. Chem. Soc.* **1988**, 18, 6255–6256.

24. Resconi L.; Cavallo L.; Fait A.; Piemontesi F., *Chem. Rev.* **2000**, 100, 1253–1345.

25. Bonduel D.; Mainil M.; Alexandre M.; Monteverde F., Dubois P., *Chem. Comm.* **2005**, 6, 781 – 783.

26. Alexandre M.; Martin E.; Dubois P.; Gracia-Marti M.; Jerome R., *Macromol. Rapid Commun.* **2000**, 21, 931.

27. Kaminsky W.; Funck A.; Wiemann K., *Macromol. Symp.* **2006**, 239, 1–6.

28. Funck A.; Kaminsky W., it Compos. Sci. Technol. **2007**, 67(5), 906–915.

29. Alexandre M.; Martin E.; Dubois P.; Marti M.G.; Jérome R., *Chem. Mat.* **2001**, 13, 236–237.

30. US 4097477 (1978), du Pont de Nemours, E. I. & Co., inv.: J. E. G. Howard, it Chem. Abstr. **1978**, 89, 164489e.

31. US 4104243 (1978), du Pont de Nemours, E. I. & Co., inv.: Howard J. E. G., *Chem. Abstr.* **1979**, 90, 39708v.

32. USSR 763379 (**1979**), Institute of Chemical Physics, Academy of Science USSR, invs.: Kostandov L.A.; Enikolopov N.S.; Dyachkovskii F.S.; Novokshonova L.A.; Gavrilov Y.A.; Kudinova O.I.; Maklakova T.A.; Akopyan L.A.; Brikenstein K.A., *Chem. Abstr.* **1981**, 94, 4588m.

33. Bergman, J.; Chen, H.; Giannelis, E.; Thomas, M.; Coates, G. *Chem. Commun.* **1999**, 21, 2179–2180

34. Johnson, L.K.; Mecking, S.M.; Brookhart, M. *J. Am. Chem. Soc.*, **1996**, 118, 267

35. Rong, J.; Jing, Z.; Li, H.; Sheng, M. *Macromol. Rapid Commun.* **2001**, 22(5), 329–334

36. Jin, Y.-H.; Park, H.-J.; Im, S.-S.; Kwak, S.-Y.; Kwak, S. *Macromol. Rapid Commun.* **2002**, 23(2), 135–140.

37. He F.; Zhang L-M.; Yang F, Chen L-S.; Wu Q., *J. Macromol. Sci. Part A* **2007**, 44(1), 11–15.

38. Yang, F.; Zhang, X.; Zhao, H.; Chen, B.; Huang, B.; Feng, Z. *J. Appl. Polym. Sci.* **2003**, 89, 3680–3684.

39. Ivanyuk, A.; Gerasin, V.; Rebrov, A.; Pavelko, R.; Antipov, E. *J. Eng. Phys. Thermophys.* **2005**, 78(5), 926–931.

40. Wang, Q.; Zhou, Z.; Song, L.; Xu, H.; Wang, L. *J. Polym. Sci.: Part A: Polym. Chem.*, **2004**, 42, 38–43

41. Wei, L.; Tang, T.; Huang, B. *J. Polym. Sci.: Part A: Polym. Chem.*, **2004**, 42, 941–949

42. Xu, J.-T.; Zhao, Y.-Q.; Wang, Q.; Fan, Z.-Q. *Macromol. Rapid Commun.* **2005**, 26, 620–625; Xu, J.-T.; Zhao, Y.-Q.; Wang, Q.; Fan, Z.-Q. *Polymer*, **2005**, 46, 11978–11985

43. Heinemann, J.; Reichert, P.; Thomann, R.; Mülhaupt, R. *Macromol. Rapid Commun*, **1999**, 20, 423–430

44. Zhang, F.; Li, S.; Karaki, T.; Adachi, M. *Jpn. Soc. Appl. Phys.* **2005**, 44(1B), 658–661

45. Ray, S.; Galgali, G.; Lele, A.; Sivaram, S. *J. Polym. Sci.: Part A: Polym. Chem.*, **2005**, 43, 304–318

46. Wang, Q.; Liu, P. *J. Polym. Sci.: Part A: Polym. Chem.*, **2005**, 43, 5506–5511

47. Xalter, R.; Halbach, T.; Mülhaupt, R. *Macromol. Symp.* **2006**, 236, 145–150

48. Wang, L.; Feng, L. X.; Yang, S.L. *J. Appl. Polym. Sci.* **1999**, 71, 2087–2090

49. Dubois, P.; Alexandre, M.; Jerome, R. *Macromol. Symp.* **2003**, 194, 13–26

50. Wang, J.; Liu, Z.; Guo, C.; Chen, Y.; Wang, D. *Macromol. Rapid Commun.* **2001**, 22(17), 1422–1426

51. Schumann, H.; Widmaier, R.; Lange, K. H. C.; Wassermann, B.C. *Z. Naturforsch.* **2005**, 60b, 614–626.

52. Scharlach, K. Dissertation, Universität Hamburg, 2008.

53. Desharun C.; Jongsomjit B.; Praserthdam P., *Catalysis Comm.* **2008**, 9 (4), 522–528.

54. Scharlach K.; Kaminsky W., *J. Zhejiang University, Sci. A* **2007**, 8(7), 987–990.

55. Alexandre M.; Pluta M.; Dubois P.; Jerome R., *Macromol. Chem. Phys.* **2001**, 202, 2239.

56. Luo Y.-W.; Cao X.-P.; Feng L.-X., *Chinese J. Polym. Sci.* **2003**, 21(3), 333.

57. Thostenson E.T.; Ren Z.; Chou T-W., *Compos. Sci.. Technol.* **2001**, 61, 1899.

58. Wiemann K.; Kaminsky W.; Gojny FH.; Schulte K., *Macromol. Chem. Phys.* **2005**, 206, 1472–1478.

59. Sandler J.; Broza K.G.; Nolte M.; Schulte K.; Lam Y.M.;, Shaffer M.S.P.; *J. Macromol. Sci. B* **2003**, 42, 479–488.

60. Assouline EL, Barber AH, Cooper CA, Klein E, Wachtel E, Wagner HD., *J. Polym. Sci. B* **2003**, 41, 520–527.

61. Müller A.J.; Trujillo M.; Arnal L.; Laredo E.; Bredeau St.; Bonduel D.; Dubois Ph., *Macromol.* **2007**, 40, 6268–6276.

62. Kaminsky W.; Funck A., *Macromol. Symp.* **2007**, 260, 1–8.

63. Peeterbroeck S.; Lepoittevin B.; Pollet E.; Benali S.; Broekaert C.; Alexandre M.; Bonduel D.; Viville P.; Lazzaroni R.; Dubois P., *Polym. Eng. Sci.* **2006,** DOI 10.1002/pen

64. Galli, P.; Vecellio, G. *J. Polym. Sci. A* **2004**, *42*, 396–415.

65. Del Duca, D.; Moore, E. P. in *Polypropylene Handbook*; Moore, E. P., Ed.; Hanser, München, **1996**.

66. Auriemma, F.; De Rosa, C. *J. Am. Chem. Soc.* **2003**, *125*, 13134–13147.

67. Dotson, D.L. ; Milliken & Company: WO03/087175 A1, **2003**

68. Bergman, J.; Chen, H.; Giannelis, E.; Thomas, M.; Coates, G. *Chem. Commun.* **1999**, 21, 2179–2180

69. Sun, T.; Garces, J.M. *Adv. Mater.* **2002**, 14(2), 128–130

70. Ma, J.; Qi, Z.; Hu, Y., *J. Appl. Polym. Sci.* **2001**, 82, 3611–3617

71. Yang K.; Huang Y.; Dong J-Y, *Polymer* **2007**, 48, 6254–6261.

72. He, A.; Hu, H.; Huang, Y.; Dong, J.-Y.; Han, C.C. *Macromol. Rapid Commun.* **2004**, 25, 2008–2013

73. He, A.; Wang, L.; Li, J.; Dong, J.; Han, C.C. *Polymer* **2006**, 47, 1767–1771

74. Ma, J.; Zhang, S.; Qi, Z.; Li, G.; Hu, Y. *J. Appl. Polym. Sci.* **2002**, 83, 1978–1985

75. Hwu, J.-M.; Jiang, G.-J. *J. Appl. Polym. Sci.* **2005**, 95, 1228–1236

76. Wang, L.; Feng, L.X.; Yang, S.L. *J. Appl. Polym. Sci.* **1999**, 71, 2087–2090

77. Garcia, M.; van Zyl, W.E.; ten Cate, M. G. J.; Stouwdam, J.W.; Verweij, H.; Pimplapure, M.S.; Weickert, G. *Ind. Eng. Chem. Res.* **2003**, *42*, 3750–3757

78. Kaminsky, W.; Wiemann, K. *Mirai Zairyo*, **2003**, 3(11), 6–12

79. W. Kaminsky, K. Wiemann, Polypropene/silica nanocomposites synthesized by *in-situ* polymerization, *Mirai Zairyo*, **2003**, 3(11), 6.

80. Ran, S.; Burger, C.; Sics, I.; Yoon, K.; Fang, D.; Kim, K.; Avila-Orta, C.; Keum, J.; Chu, B.; Hsiao, B. S.; Cokson, D.; Shultz, D.; Lee, M.; Viccaro, J.; Ohta, Y. *Colloid Polym. Sci.* **2004**, 282, 802–809.

81. Lozano, K.; Bonilla-Rios, J.; Barrera, E.V. *J. Appl. Polym. Sci.* **2001**, 80, 1162–1172.

82. Andrews, R.; Jaques, D.; Minot, M.; Rantell, T. *Macromol. Mater. Eng.* **2002**, 287, 395–403.

83. Wiemann K., Synthesis of Polypropylene Nanocomposites by *in-situ* Polymerization of Propylene with Metallocene/MAO Catalysts, Dissertation, Hambrug **2004**

84. Valentini L.; Biagiotti J.; Kenny J.M.; López Manchado M.A., *J. Appl. Polym. Sci.* **2003**, 89, 2657–2663.

# 4 Epoxy-Based Nanocomposites

*Loïc Le Pluart*

Laboratoire de Chimie Moléc ulaire et Thioorganique, UMR CNRS 6507, INC3M, FR 3038,
ENSICAEN & Unive rsité de Caen, 14050 CAEN, France
loic.le_pluart@ensicaen.fr

**Abstract**

This chapter reviews the recent advances performed in the field of epoxy-based nanocomposites. After an introduction dealing with the objectives targeted when nanostructuring epoxy networks, we present the different types
of nanofillers that have been dispersed in epoxy matrixes including nanoparticles, nanotubes, and nanoplatelets.
Section 4.2 deals with the nanostructuring process. One of the main difficulties in obtaining nanocomposites is
to avoid aggregation, which is a real limitation when particles size goes down to the nanometer range. Since
the final state of dispersion is of great importance in the composite properties, compatibilization and dispersion
procedures in the network precursors are two initial steps of the process to be studied carefully. The network
formation, and particularly the gelation, is the following step and might be greatly influenced by the presence
of nano-objects. The nanofillers often influence the polymerization of the epoxy network leading to changes in
the final structure of the network evidenced by variations in the glass transition temperature and/or the crosslink
density. Respectively, the crosslinking reaction kinetics has also an influence on the final state of dispersion of the
nanofillers and might affect the final properties of the composite. This is why the choice of the curing agent and
of the curing process is an essential parameter in the achievement of epoxy-based nanocomposites. The changes
in epoxy systems properties caused by the nanoparticles will be reviewed in details including modifications in
stiffness, toughness, thermal properties, gas permeability, and electrical conductivity. The improvements obtained
with different kinds of nanofillers are compared and discussed. As a conclusion, the actual limits of these systems
are underlined. This includes all the difficulties in characterizing, processing, and behavior modeling of these
materials, which still prevent us from designing nanocomposites with the properties we expect from their aspect
ratio and size.

**Keywords:** epoxy, clay, carbon nanotubes, nanoparticles, nanocomposites

## 4.1. Introduction

Epoxy networks are thermoset systems which are extensively used as composites matrix
in automotive, aerospace, or electronic devices. They also find numerous applications
as adhesives or coatings, which makes their overall application domain very large. This
versatility is due to the fact that their properties are mainly governed by the polymer
network structure, which can be tailored by an appropriate curing agent choice. They are
likely to present high thermomechanical properties, low density, and good adhesion with
other phases. Their main drawbacks are their brittleness when they are highly crosslinked,

their high tendency to uptake water, their polymerization shrinkage, and relatively high thermal expansion coefficient. Inorganic microscale fillers are dispersed in epoxies to reduce the polymerization shrinkage or to improve electrical conductivity or fire-retardant properties. However, the high filling rates needed to reach these objectives generally lead to the reductions in strength and toughness as well as increase in the materials density. To reduce their brittleness, core shell particles, liquid elastomers, or hyperbranched polymers can be incorporated in epoxy networks but in that case, the glass transition temperature is generally decreased, which results in a lowering of the thermomechanical stability of these materials.

From the early studies in the field of polymer nanocomposites, attention has been paid on epoxy systems since it is thought that the increased interfacial area brought by nanofillers can allow overcoming the aforementioned difficulties by reaching the desired properties with much lower quantities introduced and, therefore, limiting the negative effects. However, a real technological march has to be performed since individually dispersing nanometric objects in an epoxy network cannot be performed with the same tools and methods that were used with microscale ones. The main issues are the full understanding of network formation, nanofiller dispersion at all scale levels, and interfacial interactions at the phase boundaries. Structure–properties relationships at the nanometer level have to be established, and a bottom-up approach has to be used to tailor nanocomposites properties by domesticating the nanostructuring process.

The numbers of studies dealing with epoxy nanocomposites are very high. In this review, a description of the observed behaviors and actual trends in this field will be presented through representative examples rather than a systematic listing of all the studies performed. In a first part, focus will be made on the different kinds of nanofillers that have been dispersed in epoxy networks and on the various strategies elaborated to succeed in reaching the "properties enhancer" morphologies. The second part will deal with a specificity of epoxy nanocomposite elaboration, which is the nanocomposite self-construction. Indeed, in contrast to thermoplastic polymers in which nanofillers can be dispersed by melt blending, epoxy clay nanocomposites have to be elaborated by *in-situ* polymerization. Progress is continuously made in the field of understanding how nanofillers can affect the network formation and how this reaction can modify the fillers' state of dispersion. Through recent advances in the literature, we will describe these simultaneous structuring of the organic and inorganic parts of the nanocomposite and the consequences on the material final structure. Finally, we will present the recent advances in the enhancement of mechanical, thermal, electrical, and barrier properties that have been obtained in epoxies with nanofillers.

## 4.2. Nanofillers, Compatibilization, and Processing Strategies

Spherical inorganic particles have been widely used as fillers in polymers for long. Their incorporation in epoxies was generally targeting a reduced thermal shrinkage and a lower thermal expansion coefficient. In the fields of coatings, an increased wear resistance is also wished. Functional fillers have also been used in order to bring specific (magnetic, electric, fire-retardant) properties. Whatever the case, the loadings used are generally in the order of tens wt%, which often leads to decreases in strength and toughness of epoxies. Many works in the filled polymers domain led to the same conclusion that mechanical properties were less affected with higher interfacial adhesion and that higher specific surface would

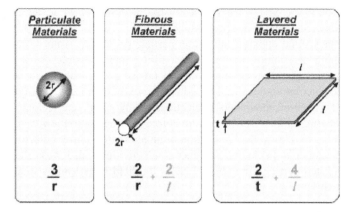

**Figure 4.1**   Surface area/volume relations for various types of nanofillers (reproduced from [1] with permission from Elsevier).

lead to higher interface quantities. The easiest way of increasing specific surface area is to decrease particle size; this is why so much research is focusing on inorganic nanofillers.

By nanofiller it is generally intended an inorganic particle having at least one of its dimensions in the nanometer range. We will not consider in this review the nanostructuring that can be made in epoxies through the incorporation of block copolymers for instance.

Among nanofillers, three categories are generally defined: spherical nanoparticles, nanoplatelets, and nanotubes or nanofibers (Figure 4.1). They have respectively 3, 2, and 1 dimensions in the nanometer range. In the case of tubes and platelets, in addition to the potential benefit arising from the high surface area available, the high anisotropy is wished to help bringing specific enhancements like barrier properties or low percolation thresholds.

The extremely high surface area of these nanofillers has, however, its drawback, since the tendency to aggregation and filler/filler interactions will be much stronger than with classical fillers. Moreover, the mechanical dispersion generally used to disperse filler in the low-viscosity reactive epoxy monomers mixture will not be able to bring the shear forces necessary to break these aggregates. This is why appropriate dispersing tools and surface modifications are necessary to control individual dispersion and interfacial adhesion [2]. This first part presents the recent advances performed in the specific strategies used to compatibilize and disperse these nanofillers in epoxies.

### 4.2.1. Spherical Nanoparticles

Improvements in the synthesis of nanometer-size particles make possible to try using the same materials as the traditional micrometer-size fillers in the elaboration of epoxy nanocomposites and evaluate the beneficial effect from this reduction in size. It is expected to reach the same properties improvements with lower filling rates and, therefore, reduce the processing difficulties arising from viscosity increases in the case of massive fillings with micron-size particles.

By nanoparticles, we intend inorganic domains whose average diameter is inferior to 100 nm. This part does not pretend to be comprehensive but through the choice of different examples (carbon black, $SiO_2$), we aim at illustrating some recent advances in spherical particles of epoxy nanocomposites processing.

In carbon black filled epoxies, individual particle dispersion is not wished since it is aimed at reaching percolation for the lowest filling rates possible. As a consequence, the latest improvements in processing focus on favoring aggregation. Schueler et al. [3] noticed that the percolation threshold is higher in epoxies than in other polymers since its dispersion in the uncured monomers is electrostatically stabilized. Helping aggregation, by moderate shearing of the reactive mixture before polymerization lowers the percolation threshold from 0.9 to 0.06 wt%. Application of an alternative current electric field during curing, led Schwarz et al. [4] to obtain higher conductivities with 1 wt% filled composites since the aggregated particles are aligned along the field lines leading to a consolidated conductive network.

When an improvement in mechanical, optical, or thermal properties is targeted, the nanoparticles have to be individualized and homogeneously dispersed to get full benefit from their high specific surface area. Various processing strategies have been tried to disperse nanoparticles [5]. It is known that covering silica surface with a coupling agent leads to improved dispersion states as demonstrated by Kang et al. [6] on silica particles having a diameter of 400 nm. However, when the particles' diameter decreases below 100 nm, mechanical stirring [7], ultrasonication [8], dissolution, and ball milling [9] cannot avoid aggregation to take place independently of the kind of nanoparticles used and despite the use of coupling agents. Homogeneous dispersions can be obtained, like Wetzel et al. [9] did with $Al_2O_3$ particles, but some unavoidable aggregates are still present due to the high specific surface of nanoparticles as shown in Figure 4.2.

Zheng et al. [10] compared the efficiency of various processing strategies to disperse silica nanoparticles in an epoxy network and obtained better properties with the combination of sonication and high-speed mixing (24,000 rpm). The optimal dispersion state obtained is, however, still strongly aggregated.

Since no shear can be applied during epoxies curing, the high specific surface area of these spherical nanoparticles makes aggregation difficult to avoid by classical processing

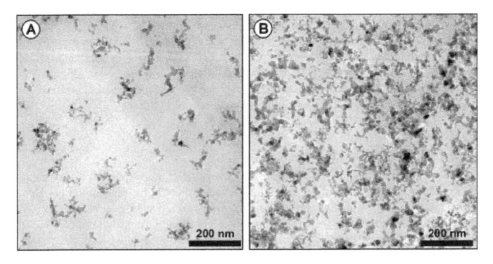

**Figure 4.2**    TEM pictures of (A) 2 vol% and (B) 10 vol% $Al_2O_3$/epoxy–amine nanocomposites (reproduced from [9] with permission from Elsevier).

techniques. This is why *in-situ* formation of the inorganic particles by the sol/gel method has been thought to be a more effective way to obtain the nanostructuring of epoxy networks.

Interpenetrated networks synthesis being detailed in another chapter of this book, we will not deeply describe the sol/gel processing method. Despite some difficulties in tailoring the size of the formed silica domains arising from the competition between inorganic and organic network formation reactions, Matejka et al. [11] successfully obtained 10–20-nm wide particles using a two-step process. It consists in a partial polymerization of the epoxy network before the inorganic precursors' introduction. This strategy allows a better control of inorganic aggregates size since their growth is stopped by organic network gelation. In a one-step process, 50–100-nm large particles were formed. More recently, Chiang et al. [12] succeeded in obtaining 25-nm wide inorganic domains, in a one-step process, using condensation polymerization of diethylphosphatoethyl-triethoxysilane. The obtained silicon- and phosphorus-rich inorganic nanoparticles are expected to improve flame retardancy of the epoxy network. The inorganic phase is grafted on the epoxy network via the epoxy hydroxyl groups formed by epoxy/amine reaction. It is, however, underlined that the tailoring of the morphologies is difficult due to complex reaction mechanisms, steric hindrance, and phase separations to control as well as interfacial interactions to optimize.

Another way of obtaining nanometric silica domains in an epoxy matrix has been developed to avoid the difficult control of morphologies arising from the sol/gel process. The introduction of polyhedral oligomeric silsesquioxanes (POSS) as nanobuilding blocks in epoxies has focused much attention. The most commonly used POSS in polymers are cage-like oligomers $(R-SiO_{1.5})_8$-octasilsesquioxanes. They can be considered as R-functionalized elementary silica nanoparticles and, therefore, can be introduced into epoxies without changing the traditional way of processing. Depending of the choice of the R groups, POSS can either be nonreactive toward the organic network, or be able to take part to the condensation polymerization when carrying epoxy or amine groups. The number of this functional groups will allow using POSS as dangling units (with monofunctional POSS) or as crosslinks whose functionality can be tailored as shown in Figure 4.3.

The nature of the nonreactive groups in the latter case will also influence the final morphology of the nanocomposites formed. Indeed, POSS can be liquid, amorphous solid, or crystallized before inclusion in epoxy. This initial state can strongly influence their final dispersion [13]. Some illustrative examples of POSS structures are shown in Figure 4.4.

Since these building blocks are functionalized, there is few to tell about their compatibilization process and/or the processing strategies used to disperse them. Up to now, the wide majority of POSS used in epoxies carry epoxy functions [13, 15–18], but more recently some authors also used amino functionalized POSS [19, 20]. Both can be dispersed at the nanometer scale. In the case of POSS used as dangling units, Matejka et al. [13] have demonstrated how crucial the choice of the seven other groups carried by the POSS is. Indeed, depending on their compatibility with the organic network, they can be responsible of aggregation as well as homogeneous dispersion. As demonstrated by Abad et al. [21], a macroscopic phase separation can also occur before polymerization, in case of incompatibility of these nonreactive groups with the epoxy/curing agent reactive mixture.

Their dispersion in the matrix is a reactive process; therefore, it will be detailed mainly in the Section 4.2, which is dedicated to the influence of nanoparticles on network formation.

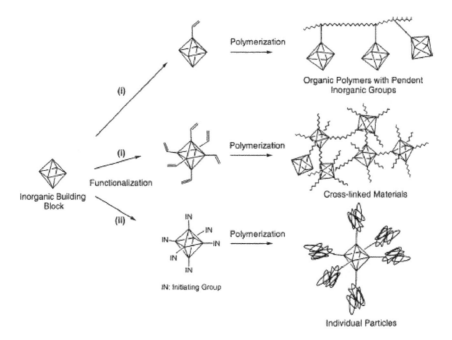

**Figure 4.3** Potential use of POSS nanobuilding blocks in epoxy networks depending on their functionality (reproduced from [14] with permission from Springer).

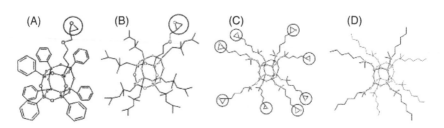

**Figure 4.4** Various POSS structures and functions: (A) crystalline dangling unit, (B) liquid dangling unit, (C) liquid crosslink, and (D) liquid nonreactive POSS (reproduced from [13] with permission from ACS).

As will be detailed in the third part, the incorporation of POSS at the molecular scale can lead to strong macroscopic properties enhancements. This is why organically modified transition-metal oxide clusters are also focusing much attention, since these well-defined nanobuilding blocks could allow giving catalytic, magnetic, or electric properties to polymers. In order to make these metal oxide nanoclusters reactive toward the matrix, organic functions are generally introduced either during the nanocluster synthesis (*in-situ* method) or by grafting on a preformed cluster [22]. The incorporation of these nanoblocks in polymers is, however, still at its beginning and examples of incorporation in epoxies are very rare.

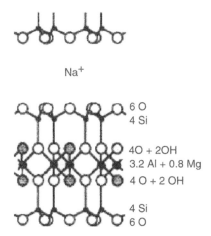

Na$^+$

6 O
4 Si

4O + 2OH
3.2 Al + 0.8 Mg
4 O + 2 OH

4 Si
6 O

**Figure 4.5**    Montmorillonite structure (reproduced from [28] with permission from Wiley).

## 4.2.2. Nanoplatelets

Among all the nanofillers introduced in epoxies, clay is the one that has been the most studied. Among all smectic clays, montmorillonite (Na-MMT) is generally chosen for reasons that will be detailed in the following paragraph. Due to the extensive number of reviews [23–27] devoted to polymer/MMT nanocomposites, its structure will not be detailed in this article. We just recall it is a 2:1 layered silicate whose crystallographic structure is presented in Figure 4.5. Isomorphic substitutions in the octahedral layer lead to negative charge density of the platelets, which is compensated by the presence of mineral cations in the interlayer gallery. The platelet thickness is around 1 nm, and its length and width are between hundreds of nanometers and 1 μm. The mineral cations are exchangeable in aqueous dispersion, which makes its compatibilization with polymers possible. Alkylammonium ions are generally used to this purpose.

Clay has not been the only platelet-shaped particle that has been dispersed in epoxy to form nanocomposites. Inspired from this strategy, recent works have been carried out with layered double hydroxides (LDH) [29–31], expanded graphite (EG) [32–35], and $\alpha$-zirconium phosphate ($\alpha$-ZrP) for instance.

LDHs (or hydrotalcite-like compounds, or anionic clays) are natural and synthetic minerals structured in positively charged layers. They are generally described by the $[M(II)_{1-x}M(III)_x(OH)_2]^{x+} [A^{n-}_{x/n} ,mH_2O]^{x-}$, where M(II) and M(III) are di- and trivalent cations and $A^{n-}$ is an exchangeable anion in the interlayer gallery. They can be intercalated by ion exchange, but their exfoliation is sometimes difficult due to the high charge density of the layers, which create strong electrostatic attractive forces. The anionic exchange is generally easier to perform with multiply charged anions and requires to be completed under inert atmosphere to avoid contamination by $CO_2$ since carbonate ions are generally easily and strongly intercalated in the LDHs galleries.

EG is obtained from natural graphite. Chemical oxidation in the presence of sulfuric and nitric acids and heating in a furnace above 600°C allows obtaining a loose and porous material structured in sheets being less than 100 nm thick and approximately 1–10 μm long and wide (Figure 4.6). Even if their aspect ratio is very high, the thickness of these graphite sheets is generally much higher than those of LDHs or clay platelets. The so-called

**Figure 4.6**    SEM micrograph of EG platelets (reproduced from [32] with permission from Elsevier).

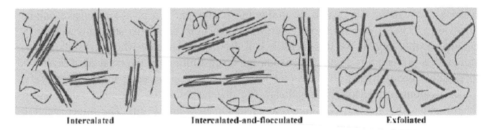

Intercalated            Intercalated-and-flocculated            Exfoliated

**Figure 4.7**    Polymer/clay nanocomposite morphologies (reproduced from [42] with permission from ACS).

nanosheets are not graphene sheets but rather consist in several layers of it [33]. The acid and temperature treatments lead to the presence of hydroxyl and carboxyl functional groups on the platelets surface.

After this short presentation of the various nanoplatelets, the strategies that have been used in order to disperse them as homogeneously as possible in epoxy matrixes will be presented. A first part is devoted to nanoclay dispersion, while the modification and processing strategies are separated in two distinct subparts. The strategies used for the other nanoplatelets are presented separately.

Many investigations have been conducted on the influence of clay platelets modification and nanocomposite processing on the final state of dispersion of these nanoparticles in epoxy matrixes. The obtained nanocomposites are generally described by the term *intercalated* or *exfoliated* depending on their morphology as presented in Figure 4.7.

The influence of the kind of clay used has been first studied by Lan et al. [36] revealing that in the layered silicates, the cation exchange capacity was the first key feature. Montmorillonite is an ideal candidate since its intermediate CEC allows alkylammonium ions to efficiently compatibilize the silicate with the epoxy monomers, without reaching the point where the steric hindrance in the galleries would prevent their diffusion, such as vermiculite does. Increasing the chain length of the alkylammonium ions used to modify the clay platelets increases the interlayer distance (*d*-spacing) of the organoclay, favoring its dispersion in an organic medium and thus leading to better dispersions in epoxy matrixes [28, 37]. Lan et al. [38] evidenced that 12 methyl groups in the alkyl chain is a minimum to get exfoliated morphologies. Kelly et al. [39] showed that solely long-chain alkylammonium was able to bring the initial swelling in the monomers, which will lead to an increase in the *d*-spacing during polymerization. Using such long-chain cations favors the cationic exchange reaction efficiency, but also leads to adsorption of excess alkylammonium cations with their counter ions causing a decrease in the hydrophobicity of the organoclay [40] as well as a decrease in its thermal stability [41].

The main ions used to compatibilize montmorillonite with epoxies are hexa and octadecylammonium (HDA and ODA-MMT), methyltallo-walkyl-bis(2-hydroxyethyl) ammonium (MT2Et-MMT), dimethyl ditallowalkylammonium (DMDT-MMT), benzyl-dimethyl-tallowalkyl ammonium (BzDMT–MMT), and trimethyloctadecylammonium (TMODA-MMT) ions. Quaternary alkylammonium ions have been mostly used in the polymer/clay nanocomposite field to favor hydrophobicity of the clay and obtain organ-oclay with higher *d*-spacing and a higher swelling potential in epoxy. Their benefit is, however, often limited with epoxies since as will be explained later, the acidity of the ions plays an important role in the polymerization gallery expansion process by catalyzing the epoxy ring opening [36, 43]. Primary alkylammonium ions have been thus observed to lead to the more desirable morphologies and properties [30, 44, 45]. Among the quaternary ammonium ions, a particular case is the methyl-tallowalkyl-bis(2-hydroxyethyl) ammo-nium, which is also able to catalyze the epoxy/amine reaction with its hydroxyl groups and, therefore, also leads to interesting morphologies in epoxies. This last alkylammonium ion has originally been introduced in epoxies to favor covalent linking between the network and the clay platelet (through a reaction between hydroxyl and epoxy groups) and thus obtain stronger clay/epoxy interfaces [46].

In order to achieve complete covalent bonding of the network to the platelet, the grafting of alkoxysilanes onto montmorillonite clay has been tried. This strategy is used to increase interaction between glass beads, glass fibers, or silica particles with epoxies, but it has not been much employed to modify clay. Nevertheless, Wang et al. [47, 48] succeeded in obtaining exfoliated epoxy nanocomposites by grafting aminopropyl- and glycidoxy-trimethylsilane directly on clay. In these studies however, a very special process, which will be detailed later, was used to elaborate the nanocomposite, which prevents us from concluding on the efficiency of the silane itself in favoring clay dispersion compared to alkylammonium ions. The low-chain length of commercial alkoxysilanes, the subsequent low increase in gallery spacing, and the fact that most of the few hydroxyls groups present on montmorillonite are located on platelets edges [41, 49] make their efficiency in a conventional processing method still unclear. A recent study performed by Xue et al. [50] with palygorskite (2:1 clay with a layered structure but which forms crosslinked ribbons instead of platelets) confirmed the grafting to proceed on the edges, and consequent difficulties in getting a properties-enhancer final state of dispersion.

Instead of using classical silanes, Liu et al. [51] tried to disperse a POSS-modified montmorillonite in epoxy. The octa($\gamma$-aminopropyl) POSS has been successfully inserted in MMT galleries after ionization of the amine groups. The dispersion of this modified MMT by a classical swelling method into epoxy monomers led to an increase in *d*-spacing, but a uniform dispersion in the crosslinked matrix could not be reached.

Recent improvements in clay platelets exfoliation in epoxies have been rather conducted through modifications of the dispersion process preceding polymerization than on clay modification strategies.

The "traditional" processing way consists in swelling the organoclay with the epoxy prepolymer. The $d$-spacing of octadecylammonium-modified montmorillonite can thus rise up to 40 Å. Further increase in $d$-spacing can solely be obtained after the hardener is added to the clay/epoxy mixture. The use of high shear mixing and sonication in order to improve this first dispersion is often reported. These dispersing tools can help breaking down big aggregates but can hardly achieve clay homogeneous dispersion at all scale levels. No clear influence of the initial swelling times or temperatures have been reported.

Recently, attempts in improving aggregates and primary particles breakdown to ensure a more homogeneous dispersion have been performed. Yasmin et al. [52] used a three-roll mill to improve organoclay dispersion in epoxy before introducing the curing agent. They observe a transition from opaque to clear and transparent solution after 3 h of mixing at 500 rpm at room temperature (to ensure high viscosity and increase shear forces). They report that this processing technique allows obtaining higher increases in epoxy nanocomposite stiffness than conventional mixing.

Koerner et al. [53] pointed out that new processes have to be imagined to get the uniform dispersion states that can be reached in thermoplastics, since a low-viscosity medium is ineffective in transferring shear stress to filler particles under a certain size. They quantify the efficiency in dispersion by measuring the number or particles by square micron on TEM micrographs. Few differences between high shear mixing and sonication of clay in epoxy are observed, but cryocompounding of the mixture before curing allows getting a uniform dispersion of tactoids containing less than five clay platelets as can be seen in Figure 4.8. The $d$-spacing of organoclay in these tactoids is not increased during processing. Cryocompounding is performed after high shear mixing, sonication, and adding of the curing agent. A bench top twin screw compounder is used at $-30°C$ to ensure high viscosity and to prevent any progress in polymerization reaction.

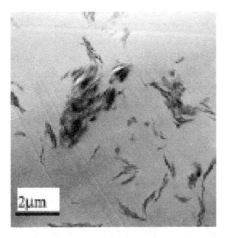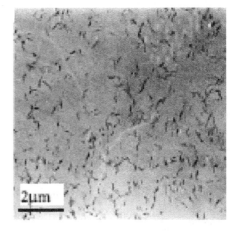

**Figure 4.8**    TEM pictures of MT2Et–MMT/epoxy nanocomposites elaborated by (a) sonication and high shear mixing, (b) cryocompounding (reproduced from [53] with permission from Elsevier).

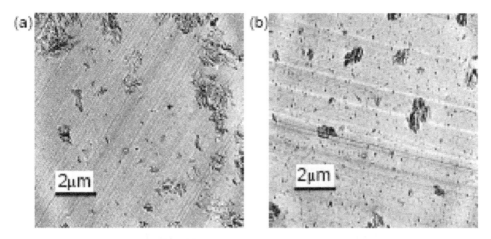

**Figure 4.9** TEM pictures of ODA–MMT/epoxy nanocomposites elaborated by (a) sonication and high shear mixing, (b) cryocompounding (reproduced from [53] with permission from Elsevier).

An interesting point is the influence of the alkylammonium ion used in this process. Octadecylammonium ions have been previously presented to be more desirable in promoting clay exfoliation. It was on the basis of an increased $d$-spacing during polymerization, which is not contradicted by this study, since it is the case with the "classical" process as well as with the cryocompounding one. However, cryocompounding seems less efficient in dispersing this organoclay than the MT2Et–MMT when Figure 4.9 is compared to Figure 4.8.

MT2Et–MMT has a 33 Å $d$-spacing in the nanocomposites, but it looks like it has to be considered as an exfoliated one. From the picture, it is clear that exfoliation should be considered as a uniform dispersion of very high aspect ratio particles, which can be achieved with platelets stacked at low distances. The interest in developing such morphologies will be presented later. On the contrary, platelet separation at high distances is of less interest if it is included in swollen spherical aggregates at the micron scale. This work points out the necessity to change the way of considering organoclay state of dispersion in an epoxy matrix and especially systematically linking increase in $d$-spacing and exfoliation.

The other way of improving platelet dispersion in epoxy/clay nanocomposites is to use organic solvents. This approach has been first considered by Brown et al. [54] who dispersed organoclay in acetone before incorporating it in the epoxy monomers. The subsequent evaporation of the low boiling point solvent allows a better dispersion of the organoclay before adding the curing agent. Luo et al. [55] have added a sonication step for the acetone/organoclay and obtained partially exfoliated structures. Still in this search for improving dispersion, Liu et al. [56] proposed to add some high-pressure dispersion (15,000 psi) to the acetone/organoclay mixture and also obtain uniform dispersion with a significant reduction in aggregates number and size compared to a classical mixing process. However, no study compares between them the different methods using solvents in order to conclude on their efficiency.

Recently Wang et al. [57] built up the so-called slurry method by first dispersing the clay in water before replacing it by acetone. This study has been conducted on silane-modified clay. The replacement of water by acetone before proceeding to silane grafting allows limiting the alkoxysilane homopolymerization since it is catalyzed by water

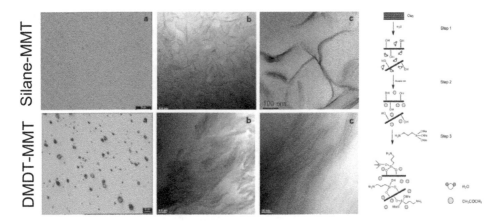

**Figure 4.10**   Right: Schematic representation of the slurry processing. Left: optical (a) and TEM (b, c) micrographs of silane–MMT/epoxy composites obtained by the "slurry method" and with DMDt–MMT/epoxy composites obtained by classical swelling/stirring method (reproduced from [57] with permission from ACS).

(cf. Figure 4.10). The acetone is removed from the mixture under vacuum at 50°C before adding the curing agent.

Getting benefit of the initial highly exfoliated structure of clay in water allows obtaining transparent mixtures up to 5 wt% of clay. The final morphologies consist in exfoliated single montmorillonite layers or thin tactoids containing 5–10 platelets much more uniformly and randomly dispersed than in the nanocomposites obtained from the conventional organoclay dispersion (Figure 4.10). According to them, most of the exfoliation and dispersion process is achieved before polymerization of the epoxy/amine network and the key feature is to succeed in transferring and maintaining the exfoliated state of clay in water and ethanol in the epoxy. Another interesting point is the lower quantity of clay modifier used, the quantity of silane grafted being around 5 wt% of the pristine clay. Chen et al. [58] also used the slurry method in order to exfoliate clay in an epoxy/anhydride network. However, instead of grafting an alkoxysilane on the pristine clay in the acetone slurry, they added a protonated catalyst of the epoxy anhydride reaction (2,4,6-*tris*-(dimethylaminomethyl)phenol) in the initial clay/water suspension. After replacement of water by acetone, the slurry containing the catalyst-modified clay is dispersed in epoxy to combine the slurry process with an intragallery catalyzed network formation. TEM micrographs of the nanocomposites reveal that the clay platelets are uniformly dispersed at all scale levels and that the remaining tactoids do not contain more than five platelets (Figure 4.11).

Even if it leads to nice morphologies, and allows decreasing the quantity of clay modifier used, the slurry method requires that the use of low boiling points solvents to achieve clay dispersion is, therefore, not completely satisfying in terms on environmental concerns.

Concerning the other types of nanoplatelets, some interesting results need to be highlighted. With LDHs, two different strategies both leading to layers exfoliation and improved properties have been exposed. Zammarano et al. [30] obtained exfoliated epoxy nanocomposites by performing an ion-exchange reaction on LDH with organosulfonate

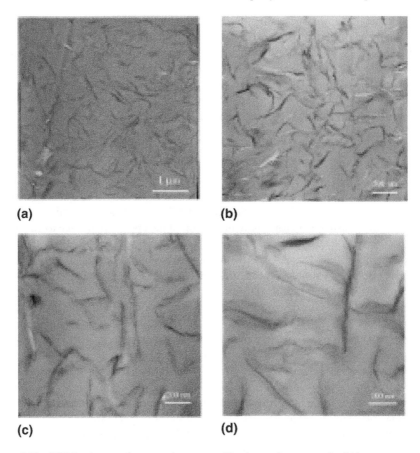

(a)

(b)

(c)

(d)

**Figure 4.11** TEM micrographs at various magnifications of epoxy anhydride nanocomposites obtained by slurry processing combined with intragallery reaction catalysis (reproduced from [58] with permission from ACS).

ions. The LDH being positively charged, the use of strong sulfonic acids such as 4-toluenesulfonic acid, 4-hydroxybenzenesulfonic acid, or 3-aminobenzene sulfonic acid allowed $d$-spacing increases and LDH compatibilization with epoxy. However, they showed with the latter that having an amine function on the anionic modifier bridges LDH platelets during epoxy/amine reaction prevents exfoliation to occur. This phenomenon has not been observed by Hsueh et al. [29], who used aminolaurate anion to compatibilize LDHs with epoxies. The exfoliation could be reached and the strong interface thus created between the nanoplatelet and the epoxy network led to much improved mechanical properties. The use of organic amino acids has been also combined to the use of organic polar solvents [HIBI05] to improve LDHs delamination, but without applying this strategy to disperse them in epoxy networks.

Recent studies have been led on EG exfoliation in epoxies. Yasmin et al. [32] showed that without any treatment surface or specific process strategies, EG could not be exfoliated in epoxies. They obtain the dispersion of 250-nm thick particles in which the $d_{002}$ interlayer distance remains constant at 3.36 Å. Li et al. [34] dispersed EG with a "slurry" process. The EG is dispersed in acetone and sonicated before being incorporated to an acetone-diluted epoxy monomer. After acetone evaporation and curing agent addition, the nanocomposites show an improved dispersion. In the same study, the influence of a UV/ozone treatment is also performed. This treatment allows graphite surface modification with higher concentrations of hydroxyl, carbonyl, and carboxyl groups and leads to enhanced nanocomposite properties, thanks to improved graphite–epoxy interfacial adhesion. However, no XRD spectrums showing an increase in graphite interlayer distance are proposed. Recently, Yasmin et al. [33] explored additional steps in the dispersing process of EG in epoxy. After EG dispersion and sonication in acetone, high shear mixing in the epoxy monomer with a three roll mill, additional sonication, and combination of both have been used. As demonstrated by XRD, all these efforts do not exfoliate the graphite platelets but lead to dispersed multilayered particles. This latter strategy leads to, however, promising results in composite properties since better improvements in stiffness than in the aforementioned studies have been obtained. If sonication provides better chance of getting exfoliated structures, it might also lead to early polymerization of the epoxy chains due to an increase in epoxy/EG mixture temperature.

As explained previously, $\alpha$-ZrP has been chosen by some authors [59–61] as a model nanoplatelet to establish structure–properties relationships in epoxy nanocomposites. The presence of a hydroxyl group of the phosphate function pointing into the interlayer space allows incorporating $n$-alkylamines in the interlayer space through acid–base reaction or hydrogen bonding, with the formation of a bilayer structure. This organic modification of $\alpha$-ZrP is performed via dispersion of the nanoplatelets in methyl ethyl ketone and progressive addition to the suspension of polyoxyalkylene monoamines. A gel is obtained preventing monoamine excess removal, but the interlayer distance is increased from 7.6 to 73.3 Å [60]. After dispersion of this gel into the epoxy prepolymer, solvent evaporation, curing agent addition, and subsequent polymerization, a uniform dispersion of individual platelets is obtained in the transparent, aggregates-free nanocomposite as can be checked in Figure 4.12. Unmodified $\alpha$-ZrP dispersion led to opaque and strongly aggregated morphologies. Boo et al. [61] precisely demonstrated that the exfoliated optimal dispersion might only be observed below 2 wt% of modified $\alpha$-ZrP since at this filling

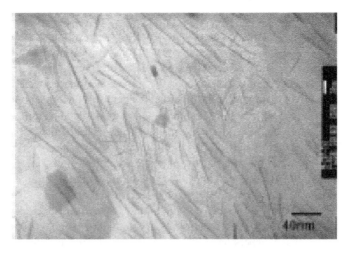

**Figure 4.12**   TEM micrograph of exfoliated of $\alpha$-ZrP layers in an epoxy matrix (reproduced from [60] with permission from ACS).

level, orientation of the platelets (attributed to excluded volume effects) is observed. This proposal is consistent with the calculations made by Celzard et al. [62]. They indicate that excluded volume concepts have to be considered to evaluate the percolation threshold of thin platelets dispersed in an epoxy matrix.

As a conclusion to this part, and in order to witness that all this issue of dispersing nanoplatelets in epoxies and in polymer in general is far from being solved, a recent remark of Pukansky (PUKA05) has to be exposed: "it is common knowledge that polyethylene and polystyrene are not miscible or compatible. Why do we expect the compatibility of a silicate layer to increase in any polymer except PE if we cover its surface with aliphatic chains?" Surface modification of inorganic nanoplatelets helps exfoliation through the decrease in interlayer bonding forces (since expanding galleries), but decreases the matrix/nanoplatelet interactions and consequently leads to low-strength composites despite large contact areas.

### 4.2.3. Nanotubes and Nanofibers

From nanotubes discovery in 1991 by Ijima [63], and subsequent evaluation of their properties, many attempts have been made to incorporate them into polymers and particularly into epoxies during the last 10 years. This tremendous research activity has already led to the publishing of comprehensive reviews on polymer/carbon nanotubes nanocomposites [1, 64, 65].

We will not detail here the different ways of synthesizing carbon nanotubes, but it is necessary for a better comprehension of the following to recall a few physical properties of these new carbon materials since various kinds of carbon nanotubes have been introduced into epoxies (Figure 4.13).

Single-wall carbon nanotubes (SWCNT) possess a diameter generally lower than 2 nm and lengths ranging from 1 to 10 μm giving an extremely high aspect ratio. Consequently, they also possess the highest specific surface area, which is estimated around 1300 m$^2$/g [67]. The longitudinal Young's modulus of these tubes is estimated to be greater than 1 TPa, which makes understandable the wish to use them as stiffness enhancers in polymers [64].

They also present high electrical conductivity and are, therefore, potential candidates for bringing conductive or at least antistatic properties to epoxy networks. Since their high price is a drawback for large commercial applications, some cheaper CNT structures have also been used. Double-wall carbon nanotubes (DWCNT) are made of two nested nanotubes. The inner and outer diameters are around 2 and 3 nm, respectively. The specific surface area is reduced to 600–800 m$^2$/g. Even cheaper are multiwall carbon nanotubes (MWCNT), whose outer diameter is between 15 and 50 nm generally. In order to limit the decrease in aspect ratio, their length is generally superior to SWCNT, going up to 50 μm.

In this part, we will also comment some results obtained with carbon nanofibers, whose diameter is generally between 50 and 200 nm. To keep high aspect ratios and specific surface area, they are also longer than nanotubes reaching 100 μm. Even if they are not made of cylindrical graphene sheets, and if their mechanical properties are intrinsically lower than those of CNTs, their much lower cost makes them a good candidate for enhancing epoxies properties at an industrial scale. In order to achieve these properties enhancements, the way of dispersing them in epoxy matrixes has been studied. Indeed, CNTs possess a strong tendency to agglomerate forming bundles or ropes due to strong van der Waals attractive forces. The higher the specific surface of the CNT, the stronger

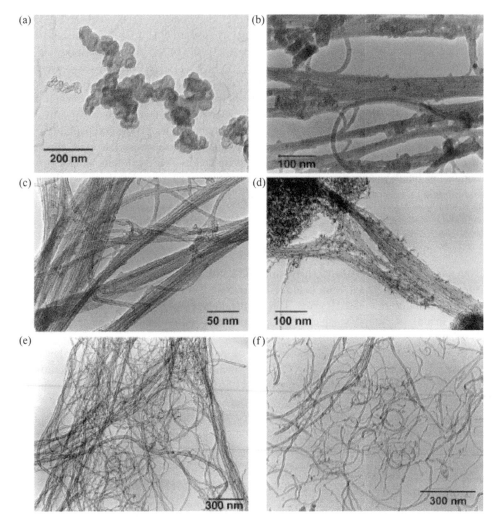

**Figure 4.13**   TEM micrographs of carbon nanofillers : (a) carbon black particles, (b) SWCNTs, (c) DWCNTs, (d) amino-functionalized DWCNTs, (e) MWCNTs, (f) amino-functionalized MWCNTs (reproduced from [66] with permission from Elsevier).

the aggregation tendency. For instance, chemical vapor deposition (CVD) synthesis process leads directly to strongly entangled nanotubes.

This is why, contrarily to what has been done with clay platelets, researchers deeply investigated first the processing strategies available to disentangle and individually disperse these nanotubes in epoxy before using chemical compatibilization. In this part, we will present the various processes independently of the final dispersion state, since, as will be detailed later, the dispersion requirement for achieving conductivity by percolation and mechanical improvements is not the same.

CNTs individualization can be achieved by dispersing them in a low boiling point solvent followed by sonication. Incorporating the epoxy prepolymer to a sonicated SW-CNT/acetone mixture, followed by solvent evaporation and curing agent addition, allowed

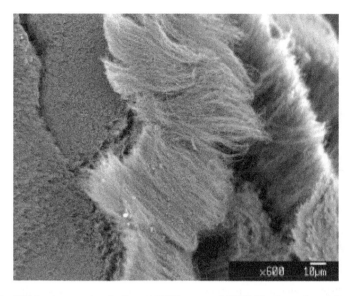

**Figure 4.14** SEM micrograph of aligned CVD grown MWCNTs (reproduced from [72] with permission from Elsevier).

Kim et al. [68] to obtain an interesting dispersion state since the nanocomposites presented an electrical conductivity of $10^{-5}$ S/cm. Precautions have to be taken with this process strategy since Lu et al. showed that too intense sonication [69] could lead to many defects in nanotubes structure and even to amorphous carbon formation. Another point is the solvent removal since an inappropriate process (too high boiling point solvent, insufficient evaporation) can dramatically affect the nanocomposite properties [70]. Recently, Song et al. [71] showed that performing sonication without using solvents was leading to lower dispersions of the carbon nanotubes.

High shear mixing has also been included in these nanocomposites elaboration process. Sandler et al. [73] have set up a dispersion process including first a sonication step in ethanol in order to disentangle the MWCNTs. The epoxy resin is then incorporated followed by ethanol evaporation. In their following studies [72, 74, 75], this first step of the process has been suppressed since they used aligned MWCNT produced with preformed substrate in the CVD process (Figure 4.14).

Shear mixing with a dissolver disk at 2000 rpm is applied to the epoxy/CNT mixture, followed by another cycle at lower temperature (ice bath used) in order to increase epoxy viscosity and submit CNTs aggregates to higher shear stresses. The addition of the curing agent is performed at higher temperatures and followed by another moderated shear mixing step. This high shear mixing process has been proven to be efficient in individually dispersing carbon nanotubes at very low filling rates, but the authors notice that partial reaggregation occurs with percentages as low as 0.0025 wt% introduced [74].

Martin et al. [72] used the same process and explain that the homogeneous dispersion of individual nanotubes obtained before adding the amine hardener is caused by repulsive force between nanotubes induced by the electrical double layer, which surrounds particles due to their negative surface charge. The dispersion state is improved when using aligned nanotubes compared to initially entangled ones since in the latter case some clusters can

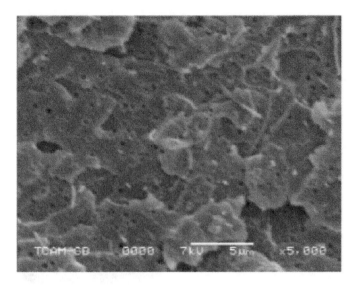

**Figure 4.15**   SEM micrographs of fracture surface of aggregates free 2 wt% carbon fibers/epoxy nanocomposite (reproduced from [76] with permission from Elsevier).

never be broken [73]. Introducing the amine curing agent modifies the ionic force of the dispersion and the CNTs aggregate. This agglomeration process is made easier by high temperature and low viscosity since it is related to the Brownian motion of the nanotubes and their diffusion in the reactive mixture [72]. Consequently, shorter nanotubes will aggregate faster than longer ones. Aggregation is slowed down if no shear is applied after the amine introduction, which is interesting if individual dispersion of nanotubes is desired. Zhou et al. [76] showed with carbon nanofibers that the efficiency of sonication in breaking aggregates is strongly reduced above 3 wt%. However, the sonication/high-speed mixing process allows obtaining individually dispersed nanofibers with a 2 wt% filling rate, as depicted by the regularity of the holes left by the fibers after debonding on the fracture surface presented in Figure 4.15.

In order to avoid agglomeration at higher filling rates, Moniruzzaman et al. [77] developed another strategy to breakdown SWCNT agglomerates by submitting them to intense shear stresses. After a classical sonication step of CNTs in DMF, incorporation of the epoxy prepolymer, and solvent evaporation, the SWCNT/epoxy mixture is sheared 1 h in a twin screw microcompounder, which allows obtaining homogeneous mixtures. The reactive blend is sheared another 30 min after hardener incorporation, and then polymerized. The resulting nanocomposites did not show any aggregates on SEM micrographs with a 0.05 wt% loading, whereas the initial epoxy/CNT mixture was strongly heterogeneous.

To limit the use of solvents which will prevent these processes to be scaled up for industrial application, Gojny et al. [78] developed a calendering approach. The direct high shear mixing of epoxy with carbon nanotubes with a three-roll calender has been inspired from the previously presented works of Yasmin et al. [52] on epoxy/clay nanocomposites. This calendering method is also efficient with CNTs. On TEM pictures of 0.1 wt% nanocomposites, CNT aggregates are still present but smaller, less compact, and more impregnated by the matrix than in nanocomposites obtained by sonication (Figure 4.16). Gojny et al. [79] also showed that the quality of CNTs dispersion is depending on their

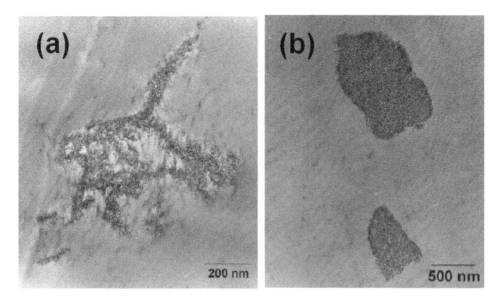

**Figure 4.16** TEM micrographs of DWCNT agglomerates in epoxy nanocomposites elaborated by (a) calendering and (b) sonication (reproduced from [78] with permission from Elsevier).

specific surface area, the DWCNT being more homogeneously dispersed in the epoxy matrix.

Recently, Thostenson et al. [80] chose to use this processing way and revealed that the gap between the rolls of the calender has to be reduced at his minimum to obtain optimal CNTs dispersion. Setting this gap to 5 μm led to a highly dispersed morphology with little or no agglomerates observed by TEM.

After having explored all the ways proposed to disperse nanotubes physically, further improvements in nanotubes dispersion in epoxies have been brought by chemistry.

Gong et al. [81] developed a surfactant approach. Covering MWCNTs with a nonreactive surfactant allowed better dispersion in epoxy according to SEM observations, but a complete homogeneous dispersion has not been achieved this way. The latest improvements have been brought by chemical modification of the nanotubes surface. Nanotubes functionalization has been reviewed by Sinnot [82] as well as by Bahr and Tour [83]. Appropriate functionalization of CNTs will increase their solubility and dispersability in epoxies, as well as the strength of the epoxy/nanotube interface [84, 85] if a covalent bonding between the nanotubes and the matrix is performed.

Recent attempts performed by Miyagawa et al. [86, 87] with fluorinated MWCNT in epoxy anhydride networks show that even if the dispersion is increased, their use should be avoided due to the detrimental effect of this modification on the epoxy/anhydride network formation and final structure. Nevertheless, the idea of incorporation fluorine atoms to reduce aggregation is not to be given up, since Park et al. [88] obtained higher dispersion and interfacial adhesion using oxyfluorinated CNTs. The presence of hydroxyl groups in addition to fluorine atoms helps the dispersion and allows covalent grafting of the epoxy on CNTs surface. Gojny et al. [89] have used amino-functionalized nanotubes (NH$_2$-CNTs). In addition to the changes in interface properties, they observed improvements in nanocomposites homogeneity, which can be seen by comparing Figure 4.16(a) with Figure 4.17.

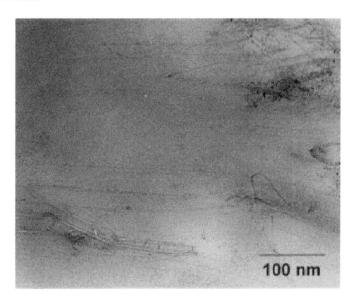

**Figure 4.17**  Homogeneous dispersion of NH$_2$-DWCNTs in an epoxy matrix (reproduced from [78] with permission from Elsevier).

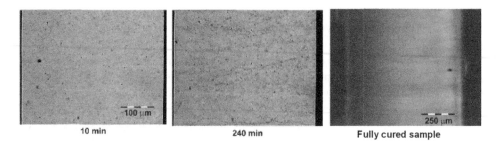

**Figure 4.18**  Evolution of dispersion of 0.01 wt% MWCNT nanocomposite when an alternative current field is applied during epoxy/amine curing (reproduced from [75] with permission from Elsevier).

Dispersion stabilization due to surface polarity variation and eventual covalent bonding with epoxy [79] is invoked to comment these morphologies.

Liu et al. [90] grafted polyoxyalkylene triamine on CNTs through a reaction with the carboxyl functions generated by an acid attack. The resulting NH$_2$-CNTs are highly dispersible in organic solvent, but the nanocomposites obtained in epoxy had aggregated morphologies probably due to the lack of high shear mixing during the process. Eitan et al. [91] have succeeded in grafting epoxy functions directly on nanotubes surface, but this kind of functionalized nanotubes has not been dispersed in epoxy matrixes to our knowledge.

The latest innovative process consists in inducing the nanotubes alignment by applying AC and DC electric fields during the polymerization reaction. This process used by Martin et al. [75], and inspired from previous studies on carbon black percolation performed by Schwarz et al. [4], aims at helping the formation of a percolating network. It strongly affects the CNTs' final dispersion as shown in Figure 4.18. Instead of waiting for a random aggregation, applying a DC field causes a dendritic progressive agglomeration from the cathode to the anode. Applying an AC field allows getting a strongly orientated percolated network, aligned along the field lines. Although the process was efficient, it did not help obtaining more appropriate dispersion states than the aforementioned one [72–74].

This first part dealt with different nanofillers preferentially incorporated to epoxy networks during this last decade, and the numerous strategies elaborated to be able to control their final state of dispersion in the matrix, since the final morphology and interfacial properties are known to rule the nanocomposite properties. A particularity of epoxy systems compared to thermoplastic nanocomposites is that the matrix is built up in the presence of the nanofillers. This is why the crosslinking reaction (or curing reaction) has to be followed carefully. This step is of great importance since nanofillers might influence the network formation and consequently the final network structure leading to the achievement of unexpected matrixes. Reciprocally, the network build up can modify the state of dispersion of the nanofillers. These two aspects are exposed in the following section.

## 4.3. Nanocomposite Self-Construction

### 4.3.1. Polymerization Reaction and Reaction-Induced Dispersion

In this part, we consider the consequences of the presence of the nanofillers on the epoxy polymerization reaction as well as the changes in nanofiller state of dispersion induced by this reaction. It is important to precise that direct comparison of results between different studies is often difficult due to the variety of nanofillers and of curing agents used with epoxies. Each system induces a specific polymerization reaction with its own kinetics; the cure temperatures used are generally different; and the final state of dispersion of the nanofiller as well as the network properties greatly depend on all these parameters.

#### 4.3.1.1. *Spherical Nanoparticles*
Incorporation of silica nanoparticles in epoxy system has been demonstrated to affect the epoxy/amine reaction. Ragosta et al. [7] showed with near infrared studies that after having reacted with the curing agent primary amines function, the epoxy groups react simultaneously with secondary amines and hydroxyls groups. The presence of hydroxyl groups on silica nanoparticles surface allows grafting of the nanofiller to the organic network. The final conversion in epoxy groups after postcuring is unaffected but the crosslink density could be lowered by this side reaction.

No catalytic or hindering effects of POSS incorporation on the epoxy/amine reactive have been reported until now. However, incorporating functionalized POSS cannot be done without having a close look at the epoxy/amine reaction since they are often involved in it. When used as dangling units, glycidyl monofunctionalized POSS have to be considered when calculating the stoichiometric amount of amine to be added to the epoxy/POSS mixture. Even if the reactivity of their glycidyl functions is probably lower than those carried by the epoxy monomers [13], reactions are successfully completed without modifying the curing cycles. However, according to Laine et al. [17], the secondary amines could not be able to react in case of strong steric hindrance caused by the presence of rings around the epoxy and amine functions. This phenomenon described in Figure 4.19

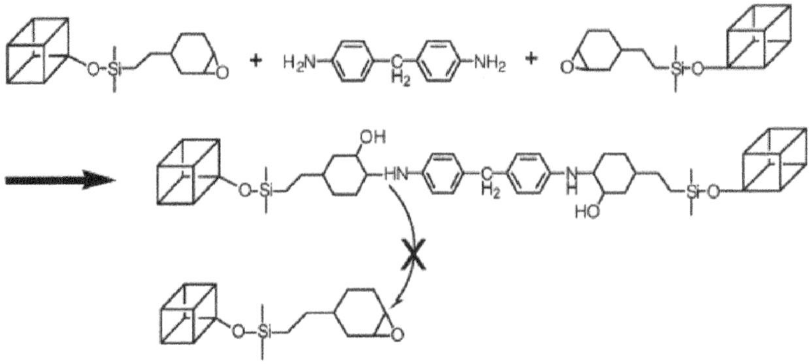

**Figure 4.19** Schematic representation of the steric hindrance leading to a decrease in diamine functionality from 4 to 2 toward POSS epoxy groups (reproduced from [93] with permission from ACS).

renders the aromatic diamines used in this work bifunctional instead of tetrafunctional. The stoichiometric ratio has, therefore, to be changed depending on the kind of POSS used.

Stoichiometric imbalance due to the phase separation between a POSS-rich phase and an epoxy-rich phase can also occur if the compatibilizing groups of monofunctional POSS are not well chosen [21]. According to Pittman et al. [92], early chemical bonding of the POSS to the epoxy network can prevent this phenomenon.

The influence of chemical network formation on POSS state of dispersion and nanocomposites morphology has been exposed by Matejka et al. [13] in a recent work.

When POSS are not functionalized, epoxy/amine reaction leads to the reaction-induced phase separation. The size of the spherical POSS domains can vary between 0.5 μm as observed by Ni et al. [19] with octanitrophenylPOSS and a few nanometers if the ligands do not favor aggregation as shown by Matejka et al. with octahexylPOSS [13].

When monofunctional reactive POSS are introduced to form dangling units in the epoxy network, the final morphologies are mostly guided by the nature of the nonreactive groups. If the POSS are initially liquid or amorphous, POSS units will probably aggregate depending on its compatibility with the epoxy network. Abad et al. [21] pointed out that with mono amino-POSS, a second phase separation occurs in the epoxy-rich phase during polymerization, after the aforementioned initial macrophase separation (Section 4.2.1). This leads to the fine amorphous POSS nodules in the epoxy-rich phase, but it demonstrates that despite the possibility of grafting the POSS monomer on the epoxy network, reaction-induced phase separation can occur in these systems as in thermoplastic/thermoset polymer blends. With a monoepoxyPOSS carrying isooctyl compatibilizing groups, Matejka et al. [13] estimated the POSS domains diameter at 3.2 nm, therefore involving around seven POSS units [13]. These domains are randomly dispersed and their size is not influenced by chemical network formation. According to the authors, this phase separation occurs because of the lower reactivity of POSS glycidyl functions compared to those carried by the epoxy monomer.

When the POSS are initially crystalline, more complex dispersion state evolutions are observed. Despite being dangling units on the chemical network, POSS are able to crystallize. The crystallographic structure in the network might be different from their initial structure due to the steric hindrance involved by their grafting. POSS crystalline

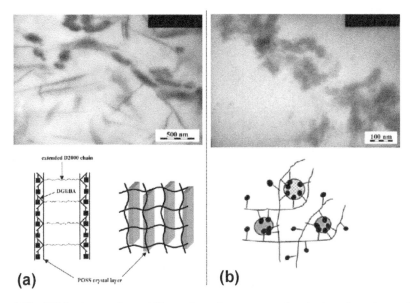

**Figure 4.20**  TEM micrographs and illustrating schemes presenting the lamellar (a), or spherical (b), silica nanophases formed in an epoxy matrix when incorporating POSS as dangling units (reproduced from [13, 94] with permission from ACS).

domains can be spherical as well as lamellar as demonstrated by the TEM pictures and relative schemes proposed by Matejka and Strachota [13, 94] grouped in Figure 4.20.

SAXS measurements during polymerization prove additionally that the organic network formation progressively reduces the initial spherical crystallized domains size by incorporating POSS to its structure and, therefore, leads to the achievement to finer POSS crystals morphologies like the lamellar morphology presented hereafter (Figure 4.20), which does not exist prior to polymerization. This reaction-induced modification of the crystalline domains is supported by observations from Abad et al. [21] who detected a decrease in crystallinity during a postcuring performed above the crystals melting temperature.

POSS have been more often used as crosslinks in the epoxy network. Most of the authors reported homogeneous molecular dispersions in that case [15, 19, 93, 95], generally based on microscopic observations. However, Matejka et al. [13] showed that despite a global good dispersion of POSS in the matrix, some aggregated amorphous domains (4 nm wide) are present and that their sizes decrease with increasing steric shielding around the POSS monomers with higher functionalities. The slow reactivity of the diamine used might be partly responsible for this aggregation. These aggregates are bigger numerous at the beginning of polymerization, and progressive involvement of POSS in the network is responsible for their improved dispersion. However, molecular dispersion of all junctions is never obtained. Moreover, a long-range ordered structure appears during polymerization and its characteristic distance increases progressively till it reaches the diamine tether length. Matejka et al. [13] proposed that when conversion increases, steric restrictions caused by the grafting of diamines around POSS monomers lead the diamine tethers, which link POSS together to be progressively more extended. Supporting this proposal, the most ordered networks are obtained with the POSS having higher functionalities.

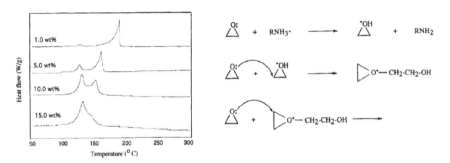

**Figure 4.21** DSC thermograms showing the growing catalytic effect of organoclay on epoxy monomer self-polymerization when increasing nanofillers loading and scheme of associated mechanism proposed (reproduced from [96] with permission from Elsevier).

### 4.3.1.2. Nanoplatelets

Concerning nanoplatelets, the most detailed results until now have been obtained with clay nanolayers. We will first focus on the influence of clay and organoclay on the polymerization reaction before exposing the various explanations proposed to justify the changes observed in the state of dispersion of the clay induced by the crosslinking reaction.

Studies from the group of Pinnavaia show that care has to be taken with the curing temperature used since epoxy prepolymer can undergo homopolymerization at quite low temperatures in the presence of organoclays. Thermally activated epoxy ring opening is usually believed not to happen below 350°C. In the presence of primary alkylammonium cations, Wang et al. [43] observed it taking place at much lower temperatures in the presence of alkylammonium-modified clays and link the catalysis of the epoxy ring opening to the acidity of the alkylammonium ions. Lan et al. [96] observed that the homopolymerization decreases down to 120°C in the presence of ODA–MMT (Figure 4.21). The mechanism proposed is that the alkylammonium cations dissociate in the clay gallery, generating a proton $H^+$, which can attack the epoxy ring and cause acid-catalyzed ring opening polymerization as described in Figure 4.21. The more acidic is the alkylammonium cation, the lower the homopolymerization temperature, which confirms the proposed mechanism.

The catalysis of the epoxy amine reaction by hydroxyl groups is a well-known phenomenon, which has also been observed with organoclay insertion in epoxies. The considered mechanism is the hydrogen bonding of the epoxy by the hydroxyl group in the transition state, followed by hydrogen transfer from donor to the epoxy. The possibility for this catalytic polymerization to happen has been observed by Messersmtih and Giannelis [46] and used to promote intragallery polymerization [97]. It has also been confirmed by Brown et al. [54] who, however, proposed that the slight effect observed when polymerizing in the presence of organoclay whose alkylammonium ions are not hydroxyl-functionalized might be solely due to the Al–OH and Si–OH on the clay layers edges.

Based on their previous results, Pinnavaia et al. [99] demonstrated that the polymerization reaction might be catalyzed in the interlayer space due to the acidity of the alkylammonium ions. These first conclusions were deduced from Brönsted acid catalysis of the epoxy amine reaction (referring to previous description by Kamon et al. [100]), and from the better dispersions obtained with the more acidic alkylammonium ions, considering (as will be exposed later) that a faster intragallery polymerization was necessary to obtain

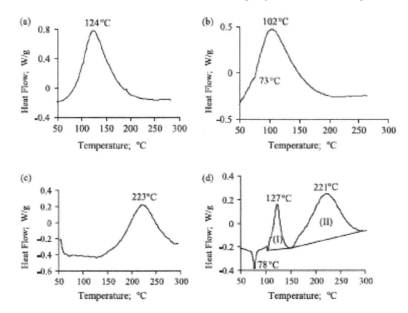

**Figure 4.22**   DSC thermograms of DGEBA epoxy monomer during condensation polymerization: (a) with Jeffamine 230, (b) with Jeffamine D230 and HDA-MMT, (c) with DDS, and (d) with DDS and HAD-MMT (reproduced from [98] with permission from Elsevier).

exfoliated nanocomposites [101]. DSC experiments by Chen et al. [102] confirmed a decrease in the epoxy-amine condensation polymerization temperature.

Recent work from Park et al. [98] reveals that this catalytic effect of acidic alkylammonium cations can or cannot be observed depending on the amine used to perform the polymerization. The amine is not directly responsible for it, but the temperature at which its reaction with epoxy starts does. Indeed alkylammonium ions are subject to thermal dissociation to form an amine and volatile HCl. This dissociation begins at around 130°C and proceeds till 162°C for hexadecylammonium chloride as observed by DSC. Consequently, the polymerization can be catalyzed solely if it starts before the alkylammonium dissociation. If the thermal dissociation happens first, the epoxy/amine polymerization will not be catalyzed as observed with the DGEBA/DDS system in Figure 4.22. Another effect of this dissociation is the formation of primary amines in the interlayer space, which might lead to a stoichiometric imbalance and consequently modified kinetics, as well as linear polymer synthesis causing a plasticization of the epoxy network.

All the results presented here up to now have been obtained with amine curing agents. Much less studies have been performed on other epoxy crosslinking reactions. However, a recent study from Liu et al. [103] reveals that some adverse effects can be observed when the crosslinking is achieved using phenol as curing agent and triphenephosphine (TPP) as a catalyst. The onset of polymerization temperature increases gradually when increasing the amount of ODA–MMT in the composite. The catalytic action of the alkylammonium-modified montmorillonite in the ring opening of epoxy prepolymer is not contradicted and is even confirmed in this study, but it is shown to be effective at much higher temperatures than TPP does. The delayed crosslinking reaction of the composites is, therefore, explained by a reaction between the $H^+$ of the alkylammonium cations and the TPP.

Increasing the organoclay loading in the composite prevents the catalyst from promoting the ring opening of epoxies. When an organoclay modified with quaternary alkylammonium ions is used, the hindered effect is no more visible, which confirms the proposed mechanism.

In summary, the main effect on the crosslinking reaction of the presence of clay platelets in epoxy systems is due to the acidic alkylammonium used to render the clay organophilic. The catalytic effect of these ions on the ring opening has been clearly demonstrated in the case of homopolymerization. Even if the catalytic effect observed on the epoxy amine reaction could also be due to the presence of hydroxyl groups on clay platelets edges, the fact that quaternary alkylammonium carrying hydroxyl groups and acidic primary alkylammonium ions modified montmorillonites lead approximately to the same catalytic power is comforting the first assumption. In any case, a careful investigation of the effects of both organoclay and organic cations used to modify it on the polymerization process has to be performed prior to the characterization of the resulting nanocomposites.

In the next part, the influence of the network crosslinking on the clay platelets state of dispersion is exposed. Various theories have been built from experimental results, but the knowledge in that field is still changing, since no theoretical model has been clearly adopted to explain why $d$-spacing is sometimes increased during curing and sometimes not.

Messersmith and Giannelis [46] first observed that during dynamic curing of epoxy amine systems, delamination of clay was happening during curing reaction. This led to the idea that clay exfoliation in epoxies might be a reaction-induced process.

From the very first studies, it has been shown that reducing the acidity of the alkylammonium cations in the organoclay galleries was deteriorating the final state of dispersion of the clay (from exfoliated to intercalated) [36]. It is important to be precise here that in order to match the authors terms, we use exfoliation to characterize a dispersion state in which the $d$-spacing has been increased to 100 Å or more, but the homogeneity of the dispersion at the micrometer scale is generally not considered. Knowing that acidity was able to catalyze the ring opening of epoxies as the hydroxyl-functionalized alkyammonium used by Giannelis did, it has been thought that exfoliation was deeply linked to the crosslinking reaction kinetics. This is why the first mechanism to a reaction-induced exfoliation brought by Pinnavaia group registered a higher polymerization rate within the clay galleries than in the bulk material. Even if the driving forces responsible for this process were not clearly identified, this idea has been the basis of the majority of the mechanisms proposed until now and led many groups to use hydroxyl-functionalized alkylammonium ions to achieve exfoliation [28, 55, 104, 105].

Kornmann et al. [106] noticed that the degree of exfoliation of clay was varying depending on the curing agent used and the curing temperature chosen. They observed that the lower was the reactivity of the curing agent, the higher the exfoliation rate. On the other hand, they showed for each different curing agent that an increase in the curing temperature (which increases the reaction speed) led to better dispersions. To explain these observations, they proposed that in addition to the intra/extragallery polymerization kinetics competition, the diffusion rate of diamine hardener had to be considered. They suggest that aliphatic amines with higher flexibility can diffuse easier in the clay galleries and participate to a faster intragallery polymerization, whereas rigid aromatic amines need higher temperature to get such a diffusion rate. The flexibility of the diamine and the cure temperature chosen will allow modifying the balance between the extragallery and intragallery reaction rates. The importance of curing temperature on the final morphologies

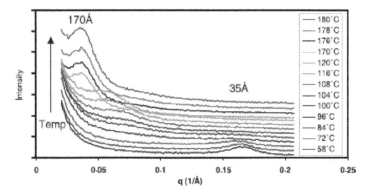

**Figure 4.23** *In-situ* SAXS measurement showing *d*-spacing increase during condensation poly-merization of a 4 wt% ODA–MMT/epoxy nanocomposite (reproduced from [102] with permission from Elsevier).

has also been underlined by Tolle et al. [107]. This theory has also been used and clearly exposed by Chen et al. [102] to comment the evolution of *d*-spacing they observed by *in-situ* SAXS measurements (Figure 4.23).

Chen et al. [97] (from the Giannelis group) followed this idea in a detailed investiga-tion using time-resolved X-ray diffraction coupled with kinetic studies of epoxy systems containing hydroxyl-functionalized alkylammonium-modified montmorillonite to ensure catalysis within the galleries. In this, an anhydride has been used to perform epoxy crosslinking, which means that the mechanisms described might not be directly compara-ble to the results obtained with diamines. A three-stage exfoliation process is proposed. In the first stage, intragallery polymerization catalyzed by the hydroxyl groups of organoclay succeeds even at room temperature and results in a first increase in the *d*-spacing. The fact that the *d*-spacing stops increasing is attributed to the presence of bridging molecules between the clay platelets. The polymerization at a higher temperature allows unreacted monomers to diffuse within the galleries, which leads to stretching and breakdown of the bridging units allowing a continuous increase in *d*-spacing to happen. The breakdown of bridging units is attributed to ester interchanges made available by the raise of the temperature. The following increase is explained, as Kornmann did, by a competition between the intra- and extragallery reaction rates as well as by the rate of diffusion of unreacted species between the silicate layers. The third stage corresponds to the cessation of the expansion. They confirm prior observations made by Ke et al. and Lu et al [108, 109] when they link the end of expansion to the gelation of the network. However, and for the first time, they correlate the exfoliation mechanism with mechanical considerations, proposing that the concomitance of network gelation and the end of the gallery expansion is due to a higher elastic modulus outside the galleries. This idea will be used to propose new mechanisms for reaction-induced exfoliation.

The latest developments in the explanation of this phenomenon have brought by Park and Jana [110]. They propose that even if a faster intragallery polymerization rate favors the gallery expansion, it is not necessary to get exfoliated nanocomposites. Moreover, they do not take into account the diffusion rate of unreacted monomers in the galleries in their explanation. According to them, the driving force is the elastic force exerted by the growing crosslinked macromolecules on the clay platelets as exposed in Figure 4.24. The molecules

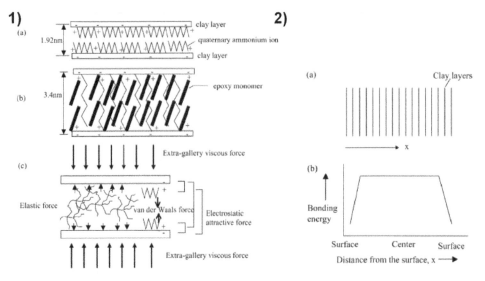

**Figure 4.24**   1. Schematic illustration of the intercalated state and exfoliation process showing the forces acting on a pair of clay layers: (a) organically modified clay, (b) epoxy monomer intercalated state, (c) forces acting on a two-particle tactoids. 2. Schematic diagram showing the relationship between the ionic bonding energy and the location of the layers in the tactoids: (a) tactoids, (b) variation of bonding energy along the thickness of the tactoid (both reproduced from [110] with permission from ACS).

confined in between clay platelets grow too fast to be able to relax and exert a higher elastic force on the platelets that those growing outside the galleries can relax easier. Moreover, considering that the electrostatic forces that maintain the platelets face to face are higher in the heart of a tactoid; they propose that solely the outermost platelet of the tactoid can be delaminated by this process (Figure 4.24). This proposal could explain why the $d$-spacings measured during an exfoliation process jump from one distance to another. According to them, the model implying continuous diffusion or unreacted monomer inside the galleries should lead to a continuous increase in the $d$-spacing. On the contrary if exfoliation occurs only with the outermost platelet, the progressive intensity decrease of the XRD peak related to intercalation $d$-spacings can be explained. Once the electrostatic forces between clay layers are overcome, the viscosity of the polymerizing system can limit exfoliation. The ability for an epoxy system to exfoliate clay during its polymerization would be linked to the ratio between its storage modulus and its viscosity in the early times of the reaction, before gelation occurs.

In another study [98], they propose that the lack of $d$-spacing increase observed when alkylammonium ions dissociate in amine and HCl is not due to the lack of catalysis in the gallery. This phenomenon is rather due to a lower crosslink density and to the plasticization of the intragallery network by the primary amines, which leads to a lower elastic modulus between the platelets. This hypothesis of network plasticization had been proposed by Shi et al. [111] to explain that tensile modulus was not increasing with increasing organomodifier chain length in rubbery epoxy nanocomposites. Its occurrence has been more recently demonstrated by Triantafillidis et al. [112]. Using protonated long aliphatic polyoxypropylene diamines as exchanged cations led to a decrease in glass transition temperature ($T_g$) of a glassy epoxy network. The model proposed by Park et al. [98] seems promising despite some imprecision in gel points measurement since it is now well known that $\tan(\delta) = 1$ is not a gelation criterion especially in systems in which cure kinetics, stoichiometry, and plasticizer amount may vary.

It is worth noting that in comparison with nanotubes for instance, very few studies have looked at the eventual percolation of clay layers. On nonepoxy thermoset systems, Meng et al. [113] observed that the percolation threshold for organoclay platelets is around 3% and that for any system with a higher filling rate, physical gels are obtained before chemical gelation. This can be related to the observations made by Chen et al. [102], who demonstrated by SAXS that the $d$-spacing in transparent nanocomposites was maximal with 1 wt% organoclay introduced and decreased when increasing loading. On an epoxy/amine system, Le Pluart et al. [114] observed a singular evolution of the viscoelastic properties during polymerization of the system in the presence of organoclay. The tan$\delta$ curves acquired during polymerization at various frequencies never crossover, preventing the gel point measurement by the classical Winter and Chambon method. This phenomenon has been attributed to potential exfoliation-induced physical gel formation before the gel point but has not been confirmed up to now. This proposal comes from the fact that organoclay dispersion in epoxy networks prepolymers has been shown to lead to physical gels formation [115] and that the nonrespect of Winter–Chambon criterion has been observed with other nanoparticles filled thermoset systems [116]. A comprehensive study of percolation and physical gel formation during epoxy network formation has not been performed yet but it could be of great interest for a better understanding of reaction-induced clay exfoliation.

As a conclusion, it has been shown that the understanding of the reaction-induced exfoliation is far from being achieved. The wide variety of the epoxy systems used render this process even harder to control and it seems that new proposals will continuously be made until a theoretical treatment of this phenomenon is performed. Despite all the mechanisms proposed in 10 years, it seems still a tough task to predict whether an unknown system will be able to exfoliate clay or not at a given temperature. For instance, the fact that the best exfoliation ratios are obtained with primary alkylammonium ions modified MMTs can be either explained by their higher catalytic activity compared to quaternary ones, or by the fact that having only one alkyl chain they less contribute to matrix plasticization. Recent experiments by Tolle et al. [117] confirmed that some progress has to be done to fully understand this gallery expansion mechanism. Indeed, they show that aging epoxy prepolymer/organoclay mixtures at room temperature affects the gallery expansion mechanism without changing the curing agent, the curing process or the organoclay used. Increasing aging time allows increasing the final nanocomposites $d$-spacing (from 42 Å for unaged samples to 110 Å for 16 weeks aged). The most striking observations being that the $d$-spacing of the uncured epoxy/organoclay mixtures remains unchanged during aging.

The main certainties obtained in that field are that $d$-spacing increase is strongly related to the polymerization of epoxy and that gelation stops the gallery expansion. The mechanical and the chemical explanations for the driving force are both interesting but none of them seems to be able to fully explain the observed behaviors.

### 4.3.1.3. Nanotubes and Nanofibers

To our knowledge, one single example concerning the influence of nanotubes on the crosslinking reaction has been reported until now. Miyagawa et al. [86] have proven that incorporating fluorinated nanotubes in an epoxy/anhydride mixture led to stoichiometric imbalance caused by fluorine release. It has been detected through a 20°C decrease in $T_g$ with an 0.1 wt% filling rate. They propose that at the high curing temperature of the nanocomposite (160°C), fluorinated nanotubes revert to CNT and the consecutive production of fluorine-free radicals leads to the observed stoichiometric imbalance. Even the imbalance corrected, the $T_g$ remains lower than in the pristine network suggesting a change in network structure, which is not detailed. This problem has not been evidenced by Zhu et al. [85] who used similarly functionalized CNTs in an epoxy cured with an aromatic diamine.

Studies performed on the evolution of CNT state of dispersion during curing have also been performed when electrical properties where targeted by the authors since it is necessary to reach percolation before epoxy gelation. Martin et al. [72] demonstrated the influence of the curing temperature of epoxy/amine networks on this percolation process. After a high shear mixing causing the individualization of the CNTs in the epoxy prepolymer, the addition of the curing agent favors the desired aggregation by modifying the ionic force of the reactive mixture. However, if the curing temperature applied is too low, the high viscosity prevents the CNTs from percolating before gelation is reached. With more elevated temperature, the aggregation process is favored by higher Brownian motion of the CNTs and lower viscosity despite the shorter time before gelation. This agglomeration process is also favored by moderate shear forces. In a following work [72], they prove via *in-situ* conductivity measurements that polymerization shrinkage is responsible for distortions in the percolated structures and that the more agglomerated the network is, the less its conductivity is affected, proving one more time that favoring aggregation during polymerization is a key feature to get optimal electric properties for CNT/epoxy nanocomposites.

### 4.3.2. Organic Network Structure

Before evaluating the improvements in epoxy properties brought by the nanofillers incorporation, it is of great importance to check whether the network structure has been modified or not. Indeed too many studies draw conclusions on eventual inorganic/organic phases interactions without carefully checking that changes in the polymerization kinetics might have modified the epoxy network structure. $T_g$ variation is the more widely used way of checking it since it reaches a maximal value when the network is fully crosslinked. Other indicators such as the evaluation of the critical exponent at the gel point, which traduces the critical gel structure, have been rarely considered.

#### 4.3.2.1. *Spherical Nanoparticles*

If the network structure is unaffected by nanoparticles introduction, $T_g$ could be either unchanged, lowered if the filler acts as a plasticizer, or increased if particle/polymer interaction hinders the network's tethers mobility acting as physical crosslinks.

The latter case has been observed by Wetzel et al. [9] with alumina nanoparticles. They assume that the nanometric dimensions of the particles increases this hindered relaxation effect compared to classical micron-size fillers. Despite considering their epoxy network unaffected by silica nanoparticles incorporation, Ragosta et al. [7] observed a decrease in $T_g$ when the filling rate reaches 10 wt%. Since silica is not known to be a plasticizer, it seems that the reaction of their high-$T_g$ epoxy with silanol groups at the silica surface (after full consumption of the primary amines) has led to a decrease in crosslink density.

They propose a plasticizing effect brought by unreacted epoxies, but it should also have been observed with the pristine epoxy since the final conversion in epoxy groups is the same in both cases. Indeed contrarily to an increase, a decrease in $T_g$ can be due to many phenomena (stoichiometric imbalance, decrease in crosslink density, plasticization...) whose influence is sometimes difficult to identify. These first examples describe well the difficulties encountered in justifying changes in $T_g$ after nanofillers incorporation since a careful kinetic study of the epoxy polymerization as well as the checking of unchanged network structure is not always performed.

Thus, as it is generally observed in epoxy nanocomposites (independently on the nanofiller used), the influence of POSS incorporation on the epoxy network structure and on the $T_g$ is subjected to contradictory observations. From studies on thermoplastic POSS nanocomposites, it is known that when they are not reactive toward the matrix, POSS act as diluents at low content causing a decrease in $T_g$, but at high loadings $T_g$ increases due to growing number POSS–polymer and POSS–POSS interactions. These POSS–polymer interactions are increased when POSS are grafted to polymeric chains.

When they are used as pendant groups on the network, Lee et al. [16] found that the onset of $T_g$ was not affected significantly but it has been broadened toward the higher temperatures. This observation has been widely confirmed [21, 94, 95, 118]. Strachota et al. [94] attributed it to a decreasing fraction of epoxy network free chains due to strong POSS–chain interactions, which can be related to the crystalline structure of POSS dangling units. Abad et al. [21] pointed out that an increase in $T_g$ should be concluded since it should be compared to an epoxy network containing the same fraction of monoepoxy monomers to keep the same crosslink density. They also attribute it to hindered chain mobility. To explain this hindered mobility, molecular dynamics studies by Bharadwaj et al. [119] and recently by Bizet et al. [120] on thermoplastic polymers demonstrate that POSS are acting as strong anchor points. This mechanism is not confirmed by Strachota et al. [94] who rather consider that POSS–POSS interactions are necessary to achieve this hindering since no lowering in chain mobility for nonaggregated POSS is observed.

When POSS are used as crosslinks in the network, the glass transition temperature is generally observed to decrease. Using octaglycidyl–POSS, Strachota et al. [94] concluded that the POSS unit is acting as a "soft" junction in the network when compared to DGEBA, since the epoxy functions are carried by flexible hexyl chains. Decreasing POSS functionality leads to even lower $T_g$ values due to lower crosslink densities and increasing plasticizing effect of nonreactive soft ligands, which is shown in Figure 4.25 through the displacement of the tan$\delta$ maximum.

Decrease in $T_g$ has been confirmed [15, 19 95], but Ni et al. [19] who used octaaminophenyl–POSS attributed it to an increased free volume due to POSS units inclusion at a molecular level, denying the eventuality of an incomplete curing reaction proposed by Li et al. [95]. Even if their results are not discussed this way, the 80°C reduction in tan$\delta$ maximum temperature obtained by Choi et al. [121] when using octa(propylglycidyl-dimethylsiloxy) POSS could also be ascribed to a higher junction "softness." Using octa(ethylcyclohexylglycidyl-dimethylsiloxy) POSS [93] with a corrected stoichiometry (because of the aforementioned change in diamine functionality) supports this hypothesis, since the measured $T_g$ is higher than in the latter case. The lower crosslink density of this network might be compensated by a lower mobility of POSS ligands.

These examples reveal that without careful variations in the parameters, it is hard to conclude on the effect of POSS incorporation into epoxy. Indeed, variations in POSS nature simultaneously affect network crosslink density, chemical nature of networks tethers,

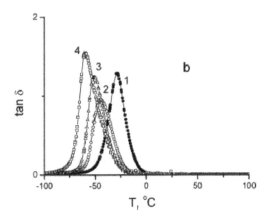

**Figure 4.25** Loss tangent versus temperature of (1) epoxy/amine, (2) octaglycidylPOSS/amine, (3) tetraglycidylPOSS/amine, (4) diglycidylPOSS/amine networks (reproduced from [94] with permission from ACS).

POSS–POSS and POSS–chain interactions, as well as the amount of inorganic cages in the polymer.

### 4.3.2.2. Nanoplatelets

In the field of epoxy/clay nanocomposites, the interrogations persist concerning the crosslink density, macromolecules mobility at the clay/epoxy interface, and the eventual plasticization of the interplatelet network by the long-chain alkylammonium cations used to modify the platelets. This is probably the domain in which the wider distribution of opinions can be found among the studies led on these nanocomposites. This diversity of opinions on the final network structure is primarily due to the fact that no clear trend in the glass transition temperature evolution has been evidenced.

Various explanations have been proposed when a decrease in the glass transition temperature of the network is measured after having dispersed organoclays in it. Kornmann et al. [122] proposed an eventual cation exchange between the alkylammonium ion and the epoxy ring, which would lead to the production of monoamines likely to react with epoxy groups, thus leading to a decrease in crosslink density. Previously, Wang et al. [123, 124] also proposed that alkylammonium could take part in the crosslinking reaction without evidencing it directly but deducing it from mechanical properties measurements. Zilg et al. [28] also measured a decrease in $T_g$ of epoxy/anhydride systems. They varied the epoxy/anhydride stoichiometric ratio to exclude a possible deviation from stoichiometric conditions and invoke the intercalation and interfacial coupling effects to explain the observed phenomenon. Another explanation has been proposed by Chen et al. [97]. According to them, the long-chain alkylammonium ions linked to the clay surface act as plasticizers of the epoxy network. This proposal is supported by the fact that the $T_g$ decreases as the interlayer spacing increases due to the growing volume fraction of intragallery plasticized network in the whole material. According to their theory, the plasticized zone should not be larger than a fully extended alkyl chain length and they indeed observe that $T_g$ does not decrease anymore when the interlayer spacing overcomes 10 nm. This allows them to evaluate the plasticized interfacial zone at the clay platelet being approximately 50 Å wide, whereas a fully extended octadecylammonium ion chain is 37 Å.

Some other author did not observe any significant effects on the glass transition temperature, concluding that the network structure was unaffected by the organoclay presence [114, 125]. Le Pluart et al. [114] checked it on rubbery as well as glassy epoxy systems, and with two different organoclays. The glass transition temperature of the fully cured networks remained unchanged. Moreover the conversion at the gel point was also unaffected, confirming the invariance of the network structure.

The work performed by Park et al. [98] may give some clues concerning the fact that decreases in glass transition temperatures can or cannot be observed. As previously explained, they think that alkylammonium ions can undergo thermal dissociation leading to the formation of free alkylamines in the organoclay gallery. If the curing temperature chosen is higher than the thermal dissociation temperature, these monoamines could react with epoxy function and lead to the lowering of the crosslink density.

On the other hand, many authors observed an increase in the glass transition. Messersmith et al. [46] explained this increase in $T_g$ and its broadening, by the restricted molecular mobility of the polymeric segments near clay particles surface. This idea of a good interfacial adhesion between clay and the epoxy network is strengthened by the fact that they use an organoclay whose alkylammonium ions carry hydroxyethyl groups, which are likely to be involved in the epoxy/anhydride crosslinking reaction. This increase has been often observed in other studies [39, 54] and the same explanation is generally given when the alkylammonium ions used were carrying reactive functions.

More surprising and interesting is the observation made by Ratna et al. [126] of a 20°C increase in the glass transition temperature of DGEBA/DETDA system containing 5 wt% of an ODA–MMT (Figure 4.26). The explanation proposed here is the confinement of the macromolecules in the galleries, which are approximately 100 Å wide. This explanation could be confirmed by a recent PALS study performed by Wang et al. [127], which tends to prove that free volume and molecular mobility are reduced in the vicinity of clay platelets. The increase in $T_g$ observed by Ratna et al. [126] is interesting since a decrease in $T_g$ and $\beta$-

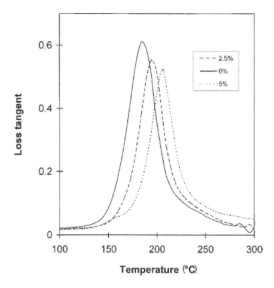

**Figure 4.26** Loss tangent versus temperature of DGEB/DETDA nanocomposites with various ODA–MMT contents (reproduced from [126] with permission from Wiley).

transition temperature has been observed by Becker et al. [128] on the same system (epoxy, curing agent organoclay, and filling rate). The curing procedure used being the same; the only noticeable difference is the 1:0.9 epoxy-to-amine ratio chosen by Becker et al. to avoid nonreacted amine groups in the final network. The glass transition temperature obtained for the unfilled epoxy system is slightly higher in the work by Ratna et al. [126]. This difference in stoichiometry might be responsible for the lower glass transition temperature of the unfilled system since it is known that the hydroxyl/epoxy etherification reaction is not negligible at high temperature when there is an excess of epoxy rings, but the change in the evolution of $T_g$ with increasing clay loading remains unclear. This last example clearly depicts the difficulties to draw some definitive conclusions on the modifications of the network structure brought by organoclays. Meng et al. [113] showed on nonepoxy thermoset systems that organoclay slightly decreases the degree of crosslinking at the gel point and modifies the gel characteristics (critical exponent). The nature of the organoclay used affects the gelation behavior of the thermoset system. This could explain why so many differences have been reported in the last 10 years concerning the trends of the network $T_g$ evolution in the presence of organoclay.

The conclusion to this part might be the one given by Koerner et al. [53]: "the impact of the organomodified montmorillonite on topology and dynamics of the networks near the montmorillonite surface is at this time beyond current standard analytical techniques." The extent of cure and the chemistry of the epoxy network being identical within the experimental resolution of DSC, FTIR, and NMR with or without organoclay, the causes of the observed $T_g$ variation can generally only be supposed and hardly be explained.

With other nanoplatelets, the examples are not sufficiently numerous to depict such various opinions. Dispersion of exfoliated LDHs has led to a 20°C increased $T_g$ epoxy network and to a broadening of the tan$\delta$ peak, which are both attributed to the restricted mobility of macromolecules at the interface due to covalent bonding since LDHs are aminolaurate modified [29]. An increase in the $\beta$-transition temperature can also be observed in this work, which supports this hypothesis. An increase in $T_g$ following platelets exfoliation is also observed in the case of EG where the use of a slurry method coupled to graphite surface treatment leads to strong nanoplatelet/polymer interfaces and subsequent reduced mobility, whereas classical dispersion only led to marginal increases [32]. When incorporating polyoxyalkylene monoamines modified $\alpha$-ZrP [61], the epoxy $T_g$ is decreased due to a deviation from stoichiometry caused by the irremovable monoamine excess.

With these platelets, there seems to be less difficulties in relating $T_g$ variations to the nanocomposites structure. This part of the review clearly demonstrates, however, how carefully the structure of the nanocomposites should be investigated compared to the pristine networks. All these possible changes in network structure have to be considered before analyzing nanocomposites properties.

### 4.3.2.3. Nanotubes and Nanofibers

The effects of CNTs dispersion on the network structure have led to fewer contradictions than in the case of nanoplatelets.

Decrease in $T_g$ has been rarely observed and it was due to unexpected processing problem as inefficient solvent removal [70] or imbalance in stoichiometry caused by defunctionalization of carbon nanotubes during curing [86].

There is more to discuss regarding the increases in $T_g$. A 10°C increase with 3 wt% carbon nanotubes [76] and +30°C with 20 wt% nanofibers [129] have been observed, as

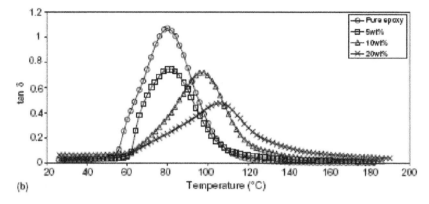

**Figure 4.27**  Loss tangent versus temperature of epoxy nanocomposites with various carbon nanofibers contents (reproduced from [129] with permission from Elsevier).

shown in Figure 4.27. As with POSS dangling units, it is attributed to lower molecular mobility of epoxy network macromolecules in the vicinity of the well-dispersed carbon nanofillers even if no effort has been devoted to cause covalent bonding between them. Strong polymer/nanofillers interaction or physical network formation could explain these observations. Gong et al. [81] observed a similar 10°C increase when dispersing 1 wt% of untreated MWCNTs. More surprising is the fact that they also report a 25°C increase when the same nanotubes are covered with a surfactant chosen to act as a dispersing agent. They suggest that the hydrophilic part of the surfactant interacts with the matrix via hydrogen bonding, whereas its hydrophobic part covers the carbon nanotubes, acting in fact as a coupling agent. Monizzuraman et al. [77] performed contradictory observations since they did not measure any change in $T_g$ after homogeneous dispersion of 0.05 wt% unmodified, high specific surface area, SWCNTs. The measured mechanical properties push the authors to suppose a chemical grafting by esterification of epoxies onto the carbon nanotubes; however, in that case the network structure is supposed to be unaffected.

The latter example testifies that the effect of CNTs dispersion on the network structure has not been clearly identified and, as in the case of platelets based nanocomposites, causes some contradictory observations and explanations to be found in the literature. However, the number of studies focused on this problem are less numerous than with clay/epoxy nanocomposites and with CNTs, full curing of the networks and infinite $T_g$ reaching are rarely discussed by the authors.

## 4.4. Nanocomposites Properties

Epoxy/clay nanocomposites have been extensively characterized during the last 10 years. Some specific dispersion states, coupled with appropriate interfacial adhesion between organic and inorganic phases, have allowed taking advantage of the nanometric dimensions of the filler and of its anisotropy to get specific properties that could not have been obtained with microsize filler without a detrimental effect on other properties. In addition to this search of physical properties improvements, more and more attention is devoted to the understanding of the structure/properties relationships of these nanocomposites particularly when mechanisms at the nanometer scale are involved.

In this part, we report the more interesting thermal, electrical, mechanical, and barrier properties improvements recently achieved with different kinds of nanofillers. We also highlight the specific choices in processing or nanofiller compatibilization that made these enhancements possible in order to give a better understanding of recent advances in structure–properties relationships in epoxy nanocomposites.

## 4.4.1. Mechanical Properties

Mechanical properties have been too widely studied in epoxy nanocomposite to comment all the results obtained in that field. Influence of nanofillers on toughness and stiffness will be presented separately, since these properties are hardly increased simultaneously with classical fillers. Stiffness improvements have often been looked for to testify a good dispersion of nanofillers at the nanometer filler, whereas we will evidence that this relationship is far from being evident. Ultimate strength generally allows getting an idea of the quality of interfacial adhesion between the phases and has been generally measured to check that incorporation of nanofillers to bring thermal or electrical properties did not weaken the epoxy too much. Finally, toughness improvements have been intensely looked for since a new way of toughening epoxies without lowering its thermomechanical stability would be a great technological improvement. A consequent objective is the identification of the toughening mechanisms involved when incorporating uniformly dispersed nanofillers.

## 4.4.2. Stiffness and Strength

Fillers being generally stiffer than polymers, the Young's modulus of composite materials is easily enhanced by their incorporation as can be understood from a simple rule of mixtures. With classical microscale fillers Nakamura et al. [130] reported an increase from 2.3 to 8 GPa for a 55 wt% silica filled epoxy. The objective with nanofillers is to reach the same range of improvements with lower quantities introduced and to avoid the lower strength observed in the latter case by controlling the interfacial interactions. Attention will also be paid on the identification of the dispersion states that help getting the best stiffness increases.

### 4.4.2.1. *Spherical Nanoparticles*
Despite the unavoidable aggregation at the nanometer level, Wetzel et al. [9] reported that improvements in stiffness and strength brought by inorganic particles incorporation up to 10 wt% in an epoxy network are not lowered when using nanoparticles instead of classical fillers. The increase of the glassy modulus and ultimate tensile strength of epoxies after nanoparticles incorporation has been often reported [7, 131, 132] and is generally attributed to the intrinsic stiffness of these inorganic particles as well as to the good interfacial properties brought by coupling agents. The aggregated state of dispersion does not lower these properties when a homogeneous dispersion is obtained and 55% modulus increases can be observed with 10 wt% of nanoparticles [7].

Using sol/gel process to obtain *in-situ* formed silica nanodomains has been shown to lead to great stiffness enhancements. Matejka et al. [11] obtained an increase by two orders of magnitude in the rubbery modulus of an epoxy containing only 10 wt% of *in-situ* formed silica. This modulus increases with increasing silica content as well as interfacial adhesion (by grafting between the two phases). Careful analysis of the stiffness values leads the authors to conclude that the obtained values do not correspond to a classical particulate

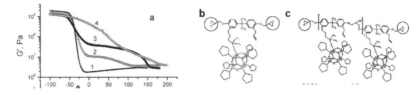

**Figure 4.28** (a) Storage modulus versus temperature of (1) DGEBA /D2000 network, (2) DGEBA+POSS grafted DGEBA monomer[b] (0.5 fct%)/ D2000 network, (3) POSS_ grafted DGEBA monomer[b]/D2000 network, and (4) POSS-grafted DGEBA oligomer[c]/D2000 network (reproduced from [94] with permission from ACS).

composite but rather to a co-continuous morphology of rubbery epoxy and glassy silica phases evidencing strong silica/silica and silica/epoxy interactions.

The stiffness enhancement observed thanks to POSS incorporation are strongly related to the way they have been introduced in the network. With POSS used as crosslinks, the glassy modulus is generally increased when increasing POSS content in the nanocomposites [15, 19] but the stronger effects are seen on the rubbery modulus since it is directly related to the network crosslink density. Strachota et al. [94] reported increases in rubbery modulus with octaglycidyl POSS, which fit with the rubber elasticity theory. Using POSS instead of DGEBA to crosslink a polyoxyalkylene diamine leads to a twice stiffer network. Moduli obtained with lower functionalities are not improved and do not fit the theory since in that case a higher aggregation prevents all epoxy functions to react and, therefore, leads to less crosslinked networks. It has to be noted that the improvements in stiffness can be much higher when the diamine tether is shorter, as proved by the more than tenfold increases in rubber modulus values that have been reported by other authors [15, 121].

However, the strongest enhancements in stiffness brought by POSS incorporation in epoxy networks have been obtained by Strachota et al. [94] when POSS are incorporated as dangling units as depicted in Figure 4.28. The well-defined and homogeneously dispersed crystallized aggregates of POSS units may act as physical crosslink in the rubbery network and, therefore, lead to huge stiffness improvements. It is increased by 1.5 orders of magnitude by replacing half of DGEBA monomers by POSS-grafted DGEBA monomer. The use of monoepoxy POSS is less efficient since it decreases the network crosslink density, and since the spherical particles morphology obtained in that case produces less chain–POSS interaction than the lamellar one. The moduli obtained are, however, still superior to the reference DGEBA/D2000 network. Another interesting point is that the reinforcement disappears with increasing temperature, which is attributed to a disordering of reinforcing crystalline domain. The higher the POSS content, the higher the disordering temperature. Once this temperature overcomes the rubbery modulus, it returns to the value planned by rubber elasticity theory, proportionally to the chemical network crosslink density.

With nonreactive POSS which are just dispersed in the epoxy amine network, or with reactive POSS which cannot crystallize, the amorphous aggregates dispersed in the network cannot act as physical crosslinks and, therefore, do not allow increasing the rubbery modulus significantly [94].

### 4.4.2.2. *Nanoplatelets*
Concerning nanoplatelets-based composites, the main parameters that have been shown to influence the epoxies stiffness are the platelets exfoliation ratio and their aspect ratio.

Lan et al. [101] first demonstrated that with layered silicates with a high aspect ratio like montmorillonite were leading to the more pronounced increases in Young modulus. In another work, Lan et al. [37] also evidenced the benefit brought by the state of dispersion of the nanoplatelets. Varying the chain length of the alkylammonium ions leads to increasing exfoliation ratios and increasing Young's moduli is consequently measured. Kornmann et al. [106] draw the same conclusions using curing agents leading to different exfoliation ratios and by comparing the relative increases in Young's modulus. Lan et al. [37] indicated that the nanoplatelets exfoliation leads to a significantly higher effect on the rubbery modulus than on the glassy modulus.

This last observation is confirmed by Messersmith et al. [46] studies who measured the storage modulus of epoxy/clay nanocomposites under and above $T_g$ and report a 60% increase in the glassy state and a 450% increase in the rubbery state with 4 vol% of exfoliated platelets. This phenomenon is illustrated in Figure 4.29 by the measurements performed by Ratna et al. [126] since similar observations are numerous in the literature [54, 104].

A first mechanism has been proposed by Lan et al. [37]. It involves the alignment of clay platelets in the rubbery state due to the stretching of macromolecules, the lack of molecular mobility in the glassy state being responsible for the lower stiffness improvements. First attempt in modeling the elastomer/clays interface to predict the Young's modulus of nanocomposite has been performed by Shia et al. [133]. In order to explain why classical models could not predict this modulus, they supposed that a reduction of the aspect ratio and of the volume fraction was caused by imperfect adhesion between clay layer and the matrix.

More recently and using a three-phase model including the epoxy matrix, the exfoliated platelets, and the intercalated clay particles, and applying the Mori–Tanaka method to calculate the modulus of the nanocomposites, Luo et al. [55] obtained interesting results on

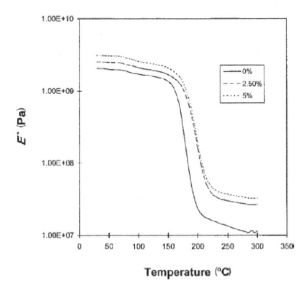

**Figure 4.29**    Storage modulus versus temperature of a DGEBA/DETDA epoxy/amine network with various amount of ODA–MMT (reproduced form [126] with permission from Wiley).

the influence of various parameters such as platelet aspect ratio, exfoliation vs. intercalation ratio, intercalated particles aspect ratio, interplatelet distance, intragallery modulus, clay loading, etc. A reasonable choice of parameters allows fitting perfectly experimental measurements. They confirmed the major influence of the exfoliation ratio on the nanocomposite stiffness. The platelets aspect ratio is also shown to be a determinant parameter. Another modeling study performed by Tsai et al. [134] supports these conclusions by demonstrating (using a shear lag model) that the load transfer efficiency in nanocomposites is excellent with exfoliated uniformly dispersed platelets due to their high aspect ratio. In case of the presence of clusters, the load transfer efficiency depends on the number of platelets in the particle and of the apparent shear stiffness between the platelets. To illustrate this modeling study, the experimental work of Yasmin et al. [52] is interesting since it links the decreasing rate of elastic modulus improvement with higher clay content to inevitable aggregation clay particles. According to Luo and Daniel [55], the presence of intercalated clay particles is not a major drawback if these particles keep a high aspect ratio. The interplatelet distance within these clusters is a less important parameter, which confirms observations made by Le Pluart et al. [114], which obtained a 60% increase in Young's modulus of a glassy epoxy with 10 wt% of a strongly intercalated organoclay. The very low exfoliation ratio of these composites was compensated by the high aspect ratio of the clay primary particles, which contained few intercalated platelets. This point has also been recently confirmed by Sheng et al. [135] in a multiscale modeling approach taking into account the morphology of nanocomposites at the micrometer and at the nanometer levels. It is shown that there is no abrupt stiffness enhancement with morphological transformation from high aspect ratio tactoids with few platelets, to completely exfoliated individual layers. These two modeling studies [55, 135] also demonstrate that the ratio between the platelet and the matrix modulus strongly affects the composite modulus enhancement, which explains why enhancements in stiffness are much higher in the rubbery state.

It has been previously exposed that some authors consider that a plasticization of the network occurs in the clay galleries. This idea originates from a work by Shi et al. [111] who proposed that one of the interaction mechanism between an organoclay and a polymer matrix was the "dissolution" of the alkylammonium ions into the polymer. To support this hypothesis, Chen et al. [97] have measured decreasing stiffness improvements of the rubbery modulus of epoxy/clay nanocomposites when the interlayer spacing was increasing. This is attributed to an increase in the volume fraction of plasticized network in the whole composite, which reaches an upper limit when the platelets interlayer overcomes 100 Å. This plasticization does not affect the stiffness values under $T_g$ since the glassy modulus is fewer dependant on the macromolecular mobility. In their modeling work, Luo et al. [55] confirmed that the influence of a lower intragallery modulus would be detrimental to the stiffness enhancement. On the contrary, the effect of higher intragallery (compared to the bulk epoxy) stiffness would only lead to limited improvements of the stiffness increase.

All these examples demonstrate that stiffness improvements are generally observed with epoxy/clay nanocomposites. It has to be noted that these stiffness enhancements have been generally observed simultaneously with ultimate tensile strength reduction [28, 47, 52, 127, 136, 137] as depicted in Figure 4.30.

This is attributed to the lower load transfer efficiency of intercalated clay clusters as proposed by Tsai et al. [134]. Wang et al. [47] evoked the possibility of flaws in the nanocomposites. They could be weak boundaries between particles, trapped bubbles, or inhomogeneities in network crosslink density and their number may increase with increasing filling rates. The negative effect on ultimate tensile strength of nanovoids in the vicinity of clay platelets (observed by TEM) has been demonstrated by Yasmin et al. [52], who showed that a stronger degassing allowed limiting this decrease. The recent observations of Wang et al. [127] confirmed these assumptions since a maximum in the ultimate strength value has been observed at the filling rate where the dispersion is optimal. Subsequent increase in the filling rate leads to an increase in clay tactoids fraction and a decrease in the ultimate tensile strength. However some examples are also reported of increasing tensile ultimate strength. For instance, Lee et al. [138] showed the strength of epoxy nanocomposites to increase with the length of the alkyl chain of the ammonium used to modify montmorillonite, this strength being always superior to that of the pure matrix.

This improvement in strength has also been noticed upon the reactive dispersion of LDH by Hsueh et al. [29], who measured a 40% increase in tensile strength with 5 wt% of aminolaurate modified LDH. This effect is attributed to the building of strong clay/epoxy interfaces and good adhesion via covalent bonding of the epoxy to the organo-LDH. Concerning other types of nanoplatelets, one can notice that the high shear mixing combined by Yasmin et al. [33] with sonication and slurry method leads to an interesting 15% increase in tensile Young' s modulus with 1wt% of EG incorporated in epoxy. This improvement is less important than those achieved with montmorillonite, but it has to be noted that in this cased, even if a solvent is used, no chemical modification of the nanofiller needs to be achieved. Li et al. [34] demonstrated that strength of epoxy/EG nanocomposite could be increased with an UV/ozone surface treatment of graphite. The decrease in strength that they observed after graphite nanosheets incorporation by a more classical process is strongly reduced through the use of this treatment.

With the same filling rate, Sue et al. [59] observed approximately similar increases in tensile modulus with exfoliated $\alpha$-ZRP. The improvement is more important if the nanocomposite modulus is compared to the plasticized network modulus since as explained before, modified $\alpha$-ZrP incorporation led to an imbalance in stoichiometric conditions.

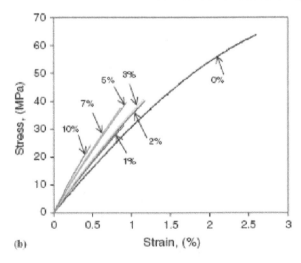

**Figure 4.30** Stress/strain curves of epoxy/anhydride nanocomposites containing various amounts of MT2Et–MMT obtained under tensile loading (reproduced from [104] with permission from Elsevier).

However the authors underline that this stiffness improvement is accompanied by a reduction of ductility. This difficulty in achieving a good stiffness/balance has been recently confirmed by Boo et al. [61], who did not measure any reinforcing effect of epoxy by $\alpha$-ZRP through $K_{Ic}$ measurements.

### 4.4.2.3. Nanotubes and Nanofibers

The first objective in studying epoxy/carbon nanotubes mechanical properties was not to achieve real improvements in stiffness and/or toughness but rather to limit the detrimental effect that incorporation of these highly conductive particles might have on them. Nowadays, progresses having been made in the dispersion strategies, more and more studies are devoted to the exploration of the possibility to induce high stiffness improvements due to the intrinsic high Young's modulus of carbon nanotubes.

The first studies often reported decreases in the strength [139] due to deficient adhesion bonding or little improvements due to difficulties in homogeneously dispersing the nanotubes [90, 140]. Gojny et al. [78] have evidenced the necessity to improve the interfacial adhesion by chemical functionalization of the tubes to succeed in improving the strength of these nanocomposites. The filling rates necessary to improve stiffness significantly were, however, higher than expected since in this work the nanotubes were still partially aggregated.

This observation confirms the first mechanical models specifically developed for polymer/CNT nanocomposites. Thostenson et al. [141] showed with a modified Halpin–Tsai model that CNT diameter is one of the key features for stiffness improvements as can be checked in Figure 4.31. Even if this model has been set up for polystyrene/CNT composites, the need to individually disperse the nanotubes (to go down from a bundle diameter to a nanotube diameter) is clearly depicted. At the same period, Fischer et al. [142]

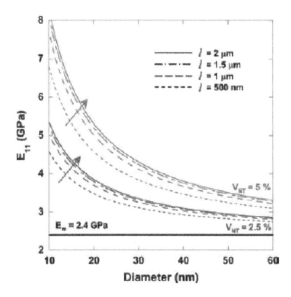

**Figure 4.31**  Influence of nanotube diameter, volume fraction, and length on the elastic properties of an aligned nanocomposite system (reproduced from [141] with permission from IOP).

used a modified Mori-Tanaka approach to evidence that nanotubes waviness in the matrix can also be the cause of moderate stiffness improvements.

Gong et al. [81] obtained a 30% increase in a glassy Young's modulus by incorporating 1 wt% of MWCNT with the surfactant dispersion process, whereas only a 10% was achieved without it proving the efficiency of the technique as well as its limitation due to the moderate enhancement observed. Almost similar improvements can be obtained with carbon nanofibers [76]. This proves that the dispersion problems are not solved at this stiffness enhancement level. Similar improvements in stiffness have also been obtained by Zhu et al. [85] with fluorinated carbon nanotubes and demonstrate the beneficial effect of functionalization, since a decrease in strength is obtained with pristine nanotubes. Gojny et al. [79] studied in detail the influence of nanotube-specific surface area, aspect ratio, volume fraction, and functionalization on the nanocomposites stiffness. The specific surface area of nanotubes has antagonist effects. It allows developing a higher interfacial surface in the nanocomposites and, therefore, promotes an efficient stress transfer but it also hinders the individualization process since it favors aggregation through particle/particle interactions. Consequently, optimal stiffness and strength improvements are not obtained with SWCNT but with DWCNT which tend less to aggregate than SWCNT and develop a higher interfacial surface than MWCNT. The positive effect of amino-functionalization of the CNTs on the efficiency of stress transfer from the epoxy matrix to the nanofiller is also demonstrated since the best improvements (+10% in strength and +15% in stiffness) are obtained with 0.5 wt% of $NH_2$-DWCNT. However, even if it does not use any solvents, and can therefore easily be upscaled to industrial level, the calendering method used does not seem to allow optimizing the individualization of nanotubes in the matrix. The difficulties obtained in individually dispersing nanotubes might be related to their aggregation process during curing underlined by Martin et al. [72]. Indeed in studies focused on mechanical properties, fewer attention is generally paid at the evolution of the dispersion during the curing process despite the efforts previously made to separate them.

Using sonication in solvent and functionalized SWCNT, Miyagawa et al. [86] obtained an interesting increase of 20% in the glassy storage modulus of an epoxy anhydride network with only 0.3 wt% nanotubes introduced. This high stiffness enhancement testifies an improvement in the homogeneity of the nanotube dispersion. Nanotubes are found to be more efficient than clay platelets, since 2.5 wt% of organoclay have to be used to reach the same nanocomposite storage modulus. However, as we try to highlight it, different nanofillers need different processing routes to reach their optimal efficiency in improving properties.

To conclude on this part, the best improvements reported have been obtained recently by Moniruzamman et al. [77] since the flexural modulus of their epoxies is increased by 17% and the flexural strength by 10% with only 0.05 wt% nanotubes introduced. The increase in strength is attributed to grafting of epoxy on the nanotubes via esterification reactions. Reaching this level of stiffness improvement is due to a more efficient process, which homogeneously and individually disperses the SWCNTs. Their strategy combines sonication in solvent and high shear mixing with a twin screw batch mixer, before and after curing agent addition, which might be the key to avoid aggregation. Indeed, Martin et al. [72] also showed that an intense shearing after introducing the curing agent was hindering CNT aggregation. Even if the strength and stiffness have been successfully increased by improving the dispersion state, most of the results obtained until now are still far from the predictions of the models based on nanotube stiffness.

### 4.4.3. Toughness

Toughening of epoxies is generally achieved through the incorporation of liquid elastomers, core–shell particles, or hyperbranched polymers, which decrease the epoxy modulus and glass transition temperature. To fight these drawbacks, toughening of epoxies has been achieved with rigid particles such as glass beads. The mechanisms involved result from a combination of interfacial debonding, void formation, and subsequent interparticle ligament shear yielding. This point has been one of the reasons that drew so much attention on toughening epoxies by nanofillers. Many questions had to be answered: will nanofillers allow toughening epoxies and if so, which kind of dispersion state is desired? Could shear yielding at the nanometer scale be induced by individual homogeneous dispersion of nanofillers? If not, what are the toughening mechanisms involved?

#### 4.4.3.1. *Spherical Nanoparticles*

Toughening of epoxies via rigid particles incorporation is generally attributed to a combination of various micromechanical mechanisms. Recent works by Lee and Yee [143, 144, 145, 146] explain that fracture surface observations of glass beads/epoxy reveal the occurrence of debonding/diffuse shear yielding, step formation, and micro shear banding. The latter deformation mechanism is considered by the authors as having the strongest influence in the observed toughening of the brittle epoxy network.

Ragosta et al. [7] with silica and Wetzel et al. [9] with alumina both observed that incorporating 10 wt% of nanoparticles (10–15 nm in diameter) could increase epoxy toughness ($K_{Ic}$ stress intensity factor) from 0.5 to 1.2 MPa m$^{1/2}$, whereas with micron-sized silica Nakamura et al. [130] needed 55 wt% to achieve such relative toughness improvements. The very comprehensive investigation of Wetzel et al. [9] reveals increases in $K_{Ic}$ and $G_{Ic}$ (critical energy release) values as well as a decrease in crack propagation rate. Many deformation mechanisms likely to occur have been identified from fracture micrographs (Figure 4.32). A combination of toughening mechanisms is invoked to explain these improvements. Crack tip blunting, which indicates plastic deformation of the epoxy matrix, is identified thanks to critical crack opening displacement measurements. The fracture surface mean roughness (measured by AFM) correlates to $G_{Ic}$ improvements and implies crack deflection to occur. They also indicate that debonding is unlikely to participate in the toughening of the epoxy due to the small dimension of fillers.

To our knowledge, the toughness of epoxy networks containing dangling POSS units has not been studied until now. Concerning networks in which POSS are used as crosslinks, the increase in crosslink density logically leads to lower toughness than with a diepoxy monomer. Choi et al. [93] observed it to fall from 0.8 to 0.5 MPa m$^{1/2}$.

#### 4.4.3.2. *Nanoplatelets*

Pioneering work in toughness improvement of glassy epoxy networks by organoclay has been performed by Zilg et al. [28]. The stress intensity factor ($K_{Ic}$) values they measured on epoxy/anhydride nanocomposites increased from 0.7 to 1.2 MPa m$^{0.5}$ with 10 wt% of organoclay. This toughness enhancement increases with clay loading and in some cases stiffness improvements were also recorded. Interestingly, this ideal situation of simultaneous stiffness/toughness improvement was not achieved with exfoliated (considering exfoliation as intercalated at very high distances) morphologies but with dispersed anisotropic intercalated clay particles. The same observations have also been made on various epoxy/amine networks [128, 137]. The reinforcing effect of intercalated clay

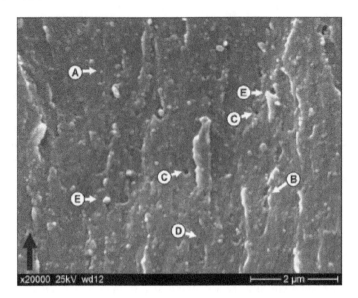

**Figure 4.32** Fracture surface of epoxy/Al2O3 nanocomposite taken in the "plastic zone" close to the initial crack. Features indicate (cautiously) (A) an apparent agglomerate, (B) plastic deformation, (C) debonding, (D) shear banding, (E) crack bridging (reproduced from [9] with permission from Elsevier).

particles observed by these authors cannot be systematically attributed to the decrease in $T_g$ they observe, since toughness improvements have also been obtained without any decrease in the glass transition temperature [47, 114, 126]. The mechanism first invoked by Zerda et al. [136] to explain epoxy toughening with these morphologies is the dissipation of energy during the crack propagation by additional surface creation when passing through intercalated clay particles. Miyagawa et al. [147] observed a correlation between $G_{Ic}$ and fracture surface roughness as Wetzel et al. [9] with Al$_2$O$_3$ spherical particles. It is also noted that intercalation morphologies lead to higher toughening than exfoliated ones confirming the previous observations. Ratna et al. [126] proposed on the basis of SEM observations that the stress concentrating nanoparticles could produce nanovoids and initiate shear yielding of the interlayer epoxy network at the tip of the propagating crack. On the other hand, Kinloch et al. [148] supposed that crack deflection was the main mechanism involved in the toughening procedure. The reinforcement they measured was lower than what is obtained with classical mica/epoxy micro composite, and the toughening was shown to decrease as platelets exfoliation increased.

The highest improvements in toughness at low clay filling rates have been obtained in epoxy/clay nanocomposites when solvents were used during the process to facilitate clay dispersion via a slurry method [47, 56, 149]. Using the slurry technique with organoclays and high pressure to process their nanocomposite, Liu et al. [56] raised the stress intensity factor from 0.75 to 1.3 MPa m$^{0.5}$ with only 1.5 phr of clay introduced. A comparison with a classical processing method reveals in Figure 4.33 the need for a good dispersion at the micron scale. Moreover $G_{Ic}$ values are improved by 300%, whereas for instance Brunner et al. [137] obtained only a 20% increase with conventionally processed nanocomposites containing 10% of organoclay.

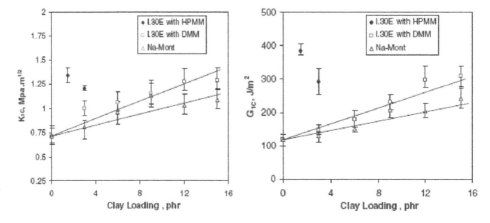

**Figure 4.33** Toughness improvement of high-pressure slurry processed ODA–MMT epoxy nanocomposites (I.30E with HPMM), swelling/melt processed ODA–MMT nanocomposites (I.30E with DMM) and Na–MMT composites (reproduced from [56] with permission from Elsevier).

Recently, Wang et al. [47] proposed some platelet nanocomposite specific toughening mechanisms. The reinforcing effect is explained by a more homogeneous dispersion at all scale levels (few clusters and clay-free epoxy zones) obtained with the slurry processing method. This good dispersion allows the formation of a large number of microcracks between clay platelets and an increase in the fracture area due to crack deflection. The clay tactoids are composed of few parallel platelets with an interlayer distance superior to WAXS detection limits. This explanation is schematically presented in Figure 4.34.

The clay tactoids act as stress concentrators upon loading. According to them, interlaminar debonding takes place since the interlayer strength is lower than the interfacial bonding strength and cohesive strength of epoxy matrix. This debonding leads to the formation of microcracks which can extend in length and even merge with further loading. When a macrocrack finally propagates through those clay tactoids, it is deflected by the presence of randomly oriented microcracks on its path and thus crack propagation is hindered. It progresses by splitting clay layers, merging microcracks, and breaking matrix ligaments between two weak points. The overall propagation path is very tortuous and the fracture surface area is considerably increased. No plastic deformation mechanisms have been observed in this study. Such a toughening mechanism requires a homogeneous random dispersion of clay tactoids through the all sample to be efficient. In a very recent work, Wang et al. [149] observed an optimal clay filling of 3%. According to them, beyond this filling level, clay aggregation in bigger tactoids would be impossible to avoid. The toughening mechanism at higher clay loading could thus be more dominated by clay clusters since they observed tail structures (generally attributed to the meeting of two crack fronts getting round a particle) on SEM micrographs. However, better comprehension of the mechanisms involved is still going on since Kinloch et al. [150] have recently observed (as Wetzel et al. [9] did on alumina nanoparticles reinforced epoxy) that the crack deflection mechanism was insufficient to explain the increases observed on epoxy nanocomposite toughness. They also noticed the occurrence of plastic deformation around clay particles thanks to SEM fracture surface observations.

Considering the few studies on epoxy/clay nanocomposites that were focused on careful toughness measurement, it will be easily understood that such studies performed with other

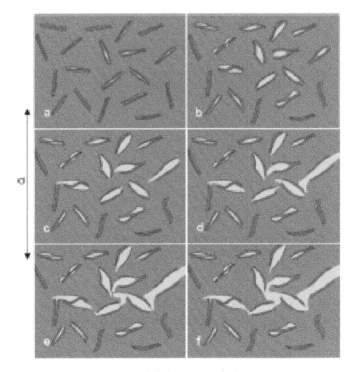

**Figure 4.34** Schematic illustration of crack initiation and propagation processes in silane–MMT epoxy nanocomposites (reproduced from [47] with permission from ACS).

nanoplatelets are pretty rare up to now. Boo et al. [61] did not see any reinforcing effect with $\alpha$-ZrP incorporation. The strong interface between amino-modified $\alpha$-ZrP and epoxy is thought to be responsible for this since neither crack deflection nor platelet debonding has been observed on fracture micrographs, the crack propagating through the material breaking down the platelets in halves.

Even if the toughening effects observed are lower than what can be obtained when incorporating liquid elastomers glass beads or hyperbranched polymers in epoxy networks, organoclay incorporation with partially intercalated structure leads to interesting improvements in both stiffness and toughness at comparatively low filling rates and without necessarily decreasing the high $T_g$ of these materials.

### 4.4.3.3. *Nanotubes and Nanofibers*

Studies on toughening epoxies with CNTs are very new. Gojny et al. [78] observed slight toughness improvements at low filling rates. Since the improvement obtained with CNTs is close to what gives carbon black at the same filling rate, it is concluded in that case that crack deflection by nanotubes aggregates is the main toughening mechanism. An optimal 0.82 MPa m$^{1/2}$K$_{Ic}$ value is obtained with 1% amino functionalized DWCNT. In a consecutive study [79], the toughening of epoxy has been improved since they obtained a fracture toughness of 0.93 MPa m$^{1/2}$ with 0.5 wt% amino-functionalized DWCNTs using exactly the same process. The optimum in toughness improvement observed at 0.5 wt% when comparing with the previous work is, however, not discussed. Nevertheless, they compare

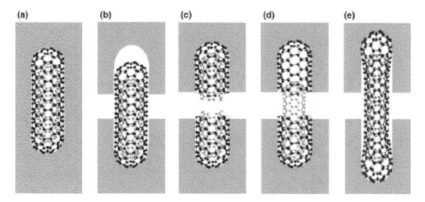

**Figure 4.35** Schematic description of possible fracture mechanisms of CNTs: (a) initial stage, (b) pull-out caused by CNT/matrix debonding in case of weak interfacial adhesion, (c) rupture of CNT-strong interfacial adhesion in combination with extensive and fast local deformation, (d) telescopic pull-out – fracture of the outer layer due to strong interfacial bonding and pull out of the inner tube, (e) bridging and partial debonding of the interface – local bonding to the matrix enables crack bridging and interfacial failure in the nonbonded regions(reproduced from [79] with permission from Elsevier).

the efficiency in toughening epoxy of SWCNTs, DWCNTs, MWCNTs (functionalized or not), and carbon black spherical particles. Micromechanical toughening mechanisms such as plastic deformation of the matrix or crack deflection are observed for all the nanofillers regardless of their shape. An additional contribution to toughening is observed with nanotubes, which is attributed to nanomechanical mechanisms and particularly crack-bridging. A proposal of the various mechanisms that might be observed depending on the nanotubes shape and on its adhesion with the matrix is presented in Figure 4.35. The last drawing presents the case the authors suppose to succeed. Functionalization of the nanotube might occur preferentially at its ends due to the process used. It results that the nanotube is covalently bonded to the matrix solely by its ends, and therefore can combine partial debonding and crack-bridging, therefore hindering the crack-opening mechanism.

Thostenson et al. [80] improved the toughness of an epoxy matrix by 75% with only 0.02 wt% of MWCNT. In their case, an optimum is also observed in the toughening of the network (Figure 4.36). They admit that interfacial debonding and pull-out of the nanotubes is very likely to occur. However, they consider crack deflection due to interaction of the crack front with agglomerated nanotubes to be more effective in toughening the epoxy matrix. Indeed the higher toughness improvements are obtained with nanocomposites in which the nanotubes dispersion is not optimal, and for which more tail-structures are observed on fracture surface micrographs. These conclusions are coherent with those made on nanoplatelet/epoxy nanocomposites. If individual dispersion of nanofillers is a key feature for stiffness improvements, micromechanical toughening mechanisms still prevail in toughening epoxies.

### 4.4.4.  Thermal Properties

#### 4.4.4.1.  *Thermal Stability and Fire Resistance*
Thermal stability and fire-retardant properties are necessary for epoxies since they are generally devoted to high-temperature technological applications. Despite their efficiency,

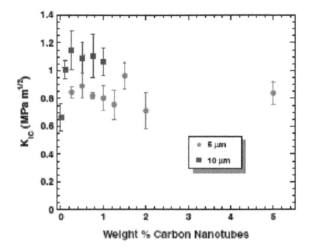

**Figure 4.36** Stress intensity factor ($K_{Ic}$) as a function of CNT loading in MWCNT–epoxy nanocomposites obtained by calendering with a 5- or 10-μm gap between the rolls (reproduced from [80] with permission from Elsevier).

halogen-containing fire retardants will have to be substituted due to the high toxicity of the gases they release upon combustion. This is why the effects of nanofillers and particularly nanoplatelets on these properties have been investigated.

*Spherical Nanoparticles*    Chia et al. [12] obtained an increase in the char residue at 800°C under nitrogen atmosphere through the dispersion of an inorganic phosphorus and silicon-rich nanophase in the epoxy network. The increased char residue is approximately twice the introduced inorganic content, which testifies an improvement in thermal degradation behavior brought by the sol/gel formed nanoparticles. However, due to incomplete condensation of inorganic precursors during nanocomposite synthesis, the degradation starts at lower temperatures than the unmodified network and even the maximum degradation temperature is decreased.

Whereas *in-situ* formation of silica nanodomains by sol/gel process leads to a decrease in the nanocomposites thermal stability, due to incomplete silanol condensation, POSS monomer units allow improvements in both thermal stability and oxidation behavior. Choi et al. [121] reported a maximum degradation temperature of octaglycidyl-POSS-based networks approximately 450°C (under nitrogen atmosphere), whereas it is approximately 400°C when the network is based on DGEBA. The char residue is not much increased since it is of 40% for the network containing around 20 wt% of silica-core particles, whereas it is 15% for an epoxy/amine network. Improvements of the thermal oxidation behavior have also been observed for POSS-based networks [15, 19, 94]. Strachota et al. [94] reported a huge increase (>150°C) in the 50 wt% degradation temperature compared to their epoxy/amine reference network (Figure 4.37) when POSS are incorporated as dangling units on DGEBA oligomer and form lamellar crystalline domains allowing physical network formation. Amorphous POSS domains or POSS used as crosslinks also increased the thermal oxidation behavior but less than the latter one. In all cases, the final residue corresponds to the quantity of silica cages introduced in the material.

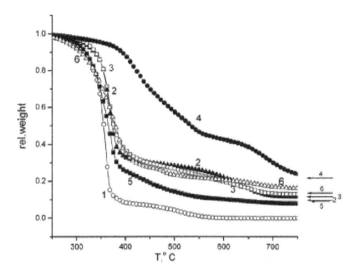

**Figure 4.37** TGA thermograms in air: (1) DGEBA/D2000, (2) DGEBA/D2000 + hepta(phenyl) monoglycidyl POSS, (3) POSS-grafted DGEBA monomer/D2000 network, (4) POSS-grafted DGEBA oligomer/D2000 network, (5) DGEBA/D2000 + hepta(octyl)monoglycidylPOSS, and (6) octaglycidylPOSS/D2000 network (reproduced from [94] with permission from ACS).

*Nanoplatelets* The efficiency of nanoplatelets in improving flame retardancy at much lower filling level than the conventional halogen-free flame retardants has been clearly evidenced with thermoplastic polymers [151], but as for the other physical properties their efficiency in epoxies strongly depends on the processing strategies used to disperse them and on the morphologies obtained.

The thermal degradation behavior of epoxy networks has recently been well described by Levchik et al. [152]. In nitrogen atmosphere, it proceeds in two steps. First, around 300°C, water elimination leads to carbon double bonds formation. This induces the breaking of carbon–oxygen weakest bonds, which leads to rearrangements in strongest structures by cyclization with moderate volatilization of small molecules. The second step of this degradation process is the complete breakdown of the network occurring around 400°C. A very low charred residue remains after this step due to poor recombination of degraded species during the network decomposition. In oxidative atmosphere, the degradation process is roughly the same, which witnesses the good resistance of epoxies to oxidation. However charring is much favored in air atmosphere since it is catalyzed by oxygen. It can reach 20% according to Camino et al. [45]. The combustion of the char takes place at above 500°C.

In nitrogen atmosphere, there are minor effects of the incorporation of clay platelets in epoxy networks. Gu et al. [153] observed a slight increase in thermal stability at low clay loadings, but this trend was inverted at higher loadings. Becker et al. [154] confirmed that the degradation begins at slightly lower temperatures (5–10°C) and that this effect is increasing at higher filling rates. They attribute this effect to the lower stability of alkylammonium ions or to the changes in matrix crosslink density that might have been caused in the composite during its polymerization in the presence of organoclay.

In a recent study, Camino et al. [45] do not mention any significant effect of clay platelets on the thermal stability in nitrogen atmosphere. However, they clearly

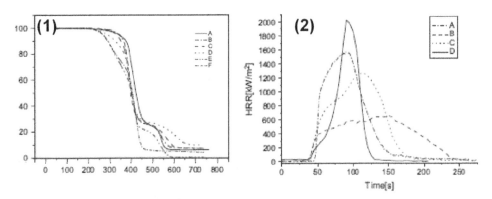

**Figure 4.38** (1) TGA curves under air flow of epoxy nanocomposites containing 10wt% of (A) Na–MMT, (B) dimethyl hydrogenated tallowalkyl (2ethylhexyl)–MMT, (C) MT2Et–MMT, and (D) ODA–MMT. TGA curves of the unfilled network under air (E) and $N_2$ (F). (2) Cone calorimeter analysis of the nanocomposites containing 10wt % of (A) dimethyl hydrogenated tallowalkyl (2ethylhexyl)-MMT, (B) MT2Et-MMT, (C) ODA-MMT, and of the unfilled network (D) (reproduced from [45] with permission from Elsevier).

demonstrate its drastic effect in oxidative atmosphere. The onset of degradation succeeds at 50 to 100°C higher in the presence of organoclay (Figure 4.38). This effect on the first step of the degradation process does not seem to be linked to the state of dispersion of the platelets since the best improvement is obtained with an unmodified clay microcomposite. The effect of organoclay incorporation in epoxy on the second step of the degradation process is the opposite with a lowering of the maximum degradation rate temperature. This is attributed to the catalytic activity of the protonated montmorillonite (resulting from Hoffman decomposition of alkylammonium ions) on the degradation process of the polymer. This effect is also linked to the exfoliation of the platelets since the better dispersed fillers lead to the lower degradation temperatures. According to them, organoclay does not modify the charring process of the matrix but a slight increase of the char degradation temperature is observed. This last effect is attributed to the oxygen sheltering effect brought by the exfoliation of clay in the matrix and is not observed with intercalated morphologies.

Concerning the fire-retardant properties of epoxy/clay nanocomposites, Camino et al. [45] obtained the best results with exfoliated morphologies. The heat release rate was reduced by 68% with 10% of MT2Et–MMT (Figure 4.38). Exfoliation is thought to promote the formation of a protective ceramic skin on the nanocomposite surface. This physical improvement of fire-retardant properties is increased by the ability of the organic modifier to catalyze the charring process during combustion without oxygen. On the other hand, the time to ignition of the epoxy matrix is not increased by organoclay incorporation probably due to their low thermal stability.

Zammarano et al. [30] have succeeded in increasing this time to ignition, obtaining self-extinguishing epoxy nanocomposites, but through the exfoliation of LDHs. These layered crystals with brucite-like octahedral structure (positively charged by cation substitutions within the structure) are organically modified by anionic exchange. The acidic molecules used to compatibilize those LDH decrease the thermal stability promoting the dehydration of the epoxy polymer. It can be concluded that acidity is responsible for this, since Hsueh et al. [29] observed a small increase in thermal stability when using nonacidic molecules to compatibilize LDHs. This increase is attributed to the reduced diffusion of volatile degradation products due to LDH barrier effect. Zammarano et al. [30] also observe catalysis of the charring reactions in nitrogen atmosphere, which is responsible for the decrease in the heat release rate (−50% with 5% of LDH) measured in cone calorimeter experiments. They also point out that these reductions can vary with the specimen thickness.

Although the results obtained by incorporating organoclays and organically modified LDH in epoxies are not directly comparables, it can be concluded that thermal stability and fire-retardant properties have to be considered separately. Thermal stability in inert atmosphere is less sensitive to nanoplatelets incorporation than in oxidative atmosphere where degradation mechanisms can be catalyzed either by the organomodifiers or by the acid sites on the layer itself. Fire-retardant properties are obtained via the formation of a protective ceramic skin by charring, which shelters the material from oxygen and helps reducing strongly the heat release rate. Self-extinguishing nanocomposites can even be obtained with exfoliated LDHs. However, as shown by Zammarano et al. [30], the results obtained with nanoplatelets are still far from what bring usual intumescent microfillers.

*Nanotubes and Nanofibers*    The thermal stability of CNT/epoxy nanocomposites has not been much studied since it is not one of the main targeted applications for these new materials. Zhou et al. [155] did not observe any significant effect of carbon nanofibers (up to 3 wt% loadings) on the thermal degradation of epoxies under nitrogen.

### 4.4.4.2. *Coefficient of Thermal Expansion*

In addition to thermal stability and fire-retardant properties, some authors have looked at the coefficient of thermal expansion (CTE) evolution after nanofillers incorporation. Indeed epoxies are widely used as coating on glass ceramic or metal substrates to bring some abrasion, photodegradation, or corrosion protection. One of the limiting points in these applications is the high CTE of epoxies compared to the substrates, which induces loss of adhesion and/or coating failure upon thermal cycling.

Kang et al. [6] showed that improved interfacial adhesion between spherical particles and the surrounding epoxy matrix could lead to significant reduction of the linear thermal expansion coefficient (CTE). Sulaiman et al. [156] recently reported that the use of octa(aminophenyl)-POSS as curing agent for high-functionality epoxy monomers could lead to the achievement of CTE values as low as 25 ppm/°C, whereas the actual incorporation of silica fillers to reach such values leads to processing difficulties. The CTE of the epoxy networks has been seen to decrease from -10% to -20% with 3 wt% of organoclay incorporated [44, 104]. This is attributed to the constrained expansion of polymer network. Koerner et al. [53] noticed a marginal effect of the dispersion state on this property, the more uniform dispersion obtained with their cryocompounding technique only slightly reduced the CTE compared to more classical dispersing strategies. The same observations and conclusions have been drawn from studies with LDHs [29] and for EG at low filling rates [32]. In the latter case however, an increase in the filling rate leads to an increase in CTE, attributed to graphite aggregation which decreases its efficiency in constraining polymer chains. It has to be noted that EG was not exfoliated in this study.

The general trend concerning nanofillers influence on CTE is that a small amount of uniformly dispersed particles leads to improvements in dimensional stability of epoxy networks.

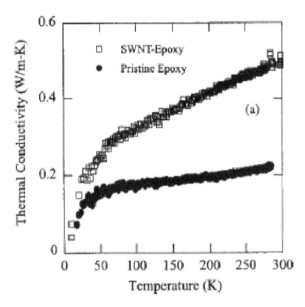

**Figure 4.39**    Thermal conductivity versus temperature of 1 wt% SWCNTs/epoxy nanocomposites (reproduced from [157] with permission from AIP).

### 4.4.4.3. *Thermal Conductivity*

Among the thermal properties, the achievement of improved thermal conductivity of epoxy nanocomposites has been studied when incorporating carbon nanotubes. The intrinsic high thermal conductivity of CNTs (estimated at 6000 W/mK for individual nanotubes) makes their dispersion in epoxies promising in order to get materials with epoxies advantages and allowing a good thermal management with a low filling rate.

Biercuk et al. [157] obtained great improvements of 125% of the thermal conductivity of epoxies at room temperature by dispersing 1 wt% of SWCNTs (Figure 4.39). This improvement is shown to be much higher than what can be obtained with vapor-grown carbon nanofibers (+45%) at the same filling rate. Consequently, the easiest percolation of these high aspect ratio nantotubes is assumed to be responsible for the measured enhancements. Noticeable improvements in thermal conductivity have also been made by Song et al. [71] with MWCNTs. They demonstrate that more aggregated morphologies were leading to reduced thermal conductivity enhancements, but it is important to precise that both morphologies led to percolation of nanotubes.

In the latest studies performed, the results obtained seem to be contradictory with the aforementioned works. Thostenson et al. [80] reported a lower increase (+60%) in an epoxy network thermal conductivity filled with 5 wt% of MWCNT. The thermal conductivity enhancement is reported to be independent of the percolation of carbon nanotubes through the epoxy matrix. In a more detailed study, Gojny et al. [66] observed a decrease in thermal conductivity with low nanotubes loadings (0.05 wt%) confirming similar observations made by Moisala et al. [158]. They demonstrate the thermal conductivity of CNT/epoxy nanocomposites to be influenced by the nanotubes aspect ratio. According to them, thermal conductivity being achieved by phonons, it can be limited due to interfacial boundary scattering, since the epoxy network is a thermal insulator. This is why the best improvements are brought by MWCNTs for which the inner layers can

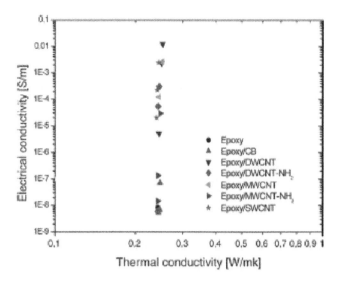

**Figure 4.40** Electrical versus thermal conductivity for a various epoxy/CNT nanocomposites (reproduced from [66] with permission from Elsevier).

act as thermal conductors without being submitted to this scattering. The SWCNTs bring the lower improvements, since their thermal conductivity is strongly hindered due to their high specific surface area. Nanotubes functionalization has also a detrimental effect since it increases the coupling of the nanotubes with the matrix. The hypothesis of thermal conductivity being independent of CNTs percolation is confirmed by plotting electrical conductivity of nanocomposites as a function of their thermal conductivity (Figure 4.40).

The improvements observed and the conclusions drawn from all these studies differ slightly, demonstrating that some progress has to be made in that field to fully understand the mechanisms engaged in increasing thermal conductivity. However, they all agree on the fact that incorporation of a small amount of carbon nanotubes improves significantly the thermal conductivity, even if the measured values are still under the predictions that can be drawn from a rule of mixture.

### 4.4.5. Electrical Conductivity

Conventional improvement in electrical conductivity of epoxies is brought by the incorporation to the matrix of micrometer scale fillers such as carbon black or carbon fibers. The filling rates of this materials can reach up to 50% in order to ensure good conductivity properties leading to an increase in composite density and sometimes to a lowering of the composites strength. This is why conductive nanofillers are of great interest since their high aspect ratios and surface areas should considerably lower the percolation threshold. Despite some recent research in this field has been performed through the incorporation of conductive nanoplatelets such as EG, the discovery of the combined high anisotropy and exceptional conductivity of carbon nanotubes has led to much numerous studies.

Even if it has been used for long in epoxies, carbon black can be considered as a nanofiller since the diameter of individual particles is around 30 nm [78]. Controlled and designed aggregation process via alternative electric field application during epoxy/amine condensation polymerization allowed Schwarz et al. [4] to obtain $10^{-2}$ S/cm conductivity value for a 1 wt% carbon black filled epoxy nanocomposite.

Li et al. [34] showed that electrical resistivity of epoxy could be lowered to $3 \times 10^3$ $\Omega$ cm after the incorporation of 2 wt% of EG. However in that first study, it is shown that in spite of higher surface areas, exfoliated morphology does not lead to the best improvements. The high aspect ratio intercalated particles might lead to lower percolation threshold and consequent higher connectivity of the percolated network at 2 wt% due to larger particle size distribution as proposed by Celzard et al. [62]. Li et al. [35] have measured the percolation threshold to be about 1 wt% (0.5 vol%) for exfoliated EG nanoplatelets. Even if it enhanced mechanical and thermal properties, a UV/ozone treatment of graphite in order to improve interfacial strength of the composite does not affect the percolation threshold. This surface treatment, however, leads to a slightly faster decrease in electrical resistivity once the threshold is reached and allows getting closer from the values obtained with the nonexfoliated morphology [34].

Since a lot of studies have been conducted on the electrical conductivity of epoxy/carbon nanotubes nanocomposites, we will try to focus here on the works that brought most interesting results and the more understanding of the mechanisms involved.

Due to the high aspect ratio of carbon nanotubes, the percolation threshold was expected to be reached with very low nanofillers loadings in the epoxy insulating matrix. Kim et al. [68] succeeded in forming a percolated network with 0.074 wt% of SWCNT using an intense sonication to disperse the nanotubes in their epoxy reactive mixture. The critical exponent $t$ of the percolated network, which represents the dimensionality of the system, has been measured at 1.3 in their study. They point out that this value of the percolation threshold is exceptionally low compared to classical values obtained from percolation theory. Since Celzard et al. [62] demonstrated that for highly anisotropic fillers, excluded volume considerations were necessary to allow good agreement between experimental values and theoretical predictions, Kim et al. [68] finally concluded that their percolation threshold value could be in agreement with a 1000 aspect ratio of carbon nanotubes. Sandler et al. [74] lowered this percolation threshold at 0.0025 wt% with MWCNT, which means with a lower aspect ratio than in the latter case. They measured the critical exponent of the percolated network to be $t = 1.2$, which concords with previously presented results. This value is, however, low for a three-dimensional network, since the values reported are more generally around 1.75 [72]. This is explained by the authors by the difference in the percolating mechanisms. Contrarily to most percolation examples reported until now, the threshold is not reached by a random percolation process, but through spontaneous aggregation of highly anisotropic particles. As previously explained, adding the curing agent to the epoxy prepolymer in which the nanotubes have been successfully individually dispersed causes an increase in the ionic force, which brings the driving force for the aggregation process. A moderate shearing of the reactive mixture as well as a low viscosity helps the percolation to proceed. A recent study by Bryning et al. [159] supports the exposed percolation mechanism. The conductivity values reported by Sandler et al. [74] in this study can reach $10^{-1}$ S/cm with 0.1 wt% loadings in nanotubes. Antistatic properties have, therefore, been successfully brought to epoxies at nanotubes loading, which should not affect their mechanical properties. The relatively low conductivity values obtained support the idea that some of the nanotubes (or of the nanotubes clusters) are covered by an insulating epoxy layer, preventing direct contact between them and leading to hopping transfer. It is also pointed out that the use of aligned nanotubes, which allows reaching a better dispersion state in the epoxy prepolymer than with entangled nanotubes, helps

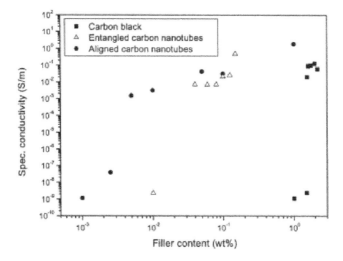

**Figure 4.41** Electrical conductivity as a function of filler content for various conductive nanofillers (reproduced from [74] with permission from Elsevier).

reaching the percolation threshold at lower rates (Figure 4.41). Three crucial steps in the achievement of percolation are evidenced in this work [74]: individualization of nanotubes in the epoxy, local cluster formation upon the addition of the amine hardener, and final aggregation to form the conductive network. Martin et al. [72] evidenced some other crucial points of this percolation process. The final conductivity of the nanocomposites is linked to the size of the clusters formed by aggregation and to the conductive network density. The bigger aggregates and the more consolidated networks lead to higher conductivity values. This has been further confirmed by the lower conductivity values obtained, when forming the percolating network through the application of an electric field [75]. The dendritic CNT network presents much more dead ends than the aggregated one previously obtained.

They also demonstrate that lower nanotubes aspect ratios (by decreasing nanotubes length) lead to lower percolation threshold (0.0021 wt% for a same filling rate) [72]. Since the opposite effect is expected based on percolation theory considerations, they attribute this phenomenon to a higher mobility of these nanotubes allowing them to aggregate more rapidly during the polymerization. In order to support this hypothesis, they compare their results to the predictions of the percolation theory taking into account the excluded volume (Figure 4.42). They justify the much lower experimental thresholds by arguing that percolation theory remains a statistical approach, which does not until now consider parameters such as diffusion of the particles in the resin (and how its viscosity affects it), or van der Waals and Coulomb interactions between the particles.

Finally, Gojny et al. [66] recently showed that even if it can help reaching a better dispersion state, the functionalization of carbon nanotubes was leading to increases in the percolation threshold. They invoke a decrease in the aspect ratio of the tubes to explain this observation, which is not consistent with the aforementioned propositions by Martin et al. [72]. Even if the full comprehension of the influence of nanotubes intrinsic properties on the percolation threshold and conductivity values is still not totally clear, all the latest studies [66, 75, 80] agree to support the idea that optimizing aggregation is the key to get the expected improvements.

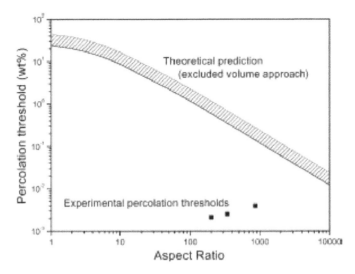

**Figure 4.42** Theoretical and experimental percolation thresholds as a function of aspect ratio (reproduced from [72] with permission from Elsevier).

### 4.4.6. Gas Permeability

From the very beginning in the studies on polymer/clay nanocomposites, montmorillonite has been considered to be able to reduce strongly the polymer permeability due to the high tortuosity in the diffusion path that would bring its high aspect ratio. However, permeability studies on epoxy/clay nanocomposites are not so numerous although it might be of great interest for specific applications such as durable liquid hydrogen tanks, or to improve the durability of coatings (or composites) through solvent or water uptake increased resistances. For instance, reducing water diffusion through epoxy matrixes is of great interest since it is the major cause of corrosion in coated electronic components.

Ogasawara et al. [160] have measured the effect of organoclay incorporation on the barrier properties to helium diffusion of an epoxy system. The diffusion coefficient is almost divided by 10 with 5 wt% of organoclay. The use of different organoclay does not show any role played by the alkylammonium used, but the effect of the filling rate is clearly evidenced. Using Hatta–Taya theory to fit their measures, the aspect ratio of clay platelets is estimated to be around 0.001. Exfoliation is necessary to improve gas barrier properties due to the huge impact of the aspect ratio on the diffusion coefficient. According to them, clay nanoplatelets are effective in decreasing the permeability and they reveal their dispersion in the epoxy matrix to be more efficient than fiber-like or spherical particles. They also notice an increase in He solubility in the nanocomposites, which is attributed to the existence of nanovoids or nanoflaws in the matrix in the vicinity of the nanoplatelets. Strong decreases in oxygen, nitrogen, and water permeabilities have also been reported by Ye et al. [161] in siloxane-modified epoxy resins membranes after the dispersion of tetradecyltrimethylammonium-exchanged montmorillonite.

Moisture barrier properties have been evaluated on epoxy/clay nanocomposites. A decrease in water uptake in the presence of organoclay is generally observed. Kim et al. [44] observed a 39% decrease in the equilibrium water uptake with an intercalated nanocomposite containing 5 wt% of ODA-MMT with an 80 Å interplatelet distance. Concerning the diffusion rate, Becker et al. [154] did not observe any change with increasing filling rate,

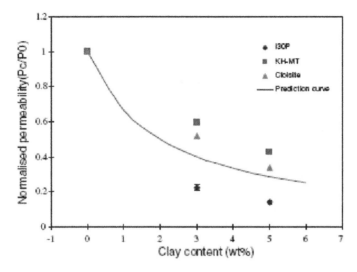

**Figure 4.43** Normalized moisture permeability for various org-MMT-based epoxy nanocomposites and tortuous path model predictions with a 200 aspect ratio (reproduced from [44] with permission from Elsevier).

while Kim et al. [44], comparing different organoclays at the same filling rate, noticed a decrease when the interplatelet distance increases. They also measured a strong decrease in the moisture permeability. Figure 4.43 presents a comparison of their measures with the prediction of a simple model taking into account the volume fraction and the aspect ratio of sheet-shaped barriers. With an aspect ratio of 200, the decrease in permeability obtained with the intercalated nanocomposite is higher than the model's prediction.

Great reductions in the diffusion rates have also been reported by Chen et al. [102], who observed an 84% decrease in the diffusion rate of acetone in an exfoliated epoxy/clay nanocomposite. Even if the measured interplatelet distances decrease from more than 200 to 150 Å with increasing the filling rate from 1 to 6 wt%, the diffusion rate is reduced with increasing organoclay content in the nanocomposite. They also point out that at the same filling rate, the better the platelets are dispersed, the more the diffusion rate decreases. Chen et al. [102] also studied the corrosion resistance of these nanocomposites. The exfoliation of 2.5% of organoclay has been shown to slightly decrease the corrosion current, which might be due to the barrier properties brought by the clay. However, the low permeability of the matrix used in this study makes it difficult to ensure a precise discrimination of the effect of the clay on the anticorrosive properties. Yeh et al. [161] recently confirmed these observations, since they reported a great increase in anticorrosion properties of siloxane-modified epoxy resins by studying the variations in corrosion potential, polarization resistance, corrosion current, and impedance spectroscopy due to organoclay dispersion.

In all these studies, the increase in the diffusion path of molecules is generally invoked to explain the barrier effects evidenced.

## 4.5. Conclusion

This review of recent advances in epoxy nanocomposites evidences that physical properties enhancement is guided by the nanofiller state of dispersion at various levels and that specific morphologies are required depending on the improvement targeted. Large progress has been made recently in the dispersion and compatibilizing of nanofillers in epoxy and has allowed improving the nanocomposites properties. The potentiality of obtaining epoxy-based materials with enhanced conductivity without decreasing mechanical strength, improved toughness without stiffness reductions, increased thermal dimensional stability, anticorrosion, barrier, or fire-retardant properties at very low filling levels has been clearly depicted.

However, theoretical predictions generally afford more potentiality to these nanocomposites. Some actual limitations are due to the multiplicity of parameters, which can be tuned to elaborate these nanocomposites. Structure–properties relationships are not easily defined and some progress has still to be done in understanding the interphase properties at the nanometer level in materials presenting such an important interfacial surface area. Tailoring the morphologies requires a better understanding of the relations between the compatibilization strategies and the final state of dispersion obtained during network formation. Researchers face new challenges at each step of the elaboration, characterization, and analysis process due to the change in the reinforcement scale. Another limitation arises from the difficulty to get pertinent parameters from materials whose properties are governed by interactions at the nanometer scale, which is not directly accessible by conventional analysis techniques. The full understanding of the elaboration–structure relationships in epoxy nanocomposites might have to wait for technological advances in characterization techniques. A good example is the potentiality of high-resolution electron microscopy used by Drummy et al. [162] recently, which allowed characterizing epoxy nanocomposites morphology at a subnanometer length scale.

Simultaneously to this research effort toward establishing elaboration–structure–properties relationships, advances are also made in the application domains of epoxy nanocomposites by introducing nanofillers in ternary systems. Nanofillers have been thus incorporated in fiber-reinforced composites [155], epoxy rubber toughened systems [105], hyperbranched toughened epoxies [163], core–shell containing epoxies [164], interpenetrated networks [165], or thermoplastic/epoxy blends [166]. These innovative materials have not been presented in this chpater since comparisons to fully understand the potentiality of these materials are made difficult by the few numbers of studies.

These third-generation composite materials and the look for synergistic effects between nanofillers incorporation and classical epoxy toughening or blending approaches will probably be the field of larger research in the forthcoming years.

## References

1. Thostenson, E. T.; Li, C.; Chou, T.-W. Compos Sci Technol 2005, 65, 491–516.
2. Pukanszky, B. Eur Polym J 2005, 41, 645–662.
3. Schueler, R.; Petermann, J.; Schulte, K.; Wentzel, H.-P. J Appl Polym Sci 1997, 63, 1741–1746.
4. Schwarz, M.-K.; Bauhofer, W.; Schulte, K. Polymer 2002, 43, 3079–3082.
5. Kickelbick, G. Prog Polym Sci 2003, 28, 83–114.
6. Kang, S.; Hong, S. I.; Choe, C. R.; Park, M.; Rim, S.; Kim, J. Polymer 2001, 42, 879–887.
7. Ragosta, G.; Abbate, M.; Musto, P.; Scarinzi, G.; Mascia, L. Polymer 2005, 46, 10506–10516.
8. Naganuma, T.; Kagawa, Y. Compos Sci Technol 2002, 62, 1187–1189.
9. Wetzel, B.; Rosso, P.; Haupert, F.; Friedrich, K. Eng Fract Mech 2006, 73, 2375–2398.
10. Zheng, W.; Wong, S.-C. Compos Sci Technol 2003, 63, 225–235.
11. Matejka, L.; Dusek, K.; Plestil, J.; Kriz, J.; Lednicky, F. Polymer 1999, 40, 171–181.
12. Chiang, C.-L.; Ma, C.-C. M. Eur Polym J 2002, 38, 2219–2224.

13. Matejka, L.; Strachota, A.; Plestil, J.; Whelan, P.; Steinhart, M.; Slouf, M. Macromolecules 2004, 37, 9449–9456.
14. Kickelbick, G.; Schubert, U. Monatsh Chem 2001, V132, 13–30.
15. Liu, Y.; Zheng, S.; Nie, K. Polymer 2005, 46, 12016–12025.
16. Lee, A.; Lichtenhan, J. D. Macromolecules 1998, 31, 4970–4974.
17. Laine, R. M.; Choi, J.; Lee, I. Adv Mater 2001, 13, 800–803.
18. Pellice, S. A.; Fasce, D. P.; Williams, R. J. J. J Polym Sci Polym Phys 2003, 41, 1451–1461.
19. Ni, Y.; Zheng, S.; Nie, K. Polymer 2004, 45, 5557–5568.
20. Choi, J.; Kim, S. G.; Laine, R. M. Macromolecules 2004, 37, 99–109.
21. Abad, M. J.; Barral, L.; Fasce, D. P.; Williams, R. J. J. Macromolecules 2003, 36, 3128–3135.
22. Schubert, U. Chem Mater 2001, 13, 3487–3494.
23. Alexandre, M.; Dubois, P. Mater Sci Eng R 2000, 28, 1–63.
24. Sinha Ray, S.; Okamoto, M. Prog Polym Sci 2003, 28, 1539–1641.
25. Tjong, S. C. Mater Sci Eng R 2006, 53, 73–197.
26. LeBaron, P. C.; Wang, Z.; Pinnavaia, T. J. Appl Clay Sci 1999, 15, 11–29.
27. Pinnavaia, T. J.; Beall, G. W. Polym. -Clay Nanocompos 2001.
28. Zilg, C.; Mülhaupt, R.; Finter, J. Macromol Chem Phys 1999, 200, 661–670.
29. Hsueh, H.-B.; Chen, C.-Y. Polymer 2003, 44, 5275–5283.
30. Zammarano, M.; Franceschi, M.; Bellayer, S.; Gilman, J. W.; Meriani, S. Polymer 2005, 46, 9314–9328.
31. Hibino, T.; Jones, W. J Mater Chem 2001, 11, 1321–1323.
32. Yasmin, A.; Daniel, I. M. Polymer 2004, 45, 8211–8219.
33. Yasmin, A.; Luo, J.-J.; Daniel, I. M. Compos Sci Technol 2006, 66, 1182–1189.
34. Li, J.; Kim, J.-K.; Lung Sham, M. Scripta Mater 2005, 53, 235–240.
35. Li, J.; Sham, M. L.; Kim, J.-K.; Marom, G. Compos Sci Technol, in press, Corrected Proof.
36. Lan, T.; Kaviratna, P. D.; Pinnavaia, T. J. Chem Mater 1995, 7, 2144–2150.
37. Lan, T.; Pinnavaia, T. J. Chem Mater 1994, 6, 2216–2219.
38. Lan, T.; Kaviratna, P. D.; Pinnavaia, T. J. Chem Mater 1994, 6, 573–575.
39. Kelly, P.; Akelah, A.; Qutubuddin, S.; Moet, A. J Mater Sci 1994, 29, 2274–2280.
40. Patzko, A.; Dekany, I. Colloid Surface 1993, 71, 299–307.
41. Le Pluart, L.; Duchet, J.; Sautereau, H.; Gerard, J.-F. J Adhesion 2002, 78, 645–662.
42. Sinha Ray, S.; Okamoto, K.; Okamoto, M. Macromolecules 2003, 36, 2355–2367.
43. Wang, M. S.; Pinnavaia, T. J. Chem Mater 1994, 6, 468–474.
44. Kim, J.-K.; Hu, C.; Woo, R. S. C.; Sham, M.-L. Compos Sci Technol 2005, 65, 805–813.
45. Camino, G.; Tartaglione, G.; Frache, A.; Manferti, C.; Costa, G. Polym Degrad Stab 2005, 90, 354–362.
46. Messersmith, P. B.; Giannelis, E. P. Chem Mater 1994, 6, 1719–1725.
47. Wang, K.; Chen, L.; Wu, J.; Toh, M. L.; He, C.; Yee, A. F. Macromolecules 2005, 38, 788–800.
48. Wang, Y.-R.; Wang, S.-F.; Chang, L.-C. Appl Clay Sci 2006, 33, 73–77.
49. Carrado, K. A.; Xu, L.; Csencsits, R.; Muntean, J. V. Chem Mater 2001, 13, 3766–3773.
50. Xue, S.; Reinholdt, M.; Pinnavaia, T. J. Polymer 2006, 47, 3344–3350.
51. Liu, H.; Zhang, W.; Zheng, S. Polymer 2005, 46, 157–165.
52. Yasmin, A.; Abot, J. L.; Daniel, I. M. Scripta Mater 2003, 49, 81–86.
53. Koerner, H.; Misra, D.; Tan, A.; Drummy, L.; Mirau, P.; Vaia, R. Polymer 2006, 47, 3426–3435.
54. Brown, J. M.; Curliss, D.; Vaia, R. A. Chem Mater 2000, 12, 3376–3384.
55. Luo, J.-J.; Daniel, I. M. Compos Sci Technol 2003, 63, 1607–1616.
56. Liu, W.; Hoa, S. V.; Pugh, M. Compos Sci Technol 2005, 65, 307–316.
57. Wang, K.; Wang, L.; Wu, J.; Chen, L.; He, C. Langmuir 2005, 21, 3613–3618.
58. Chen, B.; Liu, J.; Chen, H.; Wu, J. Chem Mater 2004, 16, 4864–4866.
59. Sue, H.-J.; Gam, K. T.; Bestaoui, N.; Clearfield, A.; Miyamoto, M.; Miyatake, N. Acta Mater 2004, 52, 2239–2250.
60. Sue, H.-J.; Gam, K. T.; Bestaoui, N.; Spurr, N.; Clearfield, A. Chem Mater 2004, 16, 242–249.
61. Boo, W. J.; Sun, L. Y.; Liu, J.; Clearfield, A.; Sue, H.-J.; Mullins, M. J.; Pham, H. Compos Sci Technol 2007, 67, 262–269.
62. Celzard, A.; McRae, E.; Deleuze, C.; Dufort, M.; Furdin, G.; Maréché, J. F. Phys Rev B 1996, 53, 6209–6214.
63. Ijima, S. Nature 1991, 354, 56.
64. Thostenson, E. T.; Ren, Z.; Chou, T.-W. Compos Sci Technol 2001, 61, 1899–1912.
65. Xie, X.-L.; Mai, Y.-W.; Zhou, X.-P. Mater Sci Eng R 2005, 49, 89–112.

66. Gojny, F. H.; Wichmann, M. H. G.; Fiedler, B.; Kinloch, I. A.; Bauhofer, W.; Windle, A. H.; Schulte, K. Polymer 2006, 47, 2036–2045.
67. Peigney, A.; Laurent, C.; Flahaut, E.; Bacsa, R. R.; Rousset, A. Carbon 2001, 39, 507–514.
68. Kim, B.; Lee, J.; Yu, I. J Appl Phys 2003, 94, 6724–6728.
69. Lu, K. L.; Lago, R. M.; Chen, Y. K.; Green, M. L. H.; Harris, P. J. F.; Tsang, S. C. Carbon 1996, 34, 814–816.
70. Lau, K.-t.; Lu, M.; Chun-ki Lam; Cheung, H.-y.; Sheng, F.-L.; Li, H.-L. Compos Sci Technol 2005, 65, 719–725.
71. Song, Y. S.; Youn, J. R. Carbon 2005, 43, 1378–1385.
72. Martin, C. A.; Sandler, J. K. W.; Shaffer, M. S. P.; Schwarz, M.-K.; Bauhofer, W.; Schulte, K.; Windle, A. H. Compos Sci Technol 2004, 64, 2309–2316.
73. Sandler, J.; Shaffer, M. S. P.; Prasse, T.; Bauhofer, W.; Schulte, K.; Windle, A. H. Polymer 1999, 40, 5967–5971.
74. Sandler, J. K. W.; Kirk, J. E.; Kinloch, I. A.; Shaffer, M. S. P.; Windle, A. H. Polymer 2003, 44, 5893–5899.
75. Martin, C. A.; Sandler, J. K. W.; Windle, A. H.; Schwarz, M.-K.; Bauhofer, W.; Schulte, K.; Shaffer, M. S. P. Polymer 2005, 46, 877–886.
76. Zhou, Y.; Pervin, F.; Rangari, V. K.; Jeelani, S. Mater Sci Eng A 2006, 426, 221–228.
77. Moniruzzaman, M.; Du, F.; Romero, N.; Winey, K. I. Polymer 2006, 47, 293–298.
78. Gojny, F. H.; Wichmann, M. H. G.; Kopke, U.; Fiedler, B.; Schulte, K. Compos Sci Technol 2004, 64, 2363–2371.
79. Gojny, F. H.; Wichmann, M. H. G.; Fiedler, B.; Schulte, K. Compos Sci Technol 2005, 65, 2300–2313.
80. Thostenson, E. T.; Chou, T.-W. Carbon 2006, 44, 3022–3029.
81. Gong, X.; Liu, J.; Baskaran, S.; Voise, R. D.; Young, J. S. Chem Mater 2000, 12, 1049–1052.
82. Sinnott, S. B. J Nanosci Nanotechnol 2002, 2, 113–123.
83. Bahr, J. L.; Tour, J. M. J Mater Chem 2002, 12, 1952–1958.
84. Dyke, C. A.; Tour, J. M. J. Phys. Chem. A 2004, 108, 11151–11159.
85. Zhu, J.; Kim, J.; Peng, H.; Margrave, J. L.; Khabashesku, V. N.; Barrera, E. V. Nano Lett 2003, 3, 1107–1113.
86. Miyagawa, H.; Rich, M. J.; Drzal, L. T. Thermochim Acta 2006, 442, 67–73.
87. Miyagawa, H.; Drzal, L. T. Polymer 2004, 45, 5163–5170.
88. Park, S.-J.; Jeong, H.-J.; Nah, C. Mater Sci Eng A 2004, 385, 13–16.
89. Gojny, F. H.; Nastalczyk, J.; Roslaniec, Z.; Schulte, K. Chem Phys Lett 2003, 370, 820–824.
90. Liu, L.; Wagner, H. D. Compos Sci Technol 2005, 65, 1861–1868.
91. Eitan, A.; Jiang, K.; Dukes, D.; Andrews, R.; Schadler, L. S. Chem Mater 2003, 15, 3198–3201.
92. Pittman, C. U.; Li, G. Z.; Ni, H. L. Macromolecul Symp 2003, 196, 301–325.
93. Choi, J.; Yee, A. F.; Laine, R. M. Macromolecules 2003, 36, 5666–5682.
94. Strachota, A.; Kroutilova, I.; Kovarova, J.; Matejka, L. Macromolecules 2004, 37, 9457–9464.
95. Li, G. Z.; Wang, L.; Toghiani, H.; Daulton, T. L.; Koyama, K.; Pittman, C. U. Macromolecules 2001, 34, 8686–8693.
96. Lan, T.; Kaviratna, P. D.; Pinnavaia, T. J. J Phys Chem Solids 1996, 57, 1005–1010.
97. Chen, J. h. S.; Poliks, M. D.; Ober, C. K.; Zhang, Y.; Wiesner, U.; Giannelis, E. Polymer 2002, 43, 4895–4904.
98. Park, J.; Jana, S. C. Polymer 2004, 45, 7673–7679.
99. Pinnavaia, T. J.; Lan, T.; Kaviratna, P. D.; Wang, Z.; Shi, H. Polym Mater Sci Eng 1996, 74, 117–118.
100. Kamon, T.; Furakawa, H. In *Epoxy Resins and Composites IV*; Dusek, K., Ed.; Springer, Berlin/Heidelberg, 1986; Vol. 80, pp 173–202.
101. Lan, T.; Kaviratna, P. D.; Pinnavaia, T. J. Polym Mater Sci Eng 1994, 71, 527–528.
102. Chen, C.; Khobaib, M.; Curliss, D. Prog Org Coat 2003, 47, 376–383.
103. Liu, D.; Shi, Z.; Matsunaga, M.; Yin, J. Polymer 2006, 47, 2918–2927.
104. Yasmin, A.; Luo, J. J.; Abot, J. L.; Daniel, I. M. Compos Sci Technol 2006, 66, 2415–2422.
105. Frohlich, J.; Thomann, R.; Mulhaupt, R. Macromolecules 2003, 36, 7205–7211.
106. Kornmann, X.; Lindberg, H.; Berglund, L. A. Polymer 2001, 42, 4493–4499.
107. Tolle, T. B.; Anderson, D. P. Compos Sci Technol 2002, 62, 1033–1041.
108. Yucai Ke, J. L., Xiaosu Yi, Jian Zhao; Zongneng Qi,. J Appl Polym Sci 2000, 78, 808–815.
109. Lü Jiankun, K. Y., Qi Zongneng, Yi Xiao-Su,. J Polym Sci Polym Phys 2001, 39, 115–120.
110. Park, J. H.; Jana, S. C. Macromolecules 2003, 36, 2758–2768.
111. Shi, H.; Lan, T.; Pinnavaia, T. J. Chem Mater 1996, 8, 1584–1587.

112. Triantafillidis, C. S.; LeBaron, P. C.; Pinnavaia, T. J. Chem Mater 2002, 14, 4088–4095.
113. Meng, J.; Hu, X.; Boey, F. Y. C.; Li, L. Polymer 2005, 46, 2766–2776.
114. Le Pluart, L.; Duchet, J.; Sautereau, H. Polymer 2005, 46, 12267–12278.
115. Le Pluart, L.; Duchet, J.; Sautereau, H.; Halley, P.; Gerard, J.-F. Appl Clay Sci 2004, 25, 207–219.
116. Chiou, B.-S.; Raghavan, S. R.; Khan, S. A. Macromolecules 2001, 34, 4526–4533.
117. Tolle, T. B.; Anderson, D. P. J Appl Polym Sci 2004, 91, 89–100.
118. Li, G. Z.; Wang, L.; Toghiani, H.; Daulton, T. L.; Pittman, J., C. U. Polymer 2002, 43, 4167–4176.
119. Bharadwaj, R. K.; Berry, R. J.; Farmer, B. L. Polymer 2000, 41, 7209–7221.
120. Bizet, S.; Galy, J.; Gerard, J.-F. Polymer 2006, 47, 8219–8227.
121. Choi, J.; Harcup, J.; Yee, A. F.; Zhu, Q.; Laine, R. M. J Am Chem Soc 2001, 123, 11420–11430.
122. Kornmann, X.; Rees, M.; Thomann, Y.; Necola, A.; Barbezat, M.; Thomann, R. Compos Sci Technol 2005, 65, 2259–2268.
123. Wang, Z.; Lan, T.; Pinnavaia, T. J. Chem Mater 1998, 8, 2200–2204.
124. Wang, Z.; Pinnavaia, T. J. Chem Mater 1998, 10, 1820–1826.
125. Lee, A.; Lichtenhan, J. D. J Appl Polym Sci 1999, 73, 1993–2001.
126. Ratna, D.; R Varley, N. R. M.; Singh Raman, R. K.; Simon, G. P. Polym Int 2003, 52, 1403–1407.
127. Wang, B.; Qi, N.; Gong, W.; Li, X. W.; Zhen, Y. P. Radiation Physics and Chemistry, 2007, 76, 146–149.
128. Becker, O.; Varley, R.; Simon, G. Polymer 2002, 43, 4365–4373.
129. Choi, Y.-K.; Sugimoto, K.-i.; Song, S.-M.; Gotoh, Y.; Ohkoshi, Y.; Endo, M. Carbon 2005, 43, 2199–2208.
130. Nakamura, Y.; Yamaguchi, M.; Okubo, M.; Matsumoto, T. J Appl Polym Sci 1992, 44, 151–158.
131. Zheng, Y.; Zheng, Y.; Ning, R. Mater Lett 2003, 57, 2940–2944.
132. Lin, J.-C.; Chang, L. C.; Nien, M. H.; Ho, H. L. Compos Struct 2006, 74, 30–36.
133. Shia, D.; Hui, C. Y.; Burnside, S. D.; Giannelis, E. P. Polym Compos 1998, 19, 608–617.
134. Tsai, J.; Sun, C. T. J Compos Mater 2004, 38, 567–579.
135. Sheng, N.; Boyce, M. C.; Parks, D. M.; Rutledge, G. C.; Abes, J. I.; Cohen, R. E. Polymer 2004, 45, 487–506.
136. Zerda, A. S.; Lesser, A. J. J Polym Sci Polym Phys 2001, 39, 1137–1146.
137. Brunner, A. J.; Necola, A.; Rees, M.; Gasser, P.; Kornmann, X.; Thomann, R.; Barbezat, M. Eng Fract Mech 2006, 73, 2336–2345.
138. Lee, J. Y.; Lee, H. K. Mater Chem Phys 2004, 85, 410–415.
139. Lau, K.-T.; Hui, D. Carbon 2002, 40, 1605–1606.
140. Penumadu, D.; Dutta, A.; Pharr, G. M.; Files, B. J Mater Res 2003, 18, 1849–1853.
141. Thostenson, E. T.; Chou, T.-W. J Phys D: Appl Phys 2003, 36, 573–582.
142. Fisher, F. T.; Bradshaw, R. D.; Brinson, L. C. Compos Sci Technol 2003, 63, 1689–1703.
143. Lee, J.; Yee, A. F. Polymer 2000, 41, 8363–8373.
144. Lee, J.; Yee, A. F. Polymer 2000, 41, 8375–8385.
145. Lee, J.; Yee, A. F. Polymer 2001, 42, 577–588.
146. Lee, J.; Yee, A. F. Polymer 2001, 42, 589–597.
147. Miyagawa, H.; Drzal, L. T. J Adhes Sci Technol 2004, 18, 1571–1588.
148. Kinloch, A. J.; Taylor, A. C. J Mater Sci Lett 2003, V22, 1439–1441.
149. Wang, L.; Wang, K.; Chen, L.; Zhang, Y.; He, C. Compos Part A: Appl Sci 2006, 37, 1890–1896.
150. Kinloch, A. J.; Taylor, A. C. J Mater Sci 2006, 41, 3271–3297.
151. Gilman, J. W. Appl Clay Sci 1999, 15, 31–49.
152. Levchik, S. V.; Camino, G.; Luda, M. P.; Costa, L.; Muller, G.; Costes, B. Polym Degrad Stab 1998, 60, 169–183.
153. Gu, A.; Liang, G. Polym Degrad Stab 2003, 80, 383–391.
154. Becker, O.; Varley, R. J.; Simon, G. P. Eur Polym J 2004, 40, 187–195.
155. Zhou, Y.; Pervin, F.; Biswas, M. A.; Rangari, V. K.; Jeelani, S. Mater Lett 2006, 60, 869–873.
156. Sulaiman, S.; Brick, C. M.; DeSana, C. M.; Katzenstein, J. M.; Laine, R. M.; Basheer, R. A. Macromolecules 2006, 39, 5167–5169.
157. Biercuk, M. J.; Llaguno, M. C.; Radosavljevic, M.; Hyun, J. K.; Johnson, A. T.; Fischer, J. E. Appl Phys Lett 2002, 80, 2767–2769.
158. Moisala, A.; Li, Q.; Kinloch, I. A.; Windle, A. H. Compos Sci Technol 2006, 66, 1285–1288.
159. Bryning, M. B.; Islam, M. F.; Kikkawa, J. M.; Yodh, A. G. Adv Mater 2005, 17, 1186–1191.
160. Ogasawara, T.; Ishida, Y.; Ishikawa, T.; Aoki, T.; Ogura, T. Compos Part A: Appl Sci, 2006, 37, 2236–2240.
161. Yeh, J.-M.; Huang, H.-Y.; Chen, C.-L.; Su, W.-F.; Yu, Y.-H. Surf Coat Tech 2006, 200, 2753–2763.

162. Drummy, L. F.; Koerner, H.; Farmer, K.; Tan, A.; Farmer, B. L.; Vaia, R. A. J Phys Chem B 2005, 109, 17868–17878.
163. Ratna, D.; Becker, O.; Krishnamurthy, R.; Simon, G. P.; Varley, R. J. Polymer 2003, 44, 7449–7457.
164. Choi, J.; Yee, A. F.; Laine, R. M. Macromolecules 2004, 37, 3267–3276.
165. Karger-Kocsis, J.; Gryshchuk, O.; Frohlich, J.; Mulhaupt, R. Compos Sci Technol 2003, 63, 2045–2054.
166. Wu, D.; Zhou, C.; Fan, X.; Mao, D.; Bian, Z. Polym Degrad Stab 2005, 87, 511–519.

# 5 Nanocomposites Based on Poly(Vinyl Chloride)

*Baris Yalcin and Miko Cakmak*

College of Polymer Engineering and Polymer Science, Polymer Engineering Department,
University of Akron Akron, OH 44325-0301, USA
byalcin@uakron.edu

Poly(vinyl chloride) (PVC) belongs to the large family of vinyl polymers. It is the one of the most frequently used thermoplastic polymer in the world and one of the most profitable thermoplastics in the history of polymer industry. It has a wide range of applications that includes vinyl siding, flooring, piping, wall coverings, and wire coatings.

The repeating unit in the polymer is $(CH_2CHCl)_n$, where $n$ can take large values. Vinyl chloride (VC) can be polymerized into PVC via bulk, solution, suspension, and emulsion polymerization. Albright [1] has reviewed the process and reactions involved in the conversions of VC into PVC polymer. A comprehensive review of PVC synthesis can be found in a recent article by Endo [2] on PVC that also includes details of block copolymers synthesis and structure and implications of tacticity and other structural parameters on morphology and crystallizability.

PVC polymer, in its original form, is not considered useful due to its limited heat stability, colorless rigid nature, and its tendency to adhere to the metallic surfaces when heated. Therefore, it is usually compounded with other ingredients to tailor its properties for a wide range of applications. A PVC compound may contain chain extenders used in the synthesis, stabilizers to prevent the degradation at the processing temperatures 150–200°C, plasticizers to soften the polymer, lubricants to aid their release from the metal surfaces of the processing equipment, pigments for coloration, polymeric processing aids, impact modifiers, and fillers for reinforcement and/or functionalization.

Enhanced functionalities can also be introduced to PVC through the use of nanofillers. These include mechanical, thermal, optical, and electrical properties. In this section, we will review the methods of preparation, structure–property relationships, and processing of PVC nanocomposites. We will mainly concentrate on the 1D silicate clay nanoplatelet reinforcement as this has been the most widely studied nanofiller for PVC so far.

## 5.1. PVC Nanocomposites with Clay Nanoplatelets

PVC has been included early in the list of potential polymer matrices in patents related to polymer clay nanocomposites. The scientific publications started to appear at the beginning of 2000 and rapidly grew as illustrated in Figure 5.1.

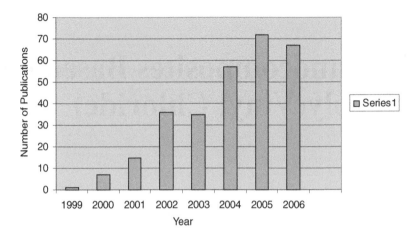

**Figure 5.1**   Number of publications related to PVC nanocomposites versus year (source SciFinder).

Mechanical properties, barrier properties, thermal stability, heat distortion, and degradation behavior of PVC nanocomposites depend on:

(a) preparation method, i.e., melt compounding or *in-situ* polymerization,
(b) type of clay and its organic modification,
(c) state of dispersion (tactoids, intercalated, exfoliated, or mixed structure),
(d) presence of the plasticizer and its type, and
(e) presence of other polymers in addition to the PVC.

In the following sections, we will review studies related to melt compounding and *in-situ* polymerization of PVC separately exploring each of the factors listed above.

## 5.1.1.  Melt Compounding of PVC Nanocomposites

Melt compounding is one of the most effective ways to obtain polymer-clay nanocomposites [3–5]. This is a preferred technique because of its ready availability and environmentally benign character due to the absence of the solvent. Melt compounding involves mixing a polymer and the clay nanofiller above the softening point (glass transition or melting) of the polymer typically in an internal mixer or twin screw extruder; though other specialized mixing equipment can also be used. During the compounding, care must be taken to optimally design the processing machines to minimize excessive shearing that can cause degradation.

In ideal preparation, the combined effects of favorable thermodynamic interactions and shear forces play important roles to diffuse polymer chains into the Van der Waals galleries between the silicate layers and push them apart leading to exfoliation. For example, Wang et al. [6–8] prepared PVC nanocomposites by melt compounding both organically modified (OMMT) (Cloisite® 30B) and unmodified montmorillonite (MMT) (Cloisite® Na+) with commercial PVC in the presence and absence of dioctyl phthalate (DOP) plasticizer in an internal mixer (Brabender) at 180°C. Cloisite® Na+ is a natural sodium montmorillonite (Na-MMT) (*d*-spacing = 1.2 nm) and 30B version is synthesized by ion-exchanging the natural sodium MMT with methyl tallow bis-2-hydroxyethyl ammonium chloride to yield a lamellar clay with a tallow core. The tallow consists mainly of C18 carbon chain. The resulting nanocomposites had mostly intercalated structure with little exfoliation. In spite of the absence of exfoliation, a low clay content of 2 wt% improved the thermal stability, flame retardancy [8], and mechanical properties [7] both with and without DOP.

Other research groups used a two-step melt compounding process to prepare PVC nanocomposites [9, 10]. In this process, the plasticizer, typically DOP at high loading level (30 per hundred PVC resin (phr)), was premixed with the clay at 80°C forming a suspension liquid. This liquid was subsequently blended with the PVC resin in a compounding machine such as a Buss Ko-Kneader or a counter-rotating twin-screw extruder at 150°C. The dimensional stability, gas permeability, and barrier properties of the PVC nanocomposites prepared by this technique were improved by more than 50% even with 1 wt% clay loading. The same group subsequently used a modified approach to their two-step melt compounding process. Instead of melt compounding the DOP/clay liquid suspension with the PVC resin, they used a mortar at room temperature forming a homogenized plastisol. The plastisol was subsequently molded at 170°C to facilitate melting of the PVC microcrystallites and interdiffusion of the PVC chains to form a uniform material with the desired shape [11]. PVC nanocomposites prepared by this method were partially intercalated and exfoliated.

Despite these favorable improvements, PVC is not very stable during melt compounding. In the presence of clay, PVC experiences strong discoloration and quickly turns into purple and then black. Unlike other polymer nanocomposite systems [12, 13], increasing rotor speed or extending compounding time is not appropriate for facilitating PVC intercalation. Since PVC is sensitive to dehydrochlorination, high temperature and strong mechanical stresses cause thermal dehydrochlorination and chain scission giving off HCl as by-product of this degradation reaction. The following conditions should be met in order to prevent degradation of the PVC/clay nanocomposites during compounding: (1) effective thermal stabilizers should be selected, (2) thermally stable organic modifiers should be used for clay treatment, and (3) proper compounding parameters should be set in order to minimize high shear stresses and local overheating.

Both the plasticizer (DOP) and the organic treatment [6, 10, 11] of the clay play important roles in discoloration and degradation of PVC. For example, the amine sites of the organo modifier can react with the active chloride atoms of the PVC and cause discoloration. To prevent this, a two-step process is used where DOP plasticizer was preintercalated prior to melt compounding solely to interact and cover the effective amine groups [6].

A detailed analysis to understand the mechanism of discoloration during melt compounding of PVC (DOP-5 phr)/OMMT nanocomposites was made by Wan et al. [14]. These authors observed that the mechanical properties and the thermal stability of discolored PVC nanocomposite batches were still superior to those of unfilled counterparts. This suggested that discoloration did not necessarily mean degradation of PVC chains. In order to prove their hypothesis, they tracked the presence of long conjugated polyene sequences indicative of PVC degradation using UV-vis spectrometry. There were no absorptions belonging to the long conjugated polyene sequences in the visible region. Instead, they observed absorptions in a region characteristic of very short polyene sequence length, which would not discolor the PVC [15, 16]. These short-length polyene sequences found in compounded PVC/OMMT nanocomposites were attributed to the early stages of degradation caused by the decomposition of the free alkyl quaternary ammonium modifier molecules at 155°C with the Hofmann elimination or an SN2 nucleophilic substitution

a) *N*-[4-(4_-aminophenyl)] phenyl phthalimide

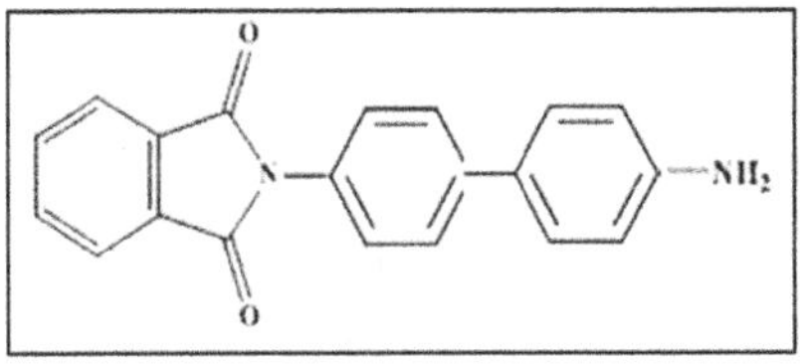

b) 1-Hexadecylamine

$$CH_3 - (CH_2)_{15} - NH_2$$

**Figure 5.2** Chemical structure of (a) thermally more stable aromatic modifier and (b) long carbon-chain amine modifier.

reaction [17]. The decomposition of alkyl quaternary ammonium resulted in numerous acidic sites, which in turn catalyzed the dehydrochlorination of PVC. However, the average polyene sequence length still remained low due to the effective organotin stabilizer. The authors concluded that the decomposition of alkyl ammonium organic modifier induces discoloration of PVC composites, but it does not lead to further PVC degradation and then deterioration of the thermal and mechanical properties. When the MMT is not organically modified, discoloration did not occur even with the addition of 10 phr MMT providing further evidence that it is the organic modifier that decomposes during melt compounding [18]. Although there are metal ions and Lewis acidic sites in unmodified MMT, they are not powerful enough to catalyze the degradation in the presence of organotin compounds. However, organo modification is still necessary for improved dispersion.

The same authors investigated the degradation behavior of the PVC and PVC/MMT (OMMT) nanocomposites by thermogravimetric analysis and calculated apparent activation and the compensation parameters by the Flynn–Wall–Ozawa and Kissinger methods [19]. The results indicated that both MMT and OMMT only affect the degradation rate of PVC and do not change its degradation mechanism. The first degradation stage in the temperature range 270–360°C was assigned to the progressive dehydrochlorination of PVC and the formation of conjugated polyene structure. PVC/MMT degraded slower than the unfilled PVC and the PVC/OMMT in the first stage. The authors suggested that the partially intercalated MMT layers were acting as a barrier to hinder the diffusion of heat and migration of degraded volatiles, and then retard the decomposition rate. The PVC/OMMT nanocomposite degraded faster due to the alkyl ammonium modifier of the OMMT.

As indicated by these studies, the type of organic modification plays an important role in the degradation behavior of PVC/MMT nanocomposites. For instance, PVC/MMT modified by the thermally stable aromatic modifier (Figure 5.2(a)) exhibits better mechanical, thermal, and flame-retardant properties than that modified by the long carbon-chain amine modifier (Figure 5.2(b)) [20]. This is due to the suppressed degradation during compounding, better interaction between the PVC matrix, and the aromatic modifier and overall better dispersion in the system.

Plasticizer type is another important factor in the melt compounding of PVC nanocomposites. Kovarova and coworkers [21] compared the mechanical properties of PVC nanocomposites containing different types of plasticizers, namely bis(2-ethylhexyl) phthalate (DOP) and bis(2-ethylhexyl) adipate (DOA) belonging to the low-molecular-weight plasticizers and Lankroflex epoxy as the agent of high-molecular-weight plasticizer. PVC/Lankroflex without any nanofiller exhibits a modulus almost 10 and 2.5 times higher than PVC/DOA and PVC/DOP, respectively.

**Table 5.1**  Clay gallery distances based on the WAXD $2\theta$ peak positions in PVC/Blendex/OMMT composites (table constructed using data from reference [22]).

| Material | Clay gallery distance (nm) |
|---|---|
| OMMT | 2.1 |
| PVC/OMMT | 3.8 |
| Blendex/OMMT | 3.7 |
| PVC/Blendex/OMMT | 4.1 |

**Table 5.2**  Mechanical properties of PVC/Blendex/OMMT nanocomposites (note: table reconstructed using data from reference [22]).

| Blendex/OMMT loading (phr) | Notched izod impact strength (J/m) | Tensile strength (MPa) | Elongation at break (%) |
|---|---|---|---|
| 0/0 | 26 | 51.3 | 17.5 |
| 30/0 | 889 | 34.9 | 76.4 |
| 0/3 | 33 | 58.7 | 61.3 |
| 30/3 | 1081 | 38.5 | 184 |

However, the addition of Cloisite 30B organoclay provided the best improvement in stiffness of the composites based on DOA plasticizer.

The same authors also investigated the influence of chain length of alkyl-amine-based modifier on the intercalation process of PVC/clay nanocomposites. They have modified MMT with octadecylamine (ODA) (18 carbons in chain), dodecylamine (DDA) (12 carbons), and octylamine (OA) (8 carbons) by the ion–dipole intercalation method. Modifiers with longer chains, ODA (18 carbons) and DDA (12 carbons), were arranged in bilayers with alkyl chains perpendicular to the aluminosilicate layer, whereas the OA (8 carbons) exhibited a monolayer arrangement. As a result of the bilayer arrangement, the PVC nanocomposites with ODA and DDA clay modifier exhibited a much better level of exfoliation than those with the OA clay modifier, as evidenced by X-ray and TEM studies.

Synergistic blends with the other polymers can easily be prepared by melt compounding route to improve the properties of PVC nanocomposites further. For instance, although the impact strength of PVC is improved by the addition of OMMT, it is still low and molded parts fracture in a brittle mode [18]. In order to improve the impact strength of PVC nanocomposites, Wan et al. [22] melt compounded PVC resin and OMMT along with an impact modifier, i.e., methyl methacrylate grafted acrylonitrile–butadiene–styrene copolymer with a high rubber content from GE Specialty Chemicals, Inc., (Blendex 338). Blendex addition expands the clay galleries evidencing intercalation of the Blendex into the OMMT galleries. Figure 5.3 shows the clay gallery distances calculated from the X-ray peak of (001) plane of the clay.

After toughening modification, the ternary blends of PVC/OMMT/Blendex exhibit substantial improvement in the Notched Izod impact strength compared to the binary blends of PVC/Blendex. This is shown in Table 5.2.

In a comprehensive morphological study with particle image analysis, Yalcin and Cakmak [23] investigated the role of DOP plasticizer on the exfoliation, dispersion, and fracture behavior of clay particles in the PVC matrix using AFM, TEM, optical

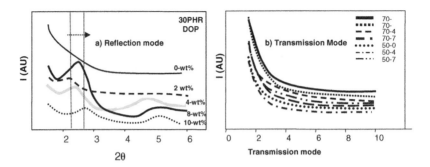

**Figure 5.3**  $2\theta$ X-ray scans of compression-molded PVC nanocomposites in (a) reflection mode and (b) transmission mode (data reconstructed from reference [23]).

microscopy, and X-ray. They prepared their nanocomposites using the two-step melt compounding procedure with the exception that OMMT (Cloisite® 30B) was premixed with the DOP plasticizer and the stabilizer *at room temperature* prior to melt compounding this suspension mixture with the PVC resin in an internal mixer at 60 rpm and 150°C for 10 min. The compounded batches were subsequently compression molded into the film form. Using AFM, the individual platy MMT particles were revealed clearly for the first time. These platelet structures were found to become preferentially aligned with their basal surfaces parallel to the broad surface of the samples especially when they are compression molded. Planar orientation of the platelets in the compression-molded films was further evidenced by $2\theta$ wide angle X-ray diffraction scans between 1.5 and $8\pi$ in the reflection and transmission mode. The $2\theta$ X-ray diffraction patterns in the transmission mode in Figure 5.3(b) did not display any peak in the entire angular region of interest, 1.5°–6°, whereas intercalation peaks around 2.5° and their higher order peaks at around 5° were clearly visible in the reflection mode (Figure 5.3(a)) indicating planar orientation of the nanoplatelets on the film plane induced by the squeezing flow of compression molding.

The AFM image of a highly loaded (10 wt%) and plasticized compound displaying both the tactoids and the single MMT crystals is shown in Figure 5.4. In these images, the stiff MMT particles appear bright, while the soft polymer appears dark. The edges of the particles are in some cases straight forming hexagonal angles and in other cases irregular. Hexagonal morphology originates from the basic unit, i.e., silicon oxide tetrahedron ($SiO_4$), of the outer tetrahedral sheets of the MMT unit crystals with 2:1 structure, i.e., a single layer of aluminum octahedral sandwiched between two layers of silicon tetrahedral. In the outer silicon oxide tetrahedron ($SiO_4$), the silicon atom is bonded to four oxygen atoms. Each tetrahedron unit shares three of its four oxygen atoms to form a sheet with hexagonal symmetry. This, in turn, gives an approximate hexagonal symmetry in the *ab*-plane of the platelets. This is an ideal morphological habit for an individual MMT crystal. Other common habits for MMT are euhedral smectite lamellae in rhombic outline and subhedral thin platelets with irregular outlines. These kinds of morphological habits usually form due to the impediment of crystallization along certain directions during growth of clay in its natural environment over the years. All these morphologies are clearly visible in the AFM image in Figure 5.4.

In mineralogy literature [24], it is well established that the layers could bend or curl not only due to external mechanical causes but also due to inherent structural features such as misfits between the component octahedral and tetrahedral sheets. Curved clay edges in

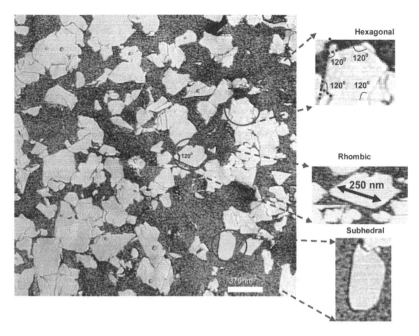

**Figure 5.4**   AFM image of PVC clay nanocomposites with 70 phr DOP and 10 wt% clay (images partly from reference [23]).

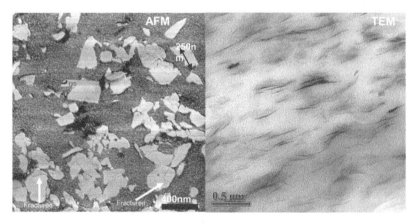

**Figure 5.5**   AFM and TEM images of plasticized PVC nanocomposites showing fracture in the left AFM image and bending in the right TEM image (images partly from reference [23]).

the TEM image in Figure 5.5 support this structure. AFM surface imaging falls short of differentiating a platelet bent in 3D. However, the AFM image in Figure 5.5 also evidences that these platelets are rather fragile and can be easily broken (fractured) most likely in bending mode during dispersion in the PVC resin in the internal mixer.

Yalcin and Cakmak [23] determined the surface area of the platelets and nearest neighbor length correlation distances between the platelets using AFM and image analysis. Dispersion and distribution of the clay platelets in the PVC matrix revealed that there is an optimum concentration of DOP for the dispersion and distribution of the particles.

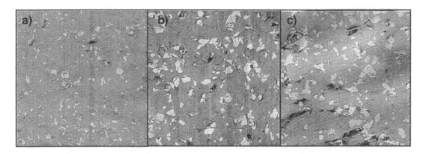

**Figure 5.6**    AFM images of PVC/OMMT at 4% clay loading and (a) 30 phr, (b) 50 phr, (c) 70 phr DOP plasticizer (images partly from reference [23]).

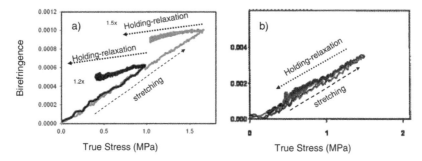

**Figure 5.7**    Birefringence–true stress behavior of (a) PVC (reference [25]), (b) amorphous PLA (reference [31]).

DOP helps the intercalation, but too high concentration adversely affects the distribution and breakage of the particles and hence the size of the platelets as can be observed in AFM images in Figure 5.6. This is related to the ability of force transfer to the clay particles through the low-viscosity polymer matrix.

In addition to preparation and characterization of PVC nanocomposites, Yalcin and Cakmak [25] have studied the real-time mechano-optical behavior of their compression-molded PVC nanocomposite films during uniaxial deformation. In fact, this is one of the few studies that focused on the structure–property relationship during processing after the preparation of PVC nanocomposites. They have investigated the effect of nanoclay loading and the amount of plasticizer as well as the rate effects on the birefringence development and true mechanical response using a custom built and instrumented uniaxial stretching system [26, 27] that allows the real-time determination of true stress, true strain, and birefringence simultaneously during deformation. In order to understand the uniaxial deformation behavior and structure as influenced by the nanoparticles during stretching, it is important to know the starting PVC network structure in the as-cast samples. PVC has a strong network of largely amorphous chains connected via small crystallites that act as physical crosslinking points [28–30]. A comparison of the birefringence–true stress behavior of PVC with fully amorphous melt cast PLA during stretching to very low strains and subsequent holding in Figure 5.7 supports this structure. During holding stage following small strains, amorphous PLA relaxes with complete birefringence recovery [31], whereas the PVC retains its birefringence substantially during stress relaxation [25]. This is because

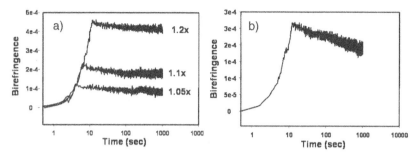

**Figure 5.8**   Birefringence–time behavior of PVC/OMMT with (a) 30 phr DOP and 10 wt% clay and (b) 70 phr DOP and 4 wt% clay (reference [25]).

the preexisting strong network with physical crosslinks, i.e., crystallites, in PVC, helps the stretched chains to remain oriented while the stress relaxes.

Addition of the OMMT nanoplatelets into the PVC structure renders even a tighter and stronger network, which results in almost complete birefringence retention during holding even after 5% strain. This is shown in Figure 5.8(a).

The physical network in PVC persists even in plasticized form [32–34]. This is because the plasticizer molecules mainly solvate the intercrystalline amorphous regions [35, 36] and leave the crystallites intact. In fact, the relaxation behavior in Figure 5.8(b) shows that birefringence decrease during relaxation increases with higher plasticizer content. At high levels of plasticizer concentration, amorphous chains have increased freedom of motion and relax back to their unoriented state more rapidly. Another factor, perhaps less effective and uncertain, could be the ability of the plasticizer to induce some melting at very high loadings. This would decrease the number of crystalline physical crosslinking junctions and form a looser structure. A reduction in crystallinity from 11% to 4% was reported by small-angle X-ray scattering as the plasticizer loading was increased from 0 to 150 phr [34]. It was shown that small amounts of DOP from 10% to 20% had no effect in crystallinity, but larger amounts could reduce the crystallinity. There were no significant changes in $d$-spacings in the unit cell or average crystallite size as determined by wide-angle X-ray scattering. However, others consider that the SAXS peak is not due to the crystallinity of PVC [37] but due to heterogeneities in the amorphous phase.

Although the amount of crystalline phase in PVC is small, it has a significant influence on the final structure. Therefore, with any mode of deformation, it is important to know what happens to the crystalline phase in PVC during processing, i.e., blow molding, tenter frame stretching, injection molding, etc. For instance, during uniaxial stretching, additional crystallites could be formed via stress-induced crystallization or the orientation could occur for the preexisting crystallites without any stress-induced crystallization. Another mechanism is the destruction of the crystallites due to stretching. The trans-conformations increase in both rigid and plasticized PVC with the development of pre-ferred chain orientation as evidenced earlier by infrared [38]. However, there was no evidence for increased crystallinity on drawing; rather, absorption peaks corresponding to the crystallinity decreased slightly on drawing. Based on Raman spectroscopy studies, Robinson et al. proposed that PVC crystallites were pulled apart and destroyed during stretching [39], especially when the temperature and the plasticizer loading was low where rigid rod rotation of the PVC crystallites in the rubbery amorphous pool was not possible.

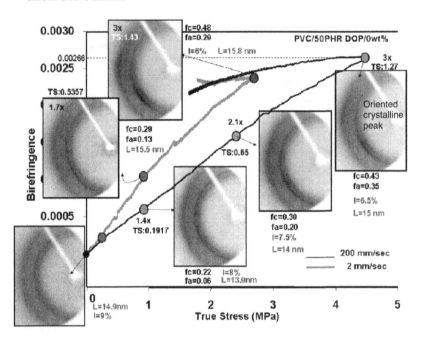

**Figure 5.9** Birefringence–true strain of PVC with 50 phr DOP during stretching at slow (2 mm/min) and fast stretching rates (200 mm/min). I, crystallinity index; L, crystal size; $f_c$, crystalline orientation factor; $f_a$, amorphous orientation factor; TS, true strain; and $x$ refers to stretch ratio

Yalcin and Cakmak determined the crystallinity index ($I$) and size ($L$) of the plasticized PVC in the rubbery stage at various strain levels with low (2 mm/min) and high (200 mm/min) stretching speeds [25]. The result is shown for PVC with 50 phr DOP stretched at 80°C in Figure 5.9. One can see that the crystallinity index denoted as "$I$" on the WAXD patterns does not increase. On the contrary, there is a slight decrease in the crystallinity index. One can also notice clearly from the WAXD patterns that the preexisting crystallites orient with stretching.

In conclusion, during stretching strain-induced crystallization in PVC is mostly prevented as the noncrystallizable atactic segments block the registry of the crystallizable syndiotactic segments [40]. Obviously, this will depend on the fraction of the syndiotactic diads present in the PVC chain architecture. The latter fraction is known to decrease with an increase in the polymerization temperature [2].

The effect of clay loading on the birefringence–true strain and birefringence–true stress curves is shown in Figure 5.10. As the nanoparticle loading is increased, PVC attains higher birefringence levels at the same strain. In fact, the presence of nanoparticle influences the orientation of both the amorphous as well as crystalline phases [25]. One way this occurs is the extension of load-bearing physical network by the inclusion of nanoplatelets as they have very strong interaction with the amorphous chains and/or crystallites. This, in turn, helps transfer the local stresses to the attached chains and increase the orientation levels of the chains. However, in doing so, filled PVC, with a tighter network structure, requires higher stress levels and the birefringence–true stress curves show decreasing birefringence with increased clay loading. However, this does not lead to indefinite increase and PVC with clay nanoparticles more than 4 wt% does not increase the orientation significantly

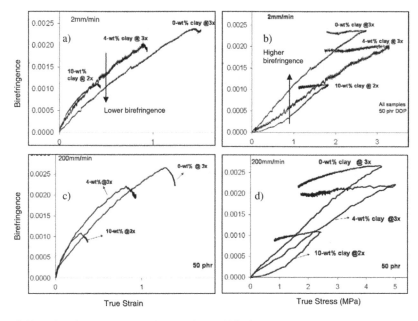

**Figure 5.10**  Birefringence–true strain (a and c) and birefringence–true stress (b and d) behavior of PVC nanocomposites (50 phr DOP) at different clay loadings. Stretching rate is 2 mm/min for (a) and (b) and 200 mm/min for (c) and (d).

during stretching. This is seen in birefringence–true strain curves in Figures 5.10(a) and (c). In fact at such high levels of clay concentration, there is a significant amount of tactoid formation, which adversely affects the continuity of the network structure.

### 5.1.2. PVC Nanocomposites by *In-Situ* Polymerization

*In-situ* polymerization is another nanocomposite preparation method successfully applied to PVC. The silicate nanoplatelets in this technique are swollen within the vinyl chloride monomer (VCM). The polymerization subsequently occurs in between the intercalated sheets.

Brittain and his coworkers have prepared PVC nanocomposites by *in-situ* suspension polymerization [41]. They polymerized PVC in the presence of initiator (AIBN) and/or comonomer-modified MMT. MMT modified with the comonomer was thermally more stable than that modified with the initiator. The initiator molecules tethered on the MMT surface was essential to initiate the polymerization but not adequate for high degree of monomer conversion. The authors had to use additional radical polymerization initiators for high MW PVC nanocomposite yield from the reaction. Others have also synthesized PVC nanocomposites by traditional PVC suspension polymerization method without making major changes to the process [42, 43]. PVC synthesized by suspension polymerization has a granular structure that is formed by the precipitation of PVC from its monomer during polymerization. Gong and his coworkers investigated the particle nature and mechanical properties of PVC/OMMT nanocomposites synthesized by *in-situ* suspension polymerization [44]. They reported that the mean particle size of the PVC nanocomposite grains decreases and the porosity of the particles increases with an increase in OMMT content from 1 to 5 wt%. This was attributed to the organophilic character of the MMT acting as a suspending agent providing coalescence resistance to the polymer droplets. Decreased mean particle size and increased porosity counterbalanced the melt viscosity increase due to clay inclusion and enhanced the cold plasticizer absorption. The nanocomposites also exhibited higher tensile strength and modulus and greater fracture toughness in comparison with unfilled PVC, while the elongation at break did not noticeably decrease.

Instead of the commonly used cationic MMT clay, Bao et al. used layered double hydroxides (LDHs) during *in-situ* suspension polymerization [45]. LDHs are anionic clays [46] that can absorb and react with hydrochloric acid produced during the thermal degradation of PVC, thus improving its thermal stability. The authors modified LDHs with dodecyl sulfate (DS) anion by coprecipitation method [47] and *in-situ* suspension polymerized VC monomer in the presence of LDH-DS. They have compared the morphology, thermal, and mechanical properties of the resulting *in-situ* polymerized nanocomposites with those prepared by direct melt compounding. Similar to the earlier work using OMMT [44], the addition of LDH-DS decreased the mean particle size of PVC nanocomposite resins and had no obvious influence on the molecular weight and molecular weight distribution of PVC. TEM and X-ray evidenced a partially intercalated and exfoliated morphology with a uniform distribution of LDH-DS particles in the PVC resin via *in-situ* polymerization method. LDH-DS was hardly intercalated by melt blending method. For the *in-situ* polymerized nanocomposite, only 4 wt% of clay loading increased the 5% weight loss temperature of PVC from 265°C to 290°C. And 10% weight loss temperatures exhibited even a larger increase from 275°C to 310°C. The tensile strength, Young's modulus, and Charpy notched impact strength of the *in-situ* polymerized PVC/LDH-DS nanocomposites were all greater than those of the unfilled PVC and the melt compounded nanocomposite. Here 3 wt% loading of LDH-DS nanoparticles was found to be an optimum amount for the PVC/LDH-DS nanocomposites.

Emulsion polymerization is another technique used for *in-situ* conversion of VCM/clay to PVC/clay nanocomposites. It is an easy and relatively inexpensive method to prepare nanocomposites. Ming Wang et al. prepared PVC/MMT hybrids by dispersing unmodified $Na^+$-MMT in a pH (9–10) controlled deionized water and subsequently introducing the VCM to polymerize the emulsion mixture at 50°C. The PVC nanocomposites showed a mixture of intercalated and well-exfoliated nanostructure. With only 2–3 wt% clay loading, the tensile modulus and the notched impact strength of the nanocomposites improved 30% and 100%, respectively. The thermal stability of the nanocomposites was also improved by the addition of MMT. Unfilled PVC experienced 25% weight loss at 300°C, whereas the nanocomposite had almost zero weight loss. With further clay loading, however, the properties started to decrease. The same authors subsequently employed a modified procedure for emulsion polymerization of VCM into PVC nanocomposites. The procedure, named "reverse feeding," differed from the earlier version by dispersing the organoclay in the VCM prior to introducing the emulsion polymerization ingredients [48]. With this method, the clays were fully wetted by the VCM and partially intercalated before the polymerization. This ensured that the silicate layers were expanded and even completely exfoliated, with the polymerization heat being released. They compared their *in-situ* emulsion polymerized nanocomposites to melt compounded ones. *In-situ* polymerized nanocomposites exhibited a much better dispersion than melt compounded ones with a much wider range of organic modifiers. This was attributed to the small volume of VCM, which could enter the galleries of the silicate layers more readily than the long PVC chains. The heat released during polymerization also helps overcome the coulombic forces of

**Table 5.3**  Electrical conductivity values of PVC/MWCNT and PVC/SWCNT nanocomposites.

| | 5 wt% | 10 wt% | 20 wt% |
|---|---|---|---|
| MWCNT/PVC (S/cm) | $5 \times 10^{-3}$ | $3.44 \times 10^{-1}$ | 1.75 |
| | 0.1 wt% | 0.2 wt% | |
| SWCNT/PVC (S/cm) | $7 \times 10^{-3}$ | $10 \times 10^{-3}$ | |

the layers. Melt-compounded nanocomposites were more prone to what kind of organic modification was used on the pristine clay. The authors suggested that the clay organic modifier molecules should possess a critical degree of polarity to allow the PVC chains to intercalate into the clay galleries.

## 5.2. **PVC Nanocomposites with Other Nanoparticles**

In this section, we will review PVC nanocomposites that were prepared with carbon nanotubes (CNT), carbon black (CB), calcium carbonate (CaCO₃), and magnetic nanoparticles. Each of these nanoinclusions improves the existing properties and/or provides novel functionality to the PVC resin.

### 5.2.1. Carbon Nanotubes

CNTs are becoming the choice of fillers in polymer nanocomposites due to their extraordinary theoretical strength, high thermal conductivity, and especially unique electrical and optical properties.

Recently, Broza et al. prepared PVC nanocomposite films using multi- (MWCNT) and single-walled carbon nanotubes (SWCNT) [49]. Homogenous dispersion of CNTs were achieved in PVC/THF (1.6/100 wt%) solution with a multistep mixing and ultrasonification procedure. Thin films were cast from CNT dispersed solution and compression molded subsequently at 175°C. As shown in Table 5.3, the PVC/CNT nanocomposites had electrical conductivity values in the semiconductor level, i.e., $10^{-6}$–$10^2$ S/cm, much higher than the insulating unfilled PVC with $10^{-16}$ S/cm. It was also shown that the sorption level in methylene chloride solvent decreased by half with 20 wt% MWCNT loading.

Jung and co-workers [50] prepared MWCNT-reinforced PVC by adsorbing the oxidized MWCNTs onto the surfaces of PVC microspheres synthesized by suspension polymerization. They have used carbon vapor deposited MWCNTs with 96% purity and oxidized them by acid treatment to achieve 99% purity before dispersing in deionized water at 0.02 wt% loading. The dispersion was performed in the presence of 0.3 wt% cetyltrimethylammonium bromide (CTAB) surfactant using sonication. PVC microspheres were subsequently added to an excess quantity of the aqueous MWCNT dispersion and further sonicated for 7 h to facilitate adsorption of the MWCNTs onto the rough and porous surfaces of suspension-polymerized PVC microspheres. Once the MWCNTs were dispersed, the aqueous dispersion turned clear indicating homogenous and stable dispersion. The stability of the final dispersion was attributed to the static interactions between the positively charged MWCNTs due to the cationic CTAB surfactant and the negatively charged PVC colloid in water. About 2.9 wt% of the MWCNTs were adsorbed onto the surface of the PVC microspheres, which resulted in an electrical conductivity increase from 10–14 to

$10^{-4}$ S/cm. The MWCNT-adsorbed PVC microspheres were later on dissolved in DMF and cast into uniform films, which exhibited $10^{-6}$ S cm$^{-1}$, which was still higher than that of pure PVC. Such PVC films have sufficient electrical conductivity for electrical charge dissipation, which requires a surface electrical conductivity in the range from $10^{-6}$ to $10^{-10}$ S/cm. In addition to electrical conductivity, the mechanical properties of these PVC nanocomposites were also improved, i.e., doubled the tensile strength and modulus.

### 5.2.2. Carbon Black

Carbon black (CB), similar to CNTs, can render the polymers electrically conductive when incorporated into the resin beyond a critical percolation concentration forming an internal conducting network. Although CB-filled polymer systems require higher percolation threshold values for electrical conductivity than CNT-filled systems, they still offer significant price advantage over CNTs. Chen et al. prepared rigid and conductive PVC by blending CB in a Haake torque rheometer set at 170°C, 60 rpm for 5 min [51]. The surface resistivity of PVC/CB composite was almost constant up to 6 phr CB. An increase from 6 to 15 phr CB resulted in a decrease of surface resistivity from $10^{10}$ to $10^4$ $\Omega$/sq, which is sufficient for electrostatic dissipation. Saad and his coworkers compared the electrical conductivity of plasticized and rigid PVC composites with fast extrusion furnace (FEF) CB [52]. They found out that PVC/CB system attains a much higher electrical conductivity at comparable CB loading than its plasticized counterpart. Percolation threshold for the rigid PVC was found to be 20 phr with FEF-CB. Others achieved a lower percolation threshold with pyrolytic CB at 10 wt% loading [53]. Noguchi et al. analyzed the structure of CB aggregates on the melt flow behavior and electrical conductivity of roll-milled PVC composites [54]. During the milling process, CB aggregates first became cylindrical in shape, then ruptured into small spheres. During this process, the electrical conductivity increased, going through a maximum with the formation of cylindrical aggregates and then decreased as the aggregates were further broken up into fine spherical particles. This phenomenon was attributed to the initial formation and subsequent elimination of electrical conductive pathways through CB aggregation and dispersion into the matrix.

### 5.2.3. Nano Calcium Carbonate (Nano-CaCO$_3$)

Pure PVC materials in general have weak impact resistance that can be improved by inclusion of solid fillers and elastomers. Among inorganic fillers, CaCO$_3$ nanoparticles have been widely used to toughen PVC. Median particle size of CaCO$_3$ nanoparticles (40–150 nm) is typically an order of smaller in magnitude than ultra ground CaCO$_3$ with 0.7–1 μm size.

Sun et al. prepared PVC/CaCO$_3$ nanocomposites in a double roll mixing machine and investigated the effect of particle size, surface treatments, and fraction of nano-CaCO$_3$ particles on the interfacial adhesion, and tensile and impact strength of nano-CaCO$_3$-filled PVC composites [55]. Increasing CaCO$_3$ content lowered the tensile strength by 15% (Figures 5.11(a) and (c)) but increased the impact strength by about 75% (Figures 5.11(b) and (d)). Titanate-treated nano-CaCO$_3$/PVC composites had superior dispersion to untreated or sodium-stearate treated CaCO$_3$/PVC composites and hence exhibited the highest tensile and impact strengths (Figures 5.11(a) and (b)). The improved dispersion with

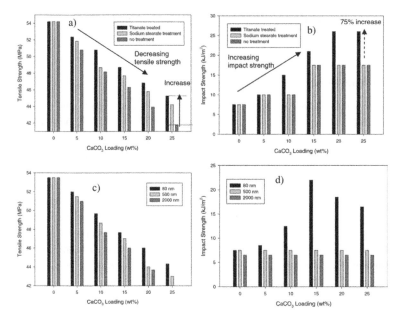

**Figure 5.11** Mechanical properties of PVC/CaCO3 nanocomposites. Upper graphs show the influence of chemical treatment and lower graphs show the influence of nanoparticles size. Data for these graphs are reconstructed using the data from reference [54]. Dimensions of the nanoparticles are shown in each graph.

titanate treatment was attributed to the formation of hydrogen-bonded sites between the PVC and the titanate treatment. The impact strength of titanate-treated nano-CaCO3/PVC composites was $26.3 \pm 1.1$ kJ m$^{-2}$, more than three times that of pure PVC materials. Both the tensile and impact strengths of CaCO3/PVC greatly increased with decreasing CaCO3 particle size (Figures 5.11(c) and (d)), which was attributed to increased interfacial contact area and enhanced interfacial adhesion between CaCO3 particles and PVC matrix.

In a similar study with nano-CaCO3 particles with 40 nm average particle size, Wu et al. prepared nanocomposites of PVC via melt blending to improve the impact strength [56]. The authors used chlorinated polyethylene (CPE) as an interfacial modifier. CPE was introduced into the PVC/CaCO3 nanocomposites through a CPE/nano-CaCO3 masterbatch, which was subsequently melt compounded with the PVC resin. Uniform dispersion of nano-CaCO3 was evidenced by TEM. TEM also evidenced that the nano-CaCO3 particles in the PVC matrix were encapsulated with the CPE layer. The impact strength increased significantly and achieved 1500% improvement with 30 wt% nano-CaCO3. Nano-CaCO3 addition increased the elongation at break monotonously from 20% to 70% along with 40% increase in Young's modulus. The toughening effect of the nano-CaCO3 was attributed to the cavitation of the PVC matrix, which consumed tremendous fracture energy.

Xie et al. prepared PVC/CaCO3 nanoparticles by *in-situ* polymerization of VC in the presence of nanosized CaCO3 particles [57]. In these nanocomposites, the CaCO3 nanoparticles stiffen and toughen PVC by acting as stress concentrators leading to interface debonding/voiding and matrix deformation. At 5 wt% loading, optimal properties were obtained in Young's modulus, tensile yield strength, elongation-at-break, and Charpy notched impact energy.

### 5.2.4. Nanomagnetites

Recently, researchers started to exploit the possibility of inducing tunable magnetic property to PVC via the incorporation of magnetic nanofillers. Magnetopolymeric nanocomposites have several application areas such as microwave absorption, recycling, and magnetic separation. These materials are also useful for the pharmaceutical and medical areas for magnetic shielding. Materials with different magnetic behaviors, i.e., ferromagnetic, paramagnetic, and superparamagnetic, are used for different applications. Ferromagnetic materials are strongly magnetic materials that are magnetized with the applied magnetic field and maintain their magnetism even when the external field is removed. This property of magnetic memory, i.e., hysteresis, makes the permanent magnets possible. Ferromagantic materials are widely used in computer disk storage and disk drives. Above a certain critical temperature, called the Curie temperature, ferromagnetic materials cease to be ferromagnetic due to deterrence of the alignment of the dipole moments with increased thermal motion of the atoms. Ferromagnetic materials become paramagnetic above their Curie temperature. Paramagnetic materials are weakly magnetic materials that are only magnetized in the presence of an applied magnetic field and loose their magnetism when the applied magnetic field is removed, i.e., zero hysteresis. Superparamagnetic materials are strong magnets similar to ferromagnets below their Curie temperature but exhibit paramagnetism even below the Curie temperature.

Sanz et al. [58] prepared PVC nanocomposites with tunable magnetic response by controlling the diameter of spherical $Co_{80}Ni_{20}$ nanoparticles around 65 nm and their concentration in the matrix from 0.5% to 50%. The authors prepared uniform $Co_{80}Ni_{20}$ nanoparticles by polyol reduction of cobalt(II) and nickel(II) tetrahydrated acetates, which were characterized by ESEM. The prepared magnetic nanoparticles were dispersed in diethyl oxalate (DEO) by ultrasonification, which was subsequently mixed with the PVC resin vigorously, ultrasonified, and heated to 120°C for gelation. The final PVC gels were molded into thin films. The authors showed that the amplitude of the magnetic response of the nanocomposites could be tailored with the suitable choice of the particle concentration and the evolution of saturation magnetization with particle concentration was fitted to a linear curve. Their nanocomposites exhibited superparamagnetic behavior except at very high nanoparticle loading the samples were ferromagnetic (100%). The temperature dependence on the magnetic characteristics has also been measured in the temperature range from 10 K to room temperature.

Flores et al. [59] prepared magnetic PVC films by casting a PVC plastisol containing ferrofluid with nanosized (15 nm) magnetite and DOP. They have used DOP plasticizer as a carrier liquid for the ferrofluid. They have functionalized their magnetite nanoparticles with oleic acid in order to prevent particle aggregation in DOP. Their PVC nanocomposites exhibited superparamagnetic behavior with no hysteresis. Mechanical properties of these PVC nanocomposites, i.e., tensile strength and elongation at break, are shown in Figure 5.12. As one can notice that the tensile strength reaches a high value of almost 3 MPa at 25 phr loading, which indicates increased strength in addition to the magnetic character of the films.

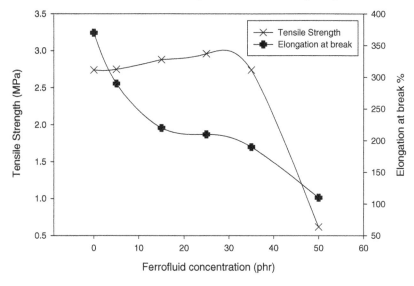

**Figure 5.12**   Mechanical properties of PVC nanocomposite films at different ferrofluid concentration (data reconstructed from reference [58]).

## 5.3.  Conclusion

PVC is a versatile polymer that can be functionalized and reinforced with nano-inclusions through chemical and physical means to render them stronger, tougher, heat resistant, electrically conducting, and even magnetic for a wide range of applications including housing, medical, electronics, as well as biosensing. It is expected to thrive in these applications in many decades to come.

## References

 1. Albright, L.F. Chem. Eng., (1967) 74, 151,
 2. Endo, K. Prog. Polym. Sci. (2002), 27(10), 2021–2054
 3. Ishida, H.; Campbell, S.; Blackwell, J.; Chem. Mater. (2000) 12(5), 1260–1267
 4. Vaia, R.A.; Ishii, H; Giannelis, E.P.; Chem. Mater. (1993) 5(12), 1694–1696
 5. Kato, M.; Usuki, A.; Okada, A. J. Appl. Polym. Sci. (1997) 66(9), 1781–1785
 6. Wang, D.; Parlow, D.; Yao, Q.; Wilkie, C. A. J. Vinyl Additive Technol. (2001) 7(4), 203
 7. Wang, D.; Parlow, D.; Yao, Q.; Wilkie, C. A. J. Vinyl Additive Technol. (2002) 8(2), 139–150
 8. Du, J.; Wang, D.; Wilkie, C. A..; Wang, J. Polym. Deg. Stab. (2003) 79, 319–324
 9. Trilica, J.; Kalendova, A.; Malacd, Z..; Simonik, J. ANTEC Conf. (2001), May 6–10, 2162.
10. Kovarova, L.; Kalendova, A.; Gerard, J-F.; Malac, J.; Simonik, J.; Weiss, Z. Macromol. Symp. (2005), 221, 105–114
11. Peprnicek, T.; Duchet, J.; Kovarova, L.; Malac, J.; Gerard, J.F.; Simonik, J. Polym. Deg. Stab. (2006) 91, 1855–1860
12. Suh, D.J.; Lim, Y.T.; Park, O.O. Polymer (2000) 41, 8557
13. Yoon, J.T.; Jo, W.H; Lee, M.S.; Ko, M.B. Polymer (2001) 42, 329
14. Wan, C.; Zhang, Y.; Zhang, Y. Polym. Test. (2004) 23, 299–306
15. Braun, D. ; Sonderhof, D. Polym. Bull. (1985) 14, 39.
16. Daniels, V.D.; Rees, N.H. J. Polym. Sci. Polym. Chem. (1974) 12, 2111

17. Xie, W.; Gao, Z.M.; Pan, W.P.; Hunter, D.; Singh, A.; Vaia, R.A. Chem. Mater. (2001) 13, 2979.

18. Wan, C.Y.; Qiao, X.Y.; Zhang, Y.; Zhang, Y.X. Polym. Test (2003) 22, 453

19. Wan, C.Y.;Tian, G.; Cui, N.; Zhang, Y.; Zhang, Y-X. J. Appl. Polym. Sci., (2004) 92, 1521–1526

20. Liang, Z.M; Wan, C.Y.; Zhang, Y.; Wei, P.;Yin, J. J. Appl. Polym. Sci. (2004) 92, 567—575

21. Kovarova, L.; Kalendova, A.; Simonik, J.; Malac, J.; Weiss, Z.; Gerard, J. F. Plast., Rubber Compos. 2004 33(7), 287

22. Wan, C. Y.; Zhang, Y.; Zhang, Y. X.; Qiao, X. Y.; Teng G. M. J. Polym. Sci: Part B: Polym. Phys., (2004) 42, 286–295

23. Yalcin, B.; Cakmak, M. Polymer (2004) 45, 6623–6638

24. Brindley,G.; Brown, W G. Crystal structures of clay minerals and their X-ray identification. Mineralogical Society Monograph, Brookfield Pub Co. (1984) Vol. 5, Chapter 2, 144

25. Yalcin, B.; Cakmak, M. J. Polym. Sci. Part B Phys. (2004) 43(6), 724–742

26. Valladares, D.; Toki, S.; Sen, T.Z; Yalcin, B., Cakmak M., Macromol. Symp. (2002) 185, 149–166.

27. Sen T.Z., Dissertation, Polymer Engineering Department, University of Akron, Akron, OH, USA, (2002).

28. Brown, H.R.; Musindi, G.M.; Stachuski, Z.H. Polymer (1982), 23, 1508

29. Ballard, D.G.H.; Burgess, A.N.; Dekonink, J.M. Polymer (1987), 28, 3

30. Summers, J. W.; Rabinovitch, E.B. J. Macromol. Sci. Phys. (1981), B20, 219

31. Mulligan, J.; Cakmak, M. Macromolecules (2005) 38(6), 2333–2344

32. Manson, J.A.; Iobst, S.A .;Acosta, R.A. J. Polym. Sci. Polym. Chem. Ed. (1972), 10, 179

33. Acosta, R.A.; Manson, J.A.; Iobst, S.A. Polym. Prepr (ACS-Polym. Chem.) (1971), 12, 745

34. Shtarkman, B.P.; Lebedev, V.P.; Yatsynina, T.L.; Kosmynin, B.P.; Gerasimov, V.I.; Genin, Y.V.; Tsvankin, D.Y. Polym. Sci. USSR (1972), 14, 1826.

35. Tabb, D.L.; Koenig, J.L. Macromolecules (1975), 8, 929

36. Taylor, R.B.; Tobolsky, A.V. J. Appl. Polym. Sci. (1964), 8, 1563

37. Straff, R.S.; Uhlmann, D.R. J. Polym. Sci., Polym. Phys. Ed. (1976), 14, 353

38. Theodorou, M.; Jasse, B. J. Polym. Sci. Polym. Phys. Ed. (1986), 24, 2543

39. Robinson, M.E.R.; Bower, D.I.; Maddams, W.F. J. Polym. Sci. Polym. Phys. Ed. (1978), 16, 2115

40. Carrega, M. Pure Appl. Chem. (1977), 49, 569

41. Xu,Y.; Malaba, D.; Huang, X.; Solis, C-A.; Brittain, W.J. Polym. Preprints. (2002) 43(2) 1312–1313

42. Gong, F.L.; Zhao, C.G.; Feng, M.; Qin, H.L. J. Mater. Sci. (2004) 39 293– 294

43. Yang, D.Y.; Liu, Q.-X.; Xie, X.L.; Zeng, F.D. J. Therm. Anal. Calorimetry, (2006) 84(2), 355–359

44. Gong, F.; Feng, M.; Zhao, C.; Zhang, S.; Yang, M. Polym. Test. (2004) 23 847–853

45. Bao, Y-Z.; Huang, Z-M.; Weng, Z-X. J. Appl. Polym. Sci., (2006) 102, 1471–1477

46. Cavani, F.; Trifiro, F.; Vaccari, A. Catal. Today (1991), 11, 173

47. Wilson, O.C.;Olorunyolemi, T.; Jaworski, A.; Borum, L.; Young, D. Appl. Clay Sci. (1999), 15, 265

48. Hu, H.; Pan, M.; Li, X.; Shi, X.; Zhang, L. Polym. Int. (2004) 53 225–231

49. Broza, G.; Piszczek, K.; Schulte, K.; Sterzynski, T. Compos. Sci. Technol. (2007) 67, 890–894

50. Jung, R. ; Kim, H-S.; Jin, H-J. Macromol. Symp. 2007, 259–264

51. Chen, C-H. ; Li, H-C.; Teng, C-C.; Yang, C-H. J. Appl. Polym. Sci., (2006) 99, 2167–2173

52. Saad, A.L.G.; Aziz, H.A..; Dimitry, O.I.H. J. Appl. Polym. Sci. (2004), 91, 1590

53. Dufeu, J.B.; Roy, C.; Ajji, A.; Choplin, L. J Appl Polym Sci (1992), 46, 2159

54. Noguchi, T. ; Nagai, T.; Seto, J. J. Appl. Polym. Sci. (1986) 31, 1913–1924

55. Sun, S.; Li, C.; Zhang, L.; Du, H.L.; Burnell-Gray, J.S. Polym. Int. (2006) 55, 158–164

56. Wu, D.; Wang, X.; Song, Y.; Jin, R. J. Appl. Polym. Sci. (2004) 92, 2714–2723

57. Xie, X-L.; Liu, Q-X.; Li, R. K-Y.; Zhou, X-P.; Zhang, Q-X.; Yu, Z-Z.; Mai,Y-W Polymer (2004) 45, 6665–6673

58. Sanz, R.; Luna, C.; Hernández-Vélez, M.; Vázquez, M.; López, D.; Mijangos, C.; Nanotechnology (2005) 16, S278–S281

59. Yáñez-Flores, I.G.; Betancourt-Galindo, R. ; Matutes Aquino, J.A.; Rodrguez-Fernández, O. J. Non-Crystal. Solids 353 (2007) 799–801

# 6 Nanocomposites Based on POSS

*Shiao-Wei Kuo[1], Chih-Feng Huang[2], and Feng-Chih Chang[2]*

[1]Department of Materials Science and Optoelectronic Engineering, Center for Nanoscience and Nanotechnology, National Sun Yat-Sen University, Kaohsiung, Taiwan

[2]Institute of Applied Chemistry, National Chiao-Tung University, Hsin-Chu, Taiwan

changfc@mail.nctu.edu.tw

**Abstract**

This chapter describes in detail the synthesis of polyhedral oligomeric silsesquioxane (POSS) compounds, the miscibility of POSS derivatives and polymers, and the preparation of monomers and polymers containing POSS, including styryl-POSS, methacrylate-POSS, norbornyl-POSS, vinyl-POSS, epoxy-POSS, phenolic-POSS, benzoxazine-POSS, amine-POSS, and hydroxyl-POSS. Both monofunctional and multifunctional monomers are used in commercial and/or high-performance thermoplastic and thermosetting polymers, the corresponding thermal, dynamic mechanical, electrical, and surface properties are also discussed in detail herein.

## 6.1. Introduction

The field of polymer nanocomposite materials has attracted great attention, imagination, and interest from polymer scientists and engineers in recent years. The simple premise of using building blocks having nanosize dimensions makes it possible to create new polymeric materials exhibiting improved physical properties, such as unprecedented flexibility. Silsesquioxanes are nanostructures having the empirical formula $RSiO_{1.5}$, where R is a hydrogen atom or an organic functional group, such as an alkyl, alkylene, acrylate, hydroxyl, or epoxide unit. Figure 6.1 illustrates the silsesquioxanes that may be formed from random, ladder, cage, and partial cage structures [1].

In 1946, Scott et al. discovered the first oligomeric organosilsesquioxane, $(CH_3SiO_{1.5})_n$, along with other volatile compounds through the thermolysis of polymeric products prepared from the methyl trichlorosilane and dimethyl chlorosilane co-hydrolysis method [2]. Interest in this field has increased dramatically in recent years, even though silsequioxane chemistry has been studied for more than half a century. In 1995, Baney et al. [3] reviewed the preparation, properties, structures, and applications of silsequioxanes, especially those of the ladder-like polysilsesquioxanes (Figure 6.1(b)). These ladder-like structures display excellent thermal stability, oxidative resistance at temperatures greater than 500°C, and good insulating and gas permeability properties [4]. More recently, attention has been concentrated on silsesquioxanes possessing the specific cage structures displayed in Figures 6.1(c)–(f). These polyhedral oligomeric silsesquioxanes are commonly abbreviated

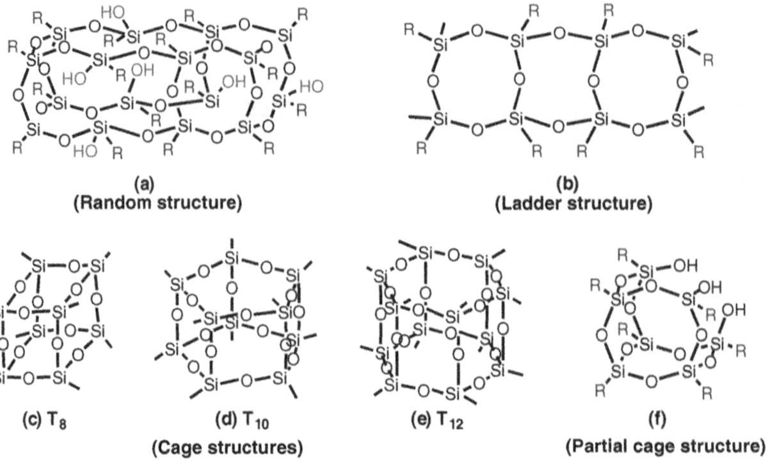

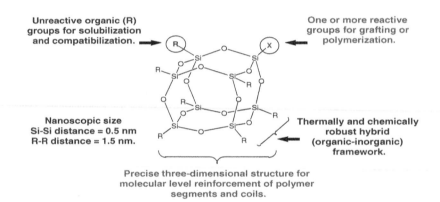

Figure 6.1    Structures of silsesquioxanes.

Unreactive organic (R) groups for solubilization and compatibilization. →

One or more reactive groups for grafting or polymerization. ←

Nanoscopic size
Si-Si distance = 0.5 nm
R-R distance = 1.5 nm.

Thermally and chemically robust hybrid (organic-inorganic) framework. ←

Precise three-dimensional structure for molecular level reinforcement of polymer segments and coils.

Figure 6.2    Chemical structure of POSS.

as "POSS." POSS compounds are true hybrid inorganic/organic chemical composites that possess an inner inorganic silicon and oxygen core $(SiO_{1.5})_n$ and external organic substituents that can possess a range of polar and nonpolar structures and functional groups.

Figure 6.2 displays POSS nanostructures having diameters ranging from 1 to 3 nm; they can be considered as the smallest possible particles of silica, i.e., molecular silica. Unlike most silicones or fillers, POSS molecules containing organic substituents on their outer surfaces are compatible or miscible with most polymers. In addition, these functional groups can be specially designed as either nonreactive or reactive to be used in polymer blending or copolymerization. POSS derivatives have been prepared with one or more covalently bonded reactive functionalities so that they become suitable for polymerization, grafting, surface bonding, or other transformations. Unlike traditional organic compounds, POSS derivatives release no volatile organic components; thus, they are odorless and environmentally friendly materials. The incorporation of POSS moieties into polymeric materials can dramatically improve the polymers' properties (e.g., strength, modulus, rigidity) as well as reduce flammability, heat evolution, and viscosity during processing. These enhancements apply to a wide range of commercial thermoplastic polymers (e.g., PE, PP, PS, PMMA, PEO, and PCL), high-performance thermoplastic polymers (e.g., polyimide and PLED), and thermosetting polymers (e.g., epoxy resin, phenolic resin, and polybenzoxazine). It is especially convenient to incorporate the POSS moieties into polymers through simple blending or copolymerization. In addition, when POSS monomers are soluble in monomer mixtures, they can be incorporated as true molecular dispersions in copolymer systems. The macrophase separation that usually occurs through the aggregation of POSS units can be avoided through copolymerization (i.e., as a result of covalent bond formation between POSS and polymers) – a significant advantage over current filler technologies. POSS nanostructures also have significant promise for use in catalyst supports and biomedical applications, such as scaffolds for drug delivery, imaging reagents, and combinatorial drug development.

In this chapter, we describe methods for synthesizing POSS compounds and preparing monomers and polymers containing POSS. We discuss the monofunctional and multifunctional POSS monomers that have been used for thermoplastic and thermosetting polymers. In addition, we compare the miscibility, phase behavior, thermal, dynamic mechanical, electric, and surface properties of polymers containing POSS.

## 6.2. General Approaches Toward Synthesizing Polyhedral Oligomeric Silsesquioxanes

POSS derivatives feature Si–O linkages in the form of a cage presenting a silicon atom at each vertex, with substituents coordinating around the silicon vertices tetrahedrally. The nature of the exo cage substituent in such compounds determines the mechanical, thermal, and other physical properties. The number of $RSiO_3$ units determines the shape of the frame, which is uniquely unstrained for 6 to 12 units. A review by Voronkov et al. published in 1982 covered the known methods of synthesizing POSS compounds [5]. In 2000, Feher and co-workers reviewed recent progress in the field of POSS synthesis [6]. There are many substituents appended to the silicon/oxygen cages $(RSiO_{1.5})_n$ (where R is an organic or inorganic group; Figure 6.3) that are suitable for polymerization or copolymerization

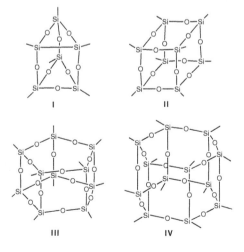

**Figure 6.3**   Cage structures of POSS where $T_6$(I), $T_8$(II), $T_{10}$(III), and $T_{12}$(IV).

between the specific POSS derivative with other monomers. The general methods of synthesizing both monofunctional and multifunctional octahedral silsesquioxanes ($T_8$), and examples of such derivatives, are described briefly below.

### 6.2.1. Monofunctional POSS

Monofunctional POSS derivatives are among the most useful compounds for polymerization or copolymerization with other monomers. Figure 6.4 summarizes the three general approaches toward synthesizing monofunctional POSS derivatives of the form $R'R_7Si_8O_{12}$.

*Route I.  Co-hydrolysis  of  trifunctional  organo-  or  hydrosilanes*: Polycondensation  of  monomers  is  the  classical  method  of  synthesizing  silsesquioxanes. When this reaction is performed in the presence of monomers possessing various R groups, mixtures of heterosubstituted compounds are obtained, including the desired monosubstituted products [7, 8].

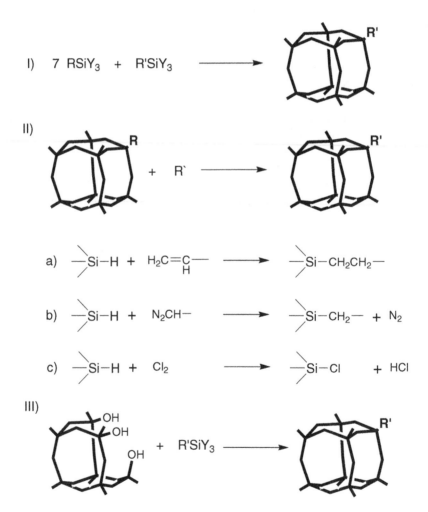

**Figure 6.4**  Three general ways of synthesizing monosubstituted octasilasesquioxane.

*Route II. Substitution reactions with retention of the siloxane cage*: Figure 6.4 presents a selection of substitution reactions using octahydro-silsesquioxane as starting materials (IIa–c) that have been applied successfully to prepare monosubstituted silsesquioxanes. By adjusting the ratio of the reactants, it is possible to obtain a considerable yield of the desired monosubstituted product [9–14].

*Route III. Corner-capping reactions*: Feher and co-workers developed this approach starting from the incompletely condensed $R_7Si_7O_9(OH)_3$ molecules ($T_7$) [15–18]. The three silanol groups are very reactive toward $RSiCl_3$, giving the fully condensed products. Variation of the R group on the silane enables the syntheses of a variety of monofunctionalized siloxane cages [19, 20]. Subsequent transformations can be performed until the desired functionality is obtained. Moreover, incompletely condensed silsesquioxanes offer a route toward the generation of hetero- and metalla-siloxanes in which a hetero main group or a transition metal element is introduced into the silicon–oxygen framework [15–17, 21–24].

## 6.2.2. Multifunctional POSS

POSS $(RSiO_{1.5})_n$ derivatives in which R is a hydrogen atom and the value of $n$ is 8, 10, or 12 are unique structures generally formed through hydrolysis and condensation of trialkoxysilanes [$HSi(OR)_3$] or trichlorosilanes ($HSiCl_3$) [5]. The hydrolysis of trimethoxysilane in cyclohexane/acetic acid mixtures in the presence of concentrated hydrochloric acid provides the octamer in low yield (13%) [25]. The hydrolytic polycondensation of trifunctional monomers of the type $RSiY_3$ leads to crosslinked three-dimensional networks and *cis*-syndiotactic (ladder-type) polymers, $(RSiO_{1.5})_n$. The reaction rates, degrees of oligomerization, and yields of the polyhedral compounds formed under these conditions are strongly dependent on several factors, including the concentration of the initial monomer in the solution, the nature of the solvent, the identity of the substituent R, the functional group Y in the initial monomer, the type of catalyst, the reaction temperature, the rate of addition of water, and the solubility of the polyhedral oligomers formed [25]. For example, POSS cages in which $n$ is 4 or 6 can be obtained in nonpolar or weakly polar solvents at 0°C or 20°C, but not in alcohols. In contrast, octa(phenylsilsesquioxane) [$Ph_8(SiO_{1.5})_8$] is more readily formed in benzene, nitrobenzene, benzyl alcohol, pyridine, or ethylene glycol dimethyl ether at high temperature (e.g., 100°C). The effects of each of these factors affecting POSS synthesis have been reviewed in depth previously [25]. Another approach toward multifunctional POSS derivatives is the functionalizing of preformed POSS cages; e.g., through Pt-catalyzed hydrosilylation of alkenes or alkynes with $(HSiO_{1.5})_8$ and $(HMe_2SiOSiO_{1.5})_8$ cages (Figure 6.5) [26–28].

## 6.3. Hydrogen Bonding and Miscibility Behavior of Polymer/POSS Nanocomposites

### 6.3.1. Hydrogen Bonding Interactions Between Polymers and POSS

The miscibility and specific interactions of polymer blends attract great attention in polymer science because of their significant potential applicability in industry. Most inorganic silicas or creamers are immiscible in most organic polymer systems because of poor specific interactions within these organic/inorganic hybrids and the negligibly small combined entropy contribution to the free energy of mixing. Specific intermolecular interactions are generally required to enhance the miscibility of polymers and inorganic particles.

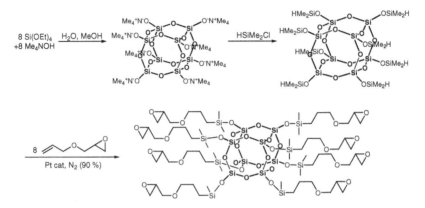

**Figure 6.5**    An example of multifunctional POSS synthesis.

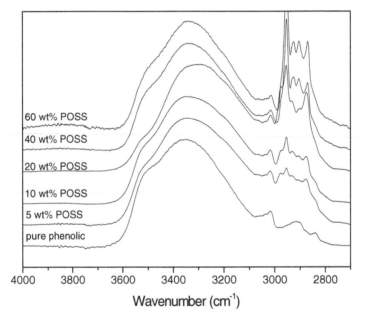

**Figure 6.6**    FT-IR spectra of phenolic/POSS hybrid with various POSS contents.

To improve properties and miscibility of hybrid materials, it is usually necessary to ensure that favorable specific interactions exist between these components, such as hydrogen bonding, dipole–dipole interactions, or acid/base complexation [29]. Determining the types and strengths of the interactions between POSS derivatives and polymers is an important challenge. For convenience, Chang et al. blended phenolic resin with a POSS derivative to investigate the miscibility, specific interactions, and microstructural behavior [30]. Figure 6.6 displays IR spectra (2700–4000 cm$^{-1}$) of pure phenolic and various phenolic/POSS hybrids measured at room temperature [30].

The spectrum of the pure phenolic polymer contains two OH components: a very broad band centered at 3350 cm$^{-1}$ that is attributed to the wide distribution of the

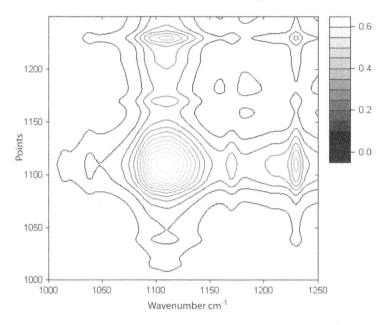

**Figure 6.7**   The synchronous 2D correlation map at 1000–1250 cm$^{-1}$ region.

hydrogen-bonded OH groups and a relatively narrow band at 3525 cm$^{-1}$ corresponding to free OH groups. Two trends are observed for the OH stretching bands in the IR spectra of the phenolic/POSS hybrids: the broad hydrogen-bonded OH band of phenolic shifts to lower wavenumber upon increasing the POSS content, approaching a minimum at 3280 cm$^{-1}$ for the hybrid containing 20 wt% POSS (Figure 6.6 (d)) [30].

This change arises from the switch of intramolecular hydroxyl–hydroxyl to intermolecular hydroxyl–siloxane interactions, i.e., hydrogen bonding between the OH groups of phenolic, and the siloxane groups of POSS. Generalized 2D IR correlation spectroscopy was used to explore the nature of the hydrogen bonding sites in the phenolic/POSS hybrid. This tool can be used to study the mechanism of interpolymer miscibility through the formation of hydrogen bonds, both qualitatively and quantitatively. Using this novel method, spectral fluctuations can be treated as a function of time, temperature, pressure, or composition to investigate the specific interactions occurring between polymer chains. 2D IR correlation spectroscopy can identify different intra- and intermolecular interactions through selected bands from a 1D vibration spectrum [31]. Figure 6.7 presents the synchronous 2D correlation maps of set A in the range from 1250 to 1000 cm$^{-1}$. The absorption bands of the POSS derivative at 1100 and 1230 cm$^{-1}$ correspond to siloxane Si–O–Si and Si–C stretching vibrations, respectively. The 1223 cm$^{-1}$ peak is due to the phenyl–OH stretching vibration of the phenolic. There are two positive cross-peaks shown in Figure 6.7, indicating hydrogen bonding interactions between the siloxane group of the POSS derivative (1100 cm$^{-1}$) and the phenyl–OH group (1223 cm$^{-1}$) of the phenolic.

## 6.3.2.  Miscibility Between Polymers and POSS Derivatives

The most important feature of a miscible polymer blend is that interassociation is stronger than self-association. Conversely, if the self-association is stronger than the interassociation, the blend tends to be immiscible or only partially miscible. According to the

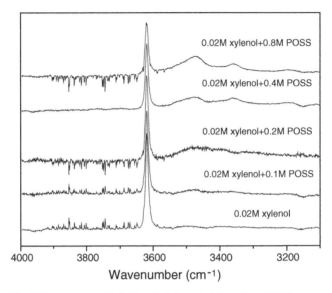

**Figure 6.8**  FT-IR spectra of 2,4-dimethylphenol with various POSS concentrations.

Painter–Coleman association model (PCAM) [29], the interassociation equilibrium constant between a noncarbonyl group component and a hydrogen bond donating component can be calculated using the classical Coggeshall and Saier (C&S) method. Figure 6.8 displays the OH group absorption of 2,4-dimethylphenol (a model compound for phenolic) in cyclohexane solutions containing various concentrations of POSS; the intensity of the free OH absorption at 3620 cm$^{-1}$ decreases with increasing POSS content. The absolute intensity of the free OH group at 3620 cm$^{-1}$ is assumed to be an indication of the content of free OH groups in the mixture [30]. Figure 6.8 indicates that the frequency of the associated OH band shifts from free OH group at 3620 to 3490 cm$^{-1}$ as a result of interassociation hydrogen bonding between 2,4-dimethylphenol and POSS [30]. The interassociation equilibrium constant, $K_A$, yielded through this procedure is 38.6, based on the classic C&S method [32], whereas the self-association equilibrium constant for the phenolic is 52.3 [33]. Clearly, the interassociation equilibrium constant from the phenolic/POSS is relatively lower compared with the self-association equilibrium constant of pure phenolic, indicating that the phenolic/POSS hybrid is partially miscible or immiscible because of its relatively poor intermolecular association [34].

For this reason, functionalization of POSS derivatives possessing pendent hydrogen bond acceptor groups is expected to improve the miscibility with phenolic resin. Functionalization of $Q_8M_8^H$ can be achieved through hydrosilylation of its Si–H groups with acetoxystyrene [34] in the presence of a Pt catalyst to form AS-POSS (Figure 6.9).

Figure 6.10 presents scaled IR spectra, recorded at room temperature, of pure phenolic and various phenolic/AS–POSS nanocomposites. Figure 6.10(a) indicates clearly that the intensity of the free OH absorption (3525 cm$^{-1}$) decreases gradually as the AS-POSS content of the blend is increased from 5 to 90 wt%. The band for the hydrogen-bonded OH units in the phenolic tends to shift to higher frequency (toward 3465 cm$^{-1}$) upon increasing the AS-POSS content. This change is due to the switch from hydroxyl–hydroxyl interactions to the formation of hydroxyl–carbonyl and/or hydroxyl–siloxane hydrogen

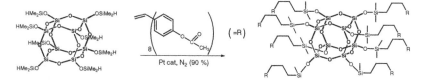

**Figure 6.9**    Chemical structures of AS-POSS.

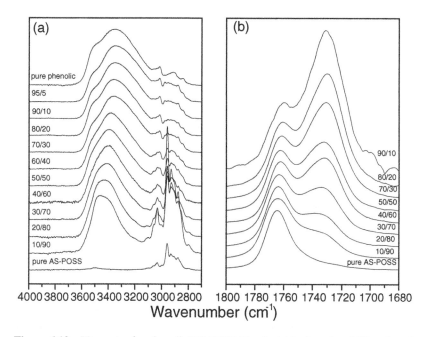

**Figure 6.10**    IR spectra for phenolic/AS-POSS blends: (a) hydroxyl and (b) carbonyl.

bonds. Figure 6.10(b) displays IR spectra (1680–1820 cm$^{-1}$) measured at room temperature for various phenolic/AS-POSS blend composites. The C=O stretching frequency is split into two bands at 1763 and 1735 cm$^{-1}$ corresponding to the free and hydrogen-bonded C=O groups, respectively. These bands are readily decomposed into two Gaussian peaks corresponding to areas of the hydrogen-bonded C=O (1735 cm$^{-1}$) and the free C=O (1763 cm$^{-1}$) peaks [35].

According to the PCAM, the interassociation equilibrium constant between a noncarbonyl group component and a hydrogen bond donating component can be calculated using the classical C&S method. To recheck the interassociation equilibrium constant between the phenolic OH groups and the POSS siloxane groups, the value of $K_A$ can be determined indirectly from a least-squares fitting procedure of the experimental fraction of hydrogen-bonded C=O groups of AS-POSS in this binary blend. Figure 6.11 indicates that the experimental values are generally lower than the predicted values when using the value of $K_A$ of 64.6 obtained from the phenolic/PAS blends [36]. This result also indicates that the OH groups of phenolic not only interact with the C=O groups of the acetoxystyrene units but also with the siloxane groups of the POSS core, which is consistent with the results

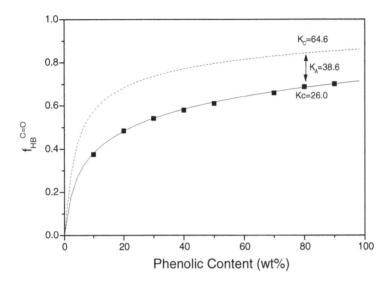

**Figure 6.11**    Fraction of hydrogen-bonded carbonyl groups versus phenolic contents.

reported from a previous study. To calculate the values of the interassociation constants $K_A$, a least-squares method was employed as described previously [36], which resulted in an interassociation equilibrium constant of 26.0 for the phenolic/AS-POSS blend. The value of $K_A$ obtained from the phenolic/PAS blend was 64.6, implying that the value of $K_A$ between the OH group of phenolic and the siloxane group of POSS is equal to 38.6 (i.e., $64.6 - 26.0 = 38.6$), which is exactly consistent with the value reported previously based on classical C&S methodology [32]. Therefore, there is a good correlation between these two different methods when determining the values of the interassociation equilibrium constants for hydroxyl–siloxane interactions. In addition, the trend of the glass transition temperature ($T_g$) of the phenolic/AS-POSS blend ($q = 25$) is greater than that of the phenolic/PAS blend ($q = -245$) system based on the Kwei equation [37].

This behavior may arise from two phenomena: (i) the star-shaped acetoxystyrene-POSS presents a larger fraction of hydrogen-bonded C=O groups than does the linear PAS, which is similar to the findings made in a previous study of a phenolic/poly(methyl methacrylate) blend system [38]; (ii) the siloxane groups of the POSS core can also form hydrogen bonds with the OH groups of the phenolic. Therefore, the presence of POSS moieties effectively increases the values of $T_g$ of the resultant organic/inorganic polymer nanocomposites.

The low molecular weight of POSS derivative is unable to effectively increase the glass transition temperature in the phenolic/AS-POSS blend system. As a result, Lin et al. synthesized a new POSS derivative containing eight phenol groups and then copolymerized it with phenol and formaldehyde to form novolac-type phenolic/POSS nanocomposites through covalent bonding, which exhibits high thermal stabilities and low surface energies (Figure 6.12) [39].

Figure 6.13 displays conventional second-run DSC and TGA thermograms of phenolic/OP–POSS nanocomposites at various weight ratios [39]. Each of these hybrids possesses essentially a single value of $T_g$, suggesting that these hybrids exhibit a single phase. The glass transition temperatures of these nanocomposites are significantly enhanced after incorporation of POSS units; however, fractions of incompletely reacted

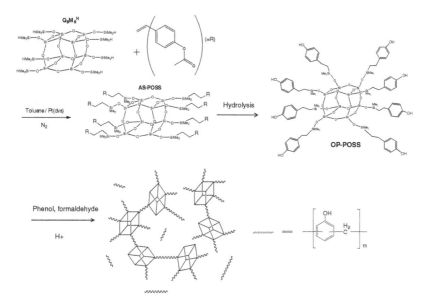

**Figure 6.12**    Synthesis procedure of phenolic/OP–POSS nanocomposites.

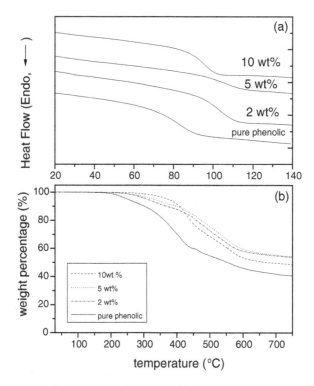

**Figure 6.13**    Thermal analyses of phenolic/OP–POSS nanocomposites containing different OP–POSS contents: (a) DSC and (b) TGA.

functional groups are still remained on the POSS and phenolic components. The enhanced glass transition temperature is resulted from the restricted motion of the polymer chains caused by the physical crosslinkage through hydrogen bonding interactions from these evenly distributed POSS units within the phenolic matrix. Figure 6.13(b) shows thermal degradation of neat phenolic resin and the POSS-containing nanocomposites [39].

The char yield increases upon increasing the POSS content except the sample incorporating 10 wt% POSS, providing further evidence that some lower molecular weight species are still present in the phenolic/octaphenol-POSS 10 wt% sample. The value of $T_d$ increases significantly upon increasing the POSS content, the phenolic/OP–POSS 10 wt% sample exhibits a value of $T_d$ 123°C higher than that of the pure phenolic resin. This phenomenon can be explained in terms of the nano-reinforcement effect of incorporating POSS moieties into polymeric matrixes. The nanoscale dispersion of POSS moieties within the matrix and their covalent and hydrogen bonds to the phenolic resin are responsible for enhancing the initial decomposition temperature [39].

## 6.4. POSS-Containing Polymers and Copolymers

POSS feedstocks, which have been functionalized with various reactive organic groups, can be incorporated into virtually any existing polymer system through either grafting or copolymerization. POSS homopolymers can also be synthesized. The incorporation of the POSS nanocluster cages into polymeric materials can result in dramatic improvements in polymer properties, including temperature and oxidation resistance, surface hardening, and reduction in flammability. Therefore, research in POSS-related polymers and copolymers has accelerated recently. Some representative systems are discussed below.

### 6.4.1. Polyolefin/POSS and Norbornyl/POSS Copolymers

A number of interesting design strategies for the preparation of polyolefin/POSS hybrid materials have evolved over the past decade. Hsiao et al. [40] used DSC to investigate a series of iPP melt-blended with nanostructured POSS molecules to study the quiescent melt crystallization behavior, and shear-induced crystallization behavior. Tabuani and co-workers [41, 42] reported the influence of the POSS substituent groups on the morphological and thermal characteristics of melt-blended PP/POSS composites. Furthermore, Coughlin et al. and Mather et al. reported polyolefin copolymers containing norbornyl-POSS macromonome [43–45]. Polyolefin-POSS copolymers incorporating a norbornylene-POSS macromonomer have been prepared using a metallocene/methyl aluminoxane (MAO) co-catalyst system (Figure 6.14) [43].

### 6.4.2. Polystyrene/POSS Nanocomposites

Haddad synthesized and characterized (Figure 6.15) a series of linear thermoplastic hybrid materials containing an organic polystyrene backbone and large inorganic silsesquioxane groups pendent to the polymer backbone [46]. The pendent inorganic groups drastically modify the thermal properties of the polystyrene, and the interchain and/or intrachain POSS–POSS interactions affected the solubility and thermal properties [46].

Couglin et al. [47] developed a synthetic route for preparing syndiotactic PS (sPS)–POSS copolymers (Figure 6.16). Copolymerizations of styrene and POSS afforded a novel nanocomposite of sPS and POSS. The rate of copolymerizations was much slower than

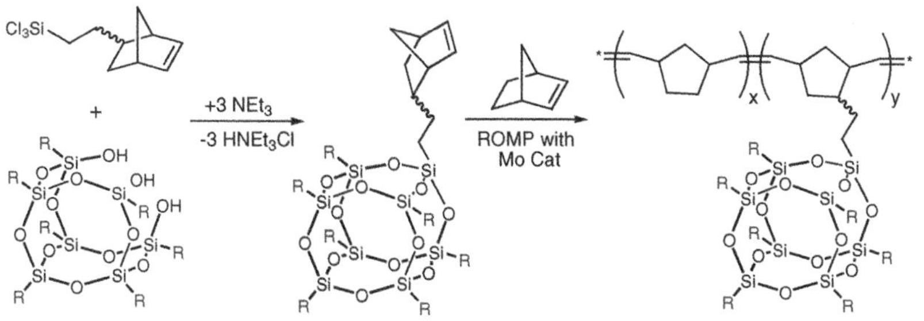
Cl3Si
+3 NEt3
-3 HNEt3Cl
ROMP with
Mo Cat

**Figure 6.14** Copolymerization of norbornylene-POSS and norbornene

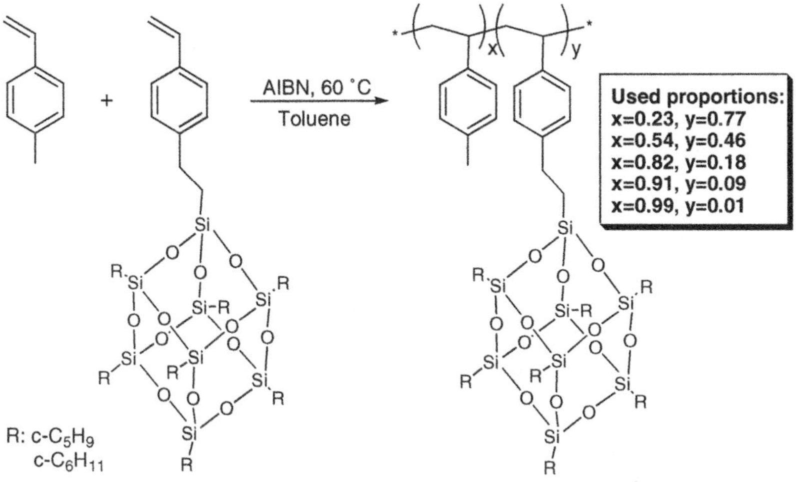
AIBN, 60 °C
Toluene
Used proportions:
x=0.23, y=0.77
x=0.54, y=0.46
x=0.82, y=0.18
x=0.91, y=0.09
x=0.99, y=0.01
R: c-C5H9
c-C6H11

**Figure 6.15** POSS styryl macromer synthesis and polymerization.

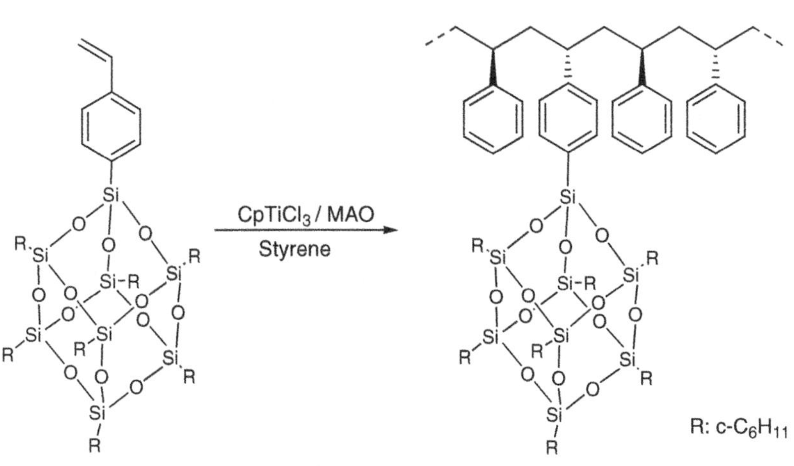
CpTiCl3 / MAO
Styrene
R: c-C6H11

**Figure 6.16** Copolymerization of styrene and POSS-styryl.

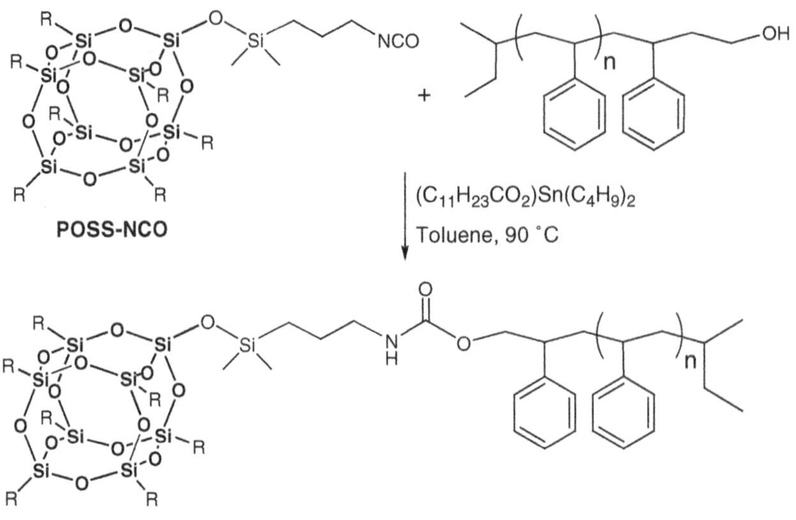

**Figure 6.17**    Synthesis of hemi-telechelic POSS–polystyrene hybrid.

that of radical polymerization, presumably because of the coordination polymerization mechanism. TGA traces of the sPS–POSS copolymers under both nitrogen and air revealed their improved thermal stability, i.e., higher degradation temperatures and char yields [47].

Couglin et al. also reported a synthetic protocol for preparing well-defined POSS–polystyrene hemi-telechelic hybrids (Figure 6.17) [48]. These model systems provide the opportunity to experimentally probe the ordering or aggregation behavior of inorganic nanoparticles within polymeric matrices.

Chang et al. synthesized and characterized a series of hybrid poly (acetoxylstyrene-co-isobutylstyryl-POSS) (PAS–POSS) systems [49, 50]. The presence of the POSS moiety effectively increases the glass transition temperatures of the resultant organic/inorganic hybrid polymers at relatively high POSS contents. Furthermore, a series of poly(hydroxystyrene-co-vinylpyrrolidone-co-isobutylstyryl-POSS) hybrid polymers incorporating various POSS contents were prepared through free radical copolymerization of acetoxystyrene, vinylpyrrolidone, and POSS, followed by selective removal of the acetyl protecting group [51, 52]. The value of $T_g$ of the PHS–PVP–POSS hybrids increases substantially upon incorporating the POSS moiety (Figure 6.18). The presence of the physically crosslinked POSS through hydrogen bonding in these hybrid polymers tends to restrict the polymer chain motion, and results in higher $T_g$.

Figure 6.19 displays the C=O stretching region (1620–1720 cm$^{-1}$) of FTIR spectra recorded at room temperature of different concentrations of ethylpyrrolidone (EPr, a model compound for PVP) in cyclohexane, PAS-co-PVP-co-POSS2.2, PAS-co-PVP, pure PVP, PVPh/PVP blend, PVPh-co-PVP, PVPh-co-PVP-co-POSS0.8, PVPh-co-PVP-co-POSS2.2, and PVPh-co-PVP-co-POSS3.2 copolymers [51, 52]. The C=O band of the EPr/cyclohexane solution is broaden and shifted to lower wavenumber upon increasing the EPr concentration because of higher probability of forming pyrrolidone/pyrrolidone interactions. These results confirm that the wavenumber and half-width are both dependent upon the specific and dipole interactions between molecules or polymer chains. In addition, the signal for the pure PVP homopolymer appeared at lower wavenumber and gives a

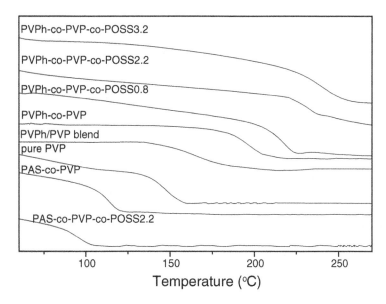

**Figure 6.18**  DSC thermograms of PAS-*co*-PVP-*co*-POSS, PAS-*co*-PVP, pure PVP, PVPh/PVP blend, PVPh-*co*-PVP, and with different POSS contents of PVPh-*co*-PVP-*co*-POSS.

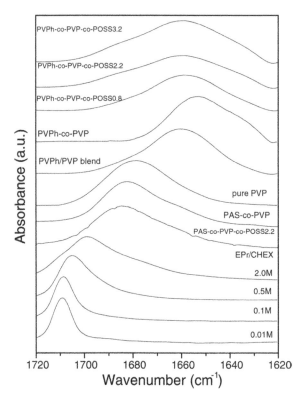

**Figure 6.19**  FTIR spectra for various EPr concentrations in cyclohexane, PAS-*co*-PVP-*co*-POSS, PAS-*co*-PVP, pure PVP, PVPh/PVP blend, PVPh-*co*-PVP, and with different POSS contents of PVPh-*co*-PVP-*co*-POSS.

broader half-width relative to that of EPr in cyclohexane because no inert diluent (nonpolar) group is present in the pure PVP homopolymer. Therefore, the probability of dipole–dipole interactions of the PVP is expected to be greater than that in the EPr/cyclohexane system. Upon incorporation of acetoxystyrene monomers into the PVP chain, the half-width of the PVP C=O band at 1680 cm$^{-1}$ decreases and shifts to higher wavenumber (1682 cm$^{-1}$). The value of $T_g$ of PAS-*co*-PVP decreases significantly (26°C) as a result of the lower degree of dipole–dipole interactions of the pyrrolidone moieties in the polymer chain. The C=O band of the PAS-*co*-PVP at 1682 cm$^{-1}$ shifts slightly higher (to 1683 cm$^{-1}$) upon incorporation of the POSS moiety into the PAS-*co*-PVP copolymer chain. It appears that the incorporation of the inert diluent group (POSS) into the polymer chain decreases the strength of its original dipole–dipole interactions [51, 52].

For this reason, the $T_g$ of PAS-*co*-PVP-*co*-POSS2.2 also decreases significantly (24°C) relative to that of the PAS-*co*-PVP copolymer. This result provides evidence for why most POSS hybrid polymer systems possess lower $T_g$, relative to those of the original polymer matrices, at lower POSS contents. Figure 6.20 displays all the shifts and interactions of the characteristic vibration bands [51, 52].

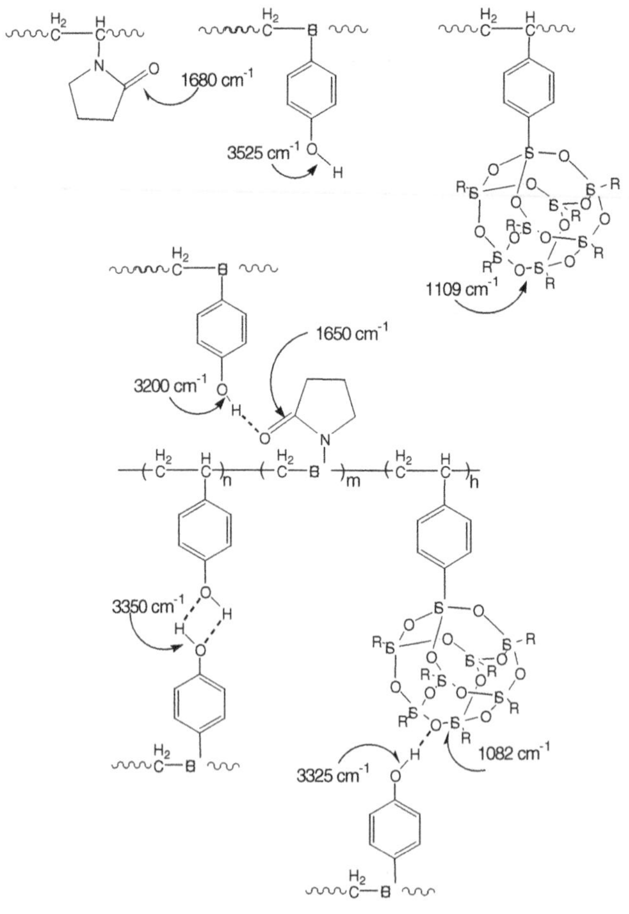

**Figure 6.20**   Interaction of POSS and PVPh.

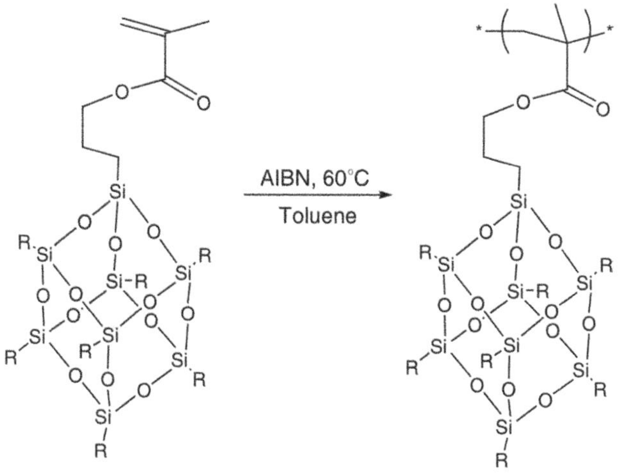

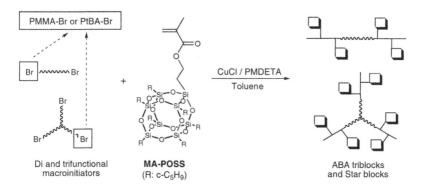

**Figure 6.21**  Synthesis of PMMA–POSS nanocomposites.

**Figure 6.22**  Synthesis of triblock and star-block copolymers.

### 6.4.3. Poly(methacrylate)/POSS Copolymers

Methacrylate-substituted POSS macromers can be prepared containing one polymerizable functional group (Figure 6.21) [53–60]. The resulting materials are generally transparent brittle plastics, i.e., the incorporation of the POSS group into linear polymers tends to prevent or reduce the segmental segregation or mobility.

Pyun and Matyjaszewski used atom transfer radical polymerization (ATRP) to synthesize an MA-POSS homopolymer and block copolymers from a cyclopentyl-substituted POSS monomer [60, 61]. They employed ATRP to prepare the block copolymer of MA-POSS with *n*-butyl acrylate; 4-(methylphenyl) 2-bromoisobutyrate as the initiator and CuCl/PMDETA as the catalyst (Figure 6.22) [60, 61]. Triblock copolymers were prepared from a difunctional (both ends) bromine-terminated poly(*n*-butyl acrylate) macroinitiator ($M_{n,NMR}$ = 7780) and MA-POSS. In the syntheses of the triblock copolymers, the macroinitiator was prepared through ATRP of *n*-BA using dimethyl-2,6-dibromoheptanedioate as the difunctional initiator. Synthesis of P(MA-POSS)-*b*-PBA-*b*-P(MA-POSS) triblock and PMA-*b*-P(MA-POSS) star-block copolymers were carried out

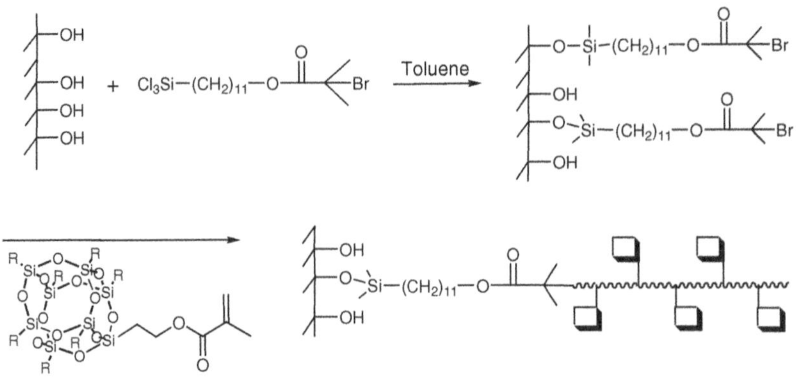

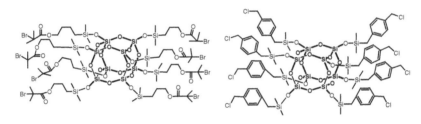

**Figure 6.23**  Surface-initiated ATRP of POSS-MA.

**Figure 6.24**  ATRP initiators of octafunctional silsesquioxane cube.

by chain extension of the difunctional bromine-terminated *n*-butyl acrylate macroinitiator using MA-POSS monomer. A star-block copolymer of methyl methacrylate and MA-POSS were also prepared through ATRP of methyl methacrylate using a trifunctional initiator. The polydispersity ($M_w/M_n$) of this star block copolymer was 1.30 [60, 61].

TEM characterization of the triblock copolymer thin film P(MA-POSS)$_{10}$-b-PBA$_{201}$-b-P(MA-POSS)$_{10}$ revealed the formation of PBA cylinders in a P(MA-POSS) matrix. Chen et al. reported the polymerization of an isobutyl-substituted MA-POSS monomer from a self-assembled monolayer (SAM) of ATRP initiators covalently immobilized on flat silicon wafers (Figure 6.23) [62]. This method is a simple and effective approach toward preparing well-defined POSS-containing polymer films from flat surfaces [62].

Laine et al. synthesized a star poly(methyl methacrylate) from an octafunctional silsesquioxane cube using the "core-first" method and ATRP (Figure 6.24) [63].

Fukuda and co-workers used an incompletely condensed POSS possessing a highly reactive trisodium silanolate group for the synthesis of several initiators for ATRP to obtain tadpole-shaped polymeric hybrids with a POSS unit at the end of the polymer chain (Figure 6.25) [64, 65].

Recently, Chang et al. reported polymer hybrids having controllable molecular weights and a chain-end-tethered POSS moiety synthesized through ATRP (Figure 6.26) [66].

Blending both PMMA–POSS and PMMA with phenolic resin revealed that the POSS terminus affected the thermal properties, miscibility behavior, and hydrogen bonding interactions [66]. A further investigation [67] of the specific interassociation interactions between the terminal siloxane units of the POSS moieties and the OH groups

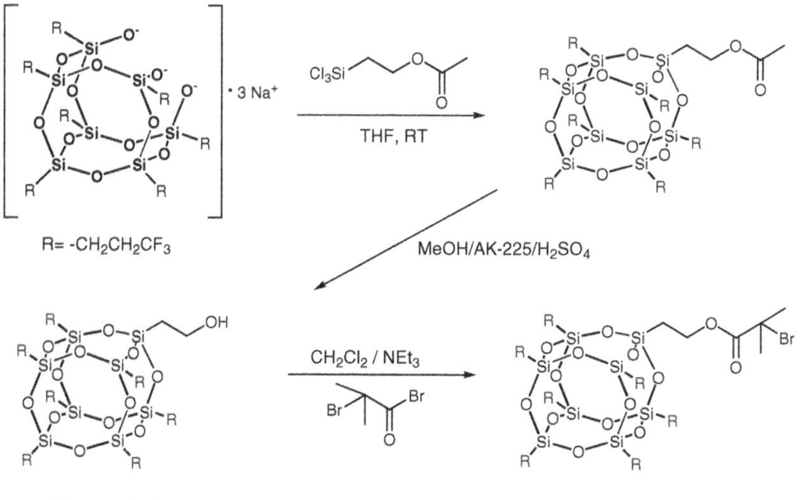

**Figure 6.25**    Synthesis of fluorinated POSS-Holding Initiator 7F-T8-BIB.

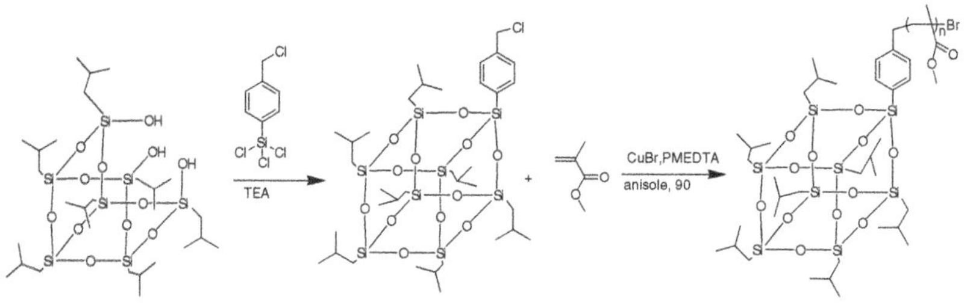

POSS-Cl                                    POSS-PMMA

**Figure 6.26**    Synthetic route to prepare PMMA–POSS and PMMA polymers.

of the phenolic revealed an interesting "screening effect" in these phenolic and low-molecular-weight PMMA–POSS blends (less than PMMA entanglement), as indicated in Figure 6.27 [66].

### 6.4.4. Poly(ethylene oxide)/POSS Nanocomposites

Poly(ethylene oxide) (PEO) is an important polymer electrolyte in lithium ion batteries because of its ability to solvate lithium ions. Nevertheless, it is a semicrystalline polymer and ionic conduction occurs mostly in the amorphous phase; the room temperature (RT) ionic conductivity of PEO is low due to its crystallization. The ion mobility increases with greater free volume upon increasing the number of chain ends; therefore, higher conductivity is expected for low-molar-mass or highly branched PEO polymers compared with linear, high-molar-mass PEO. Thus, one approach toward increasing the RT conductivity of PEO-based polymer electrolytes is the attachment of short-chain PEO oligomers as side chains to form "comb-shaped" or "hairy rod"-like polymeric structures, or as "arms" from inorganic scaffolds. Wunder et al. [67] grafted oligomeric PEO chains onto $Q_8M_8^H$ to produce the PEO functionalized silsesquioxanes as displayed in Figure 6.28.

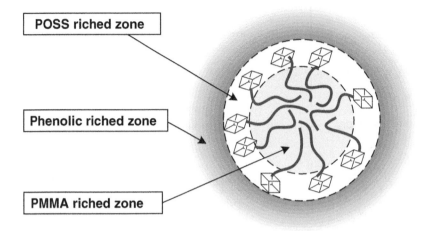

**Figure 6.27**   Proposed screening effect microstructure in phenolic/PMMA–POSS blends.

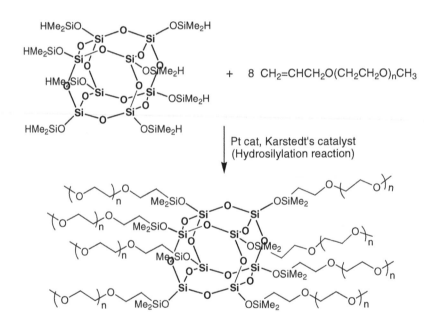

**Figure 6.28**   Reaction scheme to prepare PEO-octafunctionalized silsesquioxanes.

A strong self-supporting film prepared from 10% methyl cellulose and 90% $Q_8M_8PEO$ ($n = 8$)/LiClO$_4$ resulted in RT conductivity of $10^{-5}$ S/cm [67].

In addition, well-defined amphiphilic telechelics incorporating POSS have been synthesized through direct urethane linkage between the OH end groups of a poly(ethylene glycol) (PEG) homopolymer and the monoisocyanate groups of a POSS macromer, as indicated in Figure 6.29 [68].

They found that the synthesized amphiphilic telechelics exhibited a relatively narrow and unimodal molecular weight distribution ($M_w/M_n < 1.1$) and had close to 2.0 end

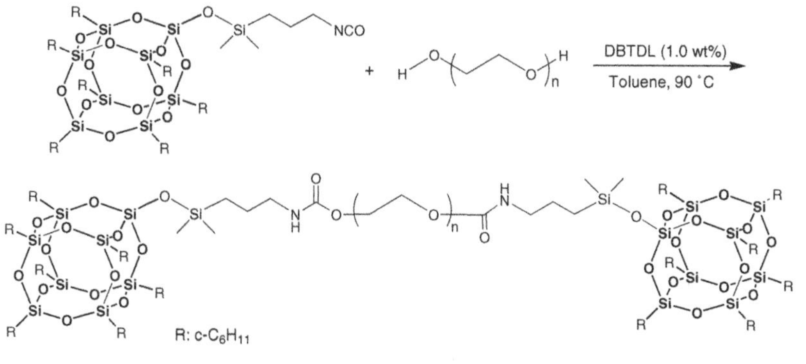

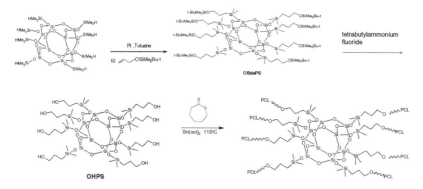

**Figure 6.29**    Reaction scheme of the amphiphilic telechelics incorporating POSS.

**Figure 6.30**    Synthesis of OHPS-cored eight-arm PCL star polymers.

groups per PEG chain [68]. The crystallinity of the PEO segments in the amphiphilic telechelics decreased dramatically when the POSS content in the amphiphilic telechelic reached 40.7%, and became amorphous at values beyond ca. 50%. As a result, they obtained several amphiphilic telechelics with a range of thermal and morphological properties through controlling the ratio of the hydrophilic PEG homopolymer and the hydrophobic and bulky POSS macromers [68].

## 6.4.5.  PCL/POSS Nanocomposites

Chang et al. synthesized a series of the organic/inorganic hybrid star PCLs through coordinated ring-opening polymerization of $\varepsilon$-caprolactone using POSS as the initiator (Figure 6.30) [69].

Similar to the linear PCL analogues reported previously, this star PCL formed inclusion complexes (ICs) with $\alpha$- and $\gamma$-CD, but not with $\beta$-CD [69]. The stoichiometries of all the ICs with $\alpha$- and $\gamma$-CD were greater than those of the corresponding CD/linear PCL ICs because of steric hindrance around the bulky POSS core and some g of the g$\varepsilon$-caprolactone units near the core were unable to form ICs. Zheng et al. [70] reported another approach for the preparation of octa(3-hydroxypropyl) POSS; they used an octafunctional initiator for the synthesis of eight-armed star-shaped PCLs. Organic/inorganic star PCLs possessing

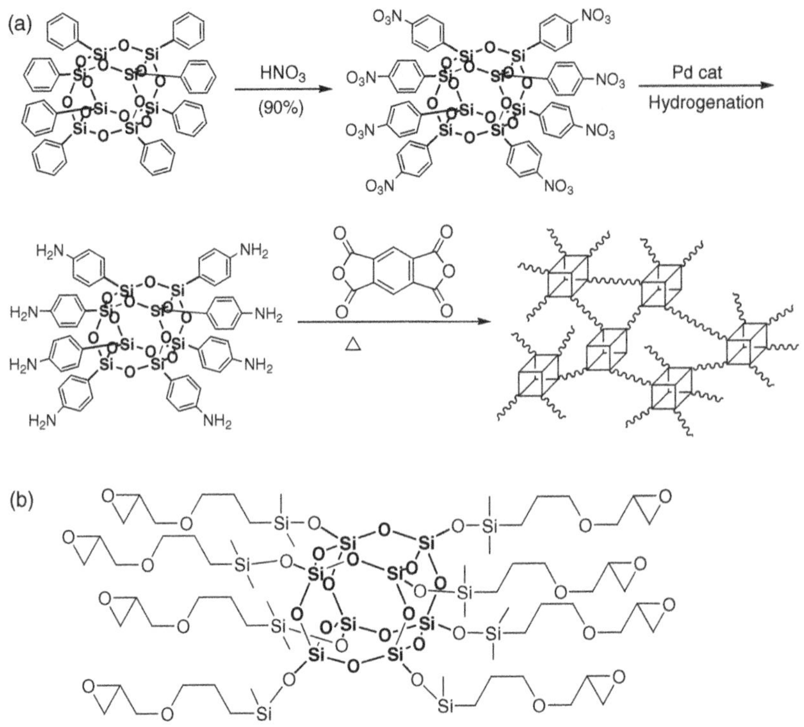

(a)

(b)

**Figure 6.31**    (a) Laine's (b) Lee's.

various degrees of polymerization were synthesized via ring-opening polymerization catalyzed by stannous(II) octanoate [Sn(Oct)$_2$] [70].

## 6.4.6.  Polyimide/POSS Nanocomposites

Polyimides (PIs) are widely used in microelectronics industries because of their outstanding characteristics, such as excellent tensile strength and modulus, good thermal stability and dielectric properties, and high resistance to organic solvents. There are two common routes for incorporating POSS units into a PI matrix. One method of achieving PI–POSS nanocomposites is to use a POSS derivative possessing eight functional groups (e.g., epoxy or amino groups) to serve as a crosslinking agent. Laine and co-workers described [71, 72] the nitration of octa-phenyl POSS, which was first reported in 1961 by Olsson [72], and subsequent production of octakis(aminophenyl)–POSS through Pd/C-catalyzed hydrogenation (Figure 6.31(a)).

Furthermore, the octa-amino POSS was employed in conjunction with dianhydrides to prepare extremely thermally resistant crosslinked PI networks [71, 72]. This amino-POSS derivative can be reacted with maleic anhydride to obtain the octa-$N$-phenylmaleimide [73], which can also serve as a crosslinking agent in maleimide polymer chemistry. He et al. [74] and Chang et al. [75] demonstrated that the PI prepolymer, polyamic acid (PAA), can react with octakis(glycidyldimethylsiloxy)-octasilsesquioxane (Figure 6.31(b)).

Using these approaches, the dielectric constants of the PI nanocomposites can be reduced, and their thermal properties modified, upon increasing the POSS content.

**Figure 6.32** SEM cross-section analysis of PI–POSS hybrid materials (a) PI-3P, (b) PI-7P, (c) PI-10P. (d) Fractured cross-section surface of the PI-10P.

Figures 6.32(a)–(d) display cross-sectional SEM images of various morphologies obtained at various epoxy-POSS contents. In this study, excess diamine (ODA) was reacted initially with BTDA and then the terminal amino groups of the polyamic acid were reacted with epoxy-POSS. By varying the equivalent ratio of the ODA, polyamic acids of various molecular weights and nanocomposites possessing various morphologies were obtained. The reduction in the dielectric constant of the PI–POSS hybrids can be explained in terms of the nanovoid volume of the POSS cores and the free volume increase resulting from the presence of the rigid, large POSS units inducing a loose PI network (Figure 6.32(d)) [75].

Wei et al. [76–78] reported the grafting reaction of a novel POSS derivative as an approach toward PI-tethered POSS nanocomposites possessing well-defined architectures (Figure 6.33). These types of PI-tethered POSS nanocomposites have both lower and tunable dielectric constants, with the lowest value of 2.3, and controllable mechanical properties, relative to those of the pure PI. The tethered POSS molecules in the amorphous PI retain a nanoporous crystal structure, but form an additional ordered architecture as a result of microphase separation. Using this approach, the dielectric constant of the film can be tuned by modifying the amount of POSS added [76–78].

## 6.4.7. Epoxy/POSS Nanocomposites

Monofunctional and multifunctional POSS–epoxies have been incorporated into the backbones of epoxy resins to improve their thermal properties [79–84]. The multifunctional

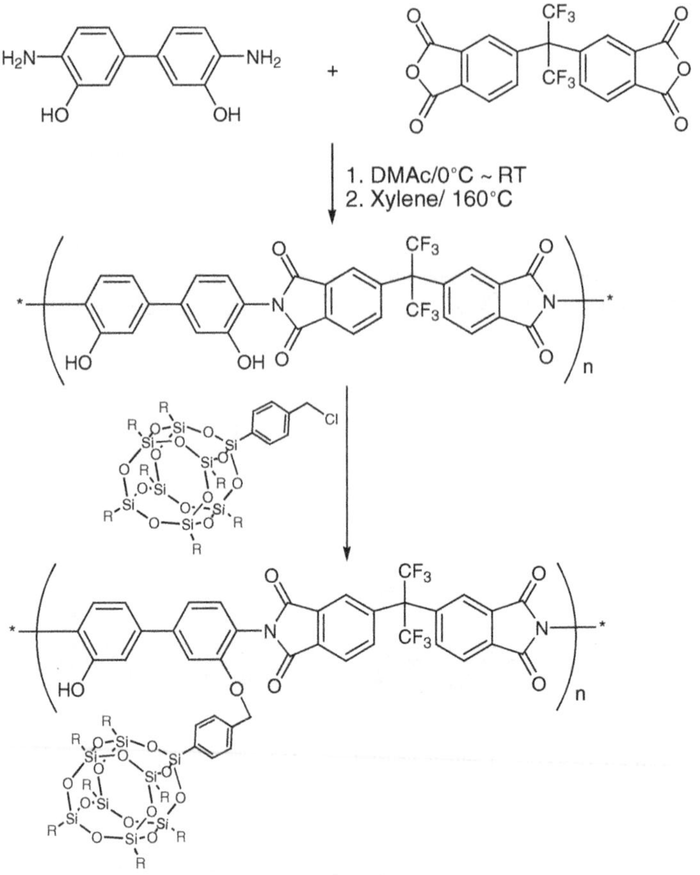

**Figure 6.33**    Polyimide-tethered POSS nanocomposites by grafting reaction.

epoxy-substituted POSS monomer was incorporated into an epoxy resin network composed of difunctional epoxies. The presence of POSS can increase the glass transition temperature of the epoxy resin because the nanoscopic size and mass of the POSS cages enhance their ability to hinder the segmental motion of molecular chains and network junctions. Chang et al. also reported a new nanomaterial (Figure 6.34) based on POSS-epoxy (OG) and *meta*-phenylenediamine (mPDA) [85]. The activation energy in curing the OG/mPDA system was higher than that of the DGEBA/mPDA system, as determined using both the Kissinger and Flynn–Wall–Ozawa methods [86, 87]. In an isothermal kinetic study based on an autocatalytic system, the activation energy for curing OG/mPDA was also higher than that of the DGEBA/mPDA system [85].

The $T_g$ of the cured OG/mPDA product is significantly higher than that of the DGEBA/mPDA material because the presence of these POSS cages is able to effectively hinder the motion of the network junctions. The cured OG/mPDA product possesses inherently higher thermal stability than the cured DGEBA/mPDA product, as evidenced by the higher maximum decomposition temperature and the higher char yield of the former system. However, the existence of a large fraction of unreacted amino groups causes a lower initial decomposition temperature of the OG/mPDA system because

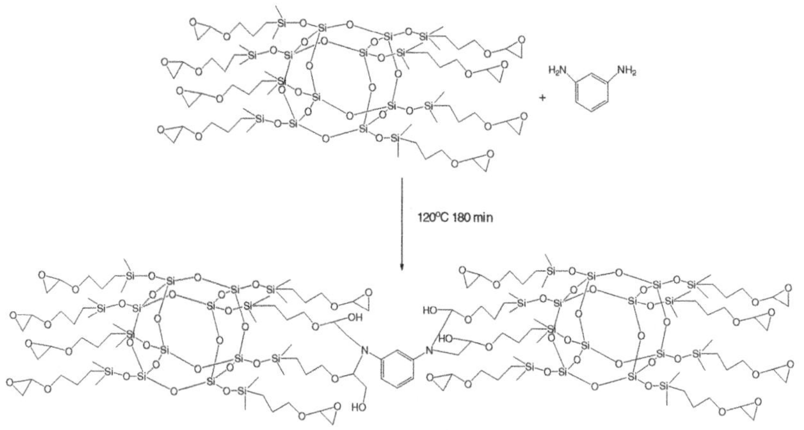

**Figure 6.34**  Curing mechanism of OG cured with mPDA.

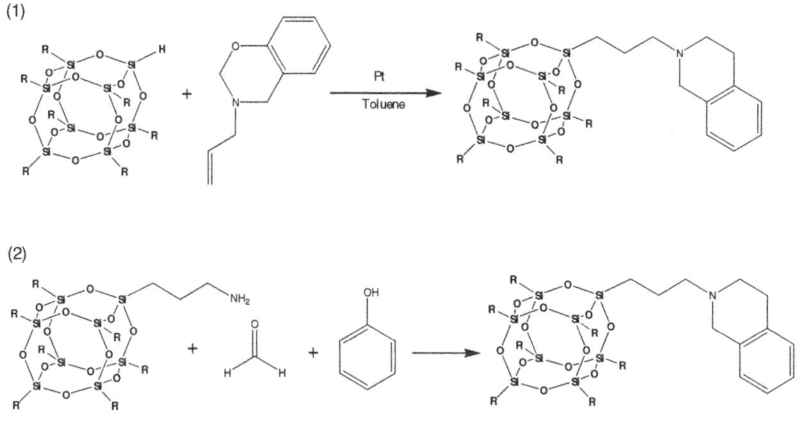

**Figure 6.35**  The chemical structure and mechanism for the synthesis of benzoxazine-POSS.

they tend to decompose or volatilize upon heating at relatively low temperatures. The dielectric constant of the OG/mPDA material (2.31) is substantially lower than that of the DGEBA/mPDA system (3.51) because of the presence of the nanoporous POSS cubes in the epoxy matrix [85].

## 6.4.8.  Polybenzoxazine/POSS Nanocomposites

Chang et al. synthesized a novel benzoxazine ring containing POSS monomer (BZ–POSS) through two routes (Figure 6.35): (i) hydrosilylation of a vinyl-terminated benzoxazine using the hydrosilane functional group of a POSS derivative (H–POSS) and (ii) the reaction of a primary amine-containing POSS (amine-POSS) with phenol and formaldehyde [88]. The BZ–POSS monomer was copolymerized with other benzoxazine monomers through ring-opening polymerization under conditions similar to those used to polymerize pure benzoxazines. The thermal properties of these POSS-containing organic/inorganic poly-benzoxazine nanocomposites were improved over those of the pure polybenzoxazine, as analyzed using DSC and TGA [88].

The chemical structure and mechanism for the synthesis of benzoxazine-POSS.

In addition, benzoxazine can be synthesized through the Mannich condensation of phenol, formaldehyde, and primary amines through ring-opening polymerization. Poly-benzoxazines are phenolic-like materials that possess dimensional and thermal stability, and release no toxic byproducts during polymerization [89]. To further improve the thermal stability of polybenzoxazines, a hydrosilane-functionalized polyhedral oligomeric silsesquioxane (H-POSS) was incorporated into the vinyl-terminated benzoxazine monomer (VB-a) followed by ring-opening polymerization. Chang et al. also prepared hybrids from a nonreactive POSS (IB-POSS) and VB-a. The value of $T_g$ of a regular polymerized VB-a (i.e., PVB-a) was 307°C, whereas for the hybrid containing 5 wt% H-POSS it was 333°C [90]. A new class of polybenzoxazine/POSS nanocomposites possessing network structures was prepared through reacting a multifunctional benzoxazine POSS (MBZ–POSS) with benzoxazine monomers (Pa and Ba) at various compositional ratios. The octafunctional cubic silsesquioxane (MBZ–POSS) used as the curing agent was synthesized from eight organic benzoxazine tethers through hydrosilylation of the vinyl-terminated benzoxazine monomer VP-a with $Q_8M_8$ using a Pt-dvs as the catalyst. Incorporation of the silsesquioxane cores into the polybenzoxazine matrix significantly hindered the mobility of the polymer chains and enhanced the thermal stability of these hybrid materials [90].

## 6.4.9.  Other Applications

### 6.4.9.1. *Polymer Light Emitting Diodes (PLEDs) Incorporating POSS Hybrid Polymers*

Heeger et al. reported the incorporation of POSS moieties into conjugated polymers [91, 92], by synthesizing (Figure 6.36) the POSS-anchored semiconducting polymers POSS-

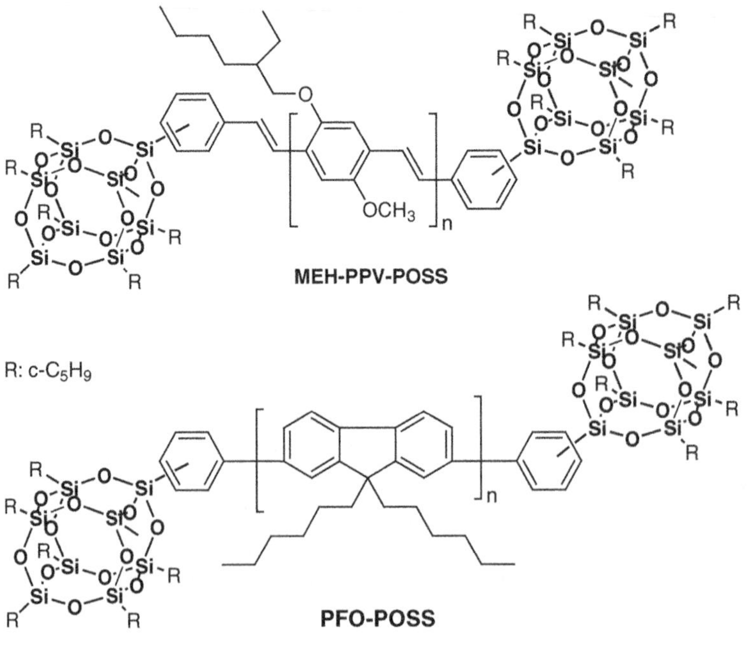

**Figure 6.36**  Molecular structure of MEH–PPV–POSS and PFO–POSS.

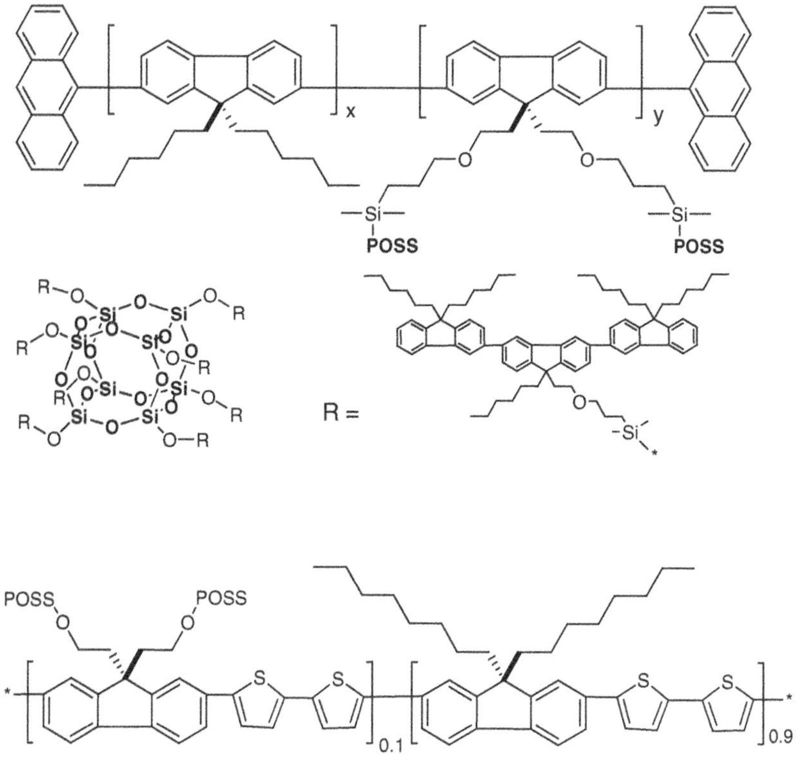

**Figure 6.37**    Molecular structure of PFO–POSS.

anchored poly(2-methoxy-5-(ethylexyloxy)-1,4-phenylenevinylene) (MEH–PPV–POSS) and POSS-anchored poly(9,9-dihexyfluorenyl-2,7-diyl) (PFO–POSS). Relative to the corresponding parent polymer, these POSS-anchored semiconducting polymers exhibited higher brightness and external quantum efficiencies [91, 92].

The role of POSS was suggested to reduce the formation of aggregation and/or excimers or to lower the concentration of conjugated defects. Shim et al. [93–96] prepared POSS-substituted polyfluorene polymers (Figure 6.37). The incorporation of the POSS groups inhibited interchain interactions and fluorenone formation and, thus, led to a reduction in the degree of undesired emission (>500 nm) of the poly(dialkylfluorene)s and an improvement in the thermal stability of the PFO–POSS systems. Devices incorporating PFO–POSS hybrids as emitting layers exhibited very stable blue light emissions and high performance [93].

Wei et al. and Hsu et al. [97–99] reported a new series of asymmetric conjugated polymers presenting POSS units on their side chains (Figure 6.38). An EL device prepared from MEH–PPV emitted a strong peak at 590 nm and a vibronic signal in the range 610–620 nm [97].

The introduction of bulky siloxane units into the PPV side chains presumably increases the interchain distance, thereby retarding interchain interactions and reducing the degree of exciton migration to defect sites. A star-like polyfluorene derivative, PFO–SQ, was synthesized through the Ni(0)-catalyzed reaction of octa(2-(4-bromophenyl) ethyl)octasilsesquioxane (OBPE–SQ) and polydioctylfluoroene (PFO), as depicted in Figure 6.39 [100]. The incorporation of the silsesquioxane core into the polyfluorene

POSS
POSS
HN
NH
OCH3
1-n
n
m


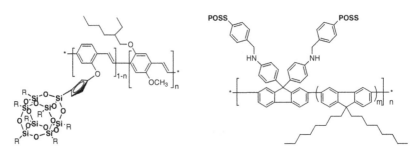

**Figure 6.38** Synthesis of MEH–PPV–POSS.

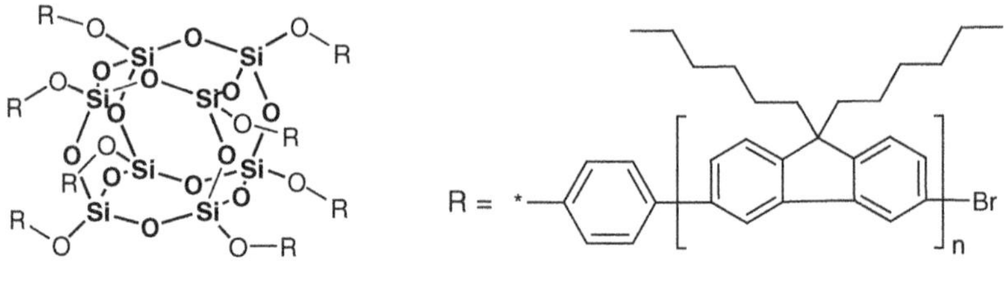

R =
Br
n


**PFO-SQ**

**Figure 6.39** Synthetic route for PFO-SQ.

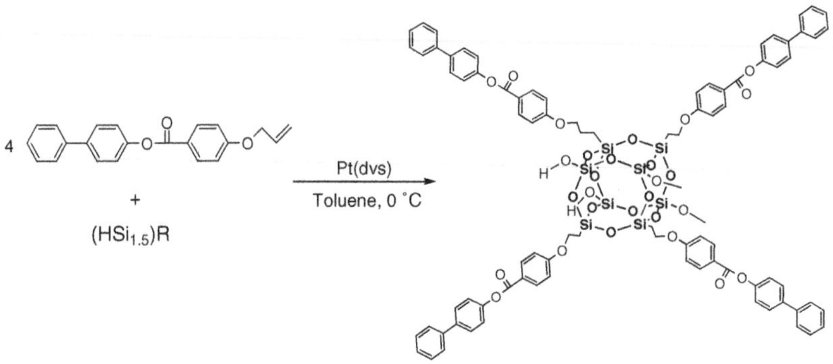

4
+
(HSi1.5)R
Pt(dvs)
Toluene, 0 °C


**Figure 6.40** Syntheses of LC–POSS.

significantly reduced the degree of aggregation and enhanced the thermal stability. The incorporation of inorganic silsesquioxane cores into polyfluorenes is a new method for preparing organic light-emitting diodes with improved thermal and optoelectronic characteristics [100].

### 6.4.9.2. *Liquid Crystal Polymers (LCPs) Incorporating POSS Hybrid Polymers*

With the goal of developing diverse building blocks for nanocomposite materials, Laine and coworkers [101] synthesized liquid crystalline materials by appending mesogenic groups to cubic silsesquioxane cores via hydrosilylation of allyloxy-functionalized mesogens with octakis(dimethylsiloxy)octasilsesquioxane ($Q_8M_8^H$). Figure 6.40 indicates that hydrosilylation leads to cubes possessing an average of five appended LC groups; this

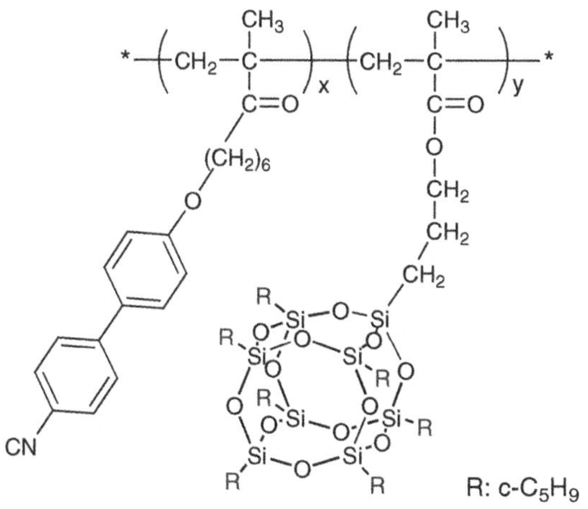

**Figure 6.41**    Synthesis of LC–POSS.

structure differs from the more "regular" fully LC-substituted analogs reported previously. Despite the structural irregularity, three of the four penta-LC-cube derivatives exhibited LC transitions, with a tendency to form SmA. One interesting observation was a redistribution of the diffuse liquid-like scattered intensity at 11–12 Å in the smectic phase upon alignment. These results provide the basis for future work on producing LC cubes as potential precursors to LC-ordered organic/inorganic nanocomposites. Although, the LC transition temperatures were reduced somewhat, they remained above those values considered useful for biologically important applications [101].

Chujo reported the preparation of a POSS macromonomer through radical copolymerization, LC hybrid copolymers incorporating various proportions of the synthesized LC monomer (Figure 6.41) [102]. The obtained LC hybrid polymers were soluble in common solvents, such as tetrahydrofuran, toluene, and chloroform. The thermal stability of the hybrid polymers increased upon increasing the ratio of POSS moieties [102].

### 6.4.9.3. *Lithographic Applications of POSS-Containing Photoresists*

Several POSS-based photoresists have been reported, including positive-tone POSS-containing photoresists. Wu et al. [103, 104] reported that the incorporation of POSS units in methacrylate-based chemically amplified photoresists influenced their reactive ion etching (RIE) behavior (Figure 6.42). Whereas polymers incorporating low POSS concentrations exhibited little improvements in their RIE resistances, the presence of 20.5 wt% POSS monomer in methacrylate-based resist significantly improved the RIE resistance in $O_2$ plasma. High-resolution transmission electron microscopy revealed that the RIE resistance improvement was due to the formation of rectangular crystallite-constituting networks of the silica cages uniformly distributed within the polymer matrix [103, 104].

Gonsalves et al. [105, 106] synthesized and characterized a series of POSS-containing positive-tone photoresists for use in both extreme ultraviolet lithography (EUVL) and electron beam lithography (Figure 6.43).

These photoresist systems exhibited the ideal combination of enhanced etch resistance and enhanced sensitivity required to satisfy both low- and high-voltage patterning applications. The photoresist sensitivity was enhanced after the direct incorporation of a

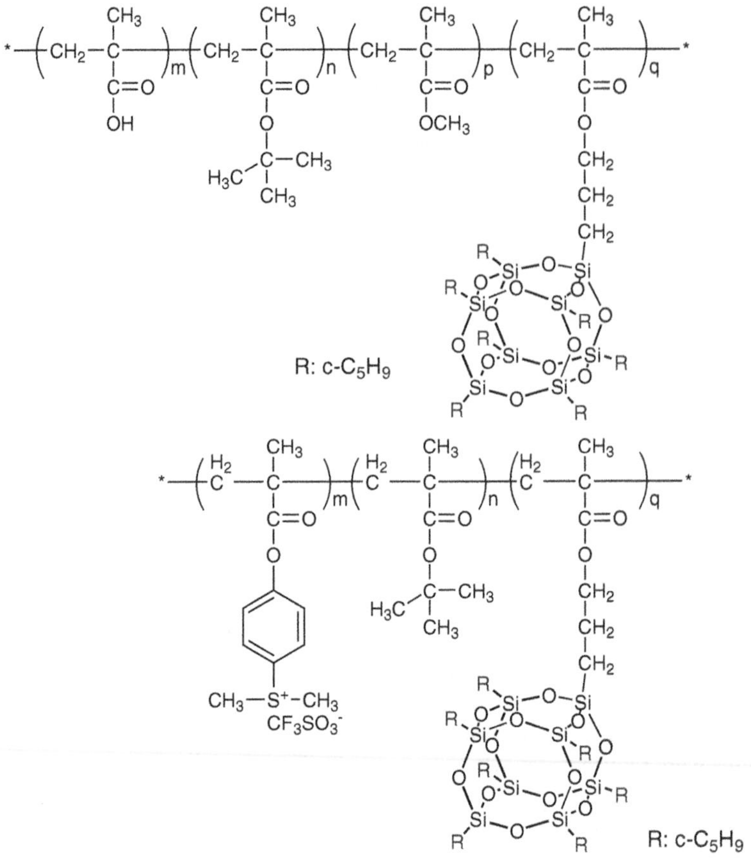

**Figure 6.42**   Microstructures of methacrylate-based chemically amplified photoresists.

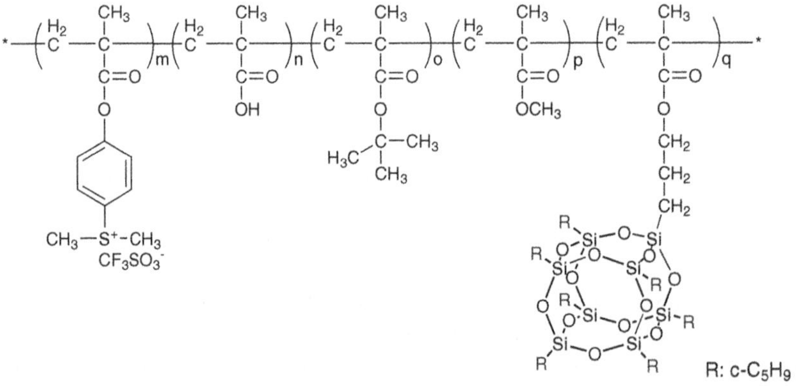

**Figure 6.43**   Microstructure of the nanocomposite resist.

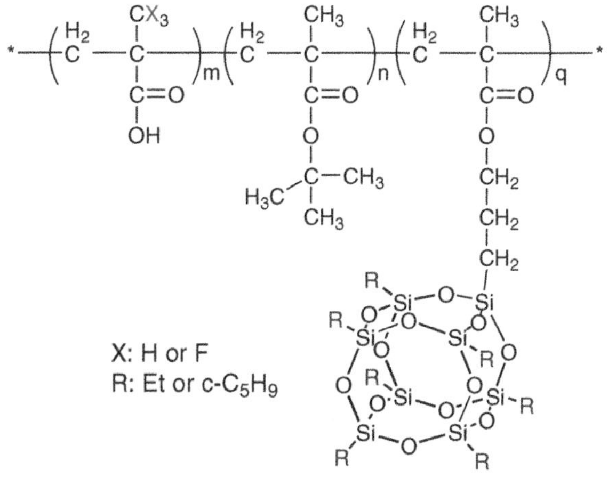

**Figure 6.44**  Microstructure of the POSS terpolymers.

photoacid-generating monomer into the resist polymer backbone, while the etch resistance of the material was improved after copolymerization with a POSS-containing monomer [105]. Argitis et al. [107–109] described the lithographic behavior and related material properties of chemically amplified, positive-tone, methacrylate-based photoresists incorporating POSS groups as the etch-resistant component (Figure 6.44).

The POSS-containing photoresists studied for 157-nm lithographic applications exhibited high sensitivity (<10 mJ/cm$^2$ under open field exposure), no silicon outgassing, and sub-100-nm resolution capabilities; indeed, 90-nm patterns in 100-nm-thick films could be resolved. Alternatively, an octavinylsesquioxane dry resist was prepared as a negative-tone resist that exhibited high sensitivity for deep-UV, electron-beam, and X-ray lithography [110]. Furthermore, Zheng et al. [111] synthesized a novel photosensitive otaccinnamoylamidophenyl POSS derivative (OcapPOSS) through the reaction of cinnamoyl chloride and octaaminophenyl POSS. The presence of the POSS cages restricted the mobility of the poly(vinyl cinnamate) macromolecular chains and, thus, hindered the formation of tetrabutane rings. DSC analysis revealed that the values of $T_g$ of the nanocomposites were significantly higher than that of the parent poly(vinyl cinnamate). Recently, Chang et al. [112] developed methacrylate-based, POSS derivative-containing photoresist materials for UV-lithography with enhanced sensitivity, higher contrast, and improved resolution as a result of the presence of hydrogen bonding interactions between the siloxane units of the POSS moieties and the OH groups [30, 39]. Hydrogen bonding interactions within these photosensitive copolymers not only formed physically crosslinked bonds but also raised the density of methacrylate olefinic units around the POSS moieties, thereby enhancing the rate of chemically crosslinked photopolymerization (Figure 6.45) [112].

### 6.4.9.4. *Low-k Applications of POSS-Containing Materials*

To decrease the dielectric constants of polymers, several research groups [113–117] have been exploring the possibility of incorporating various nanoforms into polymer matrixes to take the advantage of the low dielectric constant of air ($k = 1$). There are two typical routes employed for preparing materials with low dielectric constants: (i) nanopore formation through decomposition (e.g., thermal decomposition, photodegradation with UV irradiation, or solvent etching) of a dispersion phase within a material matrix and

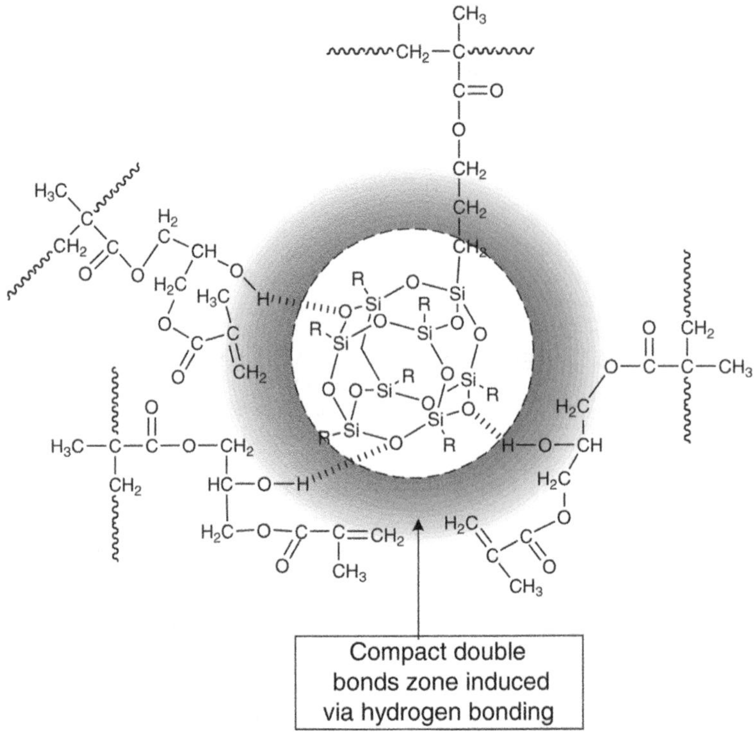

**Figure 6.45**   The proposed microstructure via hydrogen bonding interaction of POSS.

(ii) incorporation of a low-$k$ moieties (e.g., fluorinated units, supercritical fluids, or low-$k$ particles) within a material matrix. There are several drawbacks to each of these two methods, such as contaminant release, solvent effects, and compatibility of the dispersion and the matrix. Recently, a low dielectric constant material was prepared after dispersion of POSS-containing molecules into a polymer matrix [51, 53]. Chang et al. developed some POSS-containing low-$k$ materials using both of the methods mentioned above, taking advantage of hydrogen bonding interactions between the POSS moieties and OH groups to achieve homogenous nanocomposites [118, 119]. In these nanocomposite materials, the POSS moieties not only served as low-$k$ materials but also contributed to the improved thermal and mechanical properties.

## 6.5. Conclusion

The POSS nanocomposites presented here are composite materials reinforced with silica cages, i.e., an ultrafine filler of nanometer size, which is almost equal to the size of the polymer matrix. Although the POSS content can be as low as ca. 1–5 wt%, individual POSS particles can exist at distances as close as several nanometers apart. Therefore, these nanocomposites possess microstructures that do not exist in conventional composites. For this reason, the field of POSS nanocomposite research has become extremely active in recent years (Figure 6.46). In this chapter, we have reviewed only the studies that have made major contributions to this research field, focusing on the preparation of such polymer–POSS nanocomposites as styryl-POSS, methacrylate-POSS,

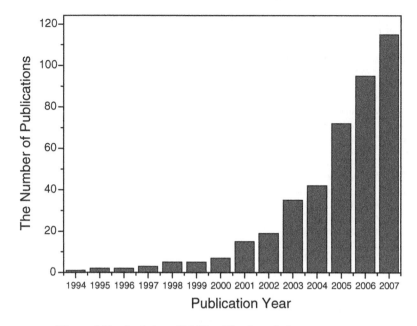

**Figure 6.46**    Statistics of POSS publications during recent years.

norbornyl-POSS, vinyl-POSS, epoxy-POSS, phenolic-POSS, benzoxazine-POSS, amine-POSS, and hydroxyl-POSS. Both monofunctional and multifunctional monomers of these types are used to prepare commercial and/or high-performance thermoplastic and thermosetting polymers. This chapter also provides details of their corresponding thermal, dynamic mechanical, electrical, and surface properties. Other important publications related to POSS are cited in the reference section [120–161].

## Acknowledgments

This work was supported financially by the National Science Council, Taiwan, Republic of China, under Contract No. NSC-96-2120-M-009-009 and NSC-96-2218-E-110-008 and Ministry of Education "Aim for the Top University" program (MOEATU program).

## References

1. Li, G.; Wang, L.; Ni, H.; Pittman Jr. C. U. *J. Inorgan. Organometal. Polym.* **2001**, 3, 123.
2. Scott, D. W. *J. Am. Chem. Soc.* **1946**,68, 356.
3. Baney, R. H.; Itoh, M.; Sakakibara, A.; Suzuki, T. *Chem. Rev.* **1995**, 95, 1409.
4. Gozdz, A. S. *Polym. Adv. Technol.* **1994**, 5, 70.
5. Voronkov, M. G.; Lavrentyev, V. I. *Topics Curr. Chem.* **1982**, 102, 199.
6. Feher, F. J.; Terroba, R.; Jin, R.; Wyndham, K. O.; Lucke, S.; Brutchey, R.; Nguyen, F. *Polym. Mater. Sci. Eng.* **2000**,82, 301.
7. Brown, I. F. *New Sci.* **1963**, 17, 304.
8. Martynova, T. N.; Chupakhina, T. I. *J. Organometal. Chem.* **1988**,345, 11.
9. Agaskar, P. A. *Inorg. Chem.* **1991**,30, 2707.
10. Chalk, A. J.; Harrod, J. F. *J. Am. Chem. Soc.* **1965**,87, 16.
11. Tsuchida, A.; Bolln, C.; Sernetz, F. G.; Frey, H.; Mulhaupt, R. *Macromolecules* **1997**,30, 2818.
12. Calzaferri, G.; Herren, D.; Imhof, R. *Helv. Chim. Acta* **1991**,74, 1278.

13. Calzaferri, G.; Imhof, R. *J. Chem. Soc., Dalton Trans.* **1992**, 3391.
14. Marcolli, C.; Imhof, R.; Calzaferri, G. *Microchim. Acta* **1997**,14, 493.
15. Feher, F. J. *J. Am. Chem. Soc.* **1986**, 108, 3850.
16. Feher, F. J.; Weller, K. J. *Organometal.* **1990**, 9, 2638.
17. Feher, F. J.; Weller, K. J . *Inorg. Chem.* **1991**, 30, 880.
18. Feher, F. J.; Newman, D. A.; Walzer, J. F. *J. Am. Chem. Soc.* **1989**,111, 1741.
19. Lichtenhan, J. D. *Comments Inorg. Chem.* **1995**,17, 115 .
20. Ruffieux, V.; Schmid, G.; Braunstein, P.; Rose, J. *Chem. Eur. J.* **1997**,3, 900.
21. Murugavel, R.; Voigt, A.; Walawalkar, M. G.; Roesky, H. W. *Chem. Rev.* **1996**,96, 2205.
22. Feher, F. J.; Budzichowski, T. A.; Weller, K. J. *J. Am. Chem. Soc.* **1989**,111, 7288.
23. Feher, F. J.; Budzichowski, T. A.*Organometal.* **1991**,10, 812.
24. Feher, F. J.; Walzer, J. F. *Inorg. Chem.* **1991**,30, 1689.
25. Frye, C. L.; Collins, W. T. *J. Am. Chem. Soc.* **1970**,92, 5586.
26. Zhang, C.; Laine, R. M. *J. Am. Chem. Soc.* **2000**, 122, 6979.
27. Sellinger, A.; Laine, R. M. *Macromolecules* **1996**, 29, 2327.
28. Sellinger, A.; Laine, R. M. *Chem. Mater.***1996**, 8, 1592.
29. Painter, P. C.; Coleman, M. C. *Prog. Polym. Sci.* **1995**, 20, 1.
30. Lee. Y. J.; Kuo, S. W.; Huang, W. J. Lee, H. Y. Chang, F. C. *J. Polym. Sci. Polym. Sci. Ed.* **2004**, 42, 1127.
31. Noda, I. *J. Am. Chem. Soc.* **1989**, 111, 8116.
32. Coggesthall, N.D.; Saier, E. L. *J. Am. Chem. Soc.* **1951**, 71, 5414.
33. Wu, H. D.; Chu, P. P.; Ma, C. C. M.; Chang, F.C. *Macromolecules* **1999**, 32, 3097.
34. Kuo, S. W. Lin, H. C. Huang, W. J. Huang, C. F. Chang, F. C. *J. Polym. Sci. Polym. Sci. Ed.* **2006**, 44, 673.
35. Moskala, E. J.; Varnell D. F.; Coleman, M. M. *Polymer*, **1985**, 26, 228.
36. Kuo, S. W. ; Chang, F. C. *Macromol. Chem. Phys.* **2002**, 203, 868.
37. Kwei, T. K. *J. Polym. Sci., Polym. Lett. Ed.*, **1984**,22, 307.
38. Huang, C. F.; Kuo, S. W.; Chen, Y. K.; Chang, F. C. *Polymer* **2004**, 45, 5913.
39. Lin, H. C.; Kuo, S. W. Huang, C. F. Chang, F. C. *Macromol. Rapid Commun.* **2006**, 27, 537.
40. Fu, B. X.; Yang, C.; Somani, R. H. Zong, S. X.; Hsiao, B. S.; Phillips, S.; Blanski, R.; Ruth, D. *J. Polym. Sci. Polym. Sci. Ed.* **2001**,39, 2727.
41. Fina, A.; Tabuani, D.; Frache, A.; Camino, G. *Polymer* **2005**, 46, 7855.
42. Pracella, M.; Chionna, D.; Fina, A.; Tabuani, D.; Frache, A.; Camino, G. *Macromol. Symp.* **2006**, 234, 59.
43. Zheng, L.; Farris, R. J.; Coughlin, E. B. *Macromolecules* **2001**,34, 8034.
44. Zheng, L.; Farris, R. J.; Coughlin, E. B. *J. Polym. Sci.: Part A: Polym. Chem.* **2001**, 39, 2920.
45. Mather, P. T.; Jeon, H. G.; Uribe, A. R. *Macromolecules* **1999**, 32, 1194.
46. Haddad, T. S.; Lichtenhan, J. D. *Macromolecules* **1996**, 29, 7302.
47. Zheng, L.; Kasi, R. M.; Farris, R. J.; Coughlin, E. B. *J. Polym. Sci.: Part A: Polym. Chem. Ed .***2002**, 40, 885.
48. Cardoen, G.; Coughlin, E. B. *Macromolecules* **2004**, 37, 5123.
49. Xu, H.; Kuo, S. W.; Huang, C. F.; Chang, F. C. *J. Polym. Res.* **2002**, 9, 239.
50. Xu, H.; Kuo, S. W.; Lee, J. S.; Chang, F. C. *Macromolecules* **2002**, 35, 8788.
51. Xu, H.; Kuo, S. W.; Chang, F. C. *Polym. Bull.* **2002**, 48, 869.
52. Xu, H.; Kuo, S. W.; Lee, J. S.; Chang, F. C. *Polymer* **2002**, 43, 5117.
53. Lichtenhan, J. D.; Otonari, Y. A.; Carr, M. J. *Macromolecules* **1995**, 28, 8435.
54. Kopesky, E. T.; Haddad, T. S.; Cohen, R. E.; Mckinley, G. H. *Macromolecules* **2004**, 37, 8992.
55. Li, G. Z.; Cho, H.; Wang, L.; Toghiani, H.; Pittman Jr. C. *J. Polym. Sci.: Polym. Chem. Ed.* **2005**, 43, 355.
56. Bizet, S.; Galy, J.; Gerard, J. F. *Macromolecules* **2006**, 39, 2574.
57. Kopesky, E. T.; McKinley, G. H.; Cohen, R. E. *Polymer* **2006**, 47, 299.
58. Castelvetro, V.; Ciardelli, F.; Vita, C. D.; Puppo A. *Macromol. Rapid Commun.* **2006**, 27, 619.
59. Patel, R. R.; Mohanraj, R.; Pittman Jr. C. U. *J. Polym. Sci.: Polym. Phy. Ed.* **2006**, 44, 234.
60. Pyun, J.; Matyjaszewski, K. *Macromolecules* **2000**, 33, 217.
61. Pyun, J.; Matyjaszewski, K. *Polymer* **2003**, 44, 2739.
62. Chen, R.; Feng, W.; Zhu, S.; Botton, G.; Ong, B.; Wu, Y. *Polymer* **2006**, 47, 1119.
63. Costa, R. O. R.; Vasconcelos, W. L.; Tamaki, R.; Laine, R. M. *Macromolecules* **2001**; 34; 5398.
64. Ohno, K.; Sugiyama, S.; Koh, K.; Tsujii, Y.; Fukuda, T.; Yamahiro, M.; Oikawa, H.; Yamamoto, Y.; Ootake, N.; Watanabe, K. *Macromolecules* **2004**, 37, 8517.
65. Koh, K.; Sugiyama, S.; Morinaga, T.; Ohno, K.; Tsujii, Y.; Fukuda, T.; Yamahiro, M.; Iijima, T.; Oikawa, H.; Watanabe, K.; Miyashita, T. *Macromolecules* **2005**, 38, 1264.

66. Huang, C. F.; Kuo, S. W.; Lin, F. J.; Huang, W. J.; Wang, C. F.; Chen, W. Y.; Chang, F. C. *Macromolecules* **2006**; 39; 300.
67. Maitra, P.; Wunder, S. L. *Chem. Mater.* **2002**, 14, 4494.
68. Kim, B. S.; Mather, P. T. *Macromolecules* **2002**, 35, 8378.
69. Chan, S. C.; Kuo, S. W.; Chang, F. C. *Macromolecules* **2005**, 38, 3099
70. Liu, Y. H.; Yang, X. T.; Zhang, W. A.; Zheng, S. X. *Polymer* **2006**, 47, 6814.
71. Laine, R. M.; Tamaki, R.; Choi, J.; Brick, C.; Kim, S. G. *Organic/Inorganic Hybrid Materials Workshop.* Sonoma, CA: ACS, 2000.
72. Tamaki, R.; Choi, J.; Laine, R. M. *Chem. Mater.* **2003**, 15, 793.
73. Olsson, K.; Gronwall, C. *Arkiv. Kemi.* **1961**, 17, 529.
74. Huang, J.; Xiao, Y.; Mya, K. Y.; Liu, X.; He, C.; Dai, J.; Siow, Y. P. *J. Mater. Chem.* **2004**, 14, 2858.
75. Lee, Y. J.; Huang, J. M.; Kuo, S. W.; Lu, J. S.; Chang, F. C. *Polymer* **2005**, 46, 173.
76. Leu, C. M.; Chang, Y. T.; Wei, K. H. *Macromolecules* **2003**, 36, 9122.
77. Leu, C. M.; Reddy, G. M.; Wei, K. H.; Shu, C. F. *Chem. Mater.* **2003**, 15, 2261.
78. Leu, C. M.; Chang, Y. T.; Wei, K. H. *Chem. Mater.* **2003**, 15, 3721.
79. Choi, J.; Harcup, J.; Yee, A. F.; Zhu,Q.; Laine, R. M. *J. Am. Chem. Soc.* **2001**, 123, 11240.
80. Liu, Y. L.; Chang, G. P. *J. Polym. Sci.: Polym. Chem. Ed.* **2006**, 44, 1869.
81. Liu, Y.; Zheng, S.; Nie, K. *Polymer* **2005**, 46, 12016.
82. Huang, J.; He, C.; Liu, X.; Xu, J.; Tay, C. S. S.; Chow, S. Y. *Polymer* **2005**, 46, 7018.
83. Strachota, A.; Kroutilova, I.; Rova, J. K.; Matejka, L. *Macromolecules* **2004**, 37, 9457.
84. Abad, M. J.; Barral, L.; Fasce, D. P.; Williams, R. J. J. *Macromolecules* **2003**, 36, 3128.
85. Chen, W. Y.; Wang, Y. Z.; Kuo, S. W.; Huang, C. F.; Tung, P. H.; Chang, F. C. *Polymer* **2004**, 45, 6897.
86. Kissinger, H. E. *Anal. Chem.* **1957**, 29, 1702.
87. Ozawa T. *Bull. Chem. Soc. Jpn.* **1965**, 38, 1881.
88. Lee, Y. J.; Kuo, S. W.; Su, Y. C.; Chen, J. K.; Tu, C. W.; Chang, F. C. *Polymer* **2004**, 45, 6321.
89. Lee, Y. J.; Huang, J. M.; Kuo, S. W.; Chen, J. K. Chang, F. C. *Polymer* **2005**, 46, 2320.
90. Lee, Y. J.; Kuo, S. W.; Huang, C. F.; Chang, F. C. *Polymer* **2005**, 47, 4378.
91. Xiao, S.; Nguyen, M.; Gong, X.; Cao, Y.; Wu, H.; Moses, D.; Heeger, A. J. *Adv. Funct. Mater.* **2003**, 13, 25.
92. Gong, X.; Soci, C.; Yang, C. Y.; Heeger, A. J.; Xiao, S. *J. Phys. D: Appl. Phys.* **2006**, 39, 2048.
93. Lee, J.; Cho, H.; Jung, B.; Cho, N.; Shim, H. *Macromolecules* **2004**,37, 8523.
94. Cho, H. J.; Hwang, D. H.; Lee, J. I.; Yung, Y. K.; Park, J. H.; Lee, J.; Lee, S. K.; Shim, H. K. *Chem. Mater.* **2006**, 18, 3780.
95. Lee, J.; Cho, H. J.; Cho, N. S.; Hwang, D. H.; Kang, J. M.; Lim, E.; Lee, J. I.; Shim, K. H. *J. Polym. Sci.: Part A: Polym. Chem.* **2006**, 44, 2943.
96. Kang, J. M.; Cho, H. J.; Lee, J.; Lee, J. I.; Lee, S. K.; Cho, N. S.; Hwang, D. H.; Shim, H. K. *Macromolecules* **2006**, 39, 4999.
97. Chen, K. B.; Chen, H. Y.; Yang, S. H.; Hsu, C. S. *J. Polym. Res.* **2006**, 13, 229.
98. Chou, C. H.; Hsu, S. L.; Yeh, S. W.; Wang, H. S.; Wei, K. H. *Macromolecules* **2005**,38, 9117.
99. Chou, C. H.; Hsu, S. L.; Dinakaran, K.; Chiu, M. Y.; Wei, K. H. *Macromolecules* **2005**,38, 745.
100. Lin, W. J.; Chen, W. C.; Wu, W. C.; Niu, Y. H.; Jen, A. K. J. *Macromolecules* **2004**, 37,2335.
101. Zhang, C.; Bunning, T. J.; Laine, R. M. *Chem. Mater.* **2001**, 13, 3653.
102. Kim, K. M.; Chujo, Y. *J. Polym. Sci.: Part A: Polym. Chem.* **2001**, 39, 4035.
103. Wu, H.; Hu, Y.; Gonsalves, K. E.; Yacaman, M. J. *J. Vac. Sci. Technol. B.* **2001**, 19, 851.
104. Wu, H.; Gonsalves, K. E. *Adv. Mater.* **2001**, 13, 670.
105. Ali, M. A.; Gonsalves, K. E.; Golovkina, V.; Cerrina, F. *Microelectron. Eng.* **2003**, 65, 454.
106. Ali, M. A.; Gonsalves, K. E.; Agrawal, A.; Jeyakumar, A.; Henderson, C. L. *Microelectron. Eng.* **2003**, 70, 19.
107. Bellas, V.; Tegou, E.; Raptis, I.; Gogolides, E.; Argitis, P.; Iatrou H.; Hadjichristidis, N.; Sarantopoulou, E.; Cefalas, A. C. *J. Vac. Sci. Technol. B* **2002**, 20, 2902.
108. Tegou, E.; Bellas, V.; Gogolides, E.; Argitis, P.; Eon, D.; Cartry, G.; Cardinaud, C. *Chem. Mater.* **2004**, 16, 2567.
109. Tegou, E.; Bellas, V.; Gogolides, E.; Argitis, P. *Microelectron. Eng.* **2004**, 73, 238.
110. Schmidt, A.; Babin, S.; Koops, H. W. P. *Microelectron. Eng.* **1997**, 35, 129.
111. Ni, Y.; Zheng, S. *Chem. Mater.* **2004**,16, 5141.
112. Lin, H. M.; Wu, S. Y.; Huang, C. F.; Kuo, S. W.; Chang, F. C. *Macromol. Rapid Commun.* **2006**, 27, 1550.
113. Carter, K. R.; McGrath, J. E. *Chem. Mater.* **1997**, 9, 105.

114. Carter, K. R.; DiPietro, R. A.; Sanchez, M. I.; Swanson, S. A. *Chem. Mater.* **2001**, 13, 213.
115. Mikoshiba, S.; Hayase, S. *J. Mater. Chem.* **1999**, 9, 591.
116. Krause, B. R.; Mettinkhof; N. F. *Macromolecules* **2001**, 34, 874.
117. Krause, B. R.; Mettinkhof, N. F. *Adv. Mater.* **2002**, 14, 1041.
118. Lee, Y. J.; Huang, J. M.; Kuo, S. W.; Chang, F. C. *Polymer* **2005**, 46, 10056.
119. Chen, W. Y.; Ho, K, S.; Hsieh, T. H.; Chang, F. C.; Wang, Y. Z. *Macromol. Rapid Commun.* **2006**, 27, 452.
120. Haddad, T. S.; Lichtenhan, J. D. *J. Inorgan. Organometal. Polym.* **1995**, 5, 237.
121. Mantz, R. A.; Jones. P. F.; Chaffee; K. P.; Lichtenhan; J. D.; Gilman; J. W.; Ismail; I. M. K; Burmeister, M. J. *Chem. Mater.* **1996**, 8, 1250.
122. Romo-Uribe A.; Mather, P. T.; Haddad, T. S.; Lichtenhan, J. D. *J. Polym. Sci.: Polym. Phys. Ed.* **1998**, 36, 1857.
123. Lee, A.; Lichtenhan, J. D. *Macromolecules* **1998**, 31, 4970.
124. Mather, P. T.; Jeon, H. G.; Romo-Uribe, A.; Haddad, T. S.; Lichtenhan, J. D. *Macromolecules* **1999**, 32, 1194.
125. Shockey, E. G.; Bolf, A. G.; Jones, P. F.; Schwab, J. J.; Chaffee, K. P.; Haddad, T. S.; Lichtenhan, J. D. *Appl. Organometall. Chem.* **1999**, 13, 311.
126. Bharadwaj, R. K.; Berry, R. J.; Farmer, B. L. *Polymer* **2000**, 41, 7209.
127. Fu, B. X.; Hsiao, B. S.; Pagola, S.; Stephens, P.; White, H.; Rafailovich, M.; Sokolov, J.; Mather, P. T.; Jeon, H. G.; Phillips, S.; Lichtenhan, J.; Schwab, J. *Polymer* **2001**, 42, 599.
128. Zheng, L.; Farris, R. J.; Coughlin, E. B. *J. Polym. Sci.: Polym. Chem. Ed.* **2001**, 39, 2920.
129. Kim, K. M.; Chujo, Y. *J. Polym. Sci.: Polym. Chem. Ed.* **2001**, 39, 4035.
130. Zheng, L.; Farris, R. J.; Coughlin, E. B. *Macromolecules* **2001**, 34, 8034.
131. Li, G. Z.; Wang, L. C.; Toghiani, H.; Daulton, T. L.; Koyama, K.; Pittman, C. U. *Macromolecules* **2001**, 34, 8686.
132. Zheng, L.; Waddon, A. J.; Farris, R. J.; Coughlin, E. B. *Macromolecules* **2002**, 35, 2375.
133. Zhang, W. H.; Fu, B. X.; Seo, Y.; Schrag, E.; Hsiao, B.; Mather, P. T.; Yang, N. L.; Xu, D. Y.; Ade, H.; Rafailovich, M.; Sokolov, J. *Macromolecules* **2002**, 35, 8029.
134. Waddon, A. J.; Zheng, L.; Farris, R. J.; Coughlin, E. B. *Nano Lett.* **2002**, 2, 1149.
135. Abad, M. J. Barral, L.; Fasce, D. P.; Williams, R. J. J. *Macromolecules* **2003**, 36, 3128.
136. Huang, J. C.; He, C. B.; Xiao, Y.; Mya, K. Y.; Dai, J.; Siow, Y. P. *Polymer* **2003**, 44, 4491.
137. Carroll, J. B.; Waddon, A. J.; Nakade, H.; Rotello, V. M. *Macromolecules* **2003**, 36, 6289.
138. Lamm, M. H.; Chen, T.; Glotzer, S. C. *Nano Lett.* **2003**, 3, 989.
139. Waddon, A. J.; Coughlin, E. B. *Chem. Mater.* **2003**, 15, 4555.
140. Huang, J. C.; Li, X.; Lin, T. T.; He, C. B.; Mya, K. Y.; Xiao, Y.; Li, J. *J. Polym. Sci.: Polym. Phy. Ed.* **2004**, 42, 1173.
141. Yei, D. R.; Kuo, S. W.; Su, Y. C.; Chang, F. C. *Polymer* **2003**, 45, 2633.
142. Baker, E. S.; Gidden, J.; Anderson, S. E.; Haddad, T. S.; Bowers, M. T. *Nano Lett.* **2004**, 4, 779.
143. Fu, B. X.; Lee, A.; Haddad, T. S. *Macromolecules* **2004**, 37, 5211.
144. Dvornic, P. R.; Hartmann-Thompson, C.; Keinath, S. E.; Hill, E. J. *Macromolecules* **2004**, 37, 7818.
145. Zheng, L.; Hong, S.; Cardoen, G.; Burgaz, E.; Gido, S. P.; Coughlin, E. B. *Macromolecules* **2004**, 37, 8606.
146. Liu, H. Z.; Zhang, W.; Zheng, S. X. *Polymer* **2005**, 46, 157.
147. Liu, H. X.; Zheng, S. X. *Macromol. Rapid Commun.* **2005**, 26, 196.
148. Anderson, S. E.; Baker, E. S.; Mitchell, C.; Haddad, T. S.; Bowers, M. T. *Chem. Mater.* **2005**, 17, 2537.
149. Turri, S.; Levi, M. *Macromolecules* **2005**, 38, 5569.
150. Liu, Y. H.; Meng, F. L.; Zheng, S. X. *Macromol. Rapid Commun.* **2005**, 26, 920.
151. Hillson, S. D.; Smith, E.; Zeldin, M.; Parish, C. A. *Macromolecules* **2005**, 38, 8950.
152. Xu, H. Yang, B.; Wang, J. Guang, S.; Li, C. Li, C. *Macromolecules* **2005**, 38, 10455.
153. Joshi, M.; Butola, B. S.; Simon, G.; Kukaleva, N. *Macromolecules* **2006**, 39, 1839.
154. Bizet, S.; Galy, J.; Gerard, J. F. *Macromolecules* **2006**, 39, 2574.
155. Soong, S. Y.; Cohen, R. E.; Boyce, M. C.; Mulliken, A. D. *Macromolecules* **2006**, 39, 2900.
156. Cui, L.; Zhu, L. *Langmuir* **2006**, 22, 5982.
157. Cui, L.; Collet, J. P.; Xu. G. Q.; Zhu, L. *Chem. Mater.* **2006**, 18, 3503.
158. Miao, J. J.; Cui, L.; Lau, H. P.; Mather, P. T.; Zhu, L. *Macromolecules* **2007**, 40, 5460.
159. Ni, Y.; Zheng, S. X. *Macromolecules* **2007**, 40, 7009.
160. Hao, N.; Boehning, M.; Schoenhals A. *Macromolecules* **2007**, 40, 9672.
161. Markovic, E.; Ginic-Markovic M.; Clarke, S.; Matisons, J.; Hussain, M.; Simon, G. P. *Macromolecules* **2007**, 40, 2694.

# 7 Carbon Nanotubes-Reinforced Thermoplastic Nanocomposites

*Rani Joseph*[1] *and Anoop Anand*[1,2]

[1] Department of Polymer Science and Rubber Technology, Cochin University
of Science and Technology, Cochin 682 022, India
rani@cusat.ac.in
[2] Composites Research Centre, Research and Development Establishment (Engineers),
DRDO – Ministry of Defence, Pune 411 015, India

**Abstract**

This chapter articulates the use of carbon nanotubes (CNTs) as a filler phase for thermoplastic polymers. The nucleation effect of CNTs in various thermoplastic matrices along with their reinforcing effect is explored. Effect of nanotubes on the rheological properties as well as CNTs-induced electrical conductivity is also detailed in this chapter.

**Keywords:** carbon nanotubes, thermoplastics, crystallization, mechanical properties, electrical conductivity

## 7.1. Introduction

Our ability to engineer novel structures has led to unprecedented opportunities in materials design. It has fueled rapid development in nanoscience and nanotechnology for the last one decade, leading to the creation of new materials with interesting nanoscale features. In the framework of this rapid development, the domain of nanocomposite materials is attracting more and more researchers – both academic and industrial. The field of nanocomposites involves the study of multiphase materials, in which at least one of the constituent phases has a minimum of one dimension of the order of nanometers [1]. The use of these nanoscale fillers to augment the properties of polymers has provided a radical alternative to conventional composites and modified polymers. The promise of nanocomposites lies in their multifunctionality, the possibility of realizing unique combinations of properties unachievable with traditional materials.

This chapter explores the potential of carbon nanotubes (CNTs) for fabricating polymer-based nanocomposites. Apart from the state-of-art research in polymer–nanotube nanocomposites, a concise introduction to CNTs, their synthesis, structure, and properties is also presented.

## 7.2. Carbon Nanotubes

The groundbreaking discovery of CNTs in 1991 followed by the realization of their amazing properties led scientists all over the world to focus their research efforts on these fascinating structures. CNTs (also known as bucky tubes) are long thin cylinders of carbon that are unique for their size, shape, and remarkable physical and electrical properties [2–5]. They can be thought of as layers of the conventional graphite structure rolled up into a cylinder such that the lattice of carbon atoms remains continuous around the circumference. Their name is derived from their size since the diameter of a nanotube is of the order of a few nanometers (approximately 50,000 times smaller than the width of a human hair), although they can be up to several micrometers in length.

The number of carbon shells in CNTs varies from 1 to as many as 50, the former being single-walled carbon nanotubes (SWNTs) and the latter multiwalled carbon nanotubes (MWNTs). These intriguing structures have sparked much excitement in recent years and a large amount of research has been dedicated to them [6–8]. They are potentially useful in a wide variety of applications in nanotechnology, electronics, optics, and other fields of materials science [9–14].

### 7.2.1. The Discovery

The discovery of CNTs dates back to the 1985 – sequence of experiments by Harry Kroto, of the University of Sussex, and Richard Smalley, of Rice University, Houston [15, 16]. During the vaporization of graphite, Kroto and Smalley were surprised by an outcome, that is, in the distribution of the resulted gas-phase carbon cluster, detected by mass spectroscopy, $C_{60}$ was by far the most dominant species. Later they realized that a *closed* cluster containing precisely 60 carbon atoms would have a structure of unique stability and symmetry, as shown in Figure 7.1. Although they had no direct evidence to support this structure, subsequent work has proved them correct. The discovery of $C_{60}$ published in

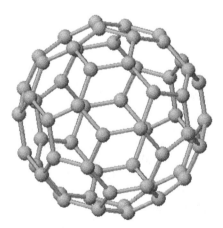

**Figure 7.1**   $C_{60}$: Buckminster fullerene.

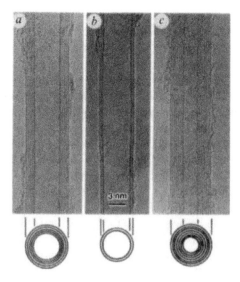

**Figure 7.2**    Iijima's electron micrographs of carbon nanotubes [20].

*Nature* in November 1985 had an impact that extended the way beyond the confines of academic chemical physics and marked the beginning of a new era in carbon science [17, 18].

However, the progress was slow in the beginning mainly because of the small quantity of $C_{60}$ produced in the Kroto–Smalley experiments. Eventually, more than a laboratory curiosity, the bulk production of $C_{60}$ was achieved by a technique developed by Wolfgang Krätschmer of the Max Planck Institute at Heidelberg, Germany and Donald Huffman of the University of Arizona, Arizona, United States. They used a simple carbon arc to vaporize graphite in an atmosphere of helium and collected the soot, which settled on the walls of the vessel. Dispersing the soot in benzene produced a red solution that could be dried down to produce beautiful plate like crystals of "fullerite": 90%, $C_{60}$ and 10%, $C_{70}$. This report appeared in *Nature* in 1990 [19].

Sumio Iijima, a Japanese electron microspopist, was fascinated by the Krätschmer–Huffman report in *Nature* paper, and decided to embark on a detailed study of the soot produced by their technique. The initial high-resolution transmission electron microscopic (HRTEM) studies were disappointing: the soot collected from the walls of the arc-evaporation vessel appeared almost completely amorphous, with little obvious long-range structures. Eventually, Iijima turned his attention to the hard cylindrical deposit, which formed on the graphite cathode after arc evaporation. This cathodic soot contained a whole range of novel graphitic structures, the most striking of which were hollow fibers, finer and more perfect than any other structure previously seen. Iijima's beautiful images of CNTs were shown first at a meeting at Richmond, Virginia in October 1991, and published in *Nature* a month later (Figure 7.2) [20].

## 7.2.2. Preparation Methods of CNTs

CNTs are generally produced by three main techniques: arc discharge, laser ablation, and chemical vapor deposition (CVD).

Arc discharge method is the most common and perhaps the easiest way to produce CNTs, as it is rather simple to undertake. In this method, the nanotubes are formed through arc vaporization of two carbon rods placed end to end, separated by approximately 1 mm, in an enclosure that is usually filled with an inert gas (He, Ar) at low pressure (between 50 and 700 mbar). A direct current of 50–100 A driven by approximately 20 V creates a high-temperature discharge between the two electrodes. The discharge vaporizes one of the carbon rods and forms a small rod-shaped deposit on the other rod. Producing nanotubes in high yield depends on the uniformity of the plasma arc and the temperature of the deposit formed on the carbon electrode [21]. Two distinct methods of synthesis can be performed with the arc discharge setup. If SWNTs are preferable, the anode has to be doped with metal catalyst, such as Fe, Co, Ni, Y, or Mo. The quantity and quality of the nanotubes obtained depend on various parameters such as the metal concentration, inert gas pressure, kind of gas, the current, and system geometry. Usually, the diameter is in the range of 1.2–1.4 nm.

In 1995, Smalley's group at Rice University introduced laser vaporization technique for the synthesis of CNTs [22]. In this process, a pulsed or continuous laser is used to vaporize a graphite target in an oven at 1200°C. The oven is filled with He or Ar gas in order to keep the pressure at 500 Torr. A very hot vapor plume forms, then expands, and cools rapidly. As the vaporized species cool, small carbon molecules and atoms quickly condense to form larger clusters, possibly including fullerenes. The catalysts also begin to condense, but more slowly at first, and attach to carbon clusters and prevent their closing into cage structures [23]. Catalysts may even open cage structures when they attach to them. From these initial clusters, tubular molecules grow into SWNTs until the catalyst particles become too large, or until conditions have cooled sufficiently that carbon no longer can diffuse through or over the surface of the catalyst particles. It is also possible that the particles become so much coated with a carbon layer that they cannot absorb more and the nanotubes stop growing. Laser ablation is almost similar to arc discharge since the optimum background gas and catalyst mix is the same as in the arc discharge process, but it results in a higher yield of SWNTs having better properties and a narrower size distribution than those produced by arc discharge.

CVD synthesis is achieved by putting a carbon source in the gas phase and using an energy source, such as plasma or a resistively heated coil, to transfer energy to gaseous carbon source. Commonly used gaseous carbon sources include $CH_4$, CO, and $C_2H_2$. The energy is used to "crack" the molecule into reactive atomic carbon. Then the carbon diffuses toward the substrate, which is heated and coated with a catalyst (usually, a first row transition metal such as Ni, Fe, or Co) where it will bind. CNTs will be formed if the proper parameters are maintained. Excellent alignment, as well as positional control on nanometer scale, can be achieved by using CVD [24, 25]. Control over the diameter as well as the growth rate of the nanotubes can also be maintained. The appropriate metal catalyst can preferentially grow SWNTs rather than MWNTs [26]. The temperatures for the synthesis of nanotubes by CVD are generally within the 650–900°C ranges [27]. Typical yields of CVD are approximately 30%. The SWNTs produced by CVD method usually have large diameters, which is poorly controlled. But on the contrary, this method is very easy to scale up, which favors commercial production.

### 7.2.3. Structure and Properties

It is the chemical versatility of carbon that it can bond in different ways to create structures with entirely different properties. Graphite and diamond, the two bulk solid phases of pure carbon, bear testimony to this. The mystery lies in the different hybridization that carbon atoms can assume. The four valence electrons in carbon, when shared equally ($sp^3$

hybridized), create isotropically strong diamond. But when only three are shared covalently between neighbors in a plane and the fourth is allowed to be delocalized among all atoms, the resulting material is graphite. The latter ($sp^2$) type of bonding builds a layered structure with strong in-plane bonds and weak out-of-plane bonding of the van der Waals type. Graphite, hence, is weak normal to its planes and is considered as a soft material due to its ability to slide along the planes. The story of fullerenes and nanotubes belongs to the architecture of $sp^2$-bonded carbon and the subtlety of a certain group of topological defects that can create unique, closed shell structures out-of-planar graphite sheets [28].

Graphite is the thermodynamically stable bulk phase of carbon up to very high temperatures under normal ranges of pressure (diamond is only kinetically stable). It is now well known that this is not the case when there are only a finite number of carbon atoms. Simply speaking, this has to do with the high density of dangling bond atoms when the size of the graphite crystallites becomes small (say, nanosize). At small sizes, the structure does well energetically by closing onto itself and removing all the dangling bonds. Preliminary experiments done in the mid-1980s, which served as the precursor to the fullerene discovery, suggested that when the number of carbon atoms is smaller than a few hundred, the structures formed correspond to linear chains, rings, and closed shells [29]. The latter, called fullerenes, are closed shell all-carbon molecules with an even number of atoms (starting at $C_{28}$) and $sp^2$ bonding between adjacent atoms.

To form curved structures (such as fullerenes) from a planar fragment of hexagonal graphite lattice, certain topological defects have to be included in the structure. To produce a convex structure, positive curvature has to be introduced into the planar hexagonal graphite lattice. This is done by creating pentagons. It is a curious consequence of the Euler's principle that one needs exactly 12 pentagons to provide the topological curvature necessary to completely close the hexagonal lattice; hence, in $C_{60}$ and all the other fullerenes ($C_{2n}$ has $(n-10)$ hexagons), there are many hexagons but only 12 pentagons. The rule of pentagon numbers will hold, however big the closed structure may be created out of hexagons and pentagons. One can thus imagine that a greatly elongated fullerene can be produced with exactly 12 pentagons and millions of hexagons. This would correspond to a CNT [30].

The structure of an SWNT can be conceptualized by wrapping a one-atom-thick layer of graphite (called graphene) into a seamless cylinder and when concentric cylinders, one inside the other is present, they are referred to as MWNTs [31]. Most SWNTs have a diameter of close to 1 nm, with a tube length that can be many thousands of times larger (Figure 7.3). SWNTs with length up to orders of centimeters have been produced.

In the figure, (a) shows a schematic of an individual SWNT, (b) shows a cross-sectional view (TEM image) of a bundle of SWNTs (transverse view shown in (d)). Each nanotube has a diameter of approximately 1.4 nm and the tube–tube distance in the bundles is 0.315 nm. Figure 7.3(c) shows the HRTEM image of a 1.5-nm-diameter SWNT, (e) is the schematic of a MWNT, and (f) shows HRTEM image of an individual MWNT. The distance between layers of the tube in (f) is 0.34 nm.

CNT is one of the strongest materials known to man, in terms of both tensile strength and modulus [32]. The strength results from the covalent $sp^2$ bonds formed between the individual carbon atoms. In the year 2000, a nanotube was tested to have a tensile strength of 63 GPa. In comparison, high-carbon steel has a tensile strength of approximately 1.2 GPa. CNTs also have very high elastic modulus, of the order of 1 TPa [33]. Since CNTs have a low density for a solid of 1.3–1.4 g/cm$^3$, its specific strength is the best of known materials.

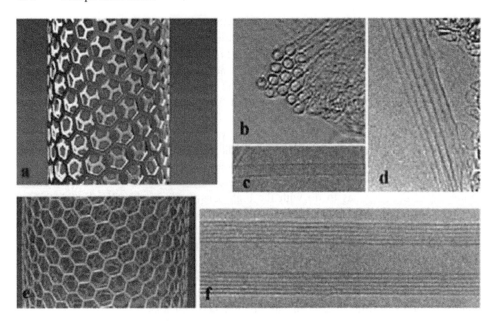

**Figure 7.3** Structure of single-walled (SWNT) (a–d) and multiwalled (MWNT) (e, f) carbon nanotubes [28].

Under excessive tensile strain, the tubes will undergo plastic deformation, which means the deformation is permanent. This deformation begins at strains of approximately 5% and can increase till the maximum strain the tube undergoes before fracture by releasing strain energy. CNTs are not nearly as strong under compression. Because of their hollow structure, they tend to undergo buckling when placed under compressive, torsional, or bending stress.

MWNTs – multiple concentric nanotubes precisely nested within one another – exhibit a striking telescoping property whereby an inner nanotube core may slide, almost without friction, within its outer nanotube shell, thus creating an atomically perfect linear or rotational bearing. This is one of the first true examples of molecular nanotechnology, the precise positioning of atoms to create useful machines. This property has already been utilized to create the world's smallest rotational motor and a nanorheostat.

## 7.3. Polymer–CNT Nanocomposites

For the last one decade, CNTs reinforced polymeric composite materials have spurred considerable attention in the materials research community, in part due to their potential to provide orders of magnitude increase in strength and stiffness when compared with virgin polymer or conventional micro- and macrocomposites. Their mechanical properties, coupled with relatively low density, make these materials ideal candidates for weight-efficient structures and have been heavily scrutinized for the same. Also for the same reason, CNTs are considered to be the ultimate reinforcement in polymeric composites. Since their discovery in early 1990s and the realization of their unique physical properties, including mechanical, thermal, and electrical, many investigators have endeavored to fabricate advanced polymer–CNT composites that exhibit one or more of these properties [34–37]. In addition to the improvements in modulus and strength, CNTs also provide enhanced electrical conductivity and heat resistance, decreased gas permeability and flammability, and increased biodegradability in the case of biodegradable polymers [38–44]. For example, as conductive filler in polymers, CNTs are quite effective compared with traditional carbon black microparticles, primarily due to their large aspect ratios. Similarly, CNTs possess one of the highest thermal conductivities known that suggests their use in composites for thermal management [45, 46]. Moreover, there has been considerable interest in theory and simulations addressing the preparation and properties of these materials, and they are also considered to be unique model systems to study the structure and dynamics of polymers in confined environments [47].

## 7.3.1.  CNTs as Nucleating Agents in Polymers

CNTs have been evaluated in recent years as additives to polymers for providing faster crystallization. For example, Probst et al. prepared polyvinyl alcohol (PVA)–SWNT nanocomposites through an ultrasonication-assisted spin casting method and studied their crystallization characteristics [48]. Differential scanning calorimetry (DSC) studies revealed that CNTs nucleate crystallization of PVA at concentrations as low as 0.1 wt%. A comparison of the crystallization exotherms of pristine PVA and PVA–SWNT nanocomposites during the first (solid line) and second (dashed line) cooling scan at 20°C/min depicting the nucleation ability of SWNTs is given in Figure 7.4. The additional thermal treatment between the first and second scans causes a shift of the crystallization temperature to lower temperatures and a lowering of the heat release associated with the

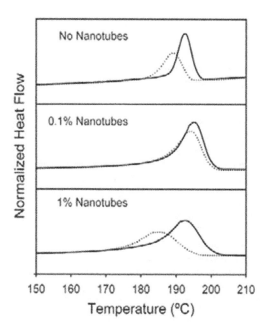

**Figure 7.4**   A comparison of the crystallization exotherms during the first (solid line) and second (dashed line) cooling scan at 20°C/min [48].

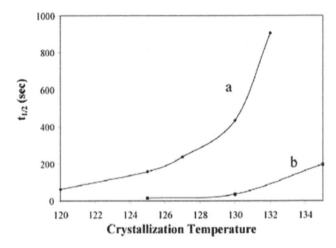

**Figure 7.5** Crystallization half-time of (a) polypropylene and (b) PP–SWNT (99.2/0.8) as a function of crystallization temperature [50].

crystallization. The authors attribute these effects to sample degradation; the effect of which was found to be worse in the sample filled with 1 wt% nanotubes.

The crystallization behavior of polypropylene (PP) in the presence of SWNTs has been reported by Grady et al. [49] They mixed modified CNTs with PP in a solvent-based process. The resulting composite had an almost negligible difference in small-strain mechanical properties. However, the introduction of CNTs changed the crystallization kinetics and was shown to nucleate crystallinity as studied using DSC. The actual amount of the nanotubes required to nucleate crystallinity was quite low; at 0.6 wt%, the number of nucleation sites increased to the point where further addition produced a very small or even negligible increase in the formation of crystallites. This extremely low value confirms that the number of nucleated crystals is proportional to the surface area of the filler.

Bhattacharyya et al. also studied the nucleation effect of melt-blended SWNTs in PP matrix using DSC [50]. It was observed that melt blending results in poor dispersion of SWNTs in the polymer matrix, but even with the poor dispersion, they act as effective nucleating agents for PP crystallization. PP containing 0.8 wt% SWNTs exhibited faster crystallization rate as compared with pristine PP. The crystallization half-time of pristine PP and PP–SWNT nanocomposites plotted against crystallization temperature depicting the nucleation efficiency of SWNTs is illustrated in Figure 7.5.

Valentini et al. also prepared PP matrix composites reinforced with SWNTs at different concentrations [51, 52]. They observed that SWNTs when melt blended with PP at a concentration of 5 wt% increased the crystallization temperature of the polymer by approximately 13°C. The DSC nonisothermal crystallization curves for neat PP and PP–SWNT nanocomposites shown in Figure 7.6 also indicate that the nucleation ability of SWNTs was not linearly dependent on the SWNT content, showing a saturation of the nucleant effect at low nanotube concentrations. The crystallization temperatures ($T_c$), the apparent melting temperatures of the crystallized samples ($T_m$), and the crystallization enthalpies ($\Delta H$) as a function of SWNT concentration are also reported in Table 7.1. The decrease of the $\Delta H$ values with increasing nanotube concentration can be directly attributed to the proportional

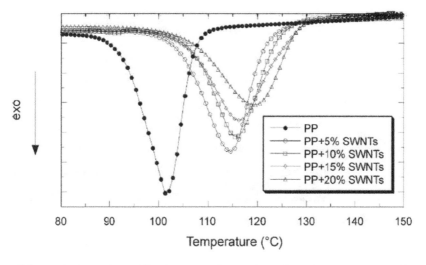

**Figure 7.6**   Nonisothermal crystallization curves for neat PP and PP–SWNT nanocomposites [52].

reduction of the PP concentration in the composite. Furthermore, no significant changes in the melting point of the PP phase were detected in the samples.

Figure 7.7 shows the microphotographs of the crystalline morphology, at the end of the crystallization process, of neat PP and PP–SWNT 5 wt% nanocomposites. For the neat PP, large spherulites are observed (average diameter $\sim$ 100 μm). In the 5 wt% concentration sample, a large number of small crystal aggregates is visible (average diameter $\sim$ 10 μm).

Anand et al. have studied the nucleation effect of SWNTs in a thermoplastic polyester, poly(ethylene terephthalate) [53]. They adopted a simple melt-compounding method for preparing PET–SWNT nanocomposites and studied the crystallization characteristics using DSC. Figure 7.8 shows the DSC cooling scans of PET–SWNT nanocomposite samples. During cooling from the melt, the SWNT containing samples showed crystallization exotherms earlier than neat PET, as also seen from the corresponding $T_c$ values given in Table 7.2. It is found that the nanocomposite sample containing SWNTs at a concentration as low as 0.03 wt% crystallizes 10°C earlier than neat PET, indicating that SWNTs are acting as effective nucleating agents for PET crystallization. The $T_c$ values continue to increase with increasing SWNT concentrations, but at a slower rate, as with the further 100-fold increase in SWNT content from 0.03 to 3.0 wt%, the additional $T_c$ increase is only about 10°C. In other words, there is a saturation of the nucleant effect at low SWNT concentrations, resulting in diminishing dependence on the increasing SWNT-induced

**Table 7.1**   DSC determined thermal parameters of PP and PP–SWNT nanocomposites [52].

| Materials | $T_c$ (°C) | $\Delta H$ (J/g) | $T_m$ (°C) |
|---|---|---|---|
| Neat PP | 101.31 | 102.41 | 166.11 |
| PP+5% SWNTs | 114.56 | 97.32 | 163.28 |
| PP+10% SWNTs | 115.41 | 92.98 | 163.44 |
| PP+15% SWNTs | 116.21 | 87.63 | 164.44 |
| PP+20% SWNTs | 119.61 | 82.79 | 165.11 |

**Figure 7.7**    Microphotographs of neat PP and PP–SWNT 5 wt% composite [52].

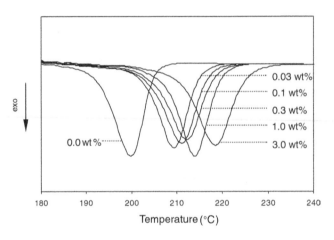

**Figure 7.8**    DSC cooling scans (20°C/min from 310°C melt) of PET–SWNT nanocomposites [53].

**Table 7.2**    DSC-determined thermal characteristics of PET–SWNT nanocomposites [53].

| Concentration of SWNTs (wt%) | $T_c$ (°C) | $\Delta H_c$ (J/g) | $T_m$ (°C) | $\Delta H_m$ (J/g) |
|---|---|---|---|---|
| 0.0 | 199.92 | 38.16 | 253.03 | 37.51 |
| 0.03 | 209.92 | 37.46 | 251.73 | 37.80 |
| 0.1 | 211.51 | 36.59 | 252.54 | 36.19 |
| 0.3 | 212.41 | 37.58 | 253.51 | 37.32 |
| 1.0 | 214.33 | 37.25 | 251.08 | 37.67 |
| 3.0 | 219.08 | 36.82 | 252.54 | 38.79 |

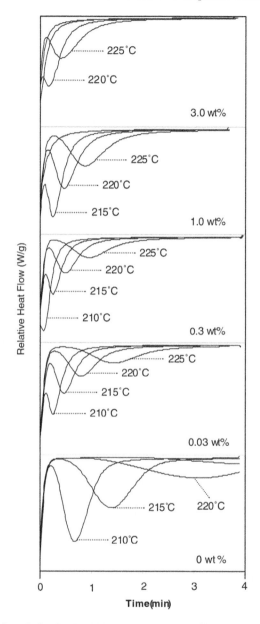

**Figure 7.9**   Heat flow during isothermal crystallization of PET–SWNT nanocomposites [53].

nucleation, possibly because of the large surface area and good dispersion of SWNTs. However, the melting temperatures $(T_m)$ and the corresponding enthalpies $(\Delta H_c$ and $\Delta H_m)$ remain almost unaltered with SWNTs.

SWNTs are also found to enhance the rate of crystallization as revealed from isothermal crystallization studies. Figure 7.9 shows the typical isothermal crystallization curves of the PET–SWNT nanocomposites at four temperatures (210, 215, 220, and 225°C). The time corresponding to the maximum in the heat flow rate (exotherm) is taken as peak time of

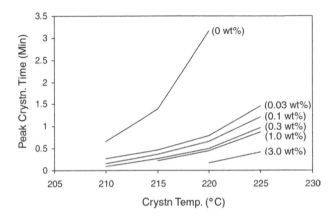

**Figure 7.10**  Peak crystallization time versus crystallization temperature for PET and PET–SWNT nanocomposites [53].

crystallization ($t_{peak}$). Such peaks are seen at each of the four isothermal crystallization temperatures for the 0.03 and 0.3 wt% SWNT-containing nanocomposites, with the earlier or faster crystallization (smaller $t_{peak}$) corresponding to lower temperature of isothermal crystallization as compared with neat PET. For the case of neat PET, no peak is seen at the highest temperature of 225°C because crystallization is very slow and would require longer time. On the other hand, for the nanocomposites with 1.0 and 3.0 wt% SWNTs, the rate of crystallization is so fast near the lowest temperatures that most of the crystallization occurs already during the cooling scan employed to reach the temperatures (210 or 215°C). This results in absence of exothermic peaks in the heat flow curves at those temperatures.

Figure 7.10 represents the peak times of crystallization for the nanocomposites, at each of the temperatures plotted against the isothermal crystallization temperature. It is noticeable that the $t_{peak}$ values for the nanocomposite samples reduce to less than 50% as compared with neat PET due to the presence of SWNTs at concentrations as low as 0.03 wt%. With the increasing SWNT concentration, there is further increase in the crystallization rate (as indicated by the decrease in $t_{peak}$), demonstrating the role of SWNTs in enhancing the rate of crystallization.

In a recent report from the same group, the authors have compared the nucleation effect of SWNTs in PET matrix when the nanotubes are incorporated to the polymer through an ultrasound-assisted solution evaporation method [54]. A comparison of the $T_c$ rise as a result of the SWNT-induced nucleation in the nanocomposites prepared via solution route and melt route is given in Table 7.3. The increase in the $T_c$ values is found to be much higher for the nanocomposites prepared via the solution route, indicating a possible improved dispersion of SWNTs by this method.

Assouline et al. reported about the nucleation ability of MWNTs in PP composites [55]. The effect of MWNTs on the nonisothermal crystallization of ethylene-vinyl acetate copolymer (EVA) was investigated by Li et al. [56]. They observed that the onset and maximum crystallization temperatures for the EVA–MWNT (10 wt%) composite were, respectively, about 10 and 5°C higher than those for the neat EVA, indicating nucleation ability of MWNTs in EVA. Kim et al. also have studied the unique nucleation of melt-compounded nanocomposites of MWNTs and poly(ethylene 2,6-naphthalate) (PEN) during nonisothermal crystallization [57]. They observed that MWNTs act as effective

**Table 7.3**  Comparison of the $T_c$ rise as a result of the SWNT-induced nucleation in PET–SWNT nanocomposites prepared via melt route and solution route [54].

| Concentration of SWNTs (wt%) | Increase in $T_c$ (°C) | |
|---|---|---|
| | Melt route | Solution route |
| 0.3 | 12.5 | 24.3 |
| 1.0 | 14.4 | 27.4 |
| 3.0 | 19.2 | 31.2 |

nucleating agents for PEN crystallization, the effect of which was more predominant at low concentrations. DSC cooling traces of the PEN–MWNT nanocomposites as a function of MWNT content at a scan rate of 10°C/min are shown in Figure 7.11. The incorporation of MWNTs into the PEN matrix increased the crystallization temperature of the PEN–MWNT nanocomposites, with this increment being greatest with lower MWNT content. In other words, the incorporation of a very small quantity of MWNTs enhanced the nucleation of PEN crystallization. The MWNTs promote the formation of heterogeneous nuclei, with lower energy consumption required to reach critical stability for crystal growth resulting in them functioning as effective nucleating agents in the PEN matrix.

## 7.3.2. CNTs for Mechanical Reinforcement

The mechanical behavior of CNTs is exciting since they are seen as the "ultimate" carbon fibers ever made. For the same reason, they hold promise as a possible reinforcing phase in polymer composites [58–61]. The traditional carbon fibers have about *50* times the specific strength (strength/density) of steel and are excellent load-bearing reinforcements

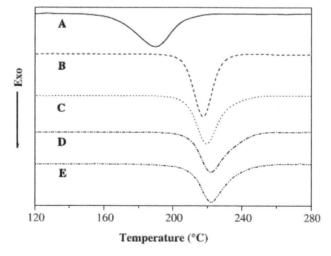

**Figure 7.11**  DSC cooling curves of PEN–MWNT nanocomposites as a function of MWNT content (a: PEN, b: PEN–MWNT 0.1%, c: PEN–MWNT 0.5%, d: PEN–MWNT 1.0%, and e: PEN–MWNT 2.0%) [57].

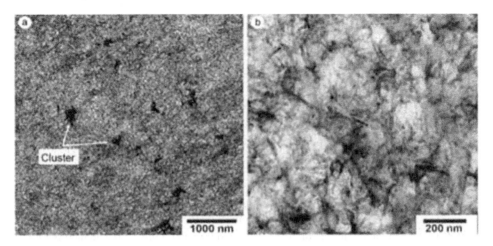

**Figure 7.12**    TEM images of PA6 filled with 5 wt% carbon nanotubes (a: 1000 nm scale bar, b: 200 nm scale bar) [64].

in composites [62]. Nanotubes should then be ideal candidates for structural applications for high-strength, light-weight, high-performance composite materials, for a series of products ranging from expensive tennis rackets to spacecraft and aircraft body parts [63]. Investigations on theeffectiveness of CNTs for providing mechanical strength to polymer matrices are briefly reviewed in this section.

Meincke et al. prepared nanocomposites of polyamide-6 (PA6) and CNTs using a corotating twin-screw extruder [64]. It was shown by TEM that the nanotubes are dispersed homogeneously in the PA matrix (Figure 7.12). Tensile tests of the composites showed a significant increase of 27% in Young's modulus; however, the elongation at break of these materials dramatically decreased due to an embrittlement of the PA6 matrix.

Zhang et al. also prepared PA6–MWNT nanocomposites with enhanced overall mechanical properties by a simple melt-compounding method [65]. Mechanical tests showed that, compared with neat PA6, the tensile modulus, tensile strength, and hardness of the composite were greatly improved by about 115, 120, and 67%, respectively, with incorporating only 1 wt% MWNTs. Typical stress–strain plots of neat PA6 and its composite are given in Figure 7.13(a). SEM observation of the fracture surfaces of the composite indicated that a homogeneous dispersion of MWNTs throughout PA6 matrix and a strong interfacial adhesion between MWNTs and the matrix had been successfully achieved, which are responsible for the significant enhancements in mechanical properties (Figure 7.13(b)).

Polymethyl methacrylate (PMMA)–nanotube nanocomposites were fabricated through an *in situ* process by Jia et al. [66], in which, chemically treated nanotubes could link with PMMA, thus obstructing the growth of PMMA, producing a C–C bond between nanotubes and the PMMA. The dispersion ratio of nanotubes in the PMMA matrix is increased and the properties of the composites are improved due to high interfacial strength. Stephan et al. also fabricated thin film of PMMA–SWNT composite by spin coating [67]. It was found that the polymer intercalated between nanotube bundles. At low concentrations, the nanotubes dispersed well, thus more uniform films were prepared. A combination of solvent casting and melt mixing has also been used to fabricate PMMA–SWNT nanocomposites [68].

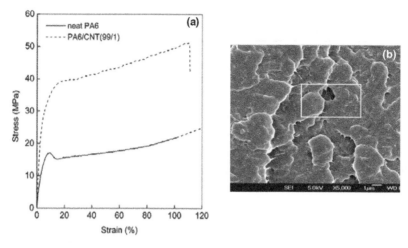

**Figure 7.13**   (a) Typical stress–strain plots of neat PA6 and PA6-MWNT 1 wt% nanocomposite. (b) SEM image showing an overall morphology of failure surface of PA6 nanocomposite containing 1 wt% MWNTs [65].

Ruan et al. reported drastically enhanced mechanical properties for ultrahigh molecular weight polyethylene (UHMWPE) films because of the addition of 1 wt% MWNTs [69]. For example, the tensile strength and modulus for the undrawn UHMWPE composite were improved by 50 and 38%, respectively. For the anisotropic UHMWPE–MWNT composite films with high draw ratios, an up to 150% increase in strain energy density has been observed together with a simultaneous increase in tensile strength of 25% and an increase in ductility up to 140%. The observations were attributed due to the chain mobility enhancement in UHMWPE induced by the MWNTs.

Several other researchers also have reported significant toughening of polymer matrices through the incorporation of CNTs. For example, Weisenberger et al. reported an 80% increase in composite fiber toughness with the incorporation of MWNTs in polyacrylonitrile (PAN) fibers [70]. Xushan et al. have incorporated MWNTs to nylon-6 and PET-fibers [71]. They observed that, when adding 0.03 wt% MWNTs into nylon-6 fiber, the strength of the nylon fiber enhanced by 33.3%. Its modulus was also enhanced by 74.4%. Adding 0.1 wt% MWNTs into PET fiber made the latter fairly conducting. The breaking strength of this conducting PET fiber showed almost no decrease and initial modulus was slightly enhanced in comparison with pure PET fiber.

Poly(p-phenylene benzobisoxazole) (PBO)-based nanocomposite fibers with SWNTs have been prepared by Kumar et al. [72]. They reported that the fibers containing 10 wt% SWNTs exhibited approximately 50% higher tensile strength as compared with the control PBO fiber. The mechanical properties of PBO and PBO–SWNT nanocomposite fibers are reported in Table 7.4. Typical stress–strain plots are also provided in Figure 7.14.

Sandler et al. also successfully explored the potential of various MWNTs and nanofibers as mechanical reinforcements in PA12 nanocomposite fibers. Using as-produced nano-materials, filler loading fractions of up to 15 wt% were realized, by following thermoplastic processing techniques to produce melt-spun nanocomposites [73]. Sreekumar et al. prepared PAN–SWNT nanocomposite fibers through solution spinning [74]. The spun nanocomposite fibers containing 10 wt% SWNTs exhibited a 100% increase in tensile modulus at room temperature, and it increased by an order of magnitude at 150°C. They

**Table 7.4**    Mechanical properties of PBO and PBO–SWNT nanocomposite fibers [72].

| Sample | Tensile modulus (GPa) | Tensile strength (GPa) | Compressive strength (GPa) |
|---|---|---|---|
| PBO | 138 ± 20 | 2.6 ± 0.3 | 0.035 ± 0.6 |
| PBO–SWNT (95/5) | 156 ± 20 | 3.2 ± 0.3 | 0.40 ± 0.6 |
| PBO–SWNT (90/10) | 167 ± 15 | 4.2 ± 0.5 | 0.50 ± 0.6 |

also observed a significant reduction in thermal shrinkage as well as polymer solubility and the glass transition temperature was increased by 40°C as compared with control PAN fiber.

Siochi et al. prepared fibers from melt-processed nanocomposites of polyimide with SWNTs [75]. They demonstrated that the fibers containing up to 1 wt% SWNTs, in which the nanotube alignment in the fiber direction was induced by shear forces during melt extrusion and fiber drawing, exhibited significantly higher tensile moduli and yield stress relative to unoriented nanocomposite films having the same SWNT concentration.

Several other groups have demonstrated improvement in properties of polystyrene, PMMA, etc., as a result of melt/wet spinning various amounts of SWNTs with the polymer matrix [76, 77].

## 7.3.3. Rheological Properties of Polymer–CNT Nanocomposites

The processability of thermoplastics and their composite materials has been a great concern to the polymer-processing industry, which is very closely related to their rheological properties in the molten state [78]. In the case of a composite system, these properties are very sensitive to the dispersion state of the filler and to a lesser extent, to the interactions between the filler and the polymer [79–81]. Because the rheological properties of the filled polymer systems are responsive to the structure, particle size, geometry, and the surface characteristics of the fillers, rheological measurements are commonly used to describe the

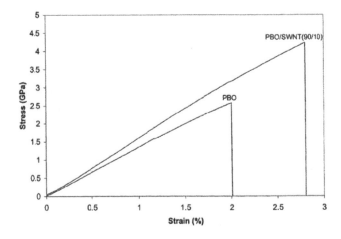

**Figure 7.14**    Typical stress–strain plots of PBO and PBO–SWNT nanocomposite fibers [72].

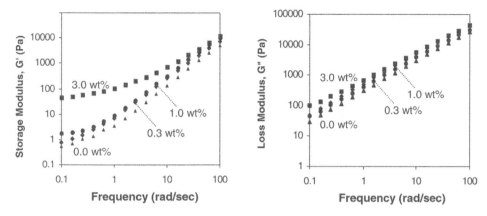

**Figure 7.15**  Effect of SWNT concentration on the storage and loss moduli of PET–SWNT nanocomposites [84].

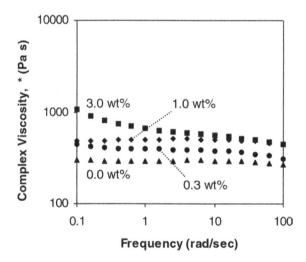

**Figure 7.16**  Effect of SWNT concentration on the complex viscosity ($T = 280°C$, strain = 10%) of PET–SWNT nanocomposites [84].

dispersion of the filler in polymer matrices [82, 83]. Critical literature reports depicting the effect of CNTs on the rheological properties of polymer matrices are briefly reviewed in this section.

The dynamic rheological properties of melt-compounded PET–SWNT nanocomposites have been reported by Anand et al. [84]. The spectra representing linear viscoelastic properties of the nanocomposites at different concentrations of SWNTs are shown in Figure 7.15.

It is noticed that, with increasing SWNT loading, the storage ($G'$) and loss ($G''$) moduli increase at all frequencies. In particular, $G'$ at low frequencies increases 100-fold on incorporation of 3.0 wt% SWNTs. These results indicate that melt-compounded SWNTs are effectively entangled in the melt state. Frequency dependence of complex viscosity ($\eta^*$) of the PET–SWNT nanocomposites is illustrated in Figure 7.16. It is noticed that neat

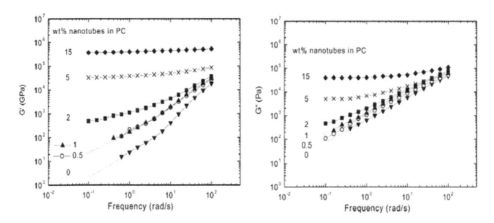

**Figure 7.17**  Storage and loss moduli of nanotube-filled polycarbonate at 260°C [86].

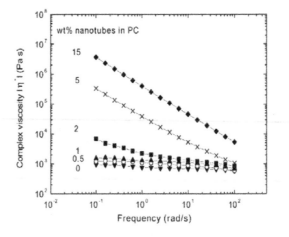

**Figure 7.18**  Complex viscosity of nanotube-filled polycarbonate at 260°C [86].

PET as well as the samples with low SWNT content displayed the expected Newtonian behavior at low frequencies and shear thinning behavior at higher frequencies. However, the 3.0 wt% SWNT sample showed a marked enhancement in the complex viscosity and strong shear thinning behavior even at low frequencies.

Li et al. also observed the decrease of shear viscosity of PET composites with MWNTs [85]. Pötschke et al. observed shear thinning behavior even at low frequencies for polycarbonate nanocomposites with CNT concentration exceeding 2.0 wt% [86, 87]. The effect of nanotubes on the dynamic moduli and complex viscosity of polycarbonate nanocomposites is shown in Figures 7.17 and 7.18, respectively.

Hu et al. have impressively reported the dynamic rheological characteristics of PET–MWNT nanocomposites [88]. It was observed that the viscosity of neat PET is almost independent of frequency, and the nanocomposite at MWNT loading of 0.5 wt% showed a weak shear thinning behavior, whereas the nanocomposites with higher MWNT loadings exhibited strong shear thinning behavior and the viscosities were of the orders of magnitude higher than that of neat PET at low frequency. However, Shin et al. reported that the addition of MWNTs to PET led to an increase in complex viscosity but the viscosity did not depend on the nanotube content up to 1.0 wt% [89].

## 7.3.4. CNTs' Induced Electrical Conductivity in Polymers

Insulating polymers can be imparted electrical conducting properties by dispersing electrically conducting particles that may form a percolative path of conducting network through the sample at concentrations exceeding certain minimum value called the percolation threshold. This approach reduces the manufacturing and maintenance costs of components as compared with those previously coated with an antistatic paint. The technology is also relevant to other applications where static electrical dissipation is needed such as computer housings or exterior automotive parts.

Although carbon black is traditionally used as a conductive filler, the small diameter and large aspect ratio of SWNTs (helps creating extensive networks that facilitate electron transport) have enabled achievement of very low percolation threshold concentrations, presumably depending on the quality of their dispersion. Earlier literature reports have revealed the percolation threshold for electrical conductivity at fairly high concentrations of nanotubes in polymers. For example, Bin et al. prepared high-density polyethylene–MWNT nanocomposites by gelation/crystallization from solutions, and revealed that the percolation occurred between 5 and 15 wt% [90]. Meincke et al. also found that CNT-filled PA6 showed an onset of electrical conductivity at nanotube loadings of 4–6 wt% [64].

Later, researchers have reported percolation threshold concentrations at around 1–2 wt% (or even less) of the nanotubes in polymer matrices. Kharchenko et al. prepared PP–MWNT nanocomposites by melt blending, and displayed that the percolation threshold was at concentrations ranging from 0.25 to 1 wt% [91]. Du et al. used coagulation method to produce PMMA–SWNT nanocomposites and disclosed percolation threshold between 0.2 and 2 wt% [92, 93]. Ounaies et al. reported that the conductivity of pristine polyimide was increased from an order of magnitude of $10^{-18}$ to $10^{-8}$ S cm$^{-1}$ at nanotube concentrations between 0.02 and 0.1 vol% [94]. Figure 7.19 illustrates the variation of electrical conductivity as a function of nanotube loading in polyimide matrix.

Coleman et al. observed that the physical doping with CNTs in a conjugated polymer matrix such as poly($p$-phenylene vinylene-$co$-2,5-dioctoxy-$m$-phenylene vinylene) could increase the conductivity of the polymer matrix by 10 orders of magnitude [95]. Regev et al. also have observed that SWNTs increased the conductivity of polystyrene by 10 orders of magnitude at very low percolation threshold concentrations [96]. Nogales et al. adopted an *in situ* polymerization method for preparing poly(butylene terephthalate)-based nanocomposites and disclosed the percolation threshold for conductivity at around 0.2 wt% of SWNT [97]. Indeed, there are several other reports in literature regarding the CNT-induced electrical conductivity in a variety of insulating polymer matrices [98].

The room temperature DC electrical conductivity results of melt-compounded PET–SWNT nanocomposites are shown in Figure 7.20 [84]. Neat PET is an excellent insulating material and has a conductivity value of the order of 10–17 S cm$^{-1}$ [99]. It is evident from the figure that SWNTs are effective in imparting electrical conductivity to the PET matrix and the conductivity reaches percolation, a value of the order of $10^{-6}$ S cm$^{-1}$ at concentrations exceeding 2.0 wt%. It is also well known that the percolation threshold is sensitive to the polymer matrix, in which the nanotubes are dispersed, and the processing

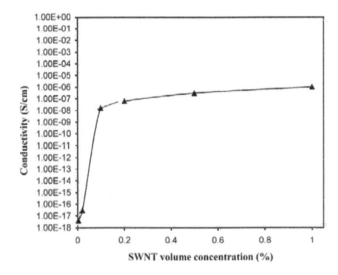

**Figure 7.19**  DC electrical conductivity variation with SWNT volume concentration in polyimide matrix [94].

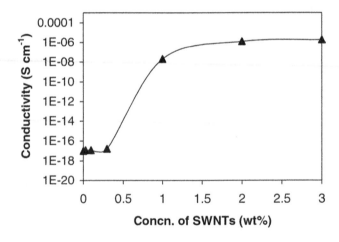

**Figure 7.20**  Electrical conductivity of melt-compounded PET–SWNT nanocomposites [84].

methods. This effect is due to the role of enhanced interfacial properties found for the nanocomposites.

After nearly a decade of research, the potential of CNTs as reinforcement for polymers has not been fully realized; the mechanical properties of at least some of the derived composites have fallen short of predicted values. Few mechanisms about adhesion, load transfer, and deformation were investigated, which make it difficult to accurately predict behaviors of nanotube–polymer composites and fabricate "ideal" nanocomposites. On the basis the overview of literature, it is found that the interaction between pristine nanotubes and polymers is dependent on the choice of the polymer matrix and also polymer conformation, thus the molecular structure may play a critical role in the interaction. Even with the "best polymer," pristine nanotube may not form strong interfaces. Yet given the magnitude of the CNT's mechanical properties, significant improvement on current composites would be possible provided that the CNT's unique attributes exhibited at the nanoscale are transferred to the macroscale. This essentially defines the fundamental challenge for applied CNT–polymer composites research [100]. A better understanding of the relationships between processing, interfacial optimization, and composite properties is a major goal of this area of research, which may lead to optimal reinforcement of polymer matrices with CNTs. The single largest impediment to use nanotubes as a filler phase for polymers is currently cost, but as nanotube production methods continue to develop, they will gradually become commercially viable fillers for multiphase materials.

# References

1. Ajayan PM, Schadler LS, Braun PV. Nanocomposite Science and Technology. Wiley-VCH; Weinheim, 2003.
2. Reich S, Thomsen C, Maultzsch J. Carbon Nanotubes: Basic Concepts and Physical Propeties. Wiley-VCH; Weinheim, 2004.
3. Endo M, Iijima S, Dresselhaus MS, editors. Carbon Nanotubes. Pergamon; Oxford, 1996.
4. Treacy MMJ, Ebbesen TW, Gibson JM. Nature 1996;381:678.
5. Wong EW, Sheehan PE, Lieber CM. Science 1997;277:1971.
6. White CT, Todorov TN. Nature 1998;393:240.
7. Qin LC, Zhao X, Hirahara K, Miyamoto Y, Ando Y, Iijima S. Nature 2000;408:50.
8. Ding W, Eitan A, Fisher FT, Chen X, Dikin DA, Andrews R, Brinson LC, Schadler LS, Ruoff RS. Nano Lett. 2003;3:1593.
9. Champion Y, Fecht HJ, editors. Nano-Architectured and Nanostructured Materials: Fabrication, Control and Properties. Wiley-VCH; Weinheim, 2004.
10. Fecht HJ, Werner M, editors. The Nano-Micro Interface: Bridging the Micro and Nanoworlds. Wiley-VCH; Weinheim, 2004.
11. čojkowski W, Blizzard JR, editors. Interfacial Effects and Novel Properties of Nanomaterials. Scitec Publications; Switzerland, 2003.
12. Kelsall R, Hamley I, Geoghegan M, editors. Nanoscale Science and Technology. Wiley; New York, 2005.
13. Schmid G, editor. Nanoparticles: From Theory to Application. Wiley-VCH; Weinheim, 2004.
14. Wolf EL, editor. Nanophysics and Nanotechnology: An Introduction to Modern Concepts in Nanoscience. Wiley-VCH; Weinheim, 2004.
15. Harris PJF. Carbon Nanotubes and Related Structures: New Materials for the Twenty First Century. Cambridge University Press; Cambridge, 1999.
16. Saito R, Dresselhaus G, Dresselhaus MS. Physical Properties of Carbon Nanotubes. Imperial College Press; London, 1998.
17. Kroto HW, Heath JR, O'Brien SC, Curl RF, Smalley RE. Nature 1985;318:162.
18. Smalley RE. Rev. Mod. Phys. 1997;69:723.
19. Krätschmer W, Lamb LD, Fostiropoulos K, Huffman DR. Nature 1990;347:354.
20. Iijima S. Nature 1991;354:56.
21. Gamaly EG, Ebbesen TW. Phys. Rev. B: 1995;52:2083.
22. Guo T, Nikolaev P, Thess A, Colbert DT, Smalley RE. Chem. Phys. Lett. 1995;243:49.
23. Scott CD, Arepalli S, Nikolaev P, Smalley RE. Appl. Phys. A, Mater. Sci. Process. 2001;72:573.
24. Ren ZF, Huang ZP, Xu JW, Wang JH, Bush P, Siegel MP, Provencio PN. Science 1998;282:1105.
25. Ren ZF, Huang ZP, Wang DZ, Wen JG, Xu JW, Wang JH, Calvet LE, Chen J, Klemic JF, Reed MA. Appl. Phys. Lett. 1999;75:1086.
26. Sinnot SB, Andrews R, Qian D, Rao AM, Mao Z, Dickey EC, Derbyshire F. Chem. Phys. Lett. 1999;315:25.
27. Masako Y, Rie K, Yoshimasa O, Etsuro O, Susumu Y. Appl. Phys. Lett. 1997;70:1817.
28. Ajayan PM. Chem. Rev. 1999;99:1787.
29. Stankevich IV, Nikerov MV, Bochvar DA. Russ. Chem. Rev. 1984;53:640.
30. Avouris P. Chem. Phys. 2002;281:429.
31. Ajayan PM, Zhou OZ. Topics Appl. Phys. 2001;80:391.

32. Srivastava D, Wei C. Appl. Mech. Rev. 2003;56:215.
33. Qian D. Appl. Mech. Rev. 2002;55:495.
34. Ma RZ, Wu J, Wei Q, Liang J, Wu DH. J. Mater. Sci. 1998;33:5243.
35. Weisenberger MC, Grulke EA, Jacques D, Rantell T, Andrews R. J. Nanosci. Nanotechnol. 2003;3.
36. Mitchell CA, Bahr JL, Arepalli S, Tour JM, Krishnamoorti R. Macromolecules 2002;35:8825.
37. Cooper CA, Ravich D, Lips D, Mayer J, Wagner HD. Compos. Sci. Technol. 2002;62:1105.
38. Curran SA, Ajayan PM, Blau WJ, Carroll DL, Coleman JN, Dalton AB, Davey AP, Drury A, McCarthy B, Maier S, Strevens A. Adv. Mater. 1998;10:1091.
39. Dalton AB, Stephan C, Coleman JN, McCarthy B, Ajayan PM, Lefrant S, Bernier P. J. Phys. Chem. B 2000;104:10012.
40. McCarthy B, Coleman JN, Curran SA, Dalton AB, Davey AP, Konya Z, Fonseca A. J. Mater. Sci. Lett. 2000;19:2239.
41. Dalton AB, Blau WJ, Chambers G, Coleman JN, Henderson K, Lefrant S, McCarthy B. Synth. Met. 2001;121:1217.
42. Andrews R, Weisenberger MC. Curr. Opin. Solid State Mater. Sci. 2004;8:31.
43. Lau T, Hui D. Composites: Part B 2002;33:263.
44. Lau T, Hui D. Carbon 2002;40:1597.
45. Kim P, Shi L, Majumdar A, McEuen PL. Phys. Rev. Lett. 2001;87:2155021.
46. Biercuk MJ, Llaguno MC, Radosavljevic M, Hyun JK, Johnson AT, Fischer JE. Appl. Phys. Lett. 2002;80:2767.
47. Pipes RB, Frankland SJV, Hubert P, Saether E. Compos. Sci. Technol. 2003;63:1349.
48. Probst O, Moore EM, Resasco DE, Grady BP. Polymer 2004;45:4437.
49. Grady BP, Pompeo F, Shambaugh RL, Resasco DE. J. Phys. Chem. 2002;B- 106:5852.
50. Bhattacharyya AR, Sreekumar TV, Liu T, Kumar S, Ericson LM, Hauge RH, Smalley RE. Polymer 2003;44:2373.
51. Valentini L, Biagiotti J, Kenny JM, Santucci S. J. Appl. Polym. Sci. 2003;87:708.
52. Valentini L, Biagiotti J, Kenny JM, Santucci S. Compos. Sci. Technol. 2003;63:1149.
53. Anand KA, Agarwal US, Joseph R. Polymer 2006;47:3976.
54. Anand KA, Agarwal US, Nisal A, Joseph R. Euro. Polym. J. 2007;43:2279.
55. Assouline E, Lustiger A, Barber AH, Cooper CA, Klein E, Wachtel E, Wagner HD. J. Polym. Sci.: Part B: Polym. Phys. 2003;41:520.
56. Li SN, Li ZM, Yang MB, Hu ZQ, Xu XB, Huang R. Mater. Lett. 2004;58:3967.
57. Kim JY, Park HS, Kim SH. Polymer 2006;47:1379.
58. Despres JF, Daguerre E, Lafdi K. Carbon 1995;33:87.
59. Iijima S, Brabec C, Maiti A, Bernholc J. J. Chem. Phys. 1996;104:2089.
60. Wong EW, Sheehan PE, Lieber CM. Science 1997;277:1971.
61. Chopra NG, Benedict L, Crespi V, Cohen M, Louie S, Zettl A. Nature 1995;377:135.
62. Dresselhaus MS, Dresselhaus G, Sugihara K, Spain IL, Goldberg HA. Graphite Fibers and Filaments. Springer; Berlin, 1988.
63. Dresselhaus MS, Dresslhous G, Avouris P. Carbon Nanotubes: Synthesis, Structure, Properties and Applications. Springer; Berlin, 2001.
64. Meincke O, Kaempfer D, Weickmann H, Friedrich C, Vathauer M, Warth H. Polymer 2004;45:739.
65. Zhang WD, Shen L, Phang IY, Liu T. Macromolecules 2004;37: 256.
66. Jia ZJ, Wang ZY, Xu C, Liang J, Wei BQ, Wu DH, Zhu SW. Mater. Sci. Eng. 1999;A271:395.
67. Stephan C, Nguyen TP, de la Chapelle, Lamy M, Lefrant S, Journet C, Bernier P. Synth. Met. 2000;108:139.
68. Haggenmueller R, Gommans HH, Rinzler AG, Fischer JE, Winey KI. Chem. Phys. Lett. 2000;30:219.
69. Ruan SL, Gao P, Yang XG, Yu TX. Polymer 2003;44:5643.
70. Weisenberger MC, Grulke EA, Jacques D, Rantell T, Andrews R. J. Nanosci. Nanotechnol. 2003;3:6.
71. Xushan G, Yan T, Shuangyan H, Zhenfu G. Chemical Fibers International 2005;55:170.
72. Kumar S, Dang TD, Arnold FE, Bhattacharyya AR, Min BG, Zhang XF, Vaia RA, Park C, Adams WW, Hauge RH, Smalley RE, Ramesh S, Willis PA. Macromolecules 2002;35:9039.
73. Sandler JKW, Pegel S, Cadek M, Gojny F, van Es M, Lohmar J, Blau WJ, Schulte K, Windle AH, Shaffer MSP. Polymer 2004;45:2001.
74. Sreekumar TV, Liu T, Min BG, Guo H, Kumar S, Hauge RH, Smalley RE. Adv. Mater. 2004;16:58.
75. Siochi EJ, Working DC, Park C, Lillehei PT, Rouse JH, Topping CC, Bhattacharyya AR, Kumar S. Composites: Part B 2004;35:439.

76. Andrews R, Jacques D, Rao AM, Rantell T, Derbyshire F, Chen Y, Chen J, Haddon RC. Appl. Phys. Lett. 1999;75:1329.
77. Qian D, Dickey EC, Andrews R, Rantell T. Appl. Phys. Lett. 2000;76:20.
78. Han CD, editor. Rheology in Polymer Processing. Academic Press; New York, 1976.
79. Ren J, Krishnamoorti R. Macromolecules 2003;36:4443.
80. Krishnamoorti R, Giannelis EP. Macromolecules 1997;30:4097.
81. Krishnamoorti R, Ren JX, Silva AS. J. Chem. Phys. 2001;114:4968.
82. Salaniwal S, Kumar SK, Douglas JF. Phys. Rev. Lett. 2002;89:258301.
83. Solomon MJ, Almusallam AS, Seefeldt KF, Somwangthanaroj A, Varadan P. Macromolecules 2001;34:1864.
84. Anand KA, Agarwal US, Joseph R. J. Appl. Polym. Sci. 2007;104:3090.
85. Li Z, Luo G, Wei F, Huang Y. Compos. Sci. Technol. 2006;66:1022.
86. Pötschke P, Fornes TD, Paul DR. Polymer 2002;43:3247.
87. Pötschke P, Bhattachryya AR, Janke A. Polymer 2003;44:8061.
88. Hu G, Zhao C, Zhang S, Yang M, Wang Z. Polymer 2006;47:480.
89. Shin DH, Yoon KH, Kwon OH, Min BG, Ik Hwang C. J. Appl. Polym. Sci. 2006;99:900.
90. Bin Y, Kitanaka M, Zhu D, Matsuo M. Macromolecules 2003;36:6213.
91. Kharchenko SB, Douglas JF, Obrzut J, Grulke EA, Migler KB. Nat. Mater. 2004;3:564.
92. Du F, Fisher JE, Winey KI. J. Polym. Sci. Part B: Polym. Phys. 2003;41:3333.
93. Du F, Scogna RC, Zhou W, Brand S, Fischer JE, Winey KI. Macromolecules 2004;37:9048.
94. Ounaies Z, Park C, Wise KE, Siochi EJ, Harrison JS. Compos. Sci. Technol. 2003;63:1637.
95. Coleman JN, Curran S, Dalton AB, Davey AP, McCarthy B, Blau W, Barklie RC. Synth. Met. 1999;102:1174.
96. Regev O, Elkati PNB, Loos J, Koning CE. Adv. Mater. 2004;16:248.
97. Nogales A, Broza G, Roslaniec Z, Schulte K, Sics I, Hsiao BS, Sanz A, Garcia-Gutierrez MC, Rueda DR, Domingo C, Ezquerra TA. Macromolecules 2004;37:7669.
98. Kymakis E, Alexandou I, Amaratunga GAJ. Synth. Met. 2002;127:59.
99. Hu G, Zhao C, Zhang S, Yang M, Wang Z. Polymer 2006;47:480.
100. Thostenson ET, Chou TW. J. Phys. D. Appl. Phys. 2003;36:573.

# 8 Specialty Rubber Nanocomposites

*Robert A. Shanks\* and Rengarajan Balaji*

Applied Sciences, RMIT University, GPO Box 2476, Melbourne 3001, Australia

## Abstract

Specialty rubbers are high-performance materials. As elastomers, they require a low modulus and high reversible elongation, but in all other properties, they must meet stringent requirements that vary depending on the application. They require properties ranging from high ultimate strength, high break strain, low permanent set, toughness, durability to cyclic strain, abrasion resistance, thermal resistance, and solvent or corrosive chemical resistance. Low hysteresis and high damping are alternatives that may be required. Fillers are added to contribute to many of these properties and nanofillers can be most effective because of large surface-area-to-volume ratio. The nanosized fillers contribute to the performance via specialized mechanisms such as resisting crack propagation, forming agglomerates, or edge bridging; having fibrous, platelet, or particulate shape; and maybe forming specific bonds with the polymer. Specialty elastomers consist of higher performing polymers each with their own particular characteristics: fluoroelastomers, polysiloxanes, and polyurethanes are typical examples. However, the use of nanofillers can increase the performance of commodity polymers for them to be considered specialty polymers as they open new applications.

## Abbreviations

| | |
|---|---|
| CNT | carbon nanotube |
| EPDM | ethylene propylene diene monomer |
| HFP | hexafluoropropylene |
| MA | maleic anhydride |
| MWNT | multiwalled carbon nanotube |
| NBR | nitrile butadiene rubber |
| OMT | organomontmorillonite |
| PNC | polymer nanocomposite |
| PP | polypropylene |
| PVDF | poly(vinylidene difluoride) |
| RTV | room temperature vulcanizing |
| S-ENP | silicone elastomeric nanoparticle |
| TEM | transmission electron microscopy |
| TFE | tetrafluoroethylene |
| TPV | thermoplastic vulcanizate |

---
\*Correspondence should be addressed to e-mail: robert.shanks@rmit.edu.au

## 8.1. Introduction

Rubber or more specifically elastomers are a unique class of materials having properties but they have commonality in the sense that they must contain a polymer to achieve high reversible elongations. Elastomers are the most widely used polymers and they often become associated with low-performance materials because of their low modulus and strength. Yet elastomers have high ultimate strength and elongation, and they are tough. Elastomers are used in many specialty applications and the aim of this chapter is to review the properties and function of elastomer nanocomposites in specialty applications. Sometimes common elastomers are used, though they may consist of special components to enhance performance. One such component, that has been traditionally used and is now recognized as an advanced material additive, is nanoparticulate fillers.

Polymer nanocomposites (PNC) are polymers (thermoplastics, thermosets or elastomers) that have been reinforced with small quantities of nanosized particles generally having high aspect ratios. PNCs represent a radical alternative to conventional filled polymers or polymer blends; a staple of the modern plastics industry. In contrast to conventional composites, where the reinforcement is of micron scale, PNCs are exemplified by discrete constituents where at least one dimensions is of the order of a few nanometers. The value of PNC technology is not solely based on the mechanical enhancement of the neat resin nor the direct replacement of current filler or blend materials. Rather, its importance comes from providing value-added properties not present in the neat resin, without sacrificing the inherent polymer processability and mechanical properties or by excessively increasing density.

Specialty rubbers depend upon the type of application and not the polymer, but some special polymers are also used are such as fluoropolymer, neoprene polymer, acrylonitrile copolymer, polysiloxane, polyurethane, and santoprene elastomers (examples are shown in Table 8.1). Specialty rubber nanocomposites made from special polymers have excellent chemical and physical properties. Some specialty rubber products are heat-resistant hose, gaskets, tire curing bladder, automotive air duct, heat-resistant engine mount, camera gaskets, sporting equipment, roll compound, and low-voltage cable insulation.

## 8.2. Fluoroelastomer Nanocomposites

### 8.2.1. Fluoroelastomers

Fluorocarbon polymer elastomers have a broad range of chemical structures that can be considered as copolymers of poly(tetrafluorethylene) with other perfluoro- or hydrocarbon monomers (Figure 8.1). They offer exceptional chemical and thermal resistance due to the strength of the carbon–fluorine bond [1]. Tetrafluoroethylene (TFE) is copolymerized with hexafluoropropylene (HFP), pentafluoropropylene, chlorotrifluoroethylene, vinylidene difluoride, ethylene, propylene, and trifluoromethoxy trifluoroethylene to form copolymers or terpolymers. Fluoroelastomers are increasingly used as high-performance materials in many industrial applications due to their excellent heat, oil, and chemical resistance. Fillers are included to increase hardness and tensile properties, though compression set is often higher. Typical fillers are carbon black, calcium silicate, clay, barium sulfate, and other minerals intended for pigmentation.

**Table 8.1**    Specialty rubber characteristics and examples of applications.

| Elastomer | Structure | Characteristics | Applications |
|---|---|---|---|
| Silicones | $CH_3$ — Si — O — $CH_3$ | Lowest $T_g$, thermal stability, and low surface energy | Sealants, O-rings and washers, electrical, and surgical implants |
| Fluorocarbons | F F / C — C / F F | Chemical and thermal resistance and low surface energy | Temperature-resistant O-rings, seals, and gaskets |
| Polysulfides | $CH_2—(S_x)—CH_2$ | Solvent, oil, and UV weather resistant | Seals, caulking compounds, and protective coatings |
| Neoprene | Cl / C = CH, $CH_2$ | Solvent, oil, and abrasion resistant | Sheets, seals, foams, hoses, and belts |
| Nitrile | CN, $CH_2$ $CH_2$ CH CH CH $CH_2$ | Oil resistant, stiffer, and less gas permeable with more nitrile | Sheets, seals, foams, and water-proofing textile finishes |
| Polyurethanes | H / N C O / $CH_2$ $CH_2$ O | Polar and rapidly reacting two-component systems | Foams and tough components, rollers |
| Thermoplastic vulcanisates | $CH_3$ / CH $CH_2$ $CH_2$ $CH_2$ | Thermoplastic, for injection molding and extrusion, and crystalline | Extruded profiles for sealing strips and wire insulation |
| Block copolymers | [ Ar / CH $CH_2$ ] [ CH CH $CH_2$ $CH_2$ ]. | Thermoplastic and glassy-rubbery two-phase material | Shoe soles, molded and extruded sections, and hot melt adhesive |
| Thermoplastic polyolefins | $CH_3—CH_2$ / CH $CH_2$ $CH_2$ $CH_2$ | Thermoplastic, compatible with other polyolefins, and crystalline | Molding, extrusions, and blending to toughen polyolefins |

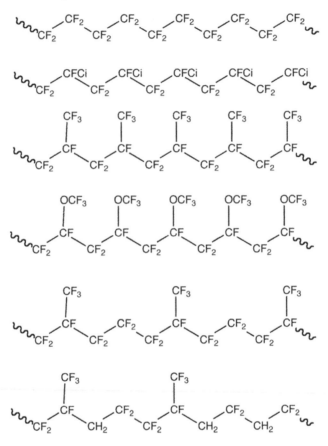

**Figure 8.1** Fluoropolymer elastomer structures, based on poly(tetrafluoroethylene) (from top): poly(tetrafluoroethylene), poly(chlorotrifluoroethylene), poly(hexafluoropropylene), poly(trifluoro-xytrifluoroethylene), poly(tetrafluoroethylene-*co*-hexafluoropropylene), and poly(vinylidenefluoride-*co*-hexafluoropropylene).

## 8.2.2. Fluoroelastomer Nanocomposites

Nanocomposites have been formed by the sol–gel approach using perfluoroalkylacrylyl ethylate, vinyl trimethyoxysilane, and tetraethylorthosilicate to increase porosity and specific surface area when used as fabric coatings [2].

In general, the PNCs with clay nanofillers have been prepared in two ways: (a) by *in-situ* polymerization, in which the polymerization occurs after mixing monomer or oligomer with organically modified clay and (b) by melt compounding, in which the organoclay is added to the polymer matrix during polymer processing. The dispersion of clay particles in a polymer matrix results in the formation of three types of nanocomposites, namely, phase-separated composites, intercalated polymer–clay nanocomposites, and the exfoliated or delaminated polymer–clay nanocomposites. Exfoliation occurs when the clay nanolayers are individually dispersed in the continuous polymer matrix. Exfoliated polymer–clay nanocomposites are especially desirable for improved properties because of the large aspect ratio of the individual clay layers, the uniform dispersion of clay, and large interfacial area between polymer chains and clay nanolayers.

The first poly(vinylidene difluoride) (PVDF)–clay nanocomposites were prepared by melt intercalation with organophilic clay. The presence of $\beta$-form of PVDF crystallized in nanocomposites in the presence of clay is interestingly reported [3].

Melt-intercalation methods have been used to prepare the nanocomposites using various fluoroelastomers such as PVDF, poly(vinylidene difluoride-*co*-tetrafluoro ethylene-*co*-perfluoromethyl vinyl ether-*co*-cure site monomer), poly(vinylidene difluoride-*co*-hexafluoropropylene [P(VDF-HFP), poly(vinylidene difluoride-*co*-hexa fluoropropylene-*co*-tetrafluoroethylene) [P(VDF-HFP-TFE)], poly(tetrafluoroethylene-*co*-hexafluoropropylene) [P(TFE-HFP)], and poly(ethylene-*co*-chlorotrifluoroethylene) [4] (Figure 8.1). The natural and organophilic clays are used as nanofillers. The organophilicity of modified clays and the polarity of the VDF containing fluoroelastomers play a significant role in the formation of fluoroelastomer nanocomposites. The intercalation of clay platelets has been proved by wide-angle X-ray scattering and transmission electron microscopy (TEM) analyses.

Exfoliation of organomodified nanoclays in a fluoroelastomer matrix is difficult to achieve, since interfacial interactions between polar nanofiller and polymer are expected to be energetically unfavorable because of the low-solubility parameters and low surface energy of fluoropolymers but some interactions are feasible due to the sufficient high polarity of the VDF repeat unit $-CF_2CH_2-$ [4]. A comparison between melt- and solution-blending processing techniques using fluoroeslatomeric terpolymer and organomodified montmorillonite clay was reported [5]. The morphology and rheological behavior of fluoroelastomer nanocomposites were affected by the preparative method. Fluoroelastomer diffused through the galleries of the organomodified nanoclays widening the gap between silicate layers during melt-blending processing. In a solution-blending method, the driving force for nanoclay exfoliation is mainly due to the desorption of the solvent molecules trapped within the silicate galleries, with a consequential large entropic gain.

### 8.2.3. Mechanical Properties

The mechanical properties of the fluoroelastomer nanocomposites depend on the nature of fillers and filler–fluoropolymer interactions. The properties of nanocomposites prepared from fluoroelastomers having different microstructure and viscosity and of nanoclays with different filler loadings have been widely investigated. Unmodified clay composite displayed an increase in modulus and ultimate stress over the unfilled polymer. Increased modulus in the case of unmodified clay composite was due to better polymer–filler interaction and filler dispersion in the matrix. As unmodified clay is polar, it is more compatible with the polar fluoroelastomers compared with organomodified clays. When the fluoroelastomer was mixed with unmodified clay, the polar clay attracted polar polymer molecules. The elastomer chains entered into the smaller gallery gap of this clay and their motion was restricted. But the modified clays had larger gallery spacings. Therefore, more polymer chains could enter the gallery gap. Hence, more polymer molecules were restrained [6].

The dynamic mechanical properties of PVDF–clay nanocomposites showed a simultaneous increase in both stiffness and toughness, leading to a situation where the overall properties were enhanced without any corresponding compromise. Most rigid fillers produce increased composite modulus compared with that of the unfilled polymer. However, this was generally associated with a significant decline in the ultimate strain. Changes in the polymer nanostructure and morphology have a profound impact on the mechanical response of fluoroelastomer nanocomposites [7].

### 8.2.4. Application of Fluoroelastomers Nanocomposites

Fluoroelastomer nanocomposites are widely used in many industrial applications because of their excellent heat and fluid resistance. They are supplanting conventional materials because they offer exceptional value during use, along with a long product lifetime. The important uses are in shaft seals, gaskets, O-rings, and fuel hose (Table 8.1). A terpolymer of vinylidene difluoride, HFP, and TFE was compounded with nanoclays to provide enhanced mechanical and physical properties of the fluoroelastomer nanocomposites produced. The products from this nanocomposite are being used in space-craft, where low-density though high strength materials are required [8].

PVDF nanocomposites have been used as piezoelectric and pyroelectric materials [9] because of their improved mechanical and electrical properties on addition of nanofillers such as clays [10]. PVDF-based nanocomposites are important engineering plastics that are used extensively in the pulp and paper industries because of their resistance to halogens and acids, in nuclear-waste processing for radiation- and hot-acid applications, and in the chemical processing industries for chemical-resistant and high-temperature applications.

## 8.3. Silicone Rubber Nanocomposites

### 8.3.1. Silicone Elastomers

Silicon elastomers consist mainly of poly(dimethylsiloxane) of varying molar mass with differing cure reactions (Figure 8.2). The cure reaction can be via peroxide initiation at elevated temperature. Other cure reactions are activated by absorbed moisture when ethoxysilane or acetoxysilane functionality is present in single-pack systems. These are called room temperature vulcanizing (RTV) silicone elastomers. A silicon–oxygen link is formed and ethanol and acetic acid are volatile by-products. A two-pack RTV system consists of a vinylsilane and a hydrosilane with chloroplatinic acid catalyst. Curing is achieved by hydrosilation of the vinyl group resulting in a silicon–carbon link and no other by-product. Silicone elastomers are one of the important types of high-temperature resistance synthetic elastomers with excellent thermal stability, low-temperature toughness, and electrical insulating properties.

Silicone elastomer nanocomposites with silica filler are frequently used. The silica can be dispersed into the uncrosslinked silicone as precipitated silica or pyrolytic silica. The sol–gel reaction using tetraethoxysilane is convenient to use either before or after curing since it is miscible with silicones.

### 8.3.2. Preparation of Silicone Rubber Nanocomposites

The typical nanofillers used are layered silicates, fumed silica, and carbon nanotubes (CNTs). The nanofillers are dispersed in silicone rubber using melt-intercalation or a sol–gel process. The synthesis of silicone elastomer nanocomposites with clay nanofiller was reported via a melt-intercalation process or an intercalation polymerization process. The melt-intercalation compounding process is one of the promising approaches of synthesizing nanocomposites by using organosilicate and conventional twin-screw extrusion

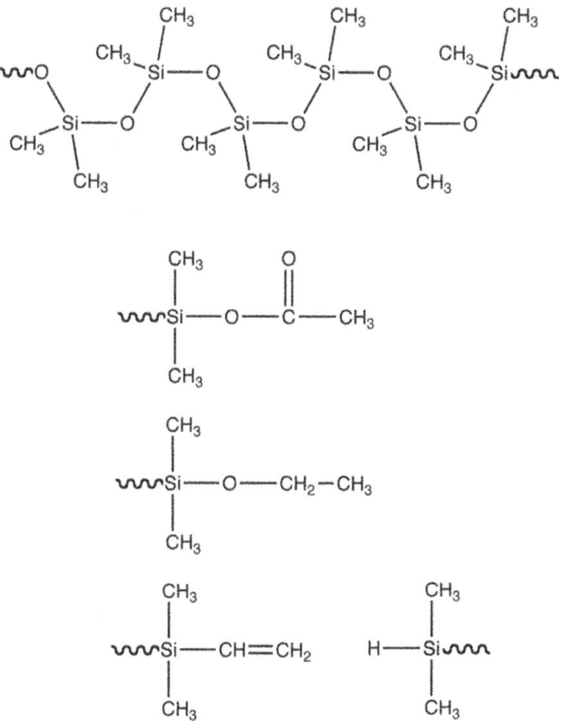

**Figure 8.2**   Polysiloxane (top) and precursors for room temperature vulcanizing systems (from top): acetoxypolysiloxane (water curable), ethoxypolysiloxane (slow water cure), vinylpolysiloxane, and hydrosilane-terminated polysiloxane (curable with hexachloroplatinic acid catalyst).

compounding equipment. Layered mica-type silicates such as montmorillonite, smectite, and vermiculite are suitable for this process [11, 12].

Silicone elastomer–organomontmorillonite (OMT) hybrid nanocomposites were prepared via melt intercalation, and the OMT particles were uniformly dispersed in the silicone elastomer matrix and exfoliated into 50 mm thickness [13].

Polydimethylsiloxane–silicate nanocomposites have been synthesized by melt processing. The synthesis involves silicate delamination in the polymer matrix followed by crosslinking. The nanocomposites exhibited decreased solvent uptake and increased thermal stability. Increased swelling resistance was attributed to strong reinforcement–matrix interactions and a large surface area attainable by exfoliation and dispersion of the silicate layers in the polymer matrix [14].

## 8.3.3. Mechanical Properties

The fire retardant poly(trimethyl vinyl silicone) elastomer–montmorillonite nanocomposites with aerosilica as synergistic reinforcement filler and magnesium hydroxide and red phosphorous as fire retardant additives showed better mechanical and flammability properties even at a low loading of fillers and additives. The addition of chemically modified filler increased the interaction between silicone elastomer matrix and filler. In the elastomer vulcanizing process, the modified filler could act as an active site to increase crosslinking and lead to a denser network-like structure. The filler particles were encapsulated in the elastomer matrix network-like structure and therefore the interaction between elastomer matrix and filler increased [15].

The silicone elastomeric nanoparticle (S-ENP), a flame retardant polyamide-6 (nylon-6) nanocomposite with high toughness, high heat resistant, high stiffness, and good flowability has been prepared. It was found that the S-ENP with average particle size of 100 nm was effective in improving flame retardancy and toughness of polyamide-6 and can also help unmodified clay to be exfoliated in polyamide-6 matrix. Through the use of clay slurry and S-ENP latex, another novel elastomeric flame retardant of S-ENP-clay (S-ENPC) was produced. It was found that the S-ENP and clay platelets in S-ENPC had a synergistic fire retardant effect on polyamide-6 that resulted in the further enhancement of fire retardancy of the polyamide-6. Polyamide-6–S-ENPC nanocomposites not only exhibited high toughness, modulus, and heat distortion temperature but also provided low melt viscosity [16].

The RTV silicone rubber–silica nanocomposites having different weight ratios of silica filler showed improved tensile strength and heat distortion temperature properties. The integrated mechanical performances of RTV silicone rubber–sol–gel silica nanocomposites are better than the RTV silicone rubber–gas phase silica nanocomposites due to more uniform distribution of sol–gel silica in silicone rubber. The improved thermal performance of RTV silicone rubber–silica nanocomposites was observed due to chemical reaction through hydroxy bonds between nanosilica and RTV silicone rubber and physical absorption, such as electrostatic and hydrogen bond absorption [17].

Silicone elastomer–clay nanocomposites were synthesized by a melt-intercalation process using synthetic Fe–OMT and natural Na–OMT. In the silicone elastomer–clay nanocomposites, especially at the application-relevant low content of clay, the mechanical properties were tested with changes in content of clay. TEM and XRD revealed a coexistence of intercalated and exfoliated silicate layers and silicone elastomer matrix were intercalated in the galleries of OMT. The effect on the tensile strength of the two kinds of clay in the silicone elastomer nanocomposites was similar. This suggested a strong interaction between the two kinds of substances and a network structure in the material. In this case, the exfoliated OMT layers combined with the intercalated agent readily interact with the copolymer chains to form a physical crosslink, in which OMT acts as a physical crosslinking junction [18].

### 8.3.4. Applications of Silicone Elastomer Nanocomposites

Nanocomposites of silicone-containing polyelectrolyte and CNTs have been used for the construction of a resistive-type humidity sensor [19]. The silicone-containing polyelectrolyte was prepared by copolymerizing the quaternary ammonium salt obtained from the reaction of dimethylaminoethyl methacrylate and $n$-bromobutane with $\gamma$-methacryloxypropyl trimethoxy silane.

Incorporation of layered clay and organoclay into silicone elastomer–poly(propylene oxide-co-ethylene oxide) blends in forming intercalated nanocomposites via blending with random nanoscale sandwich structure dampers having high stiffness and antivibration properties. Compressive vibration hysteresis evaluation indicated that the area of the hysteresis loop increased with increasing content of organically modified clay, that is, the antivibration performance was improved [20]. Silicone elastomer nanocomposites used as antivibration materials have been widely used in, for example, machine, transportation, and construction industries.

## 8.4. Polyurethane Elastomer Nanocomposites

### 8.4.1. Polyurethane Elastomers

Polyurethanes are thermoplastic or network polymers including the urethane functional group as the link between flexible prepolymer components (Figure 8.3). They are elastomers that consist of polar molecules, so they are able to provide unique properties, in

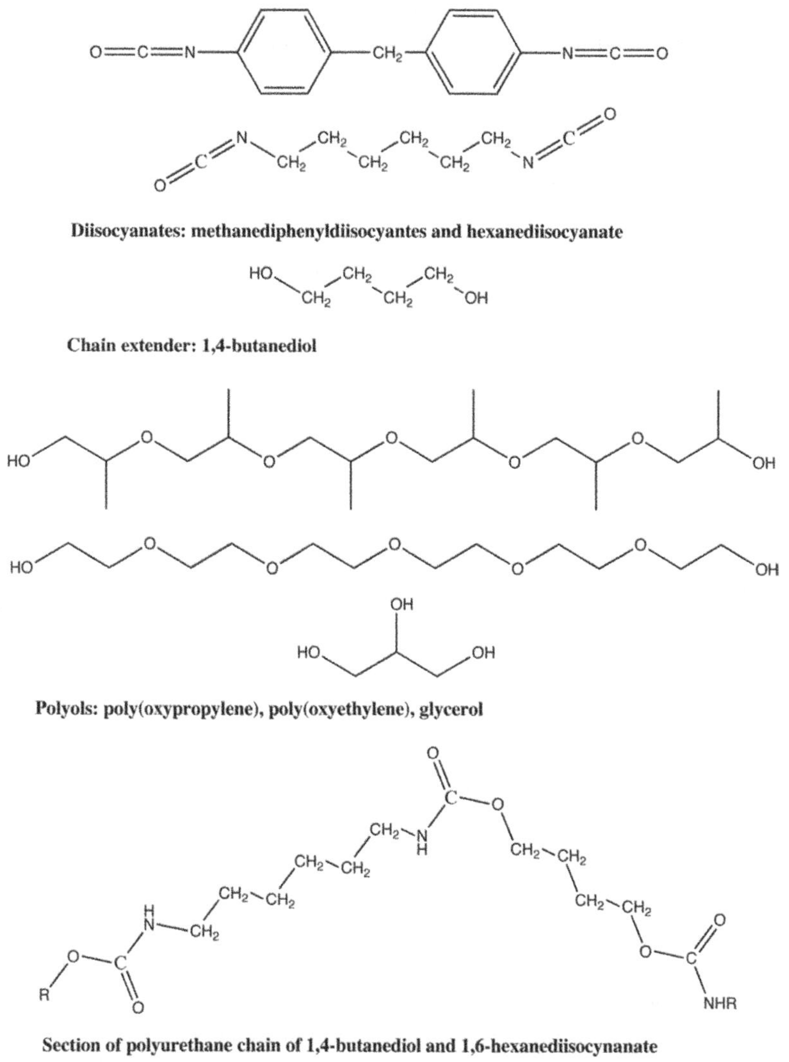

Diisocyanates: methanediphenyldiisocyantes and hexanediisocyanate

Chain extender: 1,4-butanediol

Polyols: poly(oxypropylene), poly(oxyethylene), glycerol

Section of polyurethane chain of 1,4-butanediol and 1,6-hexanediisocynanate

**Figure 8.3**  Polyurethane monomers: diisocyanates, chain extender diol, typical polyols, and an example section of polyurethane structure.

particular oil resistance. The flexible prepolymer consists of typically either an aliphatic polyether or a polyester, with the polyethers providing the greater flexibility. The chain segments containing the urethane groups are linked by a chain extender, usually derived from a diol. The urethane-containing segments are hydrogen bonded providing cohesion and a reversible elastomeric response to the structure. Alternatively, the urethane segments may be crosslinked providing additional cohesion through covalent bonds. In addition to urethane functional groups, other linking groups may be formed by related reactions of the diisocyanates. Amines added or formed during foaming reactions of diisocyanates with water yield the more rigid urea groups. Side reactions give allophonate and biuret structures. Isocyanurate ring structures can be formed, and sometimes their formation is promoted by the choice of the catalyst system.

The elastomeric segments are the polyether or polyester prepolymer polyols. These components are already polymers since they must be long enough to form reversible molecular coils. They will require functionality greater than two if network urethanes are to be formed.

### 8.4.2. *In-situ* Preparation of Network Polyurethane Elastomer Nanocomposites

The nanofiller will typically be dispersed in the polyol. This is convenient since the filler will probably contain traces of moisture that may react prematurely with an isocyanate. The polyol is usually present in the greater amount by mass due to its high molar mass. Traditionally, carbon black and pyrolytic or precipitated silicas have been used as fillers. These fillers stiffen the elastomer and increase wear resistance, damping, and cohesion. Layered clays have been increasingly used and the polyethers such as poly(oxyethylene) and poly(oxypropylene) have been found to be efficient intercalating agents. The polyol and filler can be dispersed using a high shear stirrer, disperser, or bead mill. The filler dispersion in polyol is then mixed with other components such as catalyst system, surfactant for foamed elastomers, and chain extender. The chain extender is often 1,4-butanediol or similar small molecule diol or polyol, such as glycerol. The catalyst is often a combination of tertiary amine, such as triethylenediamine, and Lewis acid, such as tin dilaurate. Water may be added for foamed elastomers.

A function of some nanofillers, such as silica particles, is to provide thixotropy while the polyurethane is being formed. This is needed for coatings and sealants to eliminate flow before cure. Thixotropy is useful during foam formation, along with surfactants, to control bubble diameter and density.

*In-situ* polymerization has been used to prepare CNT composites in polyurethanes based on polycaprolactone, using methanediphenyldiisocyanate with 1,4-butandiol as chain extender. The multiwalled carbon nanotubes (MWNTs) increased tensile strength and modulus. Strong adhesion was formed between the MWNT and the polyurethane [21].

Both the polyurethane elastomer and nanofiler can be prepared *in situ*. The polyurethane was a macrodiisocyanate prepared form poly(oxypropylene) and 2,4-toluenediisocyanate. The sodium silica–phosphate nanofiller was prepared using aqueous solutions of sodium silicate and sodium hexametaphosphate. The resulting material was an organic–inorganic hybrid polyurethane with a continuous polyurethane nanophase coexisting with a co-continuous nanophase or infinite cluster of the sodium silica–phosphate [22].

### 8.4.3. Thermoplastic Polyurethane Nanocomposites

Thermoplastic polyurethane nanocomposites can be formed by dispersing the nanofiller using a twin-screw extruder or equivalent, an internal mixer followed by extrusion or compression molding of the composite, or solution dispersion followed by processing to separate and shape the nanocomposite. Polyurethane is an efficient dispersing medium since polarity and hydrogen bonding provide compatibility with the filler. Layered clays have been found to be intercalated and exfoliated similar to their performance with nylons. The polyether or polyester component may form the predominant interactions with the filler, or the urethane groups, which are present in lower concentration, may strongly interact via hydrogen bonding and polarity.

Thermoplastic polyurethanes have been combined with various alkylammonium-intercalated montmorillonite clays using twin-screw extrusion for dispersion. The type of alkylammonium group used to treat the clay was significant in determining matrix reinforcement, similar to observations found for polyamide-6. These polar polymers have affinity for the clay surface [23].

CNTs have been used to reinforce thermoplastic polyurethane fibers. The tensile modulus and tensile strength were improved significantly, while maintaining elongation at break [24].

### 8.4.4. Phase Separation of Polyurethanes

Polyurethane elastomers whether crosslinked or thermoplastic exhibit phase separation into a dispersed phase of urethane group containing segments in a matrix phase of polyether or polyester. The matrix provides the elasticity, whereas the dispersed phase provides physical crosslinks (Figure 8.4). The thermoplastic polyurethanes are classified as thermoplastic elastomers. An important part of characterization of these materials is to

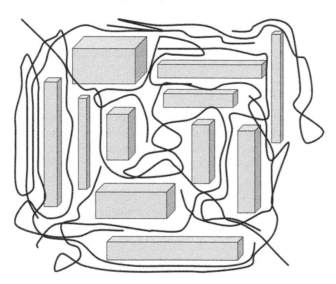

**Figure 8.4**  Schematic of a nanocomposite where the polymer–filler interactions and the polymer crystalline regions and chain entanglements provide cohesion via physical crosslinks.

establish the phase separation and its extent, usually by observation of two glass transition temperatures ($T_g$), the $T_g$ of the elastomeric matrix and the $T_g$ of the dispersed hard phase. The nanofiller particles or platelets may be associated with either or both phases depending on the interactions and concentrations. If the filler is associated mainly with the dispersed phase, then the physical crosslinks will be enhanced resulting in greater strength or yield stress, break stress, and resistance to creep. If the fillers are mainly associated with the matrix phase, then the elastomer will become stiffer and would require greater damping. If the filler is associated with both phases, then a combination of property changes will be observed.

Rheological investigation of the viscoelasticity of poly(1,4-oxybutane) polyurethanes based on toluenediisocynate and 1,4-butanediol was used to measure microphase separation. Dynamic rheological mastercurves showed that the microphase separation temperature was independent of length and type of poly(1,4-oxybutane). A dominant contribution was from the ratio of 2,4- to 2,6-isomers in the toluenediisocynate [25].

The filler will tend to average the differences between the dispersed and matrix phases. Addition of nanofillers has been shown to restrict the phase separation. If phase separation is limited, then the material will change from a typical thermoplastic elastomer to one exhibiting mainly thermoplastic properties. Phase separation is an inherent part of thermoplastic elastomer properties. Solvent resistance is mainly increased in the nanocomposites. Crosslinked polyurethane systems will be similarly affected except that they can never exhibit plastic-type deformation, so there is no need for filler-derived physical crosslinks. The duality of properties due to hard and soft phases will be reduced as the filler concentration is increased. Nanosized fillers are more effective in controlling phase separation due to their extremely large surface-area-to-volume ratio.

Polyurethane prepolymers cured with a diamine to give urea links were filled with various types of nanomaterials, including zinc oxide, aluminum oxide, and alkylammonium-treated montmorillonite. The nanofillers were dispersed in a solution of the prepolymers using ultrasonication prior to curing. The phase separation of the hard and soft segments within the polyurethane was disrupted by the addition of small amounts for nanofiller. The disruption of phase separation caused a deterioration of the mechanical properties, although properties could be improved by chemical modification of the nanoparticles [26].

## 8.4.5. Mechanical Properties

Polyurethane nanocomposites will have similar changes in mechanical properties to other PNCs, except that the interactions with the filler will be stronger, analogous to those with nylons. The complexity of changes in mechanical properties is increased by the issue of phase separation. Crosslinked polyurethanes may in addition experience changed reaction kinetics, thixotropy-modified curing, and foam formation.

The dynamic mechanical properties of polyester–polyurethanes based on methanediphenyl diisocyante showed that the entanglement modulus did not vary with hard–soft segment ratio, although the relation time in the terminal zone increased with increased hard segment [27].

Fatigue durability of polyurethanes derived from poly(oxypropylene), methanedicyclohexyldiisocyanate, and 1,4-butanediol were determined using dynamic mechanical tests in conjunction with pulsed force mode atomic force microscopy. Inclusion of alkylammonium-treated montmorillonite increased fatigue durability by about ten times. The intercalated organoclay particles showed reversible deformation that was considered to reduce stress concentration and arrest crack propagation [28]. Stress relaxation and creep response of polyurethane nanocomposites are affected by a deceased hard domain size and consequential changes in the relaxation time spectrum [29].

### 8.4.6. Applications of Polyurethane Elastomer Nanocomposites

Thermoplastic polyurethane nanocomposites have applications in extruded and injection-molded tough elastomeric components. Polyurethanes with either polyether or polyester soft segments provide elastomeric properties in conjunction with polyamide-6 microfibers for application as synthetic leather [30].

Network polyurethane nanocomposites are suitable for use in foams, sealants, and coatings where increased rigidity and damping are required. Fatigue resistance and durability are increased for layered clay nanocomposites, and crack propagation is limited by stress concentration release at the nanoparticle interfaces [31].

An important application of polyurethane elastomer nanocomposites is where flame retardance is required. Combustion and thermal behavior is enhanced by the barrier characteristics of clay layers [32]. The barrier effect of the clay lamellas inhibits the volatilization of degradation products from the polyurethane, the intimate contact between polymer and clay favors the formation of char-forming degradation products, fire-induced dripping of the polymer is decreased, and a protective char is formed by the clay and carbonaceous products [33].

Foamed polyurethane nanocomposites have been prepared and the mean cell size decreased with nanoclay concentration. The alkylammonium used to modify the montmorillonite contain hydroxy functionality and proceeded from intercalated to exfoliate during the dispersion. The clay increased the modulus of the polyurethane foam, but the hydraulic resistance was decreased due to weakened cell windows [34].

## 8.5. Other Specialty Rubbers

Neoprene, nitrile, polysulfide, and thermoplastic vulcanisate (Santoprene) are other types of specialty elastomers used for specialty rubber nanocomposites.

### 8.5.1. Neoprene

Nanocomposites of chlorine-containing polymers such as polychloroprene, chlorinated polyethylene, poly(vinyl chloride), chlorinated poly(vinyl chloride), and poly(vinylidene chloride) with organic-treated montmorillonite and natural sodium montmorillonite fillers were prepared by melt-compounding process. The chlorinated polymers are compatible with organic-treated clays but not with natural clays. The polarity of the chlorinated polymer chains and organophilicity of the organic-treated clays allow the formation of well-dispersed nanocomposites. The tensile strength and young modulus of the nanocomposites increased with increase of clay content [35]. The preparation of polychloroprene and cation-treated montmorillonite-based nanocomposites were reported. The mechanical properties of the nanocomposites are dependent not only on the filler but also on the mixing process used [36].

Neoeprene nanocomposties are used for making improved chemical protective gloves. Such gloves are having significantly improved resistance to petroleum oils and gasoline while being less flammable.

## 8.5.2. Nitrile Rubber

Nitrile rubber-unmodified montmorillonite clay nanocomposites were prepared by latex-blending method followed by melt mixing. Latex blending involves mixing of rubber latex and a clay aqueous dispersion under some specified mixing conditions, to enhance good dispersion and then coagulating the mixture by adding the electrolyte. The dynamic mechanical analysis of nanocomposites showed proportional increase in storage modulus analogous to clay loading at all temperature ranges due to confinement of polymer chains between the clay layers. The remarkable improvements in modulus, hardness, and tensile strength of nanocomposites were observed due to high level of clay dispersion [37].

Nitrile butadiene rubber (NBR)–clay nanocomposites were prepared by co-coagulating the NBR latex and clay aqueous suspension. TEM showed that the silicate layers of clay were dispersed in the NBR matrix at the nanolevel and had a planar orientation. X-ray diffraction indicated that there were some unexfoliated silicate layers in the NBR–clay nanocomposites. Stress–strain analyses showed that the silicate layers generated evident reinforcement, modulus, and tensile strength of the NBR–clay nanocomposites, which were significantly improved with an increase in the clay concentration, and strain at break was higher than that of the gum NBR vulcanizate when the amount of clay was more than 5 phr. The NBR–clay nanocomposites exhibited an excellent gas barrier property: the reduction in gas permeability in the NBR–clay nanocomposites. Compared with gum NBR vulcanisate, the oxygen index of the NBR–clay nanocomposites increased slightly [38].

The nanocomposites derived from nitrile rubber have been extensively investigated with respect to morphological and rheological properties [39], mechanical and dynamic mechanical properties [40], gas barrier properties [41], and fracture behavior and cure properties [42, 43].

Hydrogenated nitrile rubber nanocomposite has high tensile strength, low permanent set, very good abrasion resistance, and high elasticity. It shows good stability toward thermal ageing and has better properties at low temperatures compared with other heat- and oil-resistant elastomers. This combination of properties is opening up a broad range of applications for the materials, mainly in the automotive industry (Figure 8.5). Belts, seals, and hoses are important applications of nitrile rubber nanocomposites.

## 8.5.3. Thermoplastic Vulcanizate (Santoprene)

Nanocomposites comprising nanoscale platelets derived from layered silicates treated with an organic modifier in thermoplastic vulcanizates (TPVs) and PP-EPDM blends were prepared by direct melt intercalation. The interlayer spacing and dispersion of the nanoclay are greatly affected by polar forces between the nanoclay and the polymeric matrix material. The mechanical properties strongly depend on the structure and morphology of the nanocomposites, which can be modified by phase partitioning of the reinforcements. Morphology characterization provided the basis for understanding the observed structure–property relationships in this class of materials. With the increase of organoclay loading, the tensile modulus of TPV–clay nanocomposites increases with increase of organoclay loading, whereas tensile strength gradually decreases with increase of organoclay loading. In the physical blend systems, the tensile modulus increases for all PP-EPDM blend compositions and generally shows higher values in the case of selectively reinforced phases in blends having a continuous PP matrix. The tensile strength of those blends decreases

**Figure 8.5**  High-performance elastomers (temperature, oil, durable, vibration, and flex resistant) in automotive applications, such as hoses, insulation, sleeves, and coupling covers (shown with arrows), not visible gaskets and seals.

at higher nanoclay content, no matter how blending is performed. However, the tensile strength may increase when sufficient selectively reinforced EPDM phase is present [44].

The thermoplastic vulcanisate (TPV)–silica nanocomposites were prepared by the melt blending of TPV and maleic anhydride-grafted polypropylene (PP-$g$-MA) into organically modified silica ($m$-SiO$_2$), treated with $n$-hexadecyl trimethylammonium bromide as a grafting agent for TPV during the melt mixing. The dynamic mechanical analysis revealed that the glass-transition temperature of the polypropylene phase of the nanocomposites increased (in comparison to that of virgin TPV), whereas the ethylene–propylene–diene monomer phase remained almost the same. The adhesion strength between the TPV-PP-$g$-MA-$m$-SiO$_2$ nanocomposites and steel increased with increasing $m$-SiO$_2$ content [45].

The TPV–organoclay nanocomposites were prepared by melt-intercalation method. The organoclay was first treated with glycidyl methacrylate, which acts as swelling agent for the organoclays, as well as a grafting agent for the TPV during melt mixing. The X-ray diffraction analysis showed that the nanocomposite was intercalated. The dynamic mechanic analysis proved that the nanocomposite had improved mechanical properties [46]. TPV nanocomposites are widely used as body-mounted primary seal in complete dynamic sealing application.

## 8.6. Future Trends

Specialty rubber nanocomposites are now used in O-rings, seals, gaskets, lathe built components, sponge card, special laminations and assemblies, fuel hose, and sporting equipment (Figure 8.6). Several technological applications will be suited to these specialty rubber nanocomposites. The potential of nanocomposites in various sectors of research and application is promising and attracting increasing investment from governments and

**Figure 8.6**  Typical elastomer seals and gaskets, black chloroprene rubber, and translucent polydimethylsiloxane.

business in many parts of the world. While there are some niche applications in which nanotechnology has penetrated the market, the major impact will be in the future. According to several sources such as Chemical Business Newsbase and Plastic News, a significant increase in turnover of about 100%/year leading to a value of about millions of dollars (ca. 500,000 ton/year of PNCs) in 2009 is expected. Even if these estimates are optimistic, they highlight the significant technological and economical potential of polymeric nanocomposites including not only clay but other inorganic nanofillers, such as CNTs, silica, silicon carbide, and silicon nitride. It is difficult to predict which market sector would not be able to benefit from this technology. A diverse range of sectors such as aerospace, automotive, packaging (particularly food but also solar cells), electrical and electronic components, and household appliances will benefit substantially from a new range of materials that are being offered by this technology.

## 8.7. Conclusion

Specialty rubber nanocomposites are supplanting conventional materials because they offer exceptional value for use, along with a long products lifetime. They are widely used in the manufacturing of specialty products. Significant research is needed to characterize and explain the behavior of nanointerfaces, and this field can still be considered to be in its early stages. In particular, the development of accurate nanomechanical models and understanding of the properties of the polymer at the interface are required to address the outstanding issues of the polymer–nanofiller interface and interparticulate interactions and thus optimize the mechanical performance of PNCs. It is believed that one of the main issues in preparing good polymer matrix nanocomposites is uniform dispersion of the nanofiller in a polymer matrix.

# References

1. Arcella, V.; Ferro, R. Fluorocarbon elastomers, in J Scheirs (Ed.), Modern Fluoropolymers, John Wiley & Sons, London, 1998, chapter 2, pp 71–90.

2. Yeh, J.T.; Chen, C.L.; Huang, K.S. Preparation and application of fluorocarbon polymer/$SiO_2$ hybrid materials, Part 1: Preparation and properties of hybrid materials, J. Appl. Polym. Sci., 2007, 103, 1140–1145.

3. Pyria L.; Jog J.P. Poly(vinylidene fluoride/clay nanocomposites prepared by melt intercalation: crystallization and dynamic mechanical behaviour studies, J. Polym. Sci: Part B Polym. Phys., 2002, 40, 1682–1689.

4. Kim, Y.; White, J.L. Melt-intercalation nanocomposites with fluorinated polymers and a correlation for nanocomposite formation, J. Apply. Polym. Sci., 2004, 92, 1061–1071.

5. Barton, A.F.M. Handbook of Solubility Parameters and Other Cohesion Parameters, CRC Press, Boca Raton, FL, 1983, 280.

6. Valsecchi, R.; Vigano, M.; Levi, M.; Turri, S. Dynamic mechanical and rheological behaviour of fluoroelastomer-organoclay nanocomposites obtained from different preparation methods, J. Appl. Polym. Sci., 2006, 102, 4484–4487.

7. Shah, D.; Maiti, P.; Gunn, E.; Schmidt, D.F.; Jiang, D.D.; Batt, C.A.; Giannelis, E.P. Dramatic enhancements in toughness of polyvinylidene fluoride nanocomposites via nanoclay-directed crystal structure and morphology, Adv. Mater., 2004, 16, 1173–1177.

8. Maiti, M.; Bhowmick, A.K. Structure and properties of some novel fluoroelastomer/clay nanocomposite with special reference to their interaction, J. Polym. Sci: Part B Polym. Phys., 2006, 44, 162–176.

9. Zhang, Q.M.; Li, H.; Poh, M.; Xia, F.; Cheng, Y.; Xu, H. An all-organic composite actuator material with high dielectric constant, Nature, 2002, 419, 284–287.

10. Pramoda, K.P.; Mohamed, A.; Phang, I.Y.; Liu, T. Crystal transformation and thermomechanical properties of poly(vinylidene fluoride) clay nanocomposites, Polym. Int., 2005, 54, 226–232.

11. Krishnamoorthi, R.; Vaia, R.A.; Giannelis, E.P. Structure and dynamics of polymer layered silicate nanocomposites, Chem. Mater., 1996, 8, 1728–1734.

12. Vaia, R.A.; Ishii, H.; Giannelis, E.P. Synthesis and properties of two-dimensional nanostructures by direct intercalation of polymer melts in layered silicates, Chem. Mater., 1993, 5, 1694–1696.

13. Wang, S.; Long, C.; Wang, X.; Li, Q.; Qi, Z. Synthesis and properties of silicone rubber/organomontmorillonite hybrid nanocomposites, J. Apply. Polym. Sci., 1998, 69, 1557–1561.

14. Burnside, A.D.; Giannelis, E.P. Synthesis and properties of new poly(dimethylsiloxane) nanocomposites, Chem. Mater., 1995, 7, 1597–1600.

15. Yang, L.; Hu, Y.; Lu, H.; Song, L. Morphology, thermal and mechanical properties of flame-retardant silicone rubber/montmorillonite nanocomposites, J. Apply. Polym. Sci., 2006, 99, 3275–3280.

16. Dong, W.; Zhang, X.; Liu, Y.; Wang, Q;; Gui, H.; Gao, J.; Song, Z.; Lai, J.; Huang, F.; Qiao, J. Flame retardant nanocomposites of polyamide 6/clay/silicone rubber with high toughness and good flowability, Polymer, 2006, 47, 6874–6879.

17. Dengke, C.; Xishan, W., Lei, L.; Jianhui, Y. Study on RTV silicone rubber/$SiO_2$ electrical insulation nanocomposites, 2004 International Conference on Solid Dielectrics, Toulouse, France, 2004, July 5–9.

18. Kong, Q.; Hu, Y.; Song, L.; Wang, Y;; Chen, Z.; Fan, W. Influence of Fe-MMT on crosslinking and thermal degradation in silicone rubber/clay nanocomposites, Polym. Adv. Technol., 2006, 17, 463–467.

19. Li, Y.; Yang, M.J.; Chen, Y. Nanocomposites of carbon nanotubes and silicone-containing polyelectrolyte as a candidate for construction of humidity sensor, J. Mater. Sci., 2005, 40, 245–247.

20. Chiu, H.T.; Wu, J.H.; Shong, Z.J. Dynamic properties of rubber vibration isolators and antivibration performance of a nanoclay-modified silicone/poly(propylene oxide)–poly(ethylene oxide) copolymer with 20 wt% $LiClO_4$ blend system, J. Appl. Polym. Sci., 2006, 101, 3713–3720.

21. Sahoo, N.G.; Jung Y.C.; Yoo H.J.; Cho J.W. Effect of functionalized carbon nanotubes on molecular interaction and properties of polyurethane composites, Macromol. Chem. Phys., 2006, 207, 1773–1780.

22. Lebedev, E.V.; Ishchenko, S.S.; Denisenko, V.D.; Dupanov, V.O.; Privalko, E.G.; Usenko, A.A.; Privalko, V.P. Physical characterization of polyurethanes reinforced with the *in situ*-generated silica-polyphosphate nano-phase, Compos. Sci. Technol., 2006, 66, 3132–3137.

23. Chavarria, F.; Paul, D.R. Morphology and properties of thermoplastic polyurethane nanocomposites: Effect of organoclay structure, Polymer, 2006, 47, 7760–7773.

24. Chen, W.; Tao, X.; Liu, Y. Carbon nanotube-reinforced polyurethane composite fibers, Compos. Sci. Technol., 2006, 66, 3029–3034.

25. Yang, I.K.; Wang, P.J.; Tsai, P.H. Rheological investigation of microphase separation transition of polyurethane elastomer, J. Appl. Polym. Sci., 2006, 103, 2107–2112.

26. Zheng, J.; Ozisik, R.; Siegel, R.W. Phase separation and mechanical responses of polyurethane nanocomposites, Polymer, 2006, 47, 7786–7794.

27. Florez, S.; Munoz, M.E.; Santamaria, A. Novel dynamic viscoelastic measurements of polyurethane copolymer melts and their implication to tack results, Macromol. Mater. Eng., 2006, 291, 1194–1200.

28. Jin, J.; Chen, L.; Song, M.; Yao, K. An analysis on enhancement of fatigue durability of polyurethane by incorporating organoclay nanofillers, Macromol. Mater. Eng., 2006, 291, 1414–1421.

29. Xia, H.; Song, M.; Zhang Z. Richardson M; Microphase separation, stress relaxation, and creep behaviour of polyurethane nanocomposites, J. Appl. Polym. Sci., 2007, 103, 2992–3002.

30. Chen, M.; Zhou, D.L.; Chen, Y.; Zhu, P.X. Analyses of structures for a synthetic leather made of polyurethane and microfiber, J. Appl. Polym. Sci., 2007, 103, 903–908.

31. Jin, J.; Chen, L.; Song, M.; Yao, K. An analysis on enhancement of fatigue durability of polyurethane by incorporating organoclay nanofillers, Macromol. Mater. Eng, 2006, 291, 1414–1421.

32. Levchik, S.V.; Weil, E.D. Thermal decomposition, combustion and fire-retardancy of polurethanes – a review of the recent literature, Polym. Int., 2004, 53, 1585–1610.

33. Berta, M.; Lindsay, C.; Pans, G.; Camino, G. Effect of chemical structure on combustion and thermal behaviour of polyurethane elastomer layered silicate nanocomposites, Polym. Degrad. Stab., 2006, 91, 1179–1191.

34. Mondal, P.; Khakhar, D.V. Rigid polyurethane–clay nanocomposite foams: preparation and properties, J. Appl. Polym. Sci., 2007, 103, 2802–2809.

35. Kim, Y.; White, J.L. Melt intercalation nanocomposites with chlorinated polymers, J. Appl. Polym. Sci., 2003, 90, 1581–1588.

36. Yeh, M.H.; Hwang, W.S.; Chang, Y.C. Preparation and mechanical properties of polychloroprene–montmorillonite composites, Jpn. J. Appl. Phys., 2005, 44, 6847–6854.

37. Kader, M.A.; Kim, K.; Lee, Y.S.; Nah, C.; Preparation and properties of nitrile rubber/montmorillonite nanocomposites via latex mixing, J. Mater. Sci., 2006, 41, 7341–7352.

38. Wu, Y.P.; Jia, G.X.; Yu, D.S.; Zhang, L.Q. Structure and properties of nitrile rubber (NBR)–Clay nanocomposites by co-coagulating NBR latex and clay aqueous suspension, J. Apply. Polym. Sci., 2003, 89, 3855–3858.

39. Kim, J.T.; Oh, T.S.; Lee, D. Morphology and rheological properties of nanocomposites based on nitrile rubber and organophilic layered silicates, Polym. Int., 2003, 52, 1203–1208.

40. Nah, C.; Ryu, H.J.; Kim, W.D., Chang, Y.W. Preparation and properties of acrylonitrile–butadiene copolymer hybrid nanocomposites with organoclays, Polym. Int., 2003, 52, 1359–1364.

41. Kojima, Y.; Fukumori, K.; Usuki, A.; Okada, A.; Kurauchi, T. Gas permeabilities in rubber–clay hybrid, J. Mater. Sci. Lett., 1993, 12, 889–890.

42. Nah, C.; Ryu, H.J.; Han, S.H.; Rhee, J.M.; Lee, M.H. Fracture behaviour of acrylonitrile–butadiene rubber/clay nanocomposite, Polym. Int., 2001, 50, 1265–1268.

43. Kim, J.T., Oh, T.S., Lee, D.H. Curing and barrier properties of NBR/organo–clay nanocomposite, Polym. Int., 2004, 53, 406–411.

44. Lee, K.Y.; Goettler, L.A. Structure–property relationships in polymer blend nanocomposites, Polym. Eng. Sci., 2004, 44, 1103–1111.

45. Wu, T.M.; Zu, M.S. Preparation and characterization of thermoplastic vulcanizate/silica nanocomposites, J. Apply. Polym. Sci., 2005, 98, 2058–2063.

46. Mishra, J.K.; Kim, G.H.; Chung, I.J.; Ha, C.S. A new thermoplastic vulganizate (TPV)/organoclay nanocomposite: preparation, characterization and properties, J. Polym. Sci: Part B Polym. Phys., 2004, 42, 2900–2908.

# 9 Polyester Nanocomposite

*Seungsoon Im* and Wanduk Lee*

Department of Fiber & Polymer Engineering, Hanyang University,
Seongdong-gu, Seoul, 133-791, Korea

## 9.1. Introduction

The National Nanotechnology Initiative (NNI) defines nanotechnology as involving "research and technology development at the 1–100 nm range." The application of nanotechnology to polymers has produced polymeric nanocomposites in which polymers are intercalated in layered inorganic particles, such as layered silicates or admixed *via* exfoliation. Polymeric nanocomposites are prepared by admixing the different primary components in a polymer matrix, along with various secondary nanosize organic and inorganic components. Nanotechnology has grown to become wide-ranging in its applications. Of particular importance is the use of nanoparticles with a high aspect ratio that allows the structural characteristics of their molecule to be used in conjunction with their layered particle structures to maximize the strength of bonding between polymer and particles. By adding small amounts of these particles, 0.5–5 wt%, it has become possible to not only improve many of the original properties of the polymers, such as their strength, rigidity, and heat, ultraviolet, and flame resistance, but also to decrease their water absorption rates and gas permeability.

Polymer nanocomposite products first appeared in the early 1990s and consisted of nylon/layered silicates blends. Subsequently, many nanocomposites have been developed that use high-molecular-weight (high-MW) polymers. However, few nanocomposites in the polyester system, as represented by polyethylene terephthalate (PET), have come on stream commercially despite extensive research.

ASTMD 883-86b defines polyester as a high-MW polymer in which the repeat unit contains ester-bonded (–COO–) chains. Where it interacts by self-condensation with a mono hydroxyl or mono carboxyl acid or with a diol and dicarboxyl acid, it becomes a linear thermoplastic polymer. Carothers [1, 2] achieved the original synthesis of aliphatic polyester at Dupont. Schlack [3] at German Agfa and Dickson and Winfield [4, 5] in the United Kingdom developed terephthalic acid (TPA) on polyester. The initial industrial exploitation by ICI in the UK commenced in 1949, followed by Dupont in the USA in 1958. These developments provided the fiber presently used in clothing. Although PET

---

*Correspondence should be addressed to e-mail: imss007@hanyang.ac.kr

does not constitute the lion's share of commercial high-MW polymers, it is widely used in various application fields. Especially, its use in soda bottles and films has grown by 10% recently.

These developments are the result of its properties arising from its structural chemistry. For example, the crystallization rate of PET film is slow, making it easy to mould in the glassy state and hence suitable for base film of photos and bottles. PET also interacts readily. Further condensation and transesterification have allowed the production of additional commercial high-MW polymers. The related polybutylene terephthalates (PBT) are of more limited use despite being chemically similar to PET. They have faster crystallization rates and lower melting points. Following copolymerization, they yield elastic thermoplastics of use in engineering. Polyethylene naphthalate (PEN) appears to have considerable potential. It has a high glass transition temperature ($T_g$), high melting point, excellent dynamic properties, and good barrier characteristics [6, 7].

As it is outlined above, two methods are used to produce nanocomposites from thermoplastic polyester systems. First, the polymer ingredients can be added to layered smectite clays and polymerized *in situ*. Alternatively, melt compounding or solution methods can be used. Numerous treatises and reviews have been written about nanocomposites that involve polyesters [8]. This chapter summarizes the main production methods and recent research and developments (R&D) in polyester nanocomposites.

## 9.2. Nanoparticles

### 9.2.1. Layered Silicates

The commonly used layered silicates in nanocomposites are the structural family known as the 2:1 phyllosilicates. Their crystal lattice consists of two-dimensional layers in which a central edge-shared octahedral sheet of aluminum or magnesium hydroxide is fused to two external silica tetrahedral sheets. The layer thickness is approximately 1 nm, with the lateral dimensions varying from 300 Å to several microns depending on the particular silicate. These layers are stacked above each other creating a regular van der Walls gap between the layers called the interlayer or gallery. Partial isomorphic substitutions of metal cations (e.g., $Al^{3+}$ replaced by $Mg^{2+}$ or $Fe^{2+}$, or $Mg^{2+}$ replaced by $Li^+$) impart negative charges to the layers that are counterbalanced by alkali or alkaline earth cations incorporated in the galleries. This type of layered silicate is characterized by a moderate negative surface charge (known as the cation exchange capacity (CEC) and expressed in meq/100 g). The charge is not locally constant on a particular layer and varies from layer to layer, such that an average value is taken over the entire crystal. Montmorillonite (MMT), hectorite, and saponite are the most commonly used layered silicates. Their essential structure is shown schematically in Figure 9.1.

Pristine layered silicates usually contain hydrated $Na^+$ or $K^+$ cations [10]. In this pristine state, the hydrophilic layered silicates are miscible with hydrophilic polymers only. However, most polymers are hydrophobic. Therefore, the surface of a hydrophilic layered silicate must be modified out to ensure its miscibility. Generally, this is done through ion-exchange reactions with cationic surfactants. Since the forces that bond the stacked layers together in layered silicates are relatively weak, intercalation of small molecules is easy [11]. To make these hydrophilic layers more hydrophobic, the cations in the interlayer can be exchanged with cationic surfactants including primary, secondary, tertiary, and

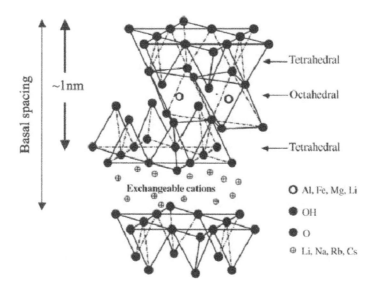

**Figure 9.1**    Structure of 2:1 phyllosilicates [9].

quaternary alkylammonium or alkylphosphonium (onium) cations. Further, alkylammonium and alkylphosphonium cations in the organosilicates lower the surface energies of the layers and improve the wettability between the organosilicates and polymer molecules. In addition, the interlayer distances increase to allow ready incorporation of polymer molecules within the interlayers. Moreover, the alkylammonium or alkylphosphonium cations provide functional groups that can react with polymer molecules and initiate monomer polymerization, thereby enhancing the interface link between the inorganic component and polymer matrix. Guan et al. [12] prepared organomodified clay using hexadecyltriphenylphosphonium bromide, 1-hexadecyl-2,3-dimethylimidazolium bromide, and 1-hexadecylprydinium bromide, resulting in an increase in the gallery spacing to 30, 21, and 19 Å, respectively. The gallery spacing during the organo-modification of clay is also dependent on the cationic exchange capacity of the pristine clay. Ou [13, 14] examined the effect of the CEC capacity of a clay on the interlayer spacing. He modified two kinds of sodium MMT with CEC values of 87 and 114 meq, respectively, using cetyltrimethylammonium chloride (CMC) and concluded that the higher the CEC, the larger the interlayer spacing.

The thermal stability of the organoclay component is also of major importance because many polymers, especially polyesters, are either melt-blended or intercalated at high temperatures to provide the corresponding nanoscale-sized composites. When the processing temperature exceeds the thermal stability of the organoclay, decomposition occurs, altering the interface between the filler and matrix polymer. Consequently, much recent research has been directed toward synthesizing organoclays that are thermally stable at high temperatures [15–20]. For example, Chang et al. [15, 16] reported synthesizing PET/clay and PBT/clay nanocomposites that had proved thermally stable using the ion-exchange reaction between sodium and dodecyltriphenylphosphonium chloride. Tables 9.1 and 9.2 list various organomodifiers and the gallery spaces of organoclays and commercial organosilicates (Clocite®, Southern clay).

**Table 9.1**    Organic modifiers and gallery spacing of various organoclays.

| Organic modifier | Gallery spacing (Å) | Reference |
|---|---|---|
| Dodecyltriphenylphosphonium | 36.1 | 23 |
| Cetyltrimethylammonium | 29.8 | 29 |
| Cetylpyridinium | 22 | 30 |
| 12-aminododecanoic acid | 17 | 31 |
| Poly(vinylpyrrolidone) | 32 | 32 |
| Hexadecyltriphenylphosphonium | 30 | |
| 1-hexadecyl-2,3-dimethylimidazolium | 21 | 20 |
| 1-hexadecylprydinium | 19 | |
| 1,2-dimethyl-3-*N*-hexadecyl imidazolium tetrafluoroborate | 18.2 | 28 |
| alkylammonium chlorohydrate (decacyl, dodecyl, tetradecyl, octadecyl) | 17.1, 17.8, 17.6, 23 | 33 |

**Table 9.2**    Structure of organic modifier and Basal spacing in Southern clay Cloisite[®][a].

| | Structure of organic modifier | Density (g/cc) | Gallery spacing (Å) |
|---|---|---|---|
| Cloisite[®] Na[+] | None | 2.86 | 11.7 |
| Cloisite[®] 30B | H₃C—N+—T with two CH₂CH₂OH groups | 1.98 | 18.5 |
| Cloisite[®] 10A | H₃C—N+—CH₂—C₆H₅ with CH₃ and HT | 1.90 | 19.2 |
| Cloisite[®] 25A | H₃C—N+—CH₂CHCH₂CH₂CH₂CH₃ with CH₃, HT, CH₂CH₃ | 1.87 | 18.6 |
| Cloisite[®] 93A | | 1.88 | 23.6 |
| Cloisite[®] 20A | H₃C—N+—HT with CH₃ and HT | 1.77 | 24.2 |
| Cloisite[®] 15A | H₃C—N+—HT with CH₃ and HT | 1.66 | 31.5 |

[a] HT, hydrogenated tallow; T, tallow (~65% C18; ~30% C16; ~5% C14).

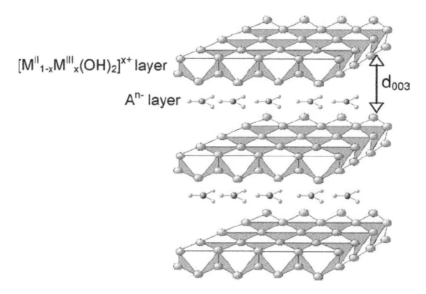

$[M^{II}_{1-x}M^{III}_x(OH)_2]^{x+}$ layer

$A^{n-}$ layer

$d_{003}$

**Figure 9.2**  Schematic representation of a typical LDH crystal structure [29].

## 9.2.2. Layered Double Hydroxides

The layered double hydroxides (LDHs) are a family of compounds that have generated increasing interest in the last few years, owing to their applications as catalysts, ion exchangers, and adsorbents [21–28]. Apart from these applications, LDHs have attracted attention for producing nanocomposites as they offer an alternative to the commonly used three-layer clay minerals. The layer charge and that of the interlayer ions are the reverse of those in common clays. A positive layer charge arises on substitution of a divalent cation by a trivalent one. In order to attain electroneutrality, an appropriate number of anions must be incorporated into the interlamellar domain. The general chemical formula of LDHs is $[M^{II}_{1-x}M^{III}_x(OH)_2]^{x+}A^{m-}_{x/m} \cdot nH_2O$, where $M^{II}$ is a divalent cation such as $Ca^{2+}$, $Mg^{2+}$, $Ni^{2+}$, $Co^{2+}$, $Zn^{2+}$, $Mn^{2+}$, or $Cu^{2+}$, $M^{III}$ is a trivalent cation such as $Al^{3+}$, $Cr^{3+}$, $Co^{3+}$, $Ni^{3+}$, $Mn^{3+}$, $Fe^{3+}$, $V^{3+}$, or $Ga^{3+}$, and A is an interlamellar anion with charge $m-$, such as $Cl^-$, $NO_3^-$, $ClO_4^-$, $CO_3^{2-}$, or $SO_4^{2-}$. Hydrotalcite is a naturally occurring LDH that contains the carbonate anion. Its formula is $Mg_6Al_2(OH)_{16}(CO_3)^{2-} \cdot 4H_2O$. A schematic representation of the LDH crystal structure is shown in Figure 9.2.

The gallery distance between each layer in an LDH such as hydrotalcite is approximately 3 Å, the layer thickness is 4.8 Å, and the long repeat period 7.8 Å. This gallery size is too narrow to accommodate intercalated monomers or polymer molecules. In addition, the hydrophilic nature of the LDH layers is largely incompatible with hydrophobic polymer molecules. For an anionic LDH to be used in a polymer nanocomposite system, the pristine LDH must be modified with suitable organic anions. Various processes are available for preparing organomodified LDHs. These include ion-exchange reactions, the rehydration of calcined oxides, and coprecipitation. Anionic exchange reactions are commonly used to provide intercalated compounds. Conventional anion exchange commonly uses aqueous solutions of the guest materials. However, when compared with cation exchange reactions involving layered polysilicates, it is difficult to use ion exchange with LDHs owing to the high selectivity of LDH for carbonate anions and the large anion exchange capacity (AEC) of LDHs. Therefore, $CO_2$ has to be excluded during the reaction and low pH values are required to achieve a high ion-exchange efficiency. Bish [30] reported efficiencies close to 100% for the exchange of carbonate in acid media. At low pH, however, the layers partially decompose.

Organomodified LDH compounds may also be prepared by rehydrating calcined oxides. The nature of LDH decomposition during calcination has been investigated in detail and it has been found that rehydration enables reconstruction of the original LDH from the calcined material [31–34]. After calcining LDH in air at 500°C, subsequent reaction with an aqueous solution of organic anions under nitrogen allows the LDH layers to reform so as to intercalate the organic anions. In this case, the calcination conditions have a fundamental effect on the purity and crystallinity of the final organic-anion-intercalated LDH compounds and must be optimized in each case. For example, Lee et al. [34] prepared PET/LDH nanocomposites using various anionic-surfactant-modified LDHs made by rehydration in which the interlayer spacing increased by up to 28 Å.

The most common method of obtaining organically modified LDH compounds is direct synthesis, so-called coprecipitation, based on the reaction of a solution containing $M^{II}$ and $M^{III}$ metal cations with a solution of organic anions in an alkaline medium [35–37]. After hydrothermal crystallization of the precipitates, organic-anion-intercalated LDH is obtained. The properties of the LDH products, such as crystallinity and particle size, are affected by various experimental parameters, such as the reaction pH, temperature, reactant solution concentrations, addition flow rates, hydrodynamic conditions in the reactor, and aging.

### 9.2.3. Spherical Nanoparticles

One of the more common classes of nanocomposites contains an organic matrix combined with an inorganic material acting as reinforcing filler. The incorporation of organic/inorganic hybrids can produce materials possessing high degrees of stiffness and strength, and with gas barrier properties, but containing far less inorganic content than conventional filled polymer composites.

Nano fumed silica (FS) [38] has been used in industry as a rheological additive and a useful reinforcement for thermosetting polymers. Hydrophilic FS can be chemically modified readily to hydrophobic FS [39] and has extremely large, smooth nonporous surfaces [38, 40, 41] that promote strong physical contact between the filler and polymer matrix. In addition, the particle–particle interactions of FSs lead to agglomerates and particle networks with pronounced shear thinning effects [42, 43]. Barthel [44] ascertained that these interactions depend heavily on the difference of polarity between the filler and liquid medium. Therefore, if the FSs have similar hydrophobic/hydrophilic properties to a polymer matrix, they are more effectively dispersed in and wetted by the matrix.

## 9.3. Preparation of Polyester Nanocomposites

Layered silicates have provided the principle components system of mainstream polymeric nanocomposites, with a wide variety of high-MW polymers used for the matrix. The manufacturing processes involved in nanocomposites now put on a point of overcoming the macro- and microphase separation between polymer matrix and clay by pre-treatment.

Three types of manufacturing process exist: solution intercalation [13, 14, 45, 46], melt blending intercalation [20, 34, 47, 48], and *in-situ* polymerization intercalation/exfoliation [12, 15, 16, 37, 49, 50–52]. Of these, the best results have been achieved with melt blending intercalation and *in-situ* intercalation/exfoliation.

Melt intercalation involves mixing the layered silicate with the molten polymer matrix. The polymer becomes incorporated in the silicate gallery where the surface of the silicate proves compatible with the polymer matrix; the greater the interlayer distance, the larger the amount polymer admitted is. Both intercalated and exfoliated nanocomposites are produced readily.

The *in-situ* polymerization method starts by widening the clay interlayers by swelling using more than one monomer, which can be polymerized linearly before undertaking further polymerization by heating. A catalyst or other monomer may be present. The final product results from chain growth within the clay galleries. A benefit of the *in-situ* process is the production of nanocomposites without the need for a physical or chemical interaction between an organic polymer, a matrix, and an inorganic nanoparticle. In the rest part of this chapter, we will describe about other methods a little by little.

## 9.4.  Nanocomposites in the PET System

Today thermoplastic PET polyester is widely used in packaging, automotive electrical and electronics, and construction industries. It is a semicrystallized high-MW polymer with excellent chemical resistance, thermal stability, melt mobility, and spinnability. It has become the most commonly produced and used synthetic fiber.

Moreover, its cost is very low owing to mass production. Despite its superior high performance, continued effort is being put into improving its effectiveness in response to market demands. In particular, research has led to the development of composite materials like PET/GF (glass fiber) and CF (carbon fiber) for use as engineering plastics. Today, both academic and industrial studies have produced improvements in the mechanical, gas barrier, and heating properties of PET/layered silicate nanocomposites. Extensive reviews of these PET/clay nanocomposites have recently been published [25].

A number of equally successful studies have examined other not-yet-commercialized high-MW polymer systems. These developments have not been exploited because of product discoloration, haze I.V. drop, side reactions, and agglomerations that develop in the PET matrix. Further difficulties have arisen when modifiers added either to facilitate polymerization or for the pre-treatment of nanoparticles of clay to aid their dispersibility have decomposed at high processing temperatures. Solutions to these and related problems, but particularly for enhancing dispersibility and avoiding decomposition, lie in further research. For example, to enhance dispersibility, it is important to obtain an understanding of the microscopic structure, which directly affects physical and mechanical properties, such as crystallinity.

### 9.4.1.  Dispersibility

The properties of nanocomposites are determined in large part by how the distribution of nanoparticles in the polymer matrix is controlled. Numerous studies have tried to improve nanoparticle distribution and the level of particle bonding interaction within the polymer matrix. The nanostructure found in polymeric nanocomposites has a primary effect on determining the polymer matrix properties. It is the nanostructure that determines how the polymer chains react and interact within the nanoparticles or even how polymer chains may differ in their behavior within layered silicate galleries, as opposed to those in LDHs.

The blending of nanoparticles with a polymer matrix generally results in the formation of an agglomerated heterogeneous nanostructure. *In-situ* polymerization generally produces a homogeneous nanostructure. Hence, to improve the nanostructure of polymeric nanocomposites, the surface modification of nanoparticles becomes necessary to ensure compatibility between the polymer chain and nanoparticle. Nevertheless, when the concentration of nanoparticles increases, agglomeration becomes unavoidable. All studies to date show that an ideal distribution is possible only with a small quantity of particles, regardless of the nature of the nanoparticles; as the quantity increases, the quality of the distribution decreases, leading to agglomeration.

Conversely when layered materials such as clay and LDH are used as nanofillers, hydrophilic, or hydrophobic swelling agents such as long-chain organic ions and water-soluble oligomers can be intercalated or absorbed in the interlayers. Modified agents or surfactants not only allow an increase in interlayer volume, but also enhance the compatibility between the polymer matrix and nanofiller. Consequently, polymer chains are readily incorporated in the interlayers during polymerization, although, depending on the preparatory methodology, differences can arise in dispersibility, as already seen in *in-situ* polymerization vs. melt blending.

Tsai et al. [53] demonstrated that individual layers of clay are not exfoliated by traditional methods, as there is no force driving the absorption of the monomer or oligomer between the silicate layers during polymerization. When they inserted antimony acetate, a catalyst for PET, as an intercalant in the gallery space of a clay, and then added bis(hydroxyethyl) terephthalate (BHET), on melt polymerization, the PET nanocomposites became evenly dispersed throughout the resulting stratiform inorganic material.

In the case of the montmorillonite clay PK805, $[Na_{0.75}(Fe_{0.18}Al_{3.33}Mg_{0.49})O_{20}(OH)_{4.3}H_2O]$, which has a CEC of 98 meq per 100 g of clay, and contains $Sb(OAc)_3$ with acidified sodium cocoamphohydroxypropylsalfonate (SB) in a sufficient molar ratio to satisfy the CEC, the interlayer space increased to 44.10 Å. Depending on chemical formula of the clay, they found that the quantity of absorbed ion can be varied to maximize the bond strength between the clay layer and modifier (Table 9.3).

**Table 9.3**  The chemical composition of PET/clay nanocomposites [53].

|  | Composition | | | |
|---|---|---|---|---|
| Sample | PET | Clay | Sb | SB |
| PET/PK-802/Sb/SB[a] | 99.29 | 0.66 | 0.038 | 0.012 |
| PET/CWC/Sb/SB[a] | 98.65 | 1.22 | 0.085 | 0.043 |
| PET/PK-805/Sb/SB[b] | 96.8 | 2.50 | 0.125 | 0.540 |
| PET/PK-805/SB[b] | 96.4 | 2.50 | 0 | 1.090 |

[a] PET/PK-802/Sb/SB and PET/CWC/Sb/SB were prepared in the same reactor under the same conditions with different ratios and types of clays.
[b] PET/PK-805/Sb/SB and PET/PK-805/SB were made from the same ratios and types of clays, but PET/PK-805/Sb/SB was treated with catalyst and PET/PK-805/SB was treated without catalyst before the polymerization.

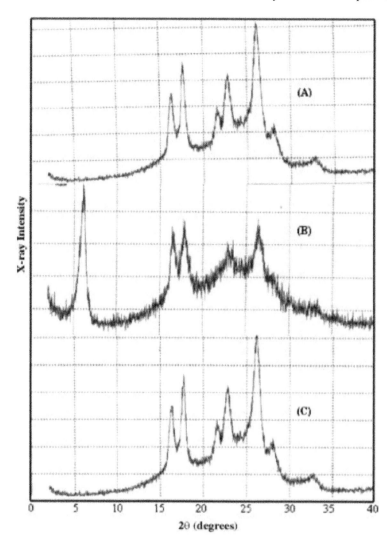

**Figure 9.3**    Powder XRD patterns of polyester/clay nanocomposites from different modified clays: (A) pure PET, (B) PET/PK-805/SB, (C) PET/PK-805/Sb/SB [53].

The dispersion of modified clay in PET/clay nanocomposites varies depending on kinds of modifier used to prepare the nanocomposites. Figure 9.3 compares the X-ray diffraction (XRD) profiles of pure PET, a polymerized PET/clay nanocomposite with SB-treated clay, and a PET/clay nanocomposite associated with both $Sb(OAc)_3$ and SB-treated clay. In the case of SB, only the (001) reflection of clay occurs at $2\theta = 6.1°$ ($d_{001}$ 14.48 Å). This (001) reflection disappears in the nanocomposite on processing with both $Sb(OAc)_3$ and SB as a result of the exfoliation clearly evident in the TEM image (Figure 9.4). These results have led to the mechanism proposed for PET/clay nanocomposite formation shown in Figure 9.5.

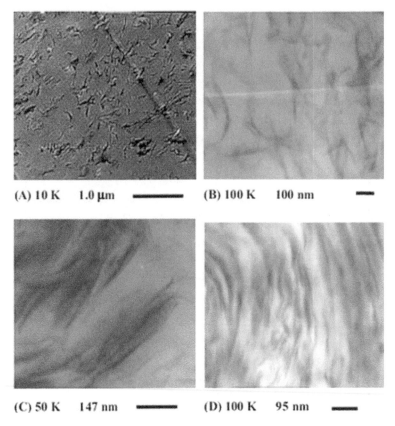

(A) 10 K    1.0 μm    ━━━━━    (B) 100 K    100 nm    ━

(C) 50 K    147 nm    ━━━━━    (D) 100 K    95 nm    ━━

**Figure 9.4** Transmission electron micrographs of samples (A) PET/PK-805/Sb/SB in low magnitude, (B) PET/PK-805/Sb/SB in high magnitude, (C) PET/PK-805/SB in low magnitude, and (D) PET/PK-805/SB in high magnitude [53].

A complementary study by Lee et al. [54] of ring-opening polymerization using the ethylene terephthalate cyclic oligomer (ETC) used different methods of processing the polycarbonate/layered silicate nanocomposites. As shown in Figure 9.6, a broad reflection occurs at around $2\theta = 2.7°$ in the ETC/OMMT mixture before polymerization, but essentially disappears after polymerization. They concluded that better dispersibility occurred where the silicate layers of clay were disrupted within the PET matrix.

There are numerous ways to promote the near-perfect exfoliation of an intercalated nanocomposite, but it has proven difficult to attain a high level of dispersibility while bringing about complete delamination of the layered silicate.

Chang et al. [15, 16] believed that the decomposition of many of the organocompounds used to modify clay was unavoidable since the polymerization and processing temperatures of PET generally exceed 280°C. They tried *in-situ* polymerization after treating clay with dodecyltriphenylphosphonium chloride, which has a very high decomposition temperature. The treated silicate layers were not exfoliated perfectly, although an intercalated or exfoliated PET/clay nanocomposite was produced.

Kim et al. [55] manufactured a PET/clay nanocomposite with an initial esterification of the trimellitic anhydride and ethylene glycol (EG) dispersed clay. This yielded the

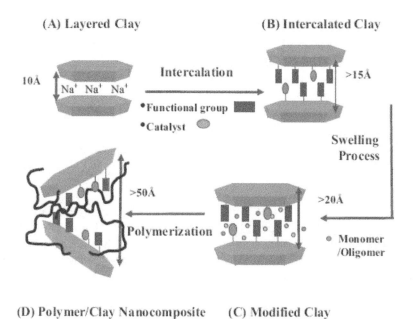

**Figure 9.5**  Proposed mechanism for the formation of polymer/clay nanocomposites using the driving force concept. (A) Sodium-type smectite clays; (B) intercalation of the catalyst/initiator and modified agent leads to the interlayer spacing expansion; (C) monomers/oligomers were driven by the catalyst/initiator to swell in the gallery of the clay lamella; (D) polymerization occurs between the adjacent silicate layers and the growth of the polymer disrupts the lamellar structure of the clay into individually disordered exfoliated layers in the polymer matrix [53].

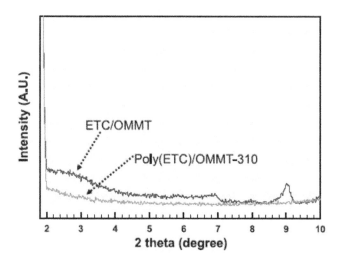

**Figure 9.6**  X-ray diffraction spectra showing the change in the *d*-spacing of OMMT resulting from the polymerization of cyclic oligomers at 310°C [54].

**Table 9.4**   The WAXD information of the various clay dispersed
TEOs [55].

| Equivalent ratio (TMA/EG) | 1/1.5 | 1/2 | 1/2.5 | 1/3 |
|---|---|---|---|---|
| First $d$-spacing (Å) | Disappeared | Disappeared | 53 | 44 |
| Second $d$-spacing (Å) | 18.1 | 18.0 | 17.2 | 17.2 |

trimellitated ester oligomer (TEO) in which the clay was dispersed and then added to polycondensed BHET formed by the esterification of TPA and EG. They reported that the dispersibility improved with an increase in the delamination of clay, as shown by the increases in the TMA/EG layer ratio in Table 9.4.

Choi et al. [51] obtained a PET/clay nanocomposite by inserting a catalyst into the clay interlayer and using an organic compound to modify the clay in a way that does not promote complete delamination of the clay layer or result in thermal degradation during polymerization. While unexfoliated layers remain, the clay-supported catalyst is well dispersed in the PET matrix.

Numerous studies have examined PET/clay nanocomposites for which *in-situ* polymerization has used organic compounds to modify the clay. For example, Hao et al. [52] prepared nanocomposites using a compatibilizer that had amino and ester groups with an 11-aminododecanoic acid (ADA)-modified clay. As shown in Figure 9.7, the XRD peak of the nanocomposite is at a greater angle than that of the ADA-treated clay. This means that the interlayer $d$-spacing of the clay decreased after *in-situ* polymerization as a result of thermal degradation. However, they found that overall, the clay was well dispersed in the PET matrix.

Recently, Hwang et al. [56] also prepared PET/Na⁺-MMT and PET/A10-MMT nanocomposites with various clay contents by *in-situ* polymerization. The TEM

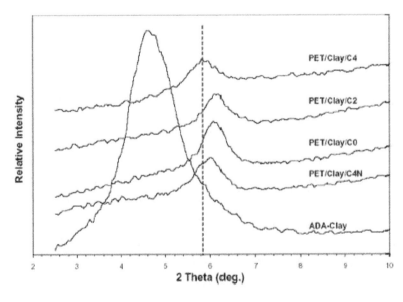

**Figure 9.7**   WAXD patterns of the PET/clay hybrids and ADA-clay [52].

(a)    (b)

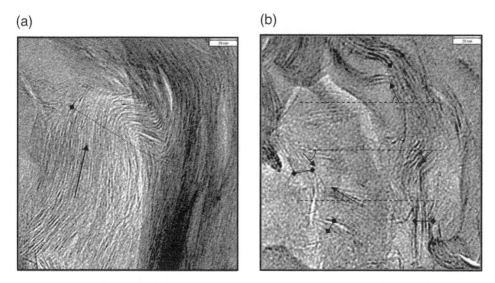

**Figure 9.8**    TEM micrographs of nanocomposite composed of (a) PET/Na$^+$-MMT 0.5 wt%, (b) PET/A10 MMT 0.5 wt% [56].

micrographs (Figures 9.8(a) and (b)), acquired at a 20-nm resolution, provide direct evidence of exfoliation of the clay particles into the PET matrix and show the effect of the alkyl-modifier on clay dispersibility. The dispersibility of clay in PET/A10-MMT nanocomposites was greater than that observed for the PET/Na$^+$–MMT nanocomposites. Thus, the resulting image in Figure 9.8(a) shows uniformly oriented Na$^+$–MMT particles with complete layer stacks containing an average of 66 silicate sheets per particle.

Conversely, A10-MMT particles in PET/A10-MMT were largely exfoliated, with a larger proportion of intercalated PET, as evidenced by significantly smaller particles composed of only four to five randomly oriented silicate sheets. The silica sheet orientation distribution was relatively broad ($\theta_1$= 44°, $\theta_2$= 63°, $\theta_3$= 47°, $\theta_4$= 30°), indicating a higher degree of dispersion for A10-MMT compared to Na$^+$–MMT in the PET matrix. In particular, neighboring circled area in Figure 9.8(b) shows an omnidirectional dispersion of the clay layer, indicating a large degree of exfoliation into the PET matrix. The greatest degree of exfoliation occurred for PET/A10-MMT 0.5 wt%.

Further, there are numerous reports on PET/clay nanocomposites prepared by melt-blending PET with clay modified using various organic compounds [20, 48, 57, 58]. Large differences can exist in melt blending and are dependent on the ingredients selected and the melt-blending conditions. This is a more economical process than *in-situ* polymerization. Davis et al. [20] synthesized 1,2-dimethyl-3-*N*-hexadecyl imidazolium tetrafluoroborate-treated MMT using melt blending to optimize the dispersion and delamination under various blending conditions. The thermal degradation temperature was 350°C. Their results yielded the highest level of MMT dispersion in PET.

Sanchez-Solis et al. [47] prepared PET/clay nanocomposites using a series of additives as compatibilizers, such as alkyl ammonium chloride, obtained from maleic anhydride (MAH), pentaerythritol (PENTA), and amines of various chain lengths. They evaluated dispersibility using rheological and mechanical properties and found that the largest separation occurred with *n*-octadecylammonium. The interlayer gap in the clay varied with

**Table 9.5**  Effect of the intercalant on
clay-interlayer distance [47].

| Clay-intercalants | $d_{001}$ (Å) |
|---|---|
| Montomorillonite | 12.4 |
| Montomorillonite-PENTA | 14.5 |
| Montomorillonite-MAH | 12.4 |
| Montomorillonite-$n$-decacylammonium | 17.1 |
| Montomorillonite-$n$-dodecylammonium | 17.8 |
| Montomorillonite-$n$-tetradecylammonium | 17.6 |
| Montomorillonite-$n$-octadecylammonium | 23.0 |

different additives, as shown in Table 9.5, as well as with the amine chain length, thereby
determining that the interlayer distance is not a linear function of the length of the chain of
an aliphatic amine. In particular, they found that all the nanocomposites treated with MAH
and PENTA were exfoliated.

Lee et al. [34] tried direct-melt compounding of PET with LDHs treated with the
anionic surfactants sodium dodecyl sulfate (DS), sodium octylsulfate (OS), and sodium
dodecyl benzenesulfonate (DBS). The XRD profiles of the parent LDH, the organic-anion-
intercalated LDHs (MGALOS, MGALDS, MGALDBS), calcinated MGALCAL, and the
PET/CDH nanocomposite at 2 wt% are shown in Figure 9.9.

The basal spacing of the parent LDH is 7.8° ($2\theta$ = 11.4°) and those of MGALOS,
MGALDS, and MGALDBS lie at 22.6, 26.8, and 28 Å, respectively. The results show that
the anionic surfactants separate the LDH interlayer. As the PET/LDH nanocomposite lacks
any peak near the basal spacing of LDHs, LDH has become well dispersed in the PET.

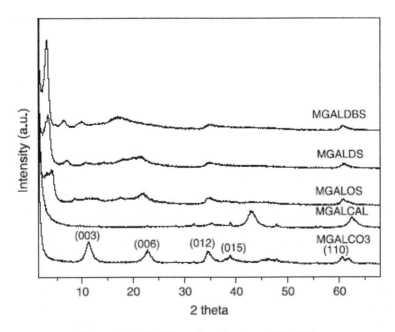

**Figure 9.9**  WAXD profiles of various LDHs [34].

There has been a recent increase in interest in carbon nanotubes, with the publication of many papers describing polymer/carbon nanotube nanocomposites prepared using melt-blending or *in-situ* polymerization in PET [59–61]. Good dispersion of carbon nanotubes occurs after they have been treated with either concentrated hydrochloric, nitric, or sulfuric acid or their mixtures, regardless of using melt-blending or *in-situ* polymerization methods. In each case, the dispersion proves far better than that obtained with untreated carbon nanotubes.

Hahm et al. [62, 63] examined the properties of PET/silica nanocomposites prepared using melt-blending and *in-situ* polymerization of FS with PET with the aim of enhancing the heat deflection temperatures and solvent diffusion barrier properties. FS nanoparticles proved to have an extremely large surface area per unit weight with numerous silanol groups (Si–OH) exposed on the surface. This causes ready aggregation via a particle–particle interaction in nonpolar liquids. The silanol groups on the surface decrease the surface energy and increase the hydrophobicity in a manner similar to that found with chemical modification using various methylsilyl groups. As a result, the dispersion of silica in the nonpolar polymer matrix is improved.

## 9.4.2. Crystallization Behavior

PET is a semicrystalline polymer with a low crystallization rate. This rate must be increased for it to be used in engineering applications. Many studies have examined the effect of nanofillers exert on the crystallization behavior of PET nanocomposites during both heating and cooling. Generally, PET nanocomposites containing nanofillers have higher crystallization rates than homogeneous PET. One reason advanced for the crystallization rate enhancement is that nanofillers act as effective heterogeneous nucleating agents in the polymer matrix. Another possible reason may lie in the nanofiller itself, especially layered materials, which might help the PET molecules stack on each other as they develop into crystallites.

Ou et al. [45] reported the nucleating effect of organoclay on crystallization of PET nanocomposite. Figure 9.10 shows the cooling and heating curves of homogeneous PET and PET/11K-M (organoclay) nanocomposites, and its data were summarized in Table 9.6. All the differential scanning calorimetry (DSC) thermograms recorded while cooling the samples from the melt contain a distinct exothermic crystallization event. The crystallization onset temperatures and crystallization temperature ($T_c$) for all four PET/11K-M nanocomposites are higher than that of homogeneous PET. The 90/10 PET/11K-M has the highest crystallization onset temperatures (220°C) and $T_c$ (208°C) of all compositions. The changes in $\Delta T_c$ and $\Delta H_c$ are related to the overall crystallization rate and crystallinity, respectively. For all the blends, $\Delta T_c$ is 12°C to 18°C narrower than that of homogeneous PET (45°C) with the 90/10 being the narrowest. Conversely, the values of $\Delta H_c$ for all the blends are larger than that of PET and increase with the 11K-M content, indicating an increased crystallinity with the organoclay content. In general, the degree of supercooling ($\Delta T = T_m - T_c$) provides a measure of a polymer's crystallizability, i.e., the smaller $\Delta T$ is, the higher the overall crystallization rate. The $\Delta T$ values for all the nanocomposites are 14°C to 21°C smaller than that of homogeneous PET with the 90/10 PET nanocomposites having the smallest $\Delta T$. These results show that the organoclay has a strong heterogeneous nucleation effect on PET owing, in large part, to the large surface area afforded by the organoclay during nucleation. The crystallization rate of PET may be accelerated

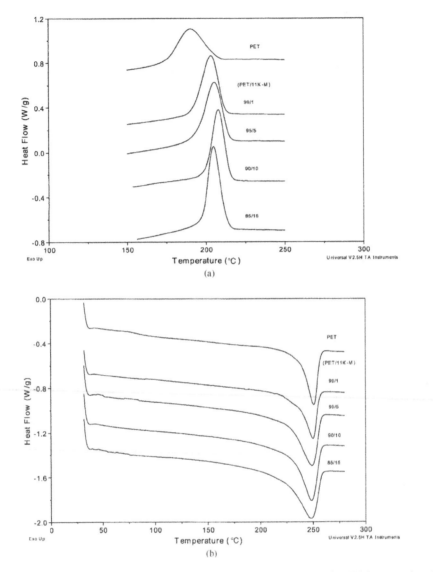

**Figure 9.10**   DSC thermograms of PET and PET blends with 1–15 wt% of 11K-M organoclay: (a) cooling scans, (b) heating scans [45].

by 1–15 wt% organoclay, and the nucleation efficiency probably reaches a maximum at 10 wt% organoclay.

Numerous investigations have examined the isothermal crystallization kinetics of PET nanocomposites, while Avrami analysis provides little insight into the structure of the lamellae or spherulites at a molecular level; the rate constant $k$ and Avrami exponent $n$ are useful for deducing the crystallization mechanism. The Avrami exponent describes the crystal growth patterns and nucleating mechanisms, while the rate constant provides information on the diffusion and nucleation rates. By plotting $\ln - \ln[1 - X_c(t)]$ as a function of $\ln t$, both the rate constant and Avrami exponent for various crystallization temperatures can be obtained by fitting straight lines to the initial portion of the curves. Figure 9.11

**Table 9.6**  DSC data of PET/11K-M nanocomposites [45].

| Composition | Melting (from heating scans) | | | | Crystallization (from cooling scans) | | | | | | |
|---|---|---|---|---|---|---|---|---|---|---|---|
| | Onset (°C) | $T_m$ (°C) | $\Delta T_m$ (°C) | $\Delta H_f$ (J/g) | $T_{on}$ (°C) | $T_c$ (°C) | $\Delta T_c$ (°C) | $t_{1/2}$ (min) | $\Delta H_c$ (J/g) | $\Delta H_c$/time (J g$^{-1}$s$^{-1}$) | $\Delta T^a$ (°C) |
| PET/11K-M | | | | | | | | | | | |
| 100/0 | 208 | 251 | 55 | 41.8 | 212 | 190 | 45 | 2.25 | 38.9 | 0.144 | 61 |
| 99/1 | 205 | 250 | 56 | 43.6 | 217 | 203 | 32 | 1.60 | 41.0 | 0.214 | 47 |
| 95/5 | 202 | 249 | 60 | 59.6 | 220 | 205 | 33 | 1.65 | 44.7 | 0.226 | 44 |
| 90/10 | 200 | 248 | 62 | 66.7 | 220 | 208 | 27 | 1.35 | 44.8 | 0.277 | 40 |
| 85/15 | 199 | 248 | 64 | 69.7 | 218 | 205 | 29 | 1.45 | 46.2 | 0.266 | 43 |

$^a\Delta = T_m - T_c$.

shows $n$ as a function of the crystallization temperature. For homogeneous PET, the average $n$ value is close to 3, confirming that crystal growth is three-dimensional (spherulitic) with a predetermined nucleation pattern. For nanocomposites, $n$ was found to average about 2. Based on additional polarized optical microscope (POM) and scanning electron microscopy (SEM) observations, led Wan et al. [49] proposed a crystallization mechanism for typical PET/clay nanocomposites. Where the polymer chain segments are adjacent to a clay surface, Wan et al. proposed that they are deposited in an ordered manner enabling lamellae growth to be initiated. These lamellae may then grow one-dimensionally and

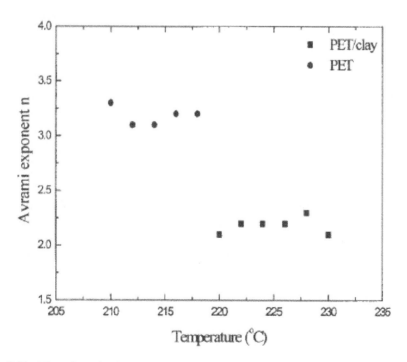

**Figure 9.11**  The value of $n$ for pure PET and the PET/clay nanocomposite as a function of the crystallization temperature during the initial stage of the isothermal process [49].

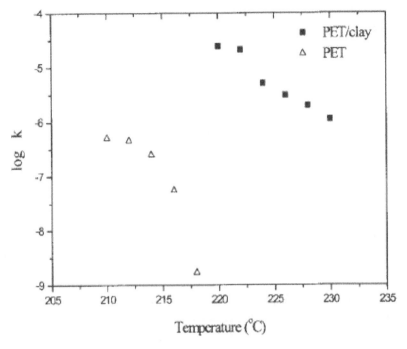

**Figure 9.12**  The log $k$ of pure PET and the PET/clay nanocomposite as a function of the crystallization temperature [49].

provide substrate for more lamellae. They can subsequently develop three-dimensionally when their nuclei are no longer subject to the geometric constraints imposed by the clay. However, crystallites fail to complete three-dimensional growth in all directions where other clays impede their growth path, leading to the formation of irregularly shaped interlocked crystallites. Figure 9.12 shows the rate constant $k$ as a function of the crystallization temperature. The values of $k$ indicate that the rate of crystallization of the PET/clay nanocomposite is much faster than that of homogeneous PET for a given temperature, demonstrating that clay can greatly enhance the overall crystallization rate in the temperature ranges studied. These results concur with those of other research groups that attribute the greatly increased crystallization rates to the heterogeneous nucleating effects of the clay.

Many nonisothermal studies of the crystallization kinetics of PET nanocomposites have also been undertaken with spherical nanofillers, which enhance the crystallization rate of PET nanocomposites.

The DSC results presented in Table 9.7 show the characteristic nonisothermal crystallization behaviors of PET/FS nanocomposites prepared using direct melt compounding. The melting peaks ($T_m$) and heats of fusion ($\Delta H_f$) of all these nanocomposites on a second heating are similar to that of pure PET regardless of the filler type or content. However, the degree of supercooling ($\Delta T$) decreases with increasing filler content in all nanocomposites, indicating that the FSs act as nucleating agents in the polymer matrix and accelerate the crystallization rate. The increase in apparent crystallinity with filler content appears to result from the incidence of very rapid crystallization due to the nucleating effect of the nanofillers, although all the samples are quenched within the maximum limit of DSC

**Table 9.7**    DSC data of PET and its nanocomposites [62].

| Samples | Actual filler content [ ][g] (wt%) | | $T_g^a$ (°C) | $T_c^b$ Peak (°C) | $\Delta H_c$ (J/g) | $T_m^c$ Peak (°C) | $\Delta H_f$(J/g) | $T_c^d$ Peak (°C) | $\Delta T^e$ (°C) | $X_c^f$ (%) |
|---|---|---|---|---|---|---|---|---|---|---|
| PET | | | 78.2 | 142.0 | −300 | 255.0 | 46.3 | 198.0 | 57.1 | 13.0 |
| PET/FS | 0.5 | | 77.4 | 136.5 | −22.5 | 255.0 | 49.1 | 207.2 | 47.9 | 21.2 |
| | 2 | | 76.8 | 130.5 | −9.5 | 255.5 | 49.2 | 210.2 | 45.3 | 31.7 |
| | 8 | | | 121.7 | −2.1 | 255.2 | 44.8 | 210.9 | 44.3 | 34.0 |
| PET/M-FS | 0.5 | [0.49] | 78.1 | 135.9 | −20.9 | 255.0 | 47.2 | 204.5 | 50.5 | 21.0 |
| | 2 | [1.95] | 76.9 | 128.4 | −7.8 | 254.9 | 45.6 | 206.1 | 48.8 | 30.1 |
| | 8 | [7.82] | | 124.4 | −1.5 | 254.9 | 43.3 | 209.6 | 45.3 | 33.3 |
| PET/O-FS | 0.5 | [0.26] | 78.3 | 140.2 | −27.0 | 255.0 | 47.1 | 202.9 | 52.1 | 16.0 |
| | 2 | [1.03] | 77.2 | 139.2 | −26.2 | 254.4 | 47.8 | 202.4 | 52.0 | 17.2 |
| | 8 | [4.25] | | 126.5 | −2.8 | 255.5 | 45.4 | 211.0 | 44.5 | 33.9 |

[a] The glass transition temperature measured on the second heating at 10.0°C/min using half $C_p$ extrapolated method.
[b] The crystallization temperature measured on the second heating at 10.0°C/min.
[c] The melting temperature measured on the second heating at 10.0°/min.
[d] The crystallization temperature measured on the second cooling at 10.0°C)/min.
[e] The degree of supercooling: ($T_c^c$ peak–$T_c^d$ peak).
[f] Apparent crystallinity: ($\Delta H_f - |\Delta H_c|)/\Delta H_f^o \times 100$, $\Delta H_f^o$: 125.5 J/g.
[g] Actual silica content determined by TGA at 800°C.

**Table 9.8**    DSC data of neat PET and its nanocomposties prepared via *in-situ* polymerization at nonisothermal crystallization [63].

| Samples | Filler content [ ][f] (wt%) | | $T_c^a$ Peak (°C) | $\Delta H_c$ (J/g) | $T_m^b$ Peak (°C) | $\Delta H_f$ (J/g) | $T_c^c$ Peak (°C) | $\Delta H_f$ (J/g) | $\Delta T^d$ (°C) | $X_c^e$ (%) |
|---|---|---|---|---|---|---|---|---|---|---|
| PET | – | | 120.3 | −27.3 | 257.9 | 49.1 | 197.2 | −47.4 | 60.7 | 17.3 |
| PET | 0.5 | [0.51] | 130.2 | −31.2 | 258.0 | 50.5 | 191.8 | −42.8 | 66.2 | 15.4 |
| /FS | 2 | [2.19] | 121.5 | −33.6 | 256.6 | 54.7 | 201.0 | −46.8 | 55.6 | 16.8 |
| PET | 0.5 | [0.49] | 125.4 | −31.1 | 258.9 | 52.0 | 186.1 | −47.6 | 72.8 | 16.6 |
| /MFS | 2 | [2.18] | 122.6 | −31.0 | 258.1 | 53.3 | 193.2 | −46.0 | 64.9 | 17.8 |

[a] The crystallization temperature measured on the second heating at 10.0°C/min.
[b] The melting temperature measured on the second heating at 10.0°C/min.
[c] The crystallization temperature measured on the second cooling at 10.0°C/min.
[d] The degree of supercooling: ($T_m^b$ peak–$T_c^c$ peak).
[e] Apparent crystallinity: ($\Delta H_f \cdot - |\Delta H_c|)/ \Delta H_f^o \times 100$, $\Delta H_f^o$: 125.5 J/g.
[f] Actual silica content determined by TGA at 750°C and air atmosphere.

before the second heating. The crystallinity of all the nanocomposites increases rapidly until a filler content of 2 wt%, and any subsequent increase is very slow. This implies that the crystallization rates and crystallinity of all of the nanocomposites are proportional to the actual filler content regardless of the filler type and critical filler amount affecting nucleation.

When PET nanocomposites filled with FS were prepared using *in-situ* polymerization, their crystallization rate slowed. Table 9.8 shows the characteristic nonisothermal crystallization behaviors of such PET nanocomposites. The crystallization temperatures ($T_c$) of all these nanocomposites shift to a higher temperature at their second heating and the degree of supercooling ($\Delta T$) increases, when compared with pure PET, except for PET nanocomposites filled with 2.0 wt% FS. This behavior implies that FSs retard the crystallization rate of PET nanocomposites, which is opposite the results obtained with nanocomposites prepared using direct melt compounding. Hahm et al. [63] explained this seeming disparity and proposed that while nanosized FSs may hinder the motion of the PET molecular chains to prevent their forming crystallites, some of the silica particles also provide initial heterogeneous nucleation sites, with the nano FSs being better dispersed in the case of *in-situ* polymerization.

Lee et al. [37] found changes in the microstructural parameters of PET/LDH nanocomposites during isothermal crystallization at 180°C and 230°C using time-resolved small-angle X-ray scattering (SAXS) analysis. As the proportion of LDH modified by dimethyl sulfoisophthalate (DMSI) increases, the long period decreases as does the extent of the amorphous development, but the average lamellar thickness remains roughly constant (Figure 9.13(a)). This behavior is most apparent at higher crystallization temperatures. In Figure 9.13(b), the long period and size of the amorphous area decrease from 28 and 18 nm to 25 and 16 nm, respectively, following a crystallization time of 20 min. It is understood that an increase in nucleation sites allows nanocomposites to crystallize more readily, leading to a reduction in the size of the average amorphous region. However, the average lamellar thicknesses remain essentially the same for all nanocomposites regardless of the LDH content, largely as a result of the dependence of the lamellar thickness at a given temperature on molecular diffusion rather than nucleation. Therefore, the decrease in the long period with increasing LDH nanofiller content arises from the reduction in the amorphous region, given the lack of change in lamellar thickness.

Nevertheless, there is a significant decrease in the long period and extent of the amorphous region with increasing LDH content at 230°C compared to the values at 180°C. In general, crystallization from the melt-state depends on both nucleation and molecular diffusion. At the lower temperature of 180°C, the LDH layers do not provide an additional effective nucleating agent for crystallization given that nucleation occurs readily at this lower crystallization temperature owing to the lower critical nucleus size and lower free-energy barrier. At the higher crystallization temperature of 230°C, molecular diffusion is significantly greater than nucleation. Consequently, the LDH nanolayers can now act as an additional nucleating agent and affect the crystallization rate, as seen in the changes observed in the microstructural parameters, such as the reduction in the long-period value and size of the amorphous region.

### 9.4.3. Thermal Stability

Where layered materials such as clay, mica, and LDH are effectively exfoliated in the polymer matrix, the thermal stability of polymer nanocomposites is improved as a result of the nanolayers providing both effective heat insulation and a mass transport barrier to the volatile gases generated during thermal decomposition. Figure 9.14 shows typical thermogravimetric analysis (TGA) thermograms of layered materials and PET nanocomposites with different nanofiller contents. The initial thermal degradation temperature ($T_D^i$) of the $C_{12}$PPh-mica/PET nanocomposites increases with nanofiller content. $T_D^i$ (2% weight loss) for PET nanocomposites ranges from 370°C to 389°C; the highest value is for 5 wt% $C_{12}$PPh-mica/PET. The increases in $T_D^i$ of these nanocomposites can result from a number

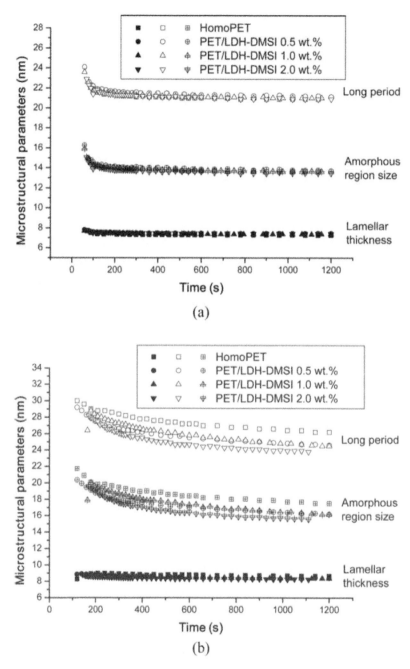

**Figure 9.13**  Changes in the microstructural parameters for PET nanocomposites during isothermal crystallization at (a) 180°C and (b) 230°C [37].

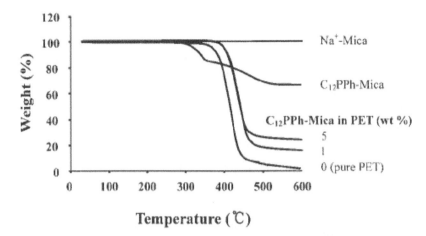

**Figure 9.14**    TGA thermograms of clay, organoclay, and PET hybrid fibers with different organoclay content [67].

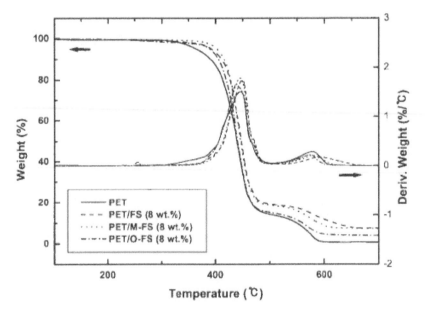

**Figure 9.15**    TGA variation of nanocomposites [62].

of different factors. Of particularly import is the high thermal stability of the clay and the interaction between the clay particles and polymer matrix, as well as the heat insulation provided by the nanolayers. Similar trends have been noted in other studies [64–66].

Spherical nanoparticles in polymer nanocomposites also enhance their thermal stability. Figure 9.15 shows the TGA records of PET/FS nanocomposites with the TGA results summarized in Table 9.9. In Figure 9.15, pure PET starts decomposing at about 300°C and the weight loss increases rapidly, whereas PET/FS and PET/methyltrichlorosilane-treated FS (M-FS) are stable to higher temperatures, but have similar weight loss curves.

**Table 9.9**  Thermogravimetry data of PET and its nanocomposites [62].

| Samples | $T_D^i (°C)^a$ | $T_D^5 (°C)$ | $wt_R^{700} (\%)^b$ |
|---|---|---|---|
| PET | 335 | 364 | 0.1 |
| PET/FS (8 wt%) | 357 | 392 | 7.6 |
| PET/M-FS (8 wt%) | 383 | 402 | 7.5 |
| PET/O-FS (8 wt%) | 334 | 390 | 3.9 |

**Table 9.10**  Heat distortion temperatures of PET/MMT composites [68].

| Samples | PET | PET1M | PET3M | PET5M |
|---|---|---|---|---|
| HDT (°C) (1.80 MPa) | 71 | 106 | 112 | 120 |

$^a$ The temperature at weight reduction of 2% and 5%.
$^b$ The weight percent of residue at 700°C.

PET/octadecyltrichlorosilane-treated FS (O-FS) shows a small weight loss from about 260°C arising from decomposition of the long octadecyl chains on the surface of the O-FS. The initial thermal degradation temperature at 2% in Table 9.9 reveals a maximum increase of 48°C when the PET is filled with M-FS, and all the PET nanocomposites are stable at a temperature higher than that of pure PET over the full range of the weight loss. Therefore, the addition of FS improves the thermal stability of PET markedly, with M-FS performing best, as a result of its better wettability and dispersibility in the matrix.

The heat distortion temperature (HDT) can be enhanced remarkably by adding layered materials. When just 1% MMT was added, the HDT was 35°C higher than that of pure PET (Table 9.10). The resistance to heat deformation under loading is increased because MMT significantly increases the crystallinity of the nanocomposite. In addition, the interaction force between the layers and polymer matrix is strong. Moreover, the polymer chain mobility between the crystallites is restricted.

# References

1.  Carothers, W.H. J Am Chem Soc 1929, 51, 2548.
2.  Carothers, W.H. Collected Papers, Interscience, New York, 1940.
3.  Schlack, P. France Patent 1953, 895,812.
4.  Whinfield, J.R.; Dickson J.T. British Patent 1946, 578,079.
5.  Whinfield, J.R. Nature, 1946, 158, 930.
6.  Fakirov, S. Handbook of Thermoplastic Polyesters, Vol. 1, John Wiley & Sons, New York, 2001
7.  Fakirov, S. Handbook of Thermoplastic Polyesters, Vol. 2, John Wiley & Sons, New York, 2001
8.  Polymer–clay nanocomposites, chapter 1, chapter 9, Wiley Series in Polymer Science, 1997.
9.  Ray, S.S.; Okamoto, M. Prog Polym Sci 2003, 28, 1539.
10. Brindly, S.W.; Brown, G. Crystal Structure of Clay Minerals and their X-Ray Diffraction; Mineralogical Society, London, 1980.
11. Theng, B.K.G. The Chemistry of Clay-Organic Reactions; Wiley, New York, 1974.
12. Guan, G.; Li, C.; Zhang, D.; Jin, Y. J Appl Polym Sci 2006, 101, 1692.
13. Ou, C.F. J Appl Polym Sci 2003, 89, 3315.
14. Ou, C.F. J Polym Sci Polym Phys Ed. 2003, 41, 2902.
15. Chang, J.H.; Kim, S.J.; Joo, Y.L.; Im, S.S. Polymer 2004, 45, 919.

16. Chang, J.H.; Kim, S.J.; Im, S.S. Polymer 2004, 45, 5171.
17. Zhu, J.; Morgan, A.B.; Lamelas, F.J.; Wilkie, C.A. Chem Mater 2001, 13, 3774.
18. Zhu, J.; Uhl, F.M.; Morgan, A.B.; Wilkie, C.A. Chem Mater 2001, 13, 4649.
19. Saujanya, C.; Imai, Y.; Tateyama, H. Polym Bull 2002, 49, 69.
20. Davis, C.H.; Mathias, L.J.; Gilman, J.W.; Shiraldi, D.A.; Shields, J.R.; Trulove, P.; Sutto, T.E.; Delong, H.C. J Polym Sci Polym Phys Ed. 2002, 40, 2661.
21. Meyn, M.; Beneke, K.; Lagaly, G. Inorg Chem 1990, 29, 5201.
22. Schmassmann, A.; Tarnawski, A.; Flogerzi, B.; Sanner, M.; Varga, L.; Halter F. Eur J Gastroenterol Hepatol 1993, 5, S111.
23. Cavani, F.; Trifiro, F.; Vaccari, A. Catal Today 1991, 11, 173.
24. Newman, S.P.; Jones, W. New J Chem 1998, 22, 105.
25. Kohjiya, S.; Sato, T.; Nakayama, T.; Yamashita, S. Macromol Rapid Commun 1981, 2, 231.
26. Schaper, H.; Berg-Slot, J.J.; Stork, W.H. J Appl Catal 1989, 54, 79.
27. Pavan, P.C.; Crepaldi, E.L.; Gomes, G.D.; Valim, J.B. Colloids Surf A 1999, 154, 399.
28. Miyata, S.; Kumura, T. Chem Lett 1973, 843.
29. Costa, F.R.; Abdel-Goad, M.; Wagenknecht, U.; Heinrich, G. Polymer 2005, 46, 4447.
30. Bish, D.L. Bull Mineral 1980, 103, 175.
31. Chibwe, K.; Jones, W. J Chem Soc Chem Commun 1989, 926.
32. Chibwe, K.; Jones, W. Chem Mater 1989, 1, 489.
33. Dimotakis, E.D.; Pinnavaia, T.J. Inorg Chem 1990, 29, 2393.
34. Lee, W.D.; Im, S.S.; Lim, H.M.; Kim, K.J. Polymer 2006, 47, 1364.
35. Hsueh, H.B.; Chen, C.Y. Polymer 2003, 44, 1151.
36. Hsueh, H.B.; Chen, C.Y. Polymer 2003, 44, 5275.
37. Lee, W.D.; Im, S.S. J Polym Sci Polym Phys Ed. 2007, 45, 28.
38. Technical Bulletin no. 6, no. 11, Degussa Corporation, Aklon, OH, 1989.
39. Donnet, J.B.; Wang, M.J.; Papirer, E.; Vidal, A. Kautsch Gummi Kunstst 1986, 39, 510.
40. Barthel, H.; Achenbach, F.; Maginot, H. Proc Int Symp on Mineral and Organic Functional Fillers in Polymers (MOFFIS 93), University de Namur, Belgium, 1993, 301.
41. Hurd, A.J.; Schaefer, D.W.; Martin, J.E. Phys Rev A 1987, 35, 2361.
42. Schreuder F.W.A.M.; Stein, H.N. Rheol Acta 1987, 26, 45.
43. Lee, G.; Murray, S.; Rupprecht, H. J Colloid Interf Sci 1985, 105, 257.
44. Barthel, H. Colloids Surf A: Physicochem Eng Aspects 1995, 101, 217.
45. Ou, C.F.; Ho, M.T.; Lin J.R. J Polym Res 2003, 10, 127.
46. Ou, C.F.; Ho, M.T.; Lin J.R. J Appl Polym Sci 2004, 91, 140.
47. Sanchez-solis, a.; Romero-Ibarra, I.; Estrada M.R.; Calderas, F.; Manero O. Polym Eng Sci 2004, 44, 1094.
48. Pegoretti, A.; Kolarik, J.; Peroni, C.; Migliaresi, C. Polymer 2004, 45, 2751.
49. Wan, T.; Chen L.; Chua Y.C.; Lu X. J Appl Polym Sci 2004, 94, 1381.
50. Guan, G.H.; Li, C.C.; Zhang D. J Appl Polym Sci 2005, 95, 1443.
51. Choi, W.J.; Kim, H.J.; Yoon, K.H.; Kwon, O.H.; Hwang C.I. J Appl Polym Sci 2006, 100, 4875.
52. Hao, J.; Lu, X.; Liu, S.; Lau S.K.; Chua Y.C. J Appl Polym Sci 2006, 101, 1057.
53. Tsai, T.Y.; Li, C.H.; Chang, C.H.; Cheng, W.H.; Hwang, C.L.; Wu R.J. Adv Mater 2005, 17, 1769.
54. Lee, S.S.; Ma, Y.T.; Rhee, H.W.; Kim J. Polymer 2005, 46, 2201.
55. Kim, S.H.; Park, S.H.; Kim, S.C. Polym B 2005, 53, 285.
56. Hwang, S.Y.; Lee, W.D.; Lim, J.S.; Park, K.H.; Im, S.S. J Polym Sci Polym Phys Ed. 2008, 46, 1022.
57. Kalgaonkar, R.A.; Jog, J.P. J Polym Sci Polym Phys Ed. 2003, 41, 3102.
58. Lam, C.K.; Cheung, H.Y.; Lau, K.T. Mater Lett 2005, 59, 1369.
59. Hu, G.; Zhao, C.; Zhang, S.; Yang, M.; Wang, Z. Polymer 2006, 47, 480.
60. Anoop Anand, K.; Agarwal, U.S.; Joseph, R. Polymer 2006, 47, 3976.
61. Shin, D.H.; Yoon, K.H.; Kwon, O.H.; Min, B.G.; Hwang, C.I. J Appl Polym Sci 2006, 99, 900.
62. Chung, S.C.; Hahm, W.G.; Im, S.S. Macromol Res 2002, 10, 221.
63. Hahm, W.G.; Myung, H.S.; Im, S.S. Macromol Res 2004, 12, 85.
64. Fischer, H.R.; Gielgens, L.H.; Koster, T.P.M. Acta Polym 1999, 50, 122.
65. Petrovic, X.S.; Javni, L.; Waddong, A.; Banhegyi, G.J. J Appl Polym Sci 2000, 76, 33.
66. Zhu, Z.K.; Yang, Y.; Yin, J.; Wang, X.; Ke, Y.; Qi, Z. J Appl Polym Sci 1999, 3, 2063.
67. Chang, J.H.; Mun, M.K.; Lee, I.C. J Appl Polym Sci 2005, 98, 2009.
68. Wang, Y.; Gao, J.; Ma, Y.; Agarwal, U.S. Composites Part B 2006, 37, 399.

# 10 Polymer Nanocomposites Based on Ionomers

*Suzana Pereira Nunes*[1]*, Dominique de Figueiredo Gomes, and Mariela Letícia Ponce*[2]

[1]King Abdullah University of Science and Technology, Thuwal, Saudi Arabia
suzana.nunes@kaust.edu.sa
[2]GKSSResearchCenter,Germany

**Abstract**

This chapter is focused on nanocomposites based on ionomers. Material developments are described mainly aiming two applications: actuators and fuel cells. For fuel cells, nanocomposites can help in obtaining membranes for direct alcohol fuel cell with low-alcohol permeation and in the preparation of polymeric membranes for use in hydrogen fuel cell operating at low humidity levels and high temperatures. An additional task for nanocomposites is to improve the membrane-electrode layers, giving a better contact and better distribution of catalysts in the membrane.

**Keywords:** nanocomposite, membranes, sulfonated polymers

## Introduction

This chapter is focused on nanocomposites, whose matrix is constituted by ionic polymers. The IUPAC [1] recommends the definition of ionomers as polymers, in which a small (typically less than 10%) but significant portion of the constitutional units carry ionic charges when dissolved in an ionizing solvent, whereas polymers with higher content of ionic charges should be referred as polyelectrolytes. Here nanocomposites of both ionic polymers will be considered without strict differentiation. Not only purely inorganic fillers but also organically modified particles and carbon nanotubes (CNTs) are also considered. Polymer nanocomposites are usually developed aiming superior mechanical, electrical, thermal, and process properties. They can even replace metals and/or other composite materials in many industrial applications. When ionic polymers are used as matrix or as a main component, more specific properties are added to the material that can then be used for applications such as actuators and fuel cell membranes.

## 10.1. Actuators

Actuators have great interest, for instance, as artificial muscle. New actuators based on CNT can provide higher work per cycle than previous actuator technologies and generate higher mechanical strength. In comparison to known ferroelectric, electrostrictive, and magnetostrictive materials, very low driving voltage for CNT actuation is a major advantage for various applications such as smart structures, multilink catheters, micropumps, flaps for microflying objects, molecular motors, or nanorobots [2, 3]. Various types of nanocomposites with polymers have been developed to be used as actuators. Yun et al. [2] developed an actuator based on a composite of carbon nanofiber polymethyl methacrylate and an ion-exchange material. Actuators based on conducting polymers having a force capacity of several tens of kilograms could be obtained [4]. An example was prepared in the form of laminated polypyrrole films, grown on Pt-coated poly(vinylidene fluoride) membranes or Pt-coated polyethylene teraphtalate fibers [4] with acrylamide gel electrolytes. Developed polymer nanocomposites have also been prepared by incorporating layered silicates, for instance, layered silicate/silica in Nafion® [5] and montmorillonite (MMT) in Nafion® [6].

There is currently a large demand for highly active actuators, which have to be flexible and miniaturized. These actuators have many potential industrial applications, including bio-microelectromechanical systems, noiseless propulsion actuators, and biomedical devices/systems [6]. Existing smart materials such as piezoelectric ceramics, polyvinylidine fluoride, and shape memory alloy actuators have been limited in application because they require high voltage or high current, or because of their brittleness, high weight, or the small strain produced [7]. Polymer gels, ionic polymer–metal composites (IPMC)s, and conducting polymers have been evaluated as potential candidates as actuators. Among them, IPMCs have been extensively studied as a new class of polymeric material exhibiting large strain capability and low-voltage operation (on the order of $10^3$ V m$^{-1}$) [6]. They are very attractive because of their capability of producing large deformations in the presence of low driving voltages. IPMCs can be ideal for operation in aqueous media because of the material affinity for water. A popular form of IPMCs is a perfluorinated sulfonate membrane, with metal composites chemically placed within the membrane [8]. The ion-exchanging capability of the base material (ion-exchange polymer, such as Nafion®) is what leads to the resulting force of an IPMC when an electric field is applied. This is due to the unique ionic nature of the fixed polymer backbone, which is permeable to cations but impermeable to anions. When an electric field is applied, hydrated cations migrate toward the cathode, resulting in an electrophoretic pressure that leads the IPMC to bend to the anode side. Figure 10.1(a) illustrates the response of an IPMC to an applied electric field.

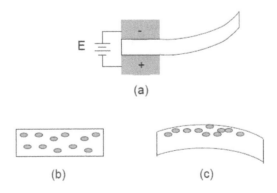

(a)

(b)                    (c)

**Figure 10.1**   (a) Bending of IPMC with an applied voltage; (b) IPMC section with no applied voltage; and (c) IPMC section with an applied voltage (adapted from [8]).

Figures 10.1(b) and (c) show the behavior of the hydrated cation species within the IPMC. When a DC voltage is applied, the IPMC tends to bend quickly to the anode side and then often slowly bends back to the cathode side (relaxation due to the existence of water backdiffusion). The mechanism for actuation can be explained by the hydrated cation of the sulfonate group in the membrane that moves with water toward the cathode just after the voltage is applied. It is possible that the weak bonds associated with the hydrated cations break after prolonged exposure to the implied electric field causing the inherent relaxation [8].

## 10.2. Fuel Cell Membranes

Probably, the most investigated application for nanocomposites with ionic polymers in the last years has been as membrane for fuel cell. The establishment of fuel cell as a competitive technology for energy conversion still depends on the development of materials with better performance. In a polymer electrolyte fuel cell, cathode and anode are separated from each other by a proton conductive membrane. Fuel oxidation in the anode (generating protons) and oxygen reduction in the anode (consuming protons) are promoted by catalysts (mainly Pt based), which are finely dispersed on the membrane surface in direct contact to the electrodes and gas diffusion layer. The requirements for good membranes are high proton conductivity, low permeability to the reactants (fuel and oxygen), and high chemical stability. While the bulk of the membrane should not allow electron transport, electron conductivity is recommended in the membrane–catalyst–electrode interface region to stimulate the electrochemical reaction. The development of highly effective fuel cell membranes is a big challenge and nanocomposites have been intensively considered as promising materials for that. Nanocomposites have been intensively investigated as part of strategies to deal with the following fuel cell key issues:

1. Optimization of the membrane–electrode–catalyst interface.
2. Preparation of membranes able to effectively operate above 100°C and under external low humidification in fuel cells fed with hydrogen.
3. Preparation of membranes with low-alcohol crossover for direct alcohol fuel cell.

### 10.2.1. Optimization of the Membrane–Electrode–Catalyst Interface

For the optimization of the fuel cell performance, the membrane-catalyst–electrode interface is as important as the membrane. The selection of effective electrodes with low-loading catalyst layers (below 0.4 mg Pt cm$^{-2}$) is decisive. Besides loading and catalyst dispersion, both ionic and electrical conductivities are required in the reactive layer. Furthermore, a strong adhesion between membrane, catalyst, and electrode must be assured [9]. Catalysts are frequently supported in carbon particles. The use of CNTs as an even higher surface area carbon support for catalysts is a more recently reported strategy. Besides the high surface area, CNTs have excellent electronic conductivity and high chemical stability [10–16]. Electrochemical measurement indicates that Pt/CNT nanocomposites display a significantly higher electrochemically active area and higher catalytic activity for the methanol oxidation reaction in comparison to a commercial Pt/C catalyst [11]. A relevant aspect regarding the preparation of CNT-based electrode nanocomposite for fuel cells is the chemical functionalization or activation of the CNT surface. Li and Hsing [12] have experimentally observed that oxygen-containing surface functionalities on CNTs can greatly affect the catalyst particle dispersion by manipulating Pt-anchoring and/or -nucleating sites. Zhao et al. [13] have shown that the addition of ionic liquids to the CNT-supported platinum catalysts contributes to the formation of small homogeneous Pt nanoparticles and to suppress the agglomeration of CNTs. In this case, the CNTs were pretreated in acid to get functional groups to promote the adherence of Pt nanoparticles. The effect of oxidation treatment of CNT surface has also been investigated by Tian et al. [11]. In the case of Pt/multiwalled carbon nanotubes (MWCNTs) prepared with as-received MWCNTs, the distribution of Pt nanoparticles is not uniform and large Pt clusters were formed. After the oxidation treatment by $H_2SO_4/H_2O_2$ solution, the dispersion of Pt particles on the MWCNTs is improved. However, agglomerates of Pt nanoparticles are clearly visible. On the contrary, the treatment of MWCNTs in $H_2SO_4/HNO_3$ significantly improved dispersion and distribution of Pt nanoparticles. Besides refluxing in oxidative acids, refluxing in aniline solution or reacting with highly active species, adsorbing functional molecules are other generally used routes for the modification of CNTs [14]. As it will be discussed later, for the preparation of nanocomposites with a good adhesion between all components, the surface modification of the additive/surface is very important.

### 10.2.2. Preparation of Membranes Able to Effectively Operate Above 100°C and Under External Low Humidification in Fuel Cells Fed with Hydrogen

At a temperature above 100°C, most of the available membranes starts to dehydrate, requiring more complex operating conditions to compensate the consequent conductivity decrease. Reviews on materials under investigation to overcome this problem have been published by Zhang et al. [17], Smitha et al. [18], Alberti and Casciola [19], Li et al. [20], and Hogarth et al. [21]. The motivations for operating at high temperatures are (1) improved reaction kinetics, (2) minimization of catalyst poisoning by CO, (3) simplification of the heat and water (humidification) management in the cell [22], (4) improved gas transport, and (5) other issues associated with catalyst and design. At temperatures above 100°C, the reaction kinetics are better and the poisoning of the catalyst by CO is lower. The adsorption of carbon monoxide on the surface of the platinum-based catalyst blocks the reaction sites (Figure 10.2). Because the adduct Pt–CO is thermoabile, by increasing the fuel operation temperature to 120–130°C, the catalyst poisoning by CO may be reduced [23]. The adsorption of CO on Pt is associated to high negative entropy, indicating that adsorption is strongly favored at low temperatures and is disfavored at higher temperatures [17].

Early reports and patent applications of Stonehart and Watanabe [24] and Antonucci and Arico [25, 26] claim the advantage of the introduction of small amounts of silica particles to

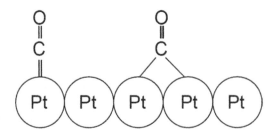

**Figure 10.2**    Adsorption of CO on Pt.

Nafion® to increase the retention of water and improve the membrane performance above 100°C. The effect is believed to be a result of the water adsorption on the oxide surface. As a consequence, the backdiffusion of the cathode-produced water is enhanced and the water electroosmotic drag from anode to cathode is reduced [21]. As a result, problems of cathode flooding can be reduced. Theoretical and experimental study has shown that polymer–inorganic nanocomposite membranes can provide additional water within the membranes for a given relative humidity [27]. A recent report of the group of Arico et al [28] confirms the effect of water retention with the inclusion of oxide particles in Nafion® and the importance of the acidity of the particle surface. An increase in both strength and amount of acid surface functional groups in the fillers enhances the water retention in the membrane: $SiO_2$-PWA (modified with phosphotungstic acid) > $SiO_2$ > neutral-$Al_2O_3$ > basic-$Al_2O_3$ > $ZrO_2$. The silica functionalization with sulfonated oligomers has been reported recently by our group [29] and will be described later in this chapter together with other strategies for the preparation of organic–inorganic membranes, which are also relevant for direct methanol fuel cell (DMFC).

### 10.2.3. Preparation of Membranes with Low-Alcohol Crossover for Direct Alcohol Fuel Cell

Instead of using hydrogen as fuel, methanol is effectively used to feed fuel cells. The DMFC has particular advantages for portable application. A recent review on membranes for DMFC has been published by Deluca and Elabd [30]. A main problem for this technology is, however, that the available commercial membranes like Nafion® have excessive methanol permeability. Methanol crossing the membrane from the anode to the cathode will react in the presence of the catalyst, competing with oxygen reduction. This competition leads to a loss of cell performance. A frequently used approach to minimize the methanol crossover has been the introduction of an inorganic phase. Besides the methanol crossover, the water transport in the membrane is at least as important. Most of the available membranes are based on sulfonated polymers. A reasonably high degree of sulfonation is required to reach enough proton conductivity. This increases also the membrane hydrophilicity and stimulates the water transport. However, if an excessive amount of water reaches the cathode, the active catalyst sites will be protected from the access to oxygen (cathode flooding), hindering the cathode reaction and again reducing the cell performance.

The simplest approach to develop organic–inorganic membranes for DMFC is to add isotropic nanosized inorganic particles like fumed silica, $TiO_2$, or $ZrO_2$. The effect can be understood using Eq. (10.1), which is based on the conventional Maxwell equation [31] and has been adapted to include effects of filler aspect ratio [32–34]:

$$P_c = \frac{P_p(1 - \phi)}{1 + (L/2W)\phi},\tag{10.1}$$

where    $P_c$ = permeability of the composite
$P_p$ = permeability of the plain membrane
$\phi$ = filler volume fraction
$L$ = filler particle length
$W$ = filler particle width.

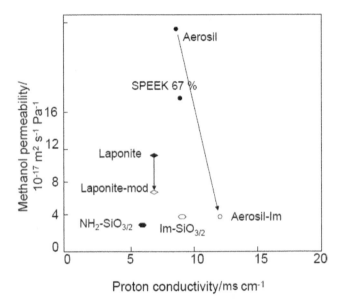

**Figure 10.3** Influence of the filler aspect ratio and surface modification on the methanol permeability and proton conductivity (adapted from [35]).

The permeability decreases with the increase of the filler volume fraction. These particles might be organically modified, mainly with silanes to incorporate groups that strengthen the interaction with the polymer matrix or even covalently bind to the matrix polymer chains. A second important consideration shown by Eq. (10.1) is that the inorganic particles with higher aspect ratios ($L/W$), like flakes, might contribute with a more effective barrier (lower permeability) effect to the membrane.

In Figure 10.3, the effect of inorganic fillers with different aspect ratios on the membrane permeability to methanol and proton conductivity is compared [35]. For the same filler concentration, the layered silicate (Laponite) was more effective in reducing the methanol permeability than spherical particles of Aerosil. Clays and layered silicates have been use by different group for the development of fuel cell membranes. Examples are Nafion®/mordenite [36] and Nafion®/MMT [37–39], sulfonated poly(ether ether ketone) (SPEEK)/MMT [40, 41], sSEBS/MMT [42], and poly(vinyl alcohol)/MMT [43].

The success of the addition of fillers with high aspect ratio depends, however, on the actual dispersion state of the multilayered silicate, which may lie between three typical situations schematized in Figure 10.4 [43, 44]. The silicate layers are not penetrated by polymer chains and a microcomposite is formed (Figure 10.4(a)), or chains are intercalated between the silicate sheets (Figure 10.5(b)) or the silicate layers are completely exfoliated and uniformly dispersed within the polymer so resulting in a nanocomposite (Figure 10.4(c)). For the barrier properties of the polymer matrix to be improved, the exfoliation of the multilayered silicate is a prerequisite for maximizing the tortuosity factor.

The surface modification of the inorganic filler is at least as important as the aspect ratio. Figure 10.4 shows how the methanol permeability decreases when the inorganic phase, independent of form or aspect, is treated with organomodified silanes. In the samples of Figure 10.4, the fillers were modified with silanes containing basic groups like amine or

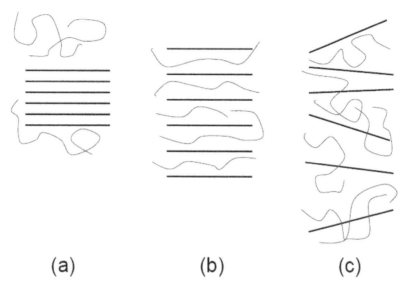

(a)                    (b)                    (c)

**Figure 10.4**    The typical dispersion states of a nanoclay within a polymer: (a) no chain penetration, (b) chain intercalation, and (c) clay exfoliation.

imidazole. The main function of the basic surface is to improve the compatibilization to the acid polymer matrix and inhibit the formation of cavities or defects between filler and matrix, which would otherwise act as a path for water and methanol transport. The methanol permeability reduced dramatically with the surface modification. Figure 10.3 also shows the examples of aminomodified silica network (polysilsesquioxane) prepared by the sol–gel process, as well as analogous networks modified with imidazole groups. In both cases, the methanol permeability as well as the proton conductivity are low. The reason is that an amino (or imidazole) group is attached to each silicon atom. The interaction of the basic groups to the sulfonic groups of the polymer matrix is strong and partially reduces the conductivity. The modification of silica or $ZrO_2$ particles with silanes, containing organic functionalized groups, opens also the possibility of really covalently linking the particles to the main chain. An example is the treatment with amino silanes and further with carbonyl-diimidazole, which then reacts with part of the sulfonic groups of the polymer matrix [45].

In order to minimize the loss of proton conductivity caused by the addition of the inorganic compound, sulfonated inorganic compounds have been used as fillers for the preparation of membranes. The approach reported by Rhee et al. [38] and Munaka et al. [46] consists of grafting the organic sulfonic acid group onto the inorganic surface by reaction of 3-mercaptopropyltrimethoxysilane (SH oxidation). First the SH group is introduced onto the surface precursor. Then, the mercapto group is converted to sulfonic acid group by oxidation with $H_2O_2$ solution. Another synthetic route reported by Munakata et al. [46] is the direct reaction of 1,3-propanesultone with the silanol surface groups. However, the thermal stability of the sulfonic acid group attached to the aliphatic propylsilane chain limit the operating temperature of the composite membrane below 100°C. Recently, sulfonated aromatic bishydroxy compounds have been grafted onto the silica surface by silanation reaction of bromophenyltrimethoxysilane for the first time [29]. First the bromophenyl group was introduced onto the silica surface, which was further reacted by nucleophilic

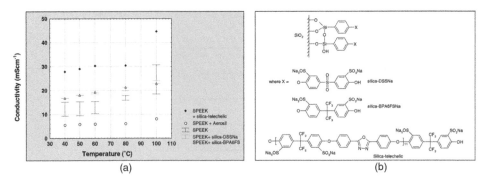

**Figure 10.5** (a) Proton conductivity of composite membranes containing 5 wt% of sulfonate filler in the temperature range of 40–100°C, (b) structure of the sulfonated fillers (adapted from [29]).

substitution reaction with sulfonate bishydroxy aromatic compounds [47]. In an effort to obtain a material with higher thermal stability, sulfonated hydroxytelechelic containing 1,3,4-oxadiazole units were also used. The resulting sulfonated silicas as fillers for the preparation of fuel cell composite membranes allow operating temperature up to 160°C and in the case of the silica-telechelic up to 160°C. The main objective of the attachment of the sulfonated aromatic bishydroxy compounds onto the silica particles was the preparation of a water-insoluble organic–inorganic filler with combined properties provided by the inorganic (keeping water in the membrane) and the sulfonated compounds (increasing proton conductivity).

Figure 10.5 shows the behavior of proton conductivity of composite membranes containing 5 wt% of sulfonated fillers in the temperature range of 40–100°C and the respective structures of the sulfonated fillers [29]. All membranes exhibited an increase in proton conductivity with increasing temperature. As it would be expected [45], the composite membrane prepared with fumed silica Aerosil 380 without any modification shows lower proton conductivity than the plain polymer SPEEK. In contrast, the composite membranes prepared with sulfonated silicas had higher proton conductivity in all range of temperature. It was expected that the addition of silica-telechelic would result in higher proton conductivity values compared with other sulfonated silicas because of the basic character of the oxadiazole rings contained in the telechelic structure. The N sites may favor additional points for proton jumps, contributing to the conductivity. Because of the better results observed for the composite membrane containing silica-telechelic, further tests at temperatures higher than 100°C were made. At temperatures higher than 80°C, proton conductivity markedly increases. The proton conductivity at 140°C for the membrane containing 5 wt% silica-telechelic was about 50–60 mS cm$^{-1}$, much higher than 33 mS cm$^{-1}$, the value for the SPEEK membrane. This result may be a consequence of the water absorption by the silica-telechelic and therefore a better water retention at high temperatures. These results are quite interesting since even when sulfonated montimorillonite was used as filler, a decrease of the proton conductivity was observed [38]. The same kind of functionalized silica have been recently used for membranes based on fluorinated polyoxadiazole also for fuel cell application [48]

The results obtained from the pervaporation experiments did not show any significant variation of methanol and water permeation [29]. On the basis of the results reported in literature [38, 41, 45, 49], a decrease of the permeability values because of the increased

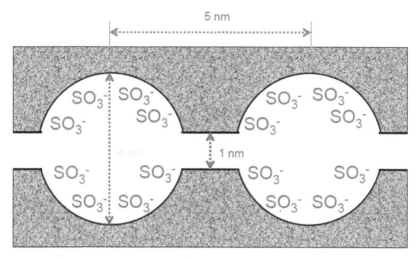

**Figure 10.6**   The proposed structure for the Nafion® ionic clusters.

barrier properties of the composite membranes should be expected. In electrolyte polymers, the bulk properties are mainly governed by the ionic aggregates, which may vary between 30 and 100 Å [50]. A structural model for the Nafion® ionic cluster in literature assumes three distinctive structural regions: the polymer network, water cores, and the interfacial domain between the two regions, where the water cores in neighboring clusters are presumably interconnected through channels (Figure 10.6) [51]. Because of its large size, the water and methanol transports may mainly occur through the ion channel. Rhee et al. [38] showed by XRD patterns that the ion channel size in the composite Nafion® membranes is reduced from 38.4 to 29.4 Å nm, resulting in significant decreased methanol and water permeabilities.

Besides decreasing the ion channel size, the addition of fillers may disrupt polymer chain packing leading to a more open polymer matrix [29, 52] or should lead to the formation of agglomerates due to the hydrophlicity of sulfonated silica. The formation of the agglomerates may lead to the formation of nonselective voids at the interface increasing flux of water and methanol [45]. For that, the modification of the filler is an important aspect. In the case of the silica-telechelic, the telechelic segments attached to the particles led to a rather good dispersion and adhesion between inorganic domains and the polymer matrix. There were some agglomerates of particles with size about 70 nm; however, the sizes are significantly lower than those formed with aerosil particles.

The composite membrane prepared using silica with sulfonated hydroxytelechelic containing 1,3,4-oxadiazole units has higher proton conductivity values in all range of temperature (40–140°C) than the membrane containing only the plain electrolyte polymer, whereas the methanol permeability determined by pervaporation experiment was unchanged. The combination of these effects may lead to significant improvement in fuel cells (fed with hydrogen or methanol) at temperatures above 100°C.

Organically modified silica particles have been widely used in different technology areas as metal ion sorbent, self-assembled monolayer, catalyst support, chromatography column packing, filler, and others [53]. The range of applications of silica particles is continuously growing due to their high specific area, large porous volume, large mechanical and thermal stability, and easy surface modification. The treatment of silica surfaces with alkylchlorosilane or organoalkoxysilane precursors such as $R'_{(4-x)}Si(OR)_x$ enables the introduction of specific functional groups ($R'$) covalently attached to the silicon–oxygen network, which can further react with other compounds acting as a network former [54].

Instead of adding fumed silica or other particles in the membrane, the inorganic phase can be *in situ* generated by sol–gel chemistry. The *in situ* generation of a silica phase in Nafion® has been early reported by Mauritz et al. [55], impregnating a manufactured membrane with silanes and inducing their hydrolysis. Recently, the same group reported self-assembled organic–inorganic membranes via sol–gel polymerization of silicon alkoxides around sulfonated blocks of polystyrene-soft block-polystyrene block copolymers [56]. By introducing silanes to a Nafion® solution and casting films, it is possible to change the morphology of the ionic clusters [57, 58]. The investigation of Nafion® or SPEEK/*in situ* generated $SiO_2$ or $ZrO_2$ membranes for fuel cell has been reported later [45, 59, 60]. The reduction of methanol and water permeability of SPEEK membranes with the generation of $ZrO_2$ in the membrane-casting solution was investigated by Silva et al. [49, 61]. A remarkable decrease of proton conductivity was, however, detected for filler contents higher than 10 wt% in the membrane. Figure 10.4 also shows the effect of introducing an *in situ* generated phase obtained by the hydrolysis of amino functionalized silanes into a SPEEK membrane. A dramatic reduction of methanol permeability was achieved. However, the amino groups had a too strong interaction to the sulfonic groups of the polymer matrix, leading also to a reduction of proton conductivity.

Without further functionalization, silica is a rather passive filler. Zirconium phosphates and phosphonates with $\alpha$- and $\gamma$-layered structures have been extensively investigated by Alberti [19, 62–65] as proton conductive materials. He has shown that their properties could be improved by introducing a suitable choice of organic groups, zirconium sulfophenylphosphate being particularly effective for proton conduction. The introduction of phosphates in polymers like SPEEK is reported to improve the proton conductivity at higher temperatures [19, 65]. Zirconium phosphate has been used in combination with Nafion® by DuPont [66, 67] and later by other groups [68–70]. Composites of zirconium phosphates and polybenzimidazole have been tested to increase the performance of the polymer in operation up to 200°C [71]. Our group has tested zirconium phosphate/SPEEK membranes for DMFC [45, 72, 73]. For this purpose, exfoliation (analogously to that shown in Figure 10.4) and further treatment with polybenzimidazole were necessary to improve the adhesion between filler and polymer matrix and reduce the methanol crossover.

Heteropolyacids, particularly Keggin-type heteropolyacids, are known for their high proton conductivity and have been considered by different groups as filler to enhance the proton conductivity of polymeric membranes for fuel cell [74–80]. Some more recent papers have been published. For instance, organic–inorganic membranes containing methacrylate covalently attached to the silica have been synthesized by the polymerization of 2-hydroxyethyl methacrylate and 3-methacryloxypropyl trimethoxysilane, and tungstophosphoric acid hydrate (WPA) was then incorporated to increase the membranes proton conductivity. The water-retention properties provided by $SiO_2$ and WPA lead to high proton conductivity (around 30 mS cm$^{-1}$) at 100–150°C [81]. Improvements in the performance of SPEEK-WC membranes by introducing heteropolyacids in the polymer matrix have been reported by Fontanova et al. [82]. The preparation of sol–gel $SiO_2$–polymer hybrid heteropolyacid-based proton exchange membranes for intermediate temperatures (80–120°C) fuel cell applications has been prepared by Pern et al. [83]. The membranes have proton conductivities of 8–15 mS cm$^{-1}$ at 80°C and 100% relative humidity (rh), of 1.5–2 mS cm$^{-1}$ at 100°C and 46% rh, and of around 0.25–0.8 mS cm$^{-1}$ at 120°C and 23% rh. A patent relative to the preparation of organic–inorganic polymer electrolyte membranes from polyethyleneimine and trialkoxysilane containing an epoxy group, trifluoromethanesulfoneimide, and at least one heteropolyacid has been published by Kim et al. [84]. Composite SPEEK/heteropolyacid membranes with different contents of heteropolyacid in the SPEEK copolymers matrix with different degree of sulfonation were investigated by Wang et al. [85]. By FTIR, they confirmed the strong interaction between the sulfonic groups on the polymer matrix and the heteropolyacid particles. High proton conductivities (95 mS cm$^{-1}$) at high temperatures (120°C) were found for the composite membranes, which was higher than those for Nafion$^{®}$ (60 mS cm$^{-1}$) at the same temperature and for the pure SPEEK (78 mS cm$^{-1}$) at 80°C. Hybrid organic–inorganic proton exchange membranes from PEG- and PVP-containing undecatungsto-cobaltoindic heteropolyacid (InCoW11) and undecatungstochromoindic heteropolyacid (InCrW11) were reported by Zhao et al. [86, 87]. The reported proton conductivity of these materials was in the order of 7 and 1 mS cm$^{-1}$ at room temperature. Composite membranes obtained by blending Nafion$^{®}$ with sulfonic-functionalized heteropolyacid–SiO$_2$ nanoparticles were reported to have a good performance in DMFC allowing operation temperatures from 80 to 200°C [88]. The power density was 33 mW cm$^{-2}$ at 80°C, 39 mW cm$^{-2}$ at 160°C, and 44 mW cm$^{-2}$ at 200°C. The function of the sulfonic-functionalized heteropolyacid–SiO$_2$ nanoparticles was to provide a proton carrier and act as water reservoir at elevated temperatures. However, for the broad use of heteropolyacids such as tungstophosphoric or molibdophosphoric acid as fillers for membranes, there are limiting problems in their application particularly if the DMFC is the aimed application. They are soluble and can be easily bled out from the membrane. In order to overcome the problem of the electrolyte dissolution, different approaches have been tried. Zaidi et al. proposed the use of heteropolyacids entrapped in zeolites [89, 90]. They also have recently reported the preparation of composite membranes from SPEEK and heteropolyacids loaded onto MCM-41 molecular sieve [91]. The highest proton conductivity of the order of 100 mS cm$^{-1}$ at room temperature was found for the membrane containing tungstophosphoric acid-loaded MCM-41 and with negligible leaching of the tungstophosphoric acid. In the catalyst literature, the entrapping of the heteropolyacid in the host material, mostly silica oxide networks, has been reported [92–95]. Usually, the heteropolyacid is fixed through covalent bonds or columbic interactions involving its free protons and the SiO2 groups of the polymer matrix resulting in a decrease of the acid strength. An alternative approach proposed by our group in order to avoid the consumption of the acid sites in the heteropolyacid was the modification of its anion structure and its fixation to an oxide network *in situ* generated by the sol–gel process from alkoxysilanes [96, 97]. Organosilyl derivatives of the heteropolyacid, for example, a divacant tungstosilicate $[\gamma\text{-SiW}_{10}\text{O}_{36}]^{8-}$, were prepared using 3-glycidoxypropyltrimethoxysilane (GPTS) (Figure 10.7). The introduction of GPTS in the anion structure of the heteropolyacid enables its attachment to a host material, by an epoxy ring-opening reaction with appropriate functional groups present in the host surface. For instance, the modified heteropolyacid was introduced into a SPEEK, containing insoluble inorganic filler (Aerosil) modified with amino groups. The amino groups react with the epoxy groups of the heteropolyacid molecules and fix them without reducing their acidity since the protons present in the heteropolyacid remain free. With this approach, an optimum balance between proton conductivity and methanol permeability [97] was achieved as shown in Figure 10.8.

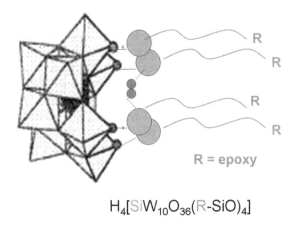

R = epoxy

$$H_4[SiW_{10}O_{36}(R\text{-}SiO)_4]$$

**Figure 10.7**    Functionalized heteropolyacid.

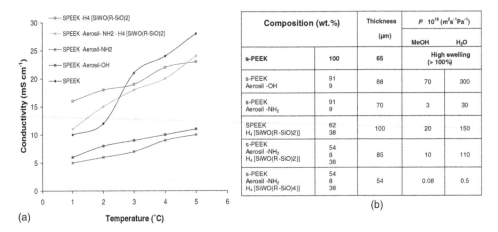

| Composition (wt.%) | | Thickness | $P\ 10^{18}\ (m^2s^{-1}Pa^{-1})$ | |
|---|---|---|---|---|
| | | (µm) | MeOH | $H_2O$ |
| s-PEEK | 100 | 65 | High swelling (> 100%) | |
| s-PEEK Aerosil -OH | 91 9 | 88 | 70 | 300 |
| s-PEEK Aerosil -NH₂ | 91 9 | 70 | 3 | 30 |
| SPEEK H₄ [SiWO(R -SiO)2)] | 62 38 | 100 | 20 | 150 |
| s-PEEK Aerosil -NH₂ H₄ [SiWO(R -SiO)2)] | 54 8 38 | 85 | 10 | 110 |
| s-PEEK Aerosil -NH₂ H₄ [SiWO(R -SiO)4)] | 54 8 38 | 54 | 0.08 | 0.5 |

(a)    Temperature (°C)

(b)

**Figure 10.8**    (a) Proton conductivity at a temperature range from 50 to 110°C and (b) methanol permeability at 55°C of composite membranes containing 38 wt% of organosilyl derivatives of the divacant tungstosilicate and 8% of the surface-modified aerosils.

Tan et al. [98] studied another approach to avoid the extraction of the heteropolyacid during fuel cell operation conditions by the preparation of composite membranes based on SPSU containing benzimidazole derivatives and heteropolyacid. The network of interactions established in these materials resulted in very high proton conductivities at 110°C.

As previously mentioned in this chapter, a detailed study of the microstructure of these organic–inorganic hybrid materials is necessary for a better understanding of the correlation between structure and their properties as proton conductors and methanol barriers for DMFC applications. By SAXS and the so-called anomalous small angle X-ray scattering (ASAXS), our group has characterized composites of ionic polymers (mainly sulfonated polyetherketone (SPEK) and SPEEK) as matrix and metal oxides, zirconium phosphates, and heteropolyacids as fillers [99]. In the case of organic–inorganic hybrid materials, the nanoparticle distribution can be well characterized by synchrotron SAXS

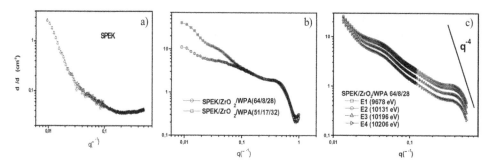

**Figure 10.9**   SAXS and ASAXS patterns of SPEK and SPEK·ZrO$_2$·WPA using energies close to the W-LIII absorption edge (10 keV) (a, c) and to the Zr–K ($\sim$ 18 keV) (b), respectively.

[100]. The ASAXS technique allows to separate the contribution of different elements to the scattering curves. The contrast between both phases can be changed and the scattering due to nanoparticles containing an specific element can be separated, by scattering measurements performed by tuning the incident beam energy to values near the absorption edge of the inorganic element under investigation (e.g., zirconium and/or tungsten).

An example of nanocomposites based on SPEK, ZrO$_2$, and WPA was recently published by our group [99] (Figure 10.9). No scattering peak was observed for the pure SPEK membrane. Scattering peaks in ionic polymers are frequently associated to phase separation or to clustering of the ionic groups [100, 101]. A reason for the absence of scattering peak is probably the fact that the sulfonic groups statistically and directly attached to the main chain of the polymer and not pendant in a longer side chain as in the case of Nafion®. They are therefore less available for clustering. In the case of the composite membranes, different scattering patterns were observed for materials with different ratio of inorganic content.

The distribution of zirconium oxospecies and WPA particles in the membranes were determined by measurements with X-ray energies close to the K-absorption edge of zirconium (17998 eV) and to the L$_{III}$-absorption edge of tungsten (10,206 eV), separating in this way the contribution of the zirconium oxospecies and the WPA to the total scattering curve. The scattering patterns of the composite membranes indicated that when both inorganic additives were added to the polymer matrix, there was a better distribution of them into the composite material, mainly as a result of interactions between WPA and zirconium oxopolymers chains. For membranes with the same WPA content, the reduction of the oxide content leads to an increase in the size of WPA aggregates.

The effect of the interaction between both inorganic components influences not only the distribution and the size of the inorganic aggregates but also the proton conductivity and the methanol crossover. Composite membranes with lower content of oxide and higher content of heteropolyacid had a still higher proton conductivity than that of the plain SPEK membrane, and the methanol and water permeabilities were, respectively, one-fifth and one-third of that of the unmodified membrane with similar thickness [96]. Aieta et al. [102] have investigated the hydration process in Nafion® and other perfluorsulfonate membranes containing various heteropolyacids (Keggin type) by SAXS. Time-resolved SAXS measurements showed that the water uptake in these membranes is a fast process and that even low concentrations of heteropolyacid changed the overall structure of the polymer matrix. Even at low relative humidity, the composite membranes had higher proton conductivity than the plain Nafion® membrane. The effect of the particle size reduction on the conductivity of Nafion®–WPA composite membranes has been studied by Ramani et al. [103]. They postulated that the reduction of additive particle size results in an enhancement of the surface/volume ratio of the additive, which leads to a more efficient proton hopping and hence larger proton conductivities.

Beside the application of heteropolyacids as proton conductors in the field of fuel cells, these compounds have found a broad application in other fields ranging from catalysis to energy storage, conversion devices (batteries, supercapacitors, and fuel cells), and sensors to biomedicine. Heteropolyanions have very interesting molecular properties, from electrochemical activity, electrochromism, photoactivity, ionic conductivity, magnetic and catalytic properties, which can be directly compared with those of related extended oxides [104]. An interesting review on the application of Keggin-type heteropolyacids has been published by Mioc et al. [105].

The field of organic–inorganic hybrids develops fast and an increasing number of new possibilities. Beyond mechanical strength and thermal and chemical stabilities, these materials are also sought for improved optical and electrical properties, luminescence, ionic conductivity, selectivity, as well as chemical or biochemical activities [104]. Some important devices based on functional materials are sensors, selective membranes, all sorts of electrochemical devices, from actuators to batteries or supercapacitors, supported catalysts, or photoelectrochemical energy conversion cells. Although in this type of materials, mechanical properties are unimportant, the emphasis is on reactivity, reaction rates, reversibility, or specificity. The hybrid approach can also be useful in this context by combining organic and inorganic species with complementary properties and reactivities. An important aspect in the nanocomposite development is the compatibilization between the ionomer matrix and the filler. Improvements on the dispersion of inorganic compounds in the network by chemical modification of the organic–inorganic surfaces are also a relevant step for developing new nanocomposite materials.

## References

1. Hess, M.; Jones, R. G.; Kahovec, J.; Kitayama, T.; Kratochvíl, P.; Kubisa, P.; Mormann, W.; Stepto, R. F. T.; Tabak, D.; Vohlídal, J.; Wilks, E. S. Pure Appl Chem 2006, 78, 2067–2074.
2. Yun, Y.-H.; Miskin, A.; Kang, P.; Jain, S.; Narasimhadevara, S.; Hurd, D; Shinde, V.; Schulz, M. J.; Shanov, V.; He, P.; Boerio, F. J.; Shi, D.; Srivinas, S. J Intel Mat Syst Str 17(3) (2006) 191–197.
3. Lee, D. Y.; Kim, K. J.; Heo, S.; Lee, M.-H.; Kim, B.-Y. KEM Bioceramics 2006, 309–311, 593–596.
4. Hutchison, A. S.; Lewis, T. W.; Moulton, S. E.; Spinks, G. M.; Wallace, G. G. Synth Met 2000, 113, 121–127.
5. Nguyen, V. K.; Yoo, Y. Sensors and Actuators B: Chemical, 2007, 120, 529–537.
6. Nam, J. D.; Choi, H. R.; Tak, Y. S.; Kim, K. J. Sens Actuators A 2003, 105, 83–90.
7. Yun, Y.-H.; Shanov, V.; Schulz, M. J.; Narasimhadevara, S.; Subramaniam, S.; Hurd, D.; Boerio, F. J. Smart Mater Struct 2005, 14, 1526–1532.
8. Paquette, J. W.; Kim, K. J.; Nam, J.-D.; Tak, Y. S. J Intel Mater Syst Str 2003, 14, 633–642.
9. Litsler, S.; McLean, G. J. Power Sources 2004, 130, 61–76.
10. Prehn, K.; Nunes, S. P.; Schulte, K. Mater Res Soc Symp Proc 2006, 885, 89–94.
11. Tian, Z. Q.; Jiang, S. P.; Liang, Y. M.; Shen, P. K. J Phys Chem B 2006, 110, 5343–5350.
12. Li, X.; Hsing, I.-M. Electrochimica Acta 51 (2006) 5250–5258.
13. Zhao, Z.W.; Guo, Z.P.; Ding, J.; Wexler, D.; Ma, Z.F.; Zhang, D.Y.; Liu, H.K. Electrochem Commun 2006, 8, 245–250.
14. Pan, D.; Chen, J.; Tao, W.; Nie, L.; Yao, S. Langmuir 2006, 22, 5872–76.

15. Ocampo, A. L.; Miranda-Hernández, M.; Morgado, J.; Montoya, J. A.; Sebastian, P.J. J Power Sources 2006, 160, 915–24.
16. Jeng, K.-T.; Chien, C.; Hsu, N.; Yen, S.; Chiou, S.; Lin, S.; Huang, W. J Power Sources 2006, 160, 97–106.
17. Zhang, J.; Xie, Z.; Zhang, J.; Tang, Y.; Song, C.; Navessin, T.; Shi, Z.; Song, D.; Wang, H.; Wilkinson, D. P.; Liu, Z.-S.; Holdcroft, S. J Power Sources 2006, 160, 872–91.
18. Smitha, B.; Sridhar, S.; Khan, A. A. J Membr Sci 2005, 259, 10–26.
19. Alberti G.; Casciola, M. Annu Rev Mater Res 2003, 33, 129–54.
20. Li, Q.; He, R.; Jensen, J. O.; Bjerrum, N. J. Chem Mater 2003, 15, 4896–4915.
21. Hogarth, W. H. J.; Diniz da Costa, J. C.; Lu, G. Q. J Power Sources 2005, 142, 223–237.
22. Faghri, A.; Guo Z. Int J Heat Mass Transfer 2005, 48, 3891–3920.
23. Baldauf, M.; Preidel, W. J Power Sources 1999, 84, 161–166.
24. Stonehart, P.; Watanabe, M. US5523181, 1996.
25. Antonucci, P. L.; Arico, A. S.; Creti, P.; Rammunni, E.; Antonucci, V. Solid States Ionics1999, 125, 431–437.
26. Antonucci, V.; Arico, A.S. EP0926754A1, 1999.
27. Choi, P.; Jalani, N. H.; Thampan, T. M. J Polym Sci Part B Polym Phys 2006, 44, 2183–2200.
28. Arico, A. S.; Baglio, V.; Di Blasi, A.; Modica, E.; Antonucci, P.L.; Antonucci, V. J Power Sources 2004, 128, 113–118.
29. Gomes, D.; Buder, I.; Nunes, S. P. J Polym Sci Part B Polym Phys 2006, 44, 2278–2298.
30. Deluca, N. W.; Elabd, Y. A. J Polym Sci Part B Polym Phys 2006, 44, 2201–2225.
31. Maxwell, J. C.; A Treatise on Electricity and Magnetism Vol 1, Clanderon Press, London, 1881.
32. Nielsen, L. E. J Macromol Sci Part A: Pure Appl Chem, 1967, 929.
33. Nunes, S. P.; Peinemann, K. V. Membrane Technology in the Chemical Industry, Wiley-VCH: Weinheim, 2001.
34. Falla, W. R.; Mulski, M.; Cussler, E. L. J Membr Sci 1996, 119, 129–138.
35. Karthikeyan, C. S.; Nunes, S. P.; Prado, L. A. S. A.; Ponce, M. L.; Silva, H., Ruffmann, B.; Schulte, K. J Membr Sci 2005, 254, 139–146.
36. Kwak, S.-H.; Yang, T.-H.; Kim, C.-S.; Yoon, K. H. Solid State Ion 2003, 160, 309–315.
37. Jung, D. H.; Cho, S. Y.; Peck, D. H.; Shin, D. R.; Kim, J. S. J Power Sources 2003, 118 205–211.
38. Rhee, C. H.; Kim, H. K.; Chang, H.; Lee, J. S. Chem Mater 2005, 17, 1691–1697.
39. Gosalawit, R.; Chirachanchai, R.; Shishatskiy, S.; Nunes, S. P. Solid State Ion 2007, 178, 1627–1635.
40. Karthikeyan, C.S.; Nunes, S. P., Schulte, K. Eur Polym J 2005, 41, 1350–1356.
41. Gaowen, Z.; Zhentao, Z. J Membr Sci 2005, 261, 107–113.
42. Won, J.; Kang, Y. S. Macromol Symp 2003, 204, 79–91.
43. Thomassin, J.-M.; Pagnoulle, C.; Caldarella, G.; Germain, A.; Jeromea, R. J Membr Sci 2006, 270, 50–56.
44. Ray, S. S.; Okamoto, M. Prog Polym Sci 2003, 28, 1539–1641.
45. Nunes, S. P.; Ruffmann, B.; Rikowski, E.; Vetter, S.; Richau, K. J Membr Sci 2002, 203, 215–225.
46. Munakata, H.; Chiba, H.; Kanamura, K. Solid State Ion 2005, 176, 2445–2450.
47. Gomes, D.; Nunes, S. P. Macromol Chem Phys 2003, 204, 2130–2141.
48. Gomes, D. and Nunes, S. P. J. Membr Sci, 2008
49. Silva, V.S.; Ruffmann, B.; Silva, H.; Gallego, Y.A.; Mendes, A.; Madeira, L.M.; Nunes, S.P. J Power Sources 2005, 140, 34–40.
50. Schlick, S. Ionomers: Characterization, Theory and Applications; CRC Press: New York, 1995.
51. Gierke, T. D.; Munn, G. E.; Wilson, F.C. J Polym Sci 1981, 19, 1687
52. Gomes, D.; Nunes, S. P.; Peinemann, K. V. J Membr Sci 2005, 246, 13–25.
53. Vansant, E.F.; Van Der Voort, P.; Vrancken, K.C. in Studies in Surface and Catalysis; Delmon, B.; Yates, J.T., Eds.; Elsevier: Amsterdam, 1995; Vol. 93, Chapters 8 and 9.
54. Sanchez, C.; Soler-Illia, G. J. A. A.; Ribot, F.; Lalot, T.; Mayer, C. R.; Cabui, V.; Chem Mater 2001, 13, 3061–3083.
55. Mauritz, K. A. Mater Sci Eng C, 1998, 6, 121–133.
56. Mauritz, K.A.; Mountz, D.A.; Reuschle, D.A.; Blackwell, R.I. Electrochim Acta 2004, 50, 565–569.
57. Zoppi, R. A.; Yoshida, I. V.P.; Nunes, S. P. Polymer 1998, 39, 1309–1315.
58. Zoppi, R. A.; Nunes, S. P. J Electroanal Chem 1998, 44, 39–45.
59. Nunes, S. P.; Rikowski, E.; Dyck, A.; Fritsch, D.; Schossig-Tiedemann, M.; Peinemann, K. V.; Euromembrane Proceedings 2000, 279–280.
60. Jones, D. J.; Roziere, J. in Handbook of Fuel Cells; Vielstich, W.; Lamm, A.; Gasteiger, H. A.; Ed.; Wiley, New York, 2003, Vol. 3, 447.

61. Silva, V.S.; Schirmer, J.; Reissner, R.; Ruffmann, B.; Silva, H.; Mendes, A.; Madeira, L.M.; Nunes, S.P. J Power Sources 2005,140, 41–49.

62. Alberti, G.; Casciola, M; Costantino, U; Vivani, R. Adv Mater 1996, 8, 291–303.

63. Alberti, G.; Casciola, M.; Massineli, L; Bauer, B. J Membr Sci 2001, 185, 73–81.

64. Alberti, G.; Casciola, M; Palombari, R. J Membr Sci 2000, 172, 233–239.

65. Tchicaya-Bouckary, L.; Jones, D.; Roziere, J. Fuel Cells 2002, 2, 40–45.

66. Grot, W. G.; Rajendran, G. US Patent 5919583, 1999.

67. Yang,C.; Costamagna, P.; Srinivasan, S.; Benziger, J.; Bocarsly, A. B. J Power Sources 2001, 103, 1–9.

68. Damay, F.; Klein, L. C. Solid State Ion 2003, 162–163, 261–267.

69. Yang, C.; Srinivasan, S.; Bocarsly, A.B.; Tulyani, S.; Benziger, J.B. J Membr Sci 2004, 237, 145–161.

70. Bauer, A.; Willert-Porada, M. J Membr Sci 2004, 233, 141–149.

71. He, R.; Li, Q.; Xiao, G.; Bierrum, N. J. J Membr Sci 2003, 226, 169–184.

72. Ruffmann, B.; Nunes, S. P. Solid State Ion 2003, 162–163, 269–275.

73. Prado, L. A. S. A.; Wittich, H.; Schulte, K.; Goerigk, G.; Garamus, V. M.; Willumeit, R.; Vetter, S.; Ruffmann, B.; Nunes, S. P. J. Polym Sci.-Phys 2003, 42, 567–575.

74. Tazi, B.; Savadogo, O. Electrochem Acta 2000, 45, 4329–4339.

75. Zaidi, S.; Mikhailenko, S.; Robertson,G.; Guiver, M.; Kaliaguine, S. J Membr Sci 2000, 173, 17–34.

76. Staiti, P.; Minutoli, M.; Hocevar, S. J Power Sources 2000, 90, 231–235.

77. Staiti, P.; Minutoli, M. J Power Sources 2001, 94, 9–13.

78. Staiti, P. J New Mat Electrochem Systems 2001, 4, 181–186.

79. Kim, Y. S.; Wang, F.; Hickner, M.; Zawodzinski, T.; Mc Grath, J. J Membr Sci 2003, 212, 263–282.

80. Honma, I.; Takeda, Y.; Bae, J.M. Solid State Ion 1999, 120, 255–264.

81. Aparicio, M.; Mosa, J.; Etienne, M.; Duran, A. J Power Sources 2005, 145, 231–236.

82. Fontananova, E.; Regina, A.; Drioli, E.; Trotta, F. Desalination 2006, 200, 658–659.

83. Pern, F.J.; Turner, J.A.; Meng, F.; Herring, A.M. Mater Res Soc Symp Proc 2006, 885, 51–57.

84. Kim, H.-K.; Balkus, K.J.; Chang, H.; Ferraris, J. P.; Yang, D.J.; Yang, Z. US2006159975, 2006

85. Wang, Z.; Ni, H.; Zhao, C.; Li, X.; Fu, T.; Na, H. J Polym Sci Part B Polym Phys 2006, 44, 1967–1978.

86. Zhao, S.; Wu, Q.; Mater Lett 2006, 60, 2650–2652.

87. Zhao, S.; Wu, Q.; Liu, Z. Polym Bull 2006, 56, 95–99.

88. Kim, H.-J.; Shul, Y.-G.; Han, H. J Power Sources 2006, 158, 137–142.

89. Zaidi, S. M. J.; AhmedAhmad, M.68. I.; Rahman, S.U. Proceedings of the ICOM 2005, Seoul, Korea, August 21–26th, p. 102.

90. Zaidi, S. M. J.; AhmedAhmad, M.68. I.; Rahman, S.U. Desalination 2006, 193, 387–397.

91. Zaidi, S. M. J.; AhmedAhmad, M.68. I. J Membr Sci 2006, 279, 548–557.

92. Misono, M.; Catal Rev Sci Eng 1987, 29, 269–321.

93. Kamada, M.; Kominami, H.; Kera, Y. J Coll Interf Sci 1996, 182, 297–300.

94. Kozhevnikov, I.V. Chem Rev 1998, 98, 171–198.

95. Mizuno, N.; Misono, M. Chem Rev 1998, 98, 199–218.

96. Ponce, M. L.; Prado, L.; Ruffmann, B.; Richau, K.; Mohr, R.; Nunes, S. P. J Membr Sci 2003, 217, 5–15.

97. Ponce, M. L.; Prado, L. A. S. A., Silva, V.; Nunes, S.P., Desalination 2004, 162, 383–391.

98. Tan, A.R.; Carvalho, L. M.; Gomes, A. S. Macromol Symp 2005, 229, 168–178.

99. Prado, L. A. S. A.; Goerigk, G.; Wittich, H.; Schulte, K.; Garamus, V. M.; Willumeit, R.; Ponce, M. L.; Nunes, S. P. J Polym Sci Part B Polym Phys 2005, 43, 2981.

100. Chu, B. In Ionomers: Characterization, Theory and Applications; Schlick, S.; Ed.; CRC Press, Boca Raton: Florida, CA, 1996; Chapter 3, pp 35–55.

101. Chu, B.; Hsiao, B.S. Chem Rev 2001, 101, 1727–1761.

102. Aieta, N.V.; Yandrasits, M. A.; Hamrock, S.J.; Stants, R.J.; Cookson, D.J. Herring, A.M. Prepr Symp – Am Chem Soc, Div Fuel Chem 2006, 51, 610–611.

103. Ramani, V.; Kunz, H. R.; Fenton, J. M. J Membr Sci 2005, 266, 110–114.

104. Gomez-Romero, P. Adv Mater 2001, 13, 163–174.

105. Mioc, U.B.; Todorovic, M.R.; Davidovic,M.; Colomban, Ph.; Holclajtner-Antunovic, I. Solid State Ion 2005, 176, 3005–3017.

# 11 Nanocomposites Characterization by Microscopy

*Rameshwar Adhikari*[*,1] *and Goerg H. Michler*[2]

[1]Central Department of Chemistry, Tribhuvan University, Kirtipur Kathmandu, Nepal
[2]Institute of Physics, Martin Luther University Halle-Wittenberg,
D-06099 Halle/Saale, Germany

## Abstract

Electron microscopy (EM), complemented by other techniques, has played a very important role in the elucidation of morphology and structure–property correlations of polymers including homopolymers, copolymers, blends, and composite materials. When employed vigilantly by experienced microscopist during sample preparation, EM most reliably characterizes every detail concerned with the morphology of polymers on various length scales. EM identifies the structure and properties of specific locations of polymers and the information is not restricted to the average values. The significance of the EM in nanocomposites science and technology arises principally from two causes. First, electron microscopic information is much more detailed than that from other sources. Using EM, information of the morphology of polymer matrix as well as of filler and the adhesion between them can be simultaneously assessed with nanometer resolution. Second, EM allows studying the response of the every structural details of the composite toward applied load (sometimes even *in situ*), enabling design of tailored materials. EM is the only technique that provides very direct evidence of intercalation and exfoliation of the filler in the polymer matrix allowing straightforward quantification of morphological features of the polymer nanocomposites (PNC). This chapter discusses the use of different microscopy techniques in the study of structure and properties of PNCs with special reference to EM and scanning force microscopy as nanoanalytical tools. It is shown that the selection of specific microscopy technique and preparation methods is crucial for the artifact-free imaging of the PNCs. The choice of the microscopic technique depends basically on the class of composite materials at hand. Besides the discussion of classical nanocomposites based on inorganic fillers, a brief account of polymer–polymer composites and bionanocomposites will be presented.

**Keywords:** polymer nanocomposite, polymer morphology, electron microscopy, filled polymers, deformation mechanisms, Nanofiller

## 11.1. Introduction

Polymers are characterized by rich variety of structural and functional diversities. The end use properties of these materials are mainly determined by their morphology on different

---

*Correspondence should be addressed to e-mail: nepalpolymer@yahoo.com

length scales, the latter being controlled by molecular structure and chain architecture as well as processing history. Therefore, it is of prime importance to understand and purposefully modify the structure of homopolymers, blends, and composites on macroscopic, microscopic, and molecular levels in order to be able to design materials having tailored property profile.

The construction of nanostructured polymers with specific mechanical and functional properties is an important facet of contemporary materials science and engineering [1]. There is a growing interest in the field of nanostructured polymeric materials due to their promising mechanical and functional properties, and these materials have stimulated in the recent years a great deal of interactions among physicists, chemists, biologists, and engineers.

The mechanical properties of polymers are determined mainly by the molecular parameters as well as the structural details (morphology) and their response toward applied load. Hence, to design polymeric materials possessing a desired property profile, it is essential to develop a suitable morphology aimed at specific micromechanical processes of deformation [2–5]. Within this context, in recent years, there has been a rapidly growing trend for the incorporation of inorganic fillers into the polymer matrix to enhance one or several physical properties. In particular, high aspect ratio fillers with platelet thickness in the range of a few nanometers (so-called nanofiller) such as layered silicates have been used to increase the stiffness and strength of amorphous and semicrystalline polymers without compromising the toughness properties [5]. Other nanofillers include alkaline earth metal compounds, alumina, silica, polyhedral oligomeric silsesquioxanes, carbon nanotubes (CNTs), and even fullerenes [6–14].

Various aspects of polymer nanocomposite (PNC) technology, including the use of microscopic techniques for their structural characterization, have been the object of many recent studies [15–20]. The prime objective of the application of nanofillers has been to "achieve large effects with the aid of small particles." It has been repeatedly verified that the properties suites comparable to those common for the traditional composites can be achieved using PNCs containing substantially less (typically less than 5 vol%) amount of filler. Thus, the nanocomposites enable greater retention of the inherent processibility, thereby retaining the toughness of the polymer matrix. The nanoscopic dimension and extreme aspect ratio inherent in the PNCs, regardless of the filler type (tubes, platelets or spheres), endow these composites with some fundamental characteristics such as very low percolation threshold, particles–particles correlations at low volume fraction, extremely large interfacial area per unit volume of filler, short interparticle distance, etc. [15–30]. There is a comparable size scale among the filler dimension, interparticle distance, and relaxation volumes of polymer chains. In the PNCs, the strong interface formed would lead not only to high composite stiffness and strength but also to the tougher products because of the nanofiller's ability to deform prior to their breakage [5, 15, 16].

Unfortunately, one often achieves an improvement in stiffness of the polymer at a cost of ductility. Experience shows that the deterioration of mechanical ductility of the polymeric materials by the addition of nanofiller results primarily from the lack of affinity between the filler and the polymer, leading to the formation of agglomerates of critical sizes that may cause a premature failure of the composites. Therefore, the common practice of polymer-based nanocomposite technology has been to modify the filler surface so as to promote interaction with the polymer matrix (see [21] for recent review).

Different microscopic techniques ranging from optical microscopy (OM) to electron microscopy (EM) and scanning probe microscopy (SPM) play a vital role in the characterization of PNC morphology and different length scales.

This book is devoted to highlight the advancement in the field of synthesis, modification, and incorporation of different kinds of nanofillers into various types of polymers. The physical properties of nanocomposites and the strategies adopted to enhance them have been the object of other chapters of this book. The unparalleled contribution of the microscopic techniques (especially of EM and scanning force microscopy (SFM)) in the development of nanocomposite science and technology can be realized by the fact that almost all of the scientific literatures devoted to PNCs make direct or indirect reference to the use of these techniques. The same is true for the chapters of this volume. Microscopic techniques play crucial role especially in the optimization of composite properties through structural characterization that provide direct clues not only about the structure and properties of the filler itself but also about the filler–matrix adhesion, filler distribution, and impact of filler on the morphology of the embedding polymer matrix. Recently, several monographs have been published that deal with characterization of nanomaterials in general [32–35] and with EM of polymers, blends, and composites along with their preparation techniques [36–38].

This aim of this chapter is to discuss the use of EM in the study of structure and properties of selected groups of PNCs with special reference to the selection of specific microscopic technique and preparation methods depending on the nature of composite materials at hand. In addition to the classical nanocomposites based on inorganic fillers, a brief account of polymer–polymer composites and bionanocomposites is presented.

## 11.2.  Experimental Techniques

The quality of the results delivered by microscopy depends much on the sample preparation. Without proper care during specimen preparation step, no conclusions free of error and artifacts can be drawn. This sections shades light on some of the important hints about sample preparation for different microscopic techniques used.

### 11.2.1.  Sample Preparation

#### 11.2.1.1.  *Film Preparation*
Samples for polymer microscopy can be prepared in a variety of manner [37, 38]. Thin films of polymers for OM, SFM, as well as EM can be prepared in different ways:

(a) Dropping the dilute polymer solution onto a substrate followed by the subsequent drying and annealing if needed. The film thickness can be generally controlled by the concentration of polymer. The thickness is, however, not uniform throughout the entire film area.

(b) Subjecting a drop of the polymer solution taken on a substrate to undergo high-speed rotation (also called spin coating). Drying and annealing of the films at elevated temperature may be needed. The film thickness, which is more uniform than in the dropping technique, can be adjusted by changing solution concentration and speed of substrate rotation.

(c) Drawing slowly a substrate (such as a rectangular plate of smooth glass, mica, or silicon wafer) dipped into a polymer solution (also called dip coating). The film thickness can be controlled by varying the solution concentration and film drawing speed.

(d) Sectioning of bulk sample by means of a microtome or an ultramicrotome. If one is interested with the morphology of "as processed" sample, the above three methods cannot be of choice because the morphology of interest is lost on dissolving the sample in a solvent. This is generally the case when the structures of the nanocomposite prepared by different routes are to be compared or exact morphology developed during particular processing route has to be evaluated. Microtomy and ultramicrotomy (sometimes equipped with cryodevices) are employed for the preparation of thin sections of the nanocomposite materials [37]. Generally, freshly prepared glass knives or diamond knives are used for this purpose.

As most of the polymeric materials used as matrix in the composites are much softer than the fillers at room temperature, the specimens should be hardened before sectioning in order to avoid local mechanical deformation of one or more of the components present in the composites. Treatment of sample with high energy radiations (such as $\gamma$-rays) or with chemical staining agent (such as osmium tetroxide) can help hardening the sample, so that it can be sectioned at room temperature. More conveniently, the samples are sectioned at cryogenic temperatures using liquid nitrogen as coolant. The sectioning temperature is generally below or close to the glass transition temperature of all the components of the polymer [37].

The thickness of the films required for OM lies in the range of a few microns to about 20 $\mu$m, whereas for the TEM investigations, it should be in the range of 50–70 nm. Films of any thickness can be studied by scanning electron microscopy (SEM) and SFM.

### 11.2.1.2. *Staining*

Because of the large mass contrast (matters comprising the elements of very different atomic masses) between the polymer and filler, nanocomposites can be easily imaged by transmission electron microscopy (TEM). Therefore, unlike in conventional TEM of heterophase polymers, no staining is required. However, the details on morphology of the polymer matrix, which is eventually modified by the presence of filler, cannot be evaluated without making the morphological details of the polymer visible. Thus, staining of one or more of the phases using heavy metal compounds (such as osmium tetroxide, ruthenium tetroxide, and uranyl acetate) might be necessary [39–42]. If the chemical staining is indispensable, it can be performed either on the bulk material prior to sectioning or on the films. As mentioned earlier, an advantage of staining the bulk sample is that the sample is hardened and can be sectioned at room temperature.

### 11.2.1.3. *Etching*

Free surfaces of the samples are studied by SEM and SFM. Smooth surfaces required for this purpose are prepared either by ultramicrotomy or by solution-casting procedures (such as spin coating). The free surface can be etched by reactive ions (called as reactive ion etching [43–45]) or by using chemicals such as in permanganic etching [46, 47].

Permanganic etching is most frequently used technique employed for the surface preparation for SEM examination of semicrystalline polymers and their composites. During etching process, the etchant destructively attacks the polymers and removes progressively the outer skin of the polymer, whereby the loosely packed amorphous phase is preferentially etched out of the sample. Therefore, this technique is useful to expose the internal structural details of the semicrystalline polymers. In case of nanocomposites, this technique is useful only if the experimentalist is interested, besides the adhesion and distribution of the filler, in filler-induced transformation of the matrix morphology.

The sample surfaces as received either can be directly dipped into the etchant solution or the fresh surface for etching can be prepared by microtomy (or ultramicrotomy). The etching period may range from a few seconds to several minutes depending on the nature of the polymer under investigation (i.e., the degradability of polymer under treatment with an etchant) and concentration of the etchant.

It should be, however, kept in mind that etching of the sample surface is not obligatory for SEM and/or SFM investigations. Nanocomposite materials can be conveniently studied by SEM using backscattered electrons (BSEs) imaging mode if the secondary electrons (SEs) mode does not provide sufficient information. For SFM, the most important requirements are the flat surfaces free of any sort of contamination and that there is sufficient difference in local mechanical properties (e.g., stiffness contrast) between the components.

## 11.2.2. Morphological Characterization by Microscopy

The wide variation of properties in polymeric materials arises from their inherent internal organization. The internal organization, often referred to as polymer morphology, is a function of various parameters including molecular structure and architecture of the polymer as well as the processing history. Polymer morphology, which is crucial for the properties of the polymeric materials, is primarily studied by microscopy. Using different microscopic techniques, structural details of polymeric materials ranging from a few Angstroms up to more than 100 mm can be evaluated. Complemented by other integral methods such as scattering methods, only microscopy can provide direct picture of structure of the polymeric materials.

Electron microscopy (EM) has played a vital role in the elucidation of morphology and structure–property correlations of polymers including homopolymers, copolymers, blends, and composites. When employed attentively by experienced microscopist with proper care during sample preparation, EM most reliably characterizes every detail concerned with the morphology of polymers on various length scales. EM identifies the structure and properties of specific locations of polymers and the information is not restricted to the average values.

The resolution powers of different microscopic techniques are schematically illustrated in Figure 11.1. It can be easily understood that the SPM and EM possess the central position among the modern nanoscale characterization techniques. Owing to the nanoscale dimension of the filler, which is a few orders of magnitude below the wavelength of light,

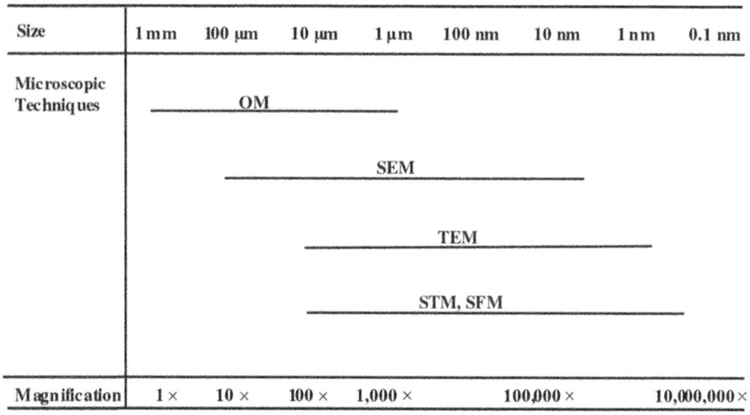

| Size | 1 mm | 100 µm | 10 µm | 1 µm | 100 nm | 10 nm | 1 nm | 0.1 nm |
|---|---|---|---|---|---|---|---|---|
| Microscopic Techniques | | OM | | | | | | |
| | | | SEM | | | | | |
| | | | | TEM | | | | |
| | | | | STM, SFM | | | | |
| Magnification | 1 × | 10 × | 100 × | 1,000 × | 100,000 × | | 10,000,000 × | |

**Figure 11.1**   Scheme showing the resolution power of different microscopic 1techniques.

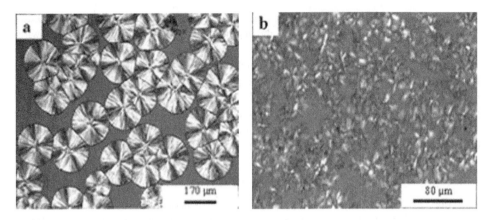

**Figure 11.2**   Polarizing optical micrographs of PP showing nucleating effect of the nanofiller; (a) neat polymer and (b) polymer containing 4 wt% layered silicate [50].

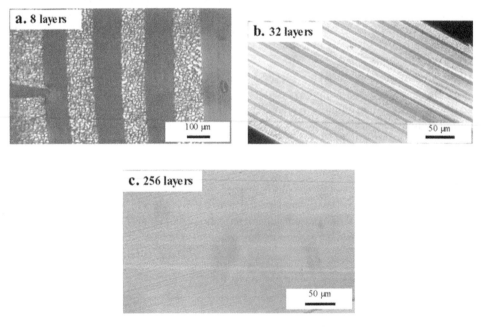

**Figure 11.3**   Polarizing optical micrographs of PET/PC multilayered composites having varying number of layers; the spherulitic texture of the PET phase gradually disappears with increasing number of layers.

the OM cannot be the method of choice for the characterization of detailed morphology of PNC. However, OM is the only technique to reliably analyze the effect of nanofiller on the structure of the polymers at larger length scales. Figures 11.2 and 11.3 illustrate the importance and the limitation of OM in the study of PNCs.

In the last decade, a large number of studies has been carried out on structure and properties of semicrystalline polymers (such as [21, 22, 48–53] and references therein) reinforced with nanofillers like nano-calcium carbonate (CaCO$_3$) and layered silicates.

In isotactic polypropylene (iPP), an important member among commodity plastics, it was found that such fillers possess strong nucleating effect leading to the formation of larger number of spherulites in presence of filler than in the neat polymer under identical processing conditions.

Figure 11.2 shows polarizing optical micrographs of neat iPP and the polymer containing 4 wt% layered silicate clay. It was found that the average spherulite diameter drastically reduced from about 200 to about 15 μm. As shown in Figure 11.2, semicrystalline polymers are not uniform solids rather they reveal polycrystalline textures under polarizing microscope. These textures, commonly known as sphurulites (see Figure 11.2(a)), are infinite in their varieties, no two species being identical [46, 47, 54]. The number of primary nuclei (which later develop as sphurulites) was increased due to the presence of nanofiller. The more the nuclei, the larger the number of sphurulites formed. As a result, the objects formed were much smaller (see Figure 11.2(b)). The nucleating effect was further enhanced on subjecting the polymer melt to undergo shear flow [50].

The spherulitic morphology of the polymers assessed by OM can be correlated with their ultimate mechanical properties. In iPP/CaCO$_3$ nanocomposites, while the crystallinity of the matrix polymer remained unaltered, the toughness of the sample was significantly increased owing to the formation of $\beta$-modification of the matrix polymer. Obviously, the presence of nanofiller was responsible for the modified deformation mechanism of the polymer. In summary, the OM is useful in determining the morphological changes occurring in semicrystalline polymers on spherulitic scale. Further, the large-size agglomerates of the fillers can be easily identified, whereby one can make preliminary evaluation on whether the surface modification of the filler was satisfactory.

Figure 11.3 shows polarizing optical micrographs of multilayered composites comprising alternating layers of polyethylene terephthalate (PET) and polycarbonate (PC). The composites were prepared by microlayer coextrusion technique [55–58]. The samples were annealed at 150°C for 30 min, so that the PET layers could undergo cold crystallization. Actually, the crystallization of PET is significantly suppressed due to the physical confinement imposed by the adjacent PC layers (Figure 11.3). For large-layer thickness (e.g. Figure 11.3(a)), the semicrystalline morphology of the PET layers consisting of spherulitic texture resembles that of the bulk PET. The development of spherulitic texture becomes weaker with increasing number of layers (or decreasing layer thickness). Finally, no spherulitic morphology appears (see Figure 11.3(c)).

Figure 11.3(c) represents the frontier where the meaningful applicability of OM ceases. Obviously, detailed morphology of the nanocomposites cannot be assessed by means of OM. In order to be able to investigate the morphological changes taking place on much smaller length scales, nanoanalytical techniques are essential.

This section can be concluded as follows. With the aid of OM, one can gain basic idea on the dispersion and distribution of the filler particles and further decide whether further high-resolution microscopic studies would be essential. One can also draw preliminary conclusions on the efficiency of the chosen processing routes.

## 11.3. Different Nanocomposite Systems

### 11.3.1. Imaging 1-, 2- and 3-Dimensional Filler Systems

Depending on the shapes of nanofillers, they can be classified as being one 1D (such as nanotubes, rods, and fibers), 2D (such as layers and platelets), and 3D (such as spherical particles). The 1D and 2D nanofillers are characterized by a large aspect ratio, whereas the 3D particles have an aspect ratio of unity. In the same way, the surface-to-volume ratio, an important parameter determining the properties of filled polymers, differs for different filler types. The choice of specimen preparation technique and that of microscopy does not depend much on the dimensionality of the filler.

### 11.3.1.1. *Overview of Filler Distribution*

After a quick inspection of a thin section, or the surface of the bulk material under an OM, the composite material may be investigated by SEM. The SEM analysis does not provide details about the dispersion, distribution, and interaction of filler with the matrix polymer but offers information on whether the filler is distributed uniformly over a large volume of the material. It further provides information on affinity of the filler toward the polymer matrix.

SEM is a surface analytical technique, in which a focused electron beam scans the sample surface in a raster pattern. In the literature, SEM has been referred to as routine method for the estimation of filler distribution in the polymer matrix. SEM imaging is carried out in two modes: SE mode and BSE mode. The former technique, which is often mentioned in the literature, reflects the topography of the sample surface. On the other hand, the contrast in the BSE mode is correlated with the atomic number of the elements, which the electron beam is interacting with, and hence enables the material specific contrast. An excellent overview of SEM application in the characterization of polymeric materials including nanocomposites can be found elsewhere [36].

A typical specimen for the SEM analysis of a bulk nanocomposite material is a cryofracture surface. Alternatively, the flat surface can be prepared by microtomy. The microtomed surfaces are occasionally etched with suitable chemical. A thin film collected on a substrate can be directly investigated. In order to avoid charging, the polymer surface is coated with a thin carbon film.

Figure 11.4 shows SEM micrographs of a nanocomposite of EPDM rubber filled with 3 wt% nanometric clay (an example of 2D filler system). The surface for the SEM imaging was prepared by breaking the rubber sample previously cooled at liquid nitrogen followed by carbon film coating. The upper picture is the common SEs image, whereas the lower one the BSEs image of the same location representing the material specific contrast.

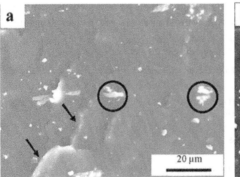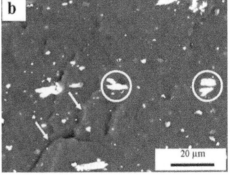

**Figure 11.4**  SEM micrograph of a cryofractured surface of layered silicate reinforced EPDM rubber; (a) SE mode image and (b) BSE mode image.

The SE image reflects the fracture surface topography. Being a rubber, the fracture surface of the sample shows flat terraces and occasionally the steps (as indicated by the arrows). The absence of plastically deformed materials (such as fibrils) indicates that the material shows elastomeric behavior. Bright spots, some of which are up to 5 μm in diameter, represent the locations pointing out of the fracture surface in SE image (Figure 11.4(a)).

A careful look at the BSE image (Figure 11.4(b)) makes it evident that the topography contrast, as observed in the SE image (see Figure 11.4(a)), does not exist; the steps/terrace structures are no more dominant. However, the contrast between the matrix polymer and dispersed filler islands (material specific contrast) is significantly increased and the particles appear brighter.

In order to determine the chemical elements present on the sample surface, energy dispersive X-ray analysis was carried out using the SEM. Modern SEMs have this facility built in the microscope. It was demonstrated that the element abundantly present in the scanned area was, besides the constituents of the polymer and the coating material, silicon (Si), a constituent of the layered silicate clay.

Limited through the resolution power of the SEM, it is not possible to image the individual clay layers (which are about 1 nm thick). The SEM images presented in Figure 11.4, nevertheless, demonstrate that a part of filler is present as agglomerates or aggregates (see the structures inside the white circles), that is, the filler is not completely exfoliated. Thus, the SEM imaging provides the first indication of whether the filler distribution is satisfactory or further optimization of processing and filler modification would be required.

The SEM technique has been extended to all sorts of polymeric materials and their nanocomposites. Figure 11.5 shows SEM micrographs of layered silicate reinforced (Figure 11.5(a)) and boehmite nanoparticles (an example of 3D filler system) reinforced (Figure 11.5(b)) polystyrene (PS) homopolymer (see [59, 60] for details). The specimens were the fracture surfaces produced by breaking the samples at liquid nitrogen temperature. SEM imaging was carried out utilizing the BSE mode.

The specimen imaged in Figure 11.5(a) is a nanocomposite of PS and 30 wt% layered silicate produced by extrusion. In the SEM micrograph, the regions with higher atomic number (i.e., the silicates) appear brighter, whereas the areas with lower atomic number

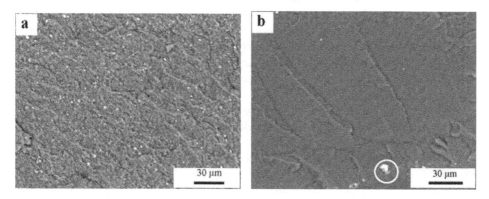

**Figure 11.5** SEM micrographs of cryofractured surfaces of polymer nanocomposites prepared by solution-casting procedure; (a) PS with 30 wt% layered silicate and (b) PS with 5 wt% boehmite; imaging was performed in BSE mode.

(i.e., the organic polymer) are darker. Thus, the white dots (up to about 1 μm in diameter), which are well dispersed in the polymer matrix, represent the aggregates of the layered silicate. Taking into consideration that 30 wt% layered silicate is mixed with the polymer, one can only see a very small amount of filler as large aggregates. Therefore, it can be concluded that the major part of the filler is efficiently dispersed, so that all the aggregates are less than 1 μm in diameter. Nevertheless, on the basis of the SEM results, it can be expected that, for an optimum amount of filler, the aggregation tendency can be reduced with an effect of improved reinforcement.

Similar experiments (using BSE mode in SEM) were conducted for a composite comprising PS homopolymer and organophilic modified boehmite nanoparticles (3D filler system). As in the previous case, the boehmite nanoparticles, owing to the higher atomic number of the constituent elements than that of the matrix polymer, appear brighter than the surrounding matrix (see Figure 11.5(b)). At low filler concentration, bright spots (agglomerates of filler particles) are visible only at few locations (for instance, see white circle in Figure 11.5(b)), attesting a very good compatibility of the filler with the polymer matrix.

Here the marked variations in the production of BSEs by different phases reflect the material contrast in the electron micrographs. One of the advantages in using SEM is that it allows the inspection of the filler dispersion and morphology of much larger sample areas than in the case of TEM.

### 11.3.1.2.  *Details of Nanostructured Morphology*
In Section 11.3.1.1, we addressed the analysis of nanoparticle dispersion in polymer matrix with the help of SEM. However, nanoscale-sensitive microscopic technique is required in order to characterize the details about the dispersion state of the nanoparticles. For this purpose, TEM and SFM are the preferred techniques. In this section, we introduce the application of the TEM in the evaluation morphology of PNCs based on representative 1D, 2D, and 3D fillers. The thin films for the TEM studies were prepared by cryo-ultramicrotomy and had a thickness in the ranger of 50–70 nm. TEM has played very important role in the straightforward evaluation of the composite morphology of the resulting materials. There is no technique, except TEM, that can provide so direct information with high precision about structure, dispersion state, and adhesion of those fillers.

CNTs and nanofibers are the typical 1D fillers that are characterized by exceptionally large surface area of about 1000 $m^2$ $g^{-1}$. Provided that the filler surfaces are modified in order to make them compatible with polymer matrix, the resulting composites will have a very strong interface leading to a potentially strong and tough materials. Melt mixing undershearing, solution-mediated processes, and *in-situ* polymerizations are the commonly practiced techniques employed for the incorporation of CNTs and carbon nanofibers into the polymer matrix [51–53, 61–64].

Right after their invention in 1991 [65], the CNTs attracted the interest of the materials scientists as a potential candidate for nanoscale filler. Especially polymer composites containing CNTs possess an enormous innovation potential. In the polymer/CNT composites, thanks to the excellent filler mechanical properties (such as moduli 200–1000 GPa, strength 200–900 MPa), one would expect high reinforcing effects at low CNT content without loss of properties such as processability and surface gloss. On the contrary, one could also exploit the electrical conductivity of the CNTs and achieve the antistatic or conducting composite materials that have added advantages for the applications such as in electrostatic painting (such as saving of paints, environmental protection, etc.).

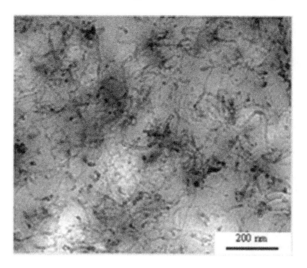

**Figure 11.6**   TEM image of a nanocomposite comprising a PC and 5 wt% MWNTs (reproduced with permission from [66]).

In Figure 11.6, typical morphology of a nanocomposite comprising PC and multiwalled carbon nanotubes (MWNT) [66] is displayed. The sample for TEM studies was prepared by ultramicrotomy. No staining is required as sufficient density contrast between inorganic nanotubes and organic polymer exists. Thus, the CNTs appear dark in the TEM image. In the TEM micrograph, even the individual CNT frequently dispersed well in the polymer matrix can be noticed. Nevertheless, the nanotubes are not homogeneously distributed as single entities rather they form small (and sometimes larger) aggregates.

It was demonstrated that surface modification of the CNTs plays an important role for adhesion, distribution, and reinforcement effect in the polymer–CNT composites. As example, the incorporation of the virgin CNT in to the polymer could not bring any positive effect [67], whereas the composites with surface modified CNT [68–70] or *in-situ* polymerized composites [71] endowed the polymers with an improvement in mechanical performances. In all those studies, EM has played vital role in the structural characterization of the nanocomposites.

TEM has been used to characterize the morphology of nanocomposites containing 2D and 3D fillers such as layered silicates and alumina (boehmite) nanoparticles, respectively (see Figures 11.7 and 11.8). The thin section of each sample was investigated by TEM without any chemical treatment.

Figure 11.7 shows the detailed morphology of PS/4 wt% layered silicate nanocomposite. The overview image of similar composite with higher (30 wt%) amount of filler was presented in Figure 11.5(a). The two pictures in Figure 11.7 were taken from two different locations of the sample. Clearly, the sample consisted of regions having nonuniform distribution of the filler in the polymer matrix, the information that was not assessable via SEM investigation alone. Figure 11.7(a) corresponds to the sample location where nothing could be observed in the SEM images (due to extremely small thickness of the dispersed filler) and Figure 11.7(b) to the locations having agglomerations.

Figure 11.7(a) shows that the layered silicate filler has been exfoliated to some extent as demonstrated by several individual nanolayers (dark lines). Most of the layers are present

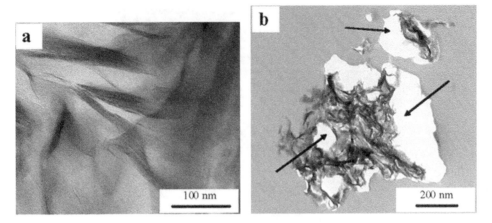

**Figure 11.7**  TEM micrograph of a composite comprising PS and organophilic modified layered silicate; filler content 4 wt%.

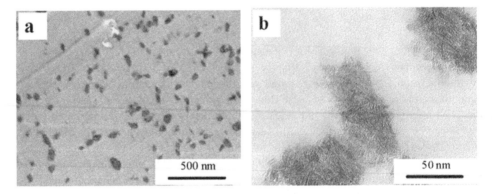

**Figure 11.8**  Lower (a) and higher (b) magnifications of TEM micrographs showing the morphology of a composite comprising PS and organophilic modified boehmite; filler content 5 wt%.

in the form of bundles (dark strips), in which the layers appear as being separated from each other but still exist in association with other layers (the intercalated layers). By measuring the spacing of the layers in the composites, the scattering techniques (such as wide-angle X-ray diffraction) may also provide hints on intercalation and exfoliation of the layered silicate in the PNCs. However, TEM is the only method that can provide very direct evidence of these nanoscale phenomena.

The TEM imaging provides additionally the information on the influence of filler agglomeration on the mechanical performance of the composite material. Figure 11.7(b) shows, for instance, that the stress applied during specimen preparation by ultramicrotomy (i.e., sectioning) is concentrated in and around the filler particle leading to the formation of microcracks (as shown by arrows), an observation that may lead to the deterioration the properties of the composite. Note that under identical sectioning conditions, such voids were not formed in the regions where the filler has been finely and uniformly distributed following the intercalation and exfoliation (see Figure 11.7(a)). This demonstrates how important the exfoliation and uniform distribution of nanofiller are for the mechanical properties enhancement of the nanocomposites.

It is worth mentioning that, unlike many integral techniques, the structural information obtained by the TEM includes very local properties of the filler and the filler matrix interfaces.

Figure 11.8 shows the TEM micrographs of PS/5 wt% boehmite nanocomposite, a 3D filler system [60]. As in the nanocomposites with CNT (a 1D filler system) and layered silicate (a 2D filler systems), the boehmite nanoparticles, owing to their higher atomic number than that of polymer matrix, scatter electrons to higher extent and thus appear dark in the TEM images. In the low magnification TEM micrograph (Figure 11.8(a)), one can see well-dispersed spherical boehmite particles whose size ranges from about 10 to 100 nm. Therefore, the particles can be regarded as 3D nanoparticles. On the other hand, the higher magnification micrograph (Figure 11.8(b)) shows the details of structure of boehmite particles and their adhesion to the polymer matrix. It can be noticed that the particles are not necessarily perfectly spherical in shape and, as expected, have well-defined interface with the polymer matrix.

Each boehmite nanoparticle, similar to that of native boehmite, depicts a crystalline texture comprising lamellar crystals arranged in a staple-like fashion. The thickness of the crystal lies in the range of 3–5 nm.

Obviously, good dispersion of the nanoparticles in the polymer matrix can be directly linked with the optimum surface modification of the filler particles. The nanofiller in this case was coated with aliphatic carboxylic acid (containing 10–14 carbon atoms in main chain). There have been several electron microscopic studies on the boehmite-based nanocomposites [22, 70] including those prepared by *in-situ* polymerization [71, 72].

A common drawback for CNTs, layered silicates, or boehmite as nanosized fillers is the difficulty to achieve molecular-level dispersion, or exfoliation of these fillers even at low loadings, due to their incompatibility with the polymer matrix. The result is that the reinforcement of the polymer predicted by theoretical consideration is hardly achieved. Therefore, the researchers are looking for alternative means to reach the goal of fully exfoliated nanocomposites by developing so-called molecular composites such as those based on polyhedral oligomeric silsesquioxanes (for instance, see [10] and the references therein). Other attempts are concentrated in incorporating zero-dimensional filler based on metal nanoclusters [73, 74], which are meant for specific functional applications. Noteworthy in this regard is that high-resolution microscopic techniques such as TEM and SFM have been utilized not only for the straightforward characterization of these nanocomposites but also for the understanding of the functional properties associated with those filler.

## 11.3.2.  Block Copolymer-Based Nanocomposites

Block copolymers constitute a special class of heterogeneous polymers, in which two or more homopolymer chains (blocks) exist in a single molecule. The incompatibility of the constituent blocks leads to intramolecular phase separation but the chemical connectivity restricts the spatial dimension of phase segregation to nanoscale [29, 75–79]. As a result, at sufficiently high molecular weight, monodisperse block copolymers form an array of periodic nanostructures (periodicity 10–100 nm) commonly referred to as microphase-separated structures. In fact, the individual nanodomains thus formed preserve the properties of respective homopolymers. By changing the relative composition, the compatibility between different components, and the architecture of the copolymer

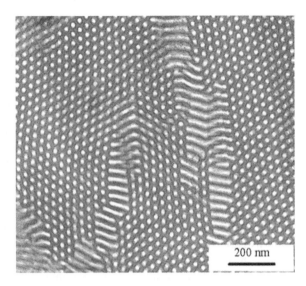

**Figure 11.9**   TEM image of an SBS block copolymer blend showing coexisting cylindrical (1D) and lamellar (2D) structures; the rubbery butadiene phase was stained with osmium tetroxide vapor.

molecules, the size and the type of nanostructures can be precisely controlled [75–79]. Because of the presence of nanostructured morphology with periodic composite texture, these copolymers can be regarded as nanocomposites. The main reason of continuously increasing interest for these materials over the last few decades is their potential application in nanotechnology such as templating of various nanoobjects [77].

Since the synthesis of first block copolymers by living anionic polymerization in the mid-fifties of the last century, the EM has been providing invaluable contribution to the understanding of their structure–property correlations, and thus, the EM has been established as the most powerful analytical tool in the block copolymer research. Today, different scattering techniques, SFM, and computer simulations provide valuable supplement to the structural characterization achieved by the EM.

The imaging of block copolymer nanostructures, which is carried out preferably with the aid of TEM, is generally accomplished by staining one of the phases by heavy metal compounds as in the case of semicrystalline homopolymers, blends, and copolymers [39–42]. Figure 11.9 shows TEM image of a blend of two different styrene/butadiene/styrene (SBS)-based triblock copolymers. The butadiene phase of the copolymer was stained by osmium tetroxide prior to the TEM imaging. The image shows 1D arrangement of the cylindrical microdomains having uniform diameter and 2D lamellar microdomains, which exist in the copolymer blend side by side. As in the case of PNCs comprising inorganic filler, it is the TEM that offers direct evidence of this morphology without ambiguity.

However, the imaging of block copolymers by TEM is not limited to the stained specimens. The experiments can also be carried out using special techniques such as holography and electron energy loss spectroscopy (EELS) (for instance, see Ref. [80]) or by means of SFM.

Some block copolymers even contain intrinsic contrast due to presence of transition metal atoms in a block and hence can be imaged without artificial staining [81]. Electron microscopy, especially the TEM, has been intensively employed to understand the

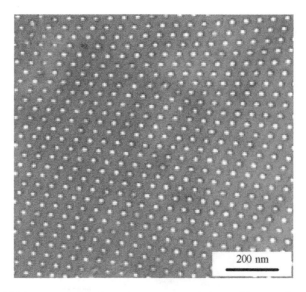

**Figure 11.10**   TEM image showing the nanoporous morphology of PtBA–PCEMA block copolymer formed by hydrolysis of the minority PtBA phase (reproduced with permission from [87]).

morphology formation in several block-copolymer-based hybrid nanocomposites [82, 83] and quantum dots [84, 85].

Block copolymers containing the degradable (etchable) minority component offer a rather convenient route to nanoporous materials, in which the pores have exactly the same configuration as in the parent block copolymer. For example, selective etching of the cylinder-forming phase in an ordered block copolymer results in the formation of nanoscopic channels, provided the matrix material can support such a structure [86].

Nanoporous materials including porous organic polymers have been shown to be useful for many applications including separations, catalysis, and templating. The applications of block copolymer-based nanoporous materials may include high surface area supports, nanomaterial templates, and size-specific separation media.

An example of such a nanoporous material is illustrated in Figure 11.10. The polymer used was a diblock copolymer based on poly($t$-butylacrylate) (PtBA) and poly(2-cinnamoylethyl methacrylate) (PCEMA) [87]. The matrix PCEMA was cross-linked by exposing the polymer in ultraviolet light followed by selective hydrolysis and subsequent removal of the cylindrical microdomains. TEM imaging showed that the dispersed phase was nothing but the hollow cylinders embedded in the cross-linked matrix of PCEMA. The hollow cylinders appear bright in the TEM images due to the fact that the electrons can pass through the pores owing to nil electron density. Obviously, it was not necessary to stain the sample, unlike the neat block copolymer sample presented in Figure 11.9, in order to visualize the nanostructures of the sample.

In Figure 11.8, it was noted that the thickness of the boehmite crystals lies in the range of 2–5 nm. Thus, the size scale of these nanostructures is approximately one order lower than that of the block copolymer nanostructures [72]. Therefore, by optimizing the processing parameters, one might be able to introduce the particles selectively into one of the block copolymer phases to achieve the so-called nano-reinforcement effect. In an attempt to incorporate boehmite nanoparticles into the block copolymer lamellae in

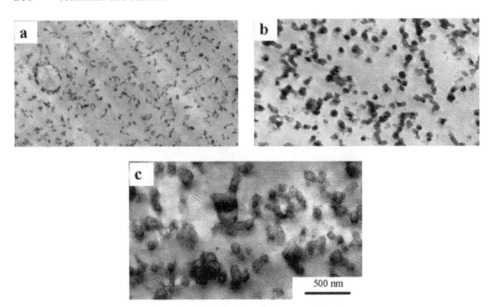

**Figure 11.11**    TEM micrographs showing the morphology of SBS–boehmite nanocomposites containing different weight fraction of boehmite: (a) 5%, (b) 20%, and (c) 40% (reproduced with permission from [60]).

a selective manner, their nanocomposites with lamellae forming SBS trilock copolymers were prepared and studied with different microscopic techniques including TEM.

Figure 11.11 illustrates the representative TEM micrographs of SBS–boehmite nanocomposites containing various amounts of nanofiller [60]. On going from Figures 11.11(a) to (c), the concentration of the nanofiller increases. In the same order, the number of aggregates decreases and their size increases. For the composite containing 5 wt% boehmite nanoparticles, the size of the particles is in the same range as for PS homopolymer (compare Figures 11.8(a) and 11.11(a)), although the filler in the block copolymer seem to be smaller. The increasing aggregate size with increasing filler content results from the fact that, in spite of the organophilic modification, the particles have their affinity toward themselves rather than to the polymers. At higher filler concentration, some of the particles reach a diameter of more than 200 nm (see Figure 11.11(c)). It suggests that the tendency of the particle agglomeration, which is not necessarily favorable for nano-reinforcement of the polymers, increases with higher feeding of the filler into the polymer matrix.

It is to be noted that the internal particle morphology remains unchanged independent of the composite composition, that is, the particles have structures identical to that observed in Figure 11.8(b), regardless of how much filler is introduced in the polymer.

The TEM micrograph in Figure 11.11(c) shows both boehmite particles and the morphology of the surrounding matrix. Prior to the TEM studies, the thin sections prepared by ultramicrotomy was stained with $OsO_4$ vapor to expose the morphological details of the polymer matrix. As expected, the lamellar morphology of the matrix remains unaffected by the incorporated nanoparticles.

To improve the mechanical performance of the SBS block copolymers, layered silicate-based nanocomposites were prepared using two-step processing route, namely solvent casting followed by melt extrusion [59, 88]. Excellent exfoliation of the nanofiller was revealed by the TEM that formed the basis of enhanced mechanical performance of the nanocomposite [59, 89].

### 11.3.3. Special Coextruded Multilayered Composites and Monocomposites

So far, we have discussed the electron microscopic studies of nanocomposites, in which the filler is made up of an inorganic material. However, there have been significant efforts and subsequent progresses in the development of composite materials comprising only polymers [57, 90–92] and sometimes only single polymer (so-called green composites) [90–95].

Being stimulated by the fact that the living beings in nature have synthesized some very sturdy materials with surprising mechanical properties through millions of years of evolution and natural selection, the materials scientists have turned their attention to develop new synthetic materials that mimic the fundamental structure and properties of natural biocomposites. The development of layered composites of metal laminates [96] and of polymers [90] is the attempt toward designing biomimetic functional composites.

Coextrusion is one way in which two or more polymers can be physically combined for the purpose of achieving the polymer composites possessing improved mechanical, electrical, and barrier properties [90]. Via this technique, polymer composites containing thousands of alternating laminates of different polymers can be produced, in which the thickness of individual layers goes down to several nanometers.

Figure 11.12 shows TEM micrograph of a coextruded multilayered composite comprising alternating 2D laminates of PET and PC [96]. The ultramicrotomed sections of the samples were treated with ruthenium tetroxide ($RuO_4$) in order to selectively stain less-dense PC phase. Thus, the PC layers appear dark in the TEM images.

In the nanolayered composite depicted in Figure 11.12, both the polyesters PC and PET are compatible to each other. Therefore, the adhesion between the polymers is good enough to ensure the integrity of the layers. In contrast, even for the strongly incompatible polymers such as PS and polymethyl methacrylate (PMMA), the alternating layers keep mechanical integrity surprisingly well in the multilayered composites [91]. Particularly, the adhesion between the layers was found to improve with the decreasing thickness of individual polymers, thereby enhancing the mechanical performance of the composites.

Nanolayer processing via multilayer coextrusion facilitates the creation of new hierarchical systems. Although this route is far away from the unique behavior of the living cells, in which the complex biosynthesis and nanoprocessing take place concurrently, this is highly flexible tool for the fabrication of unique architectures of otherwise incompatible polymers [90].

Besides TEM, other microscopic techniques such as OM, SEM, and SFM are often used for the elucidation of structure and properties of multilayered composites. The results are often supplemented by scattering studies [58].

A notable disadvantage of composite polymeric materials is that they are often problematic for the common processing conditions (such as thermoforming) and are mostly nonrecyclable [98]. In order to produce the reinforced polymeric systems while retaining their recyclability and the ease of thermoforming, new kinds of *self-reinforced* polymer compositions (popularly called as "green composites" or monocomposites) have been developed [93–95]. This route, often referred to as self-reinforcement, uses the single polymer having different orientational properties for designing the composite structures.

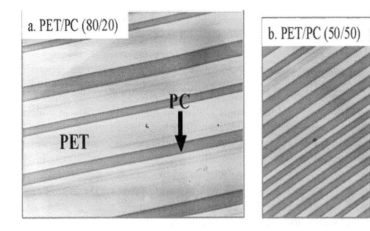

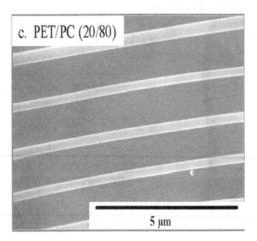

**Figure 11.12**    TEM micrograph showing the morphology of a multilayered PET–PC composite having different PET–PC compositions; the PC phase was stained with ruthenium tetroxide (reproduced with permission from [97]).

These composites are still in the phase of development and have not reached the nanometer dimension in the internal organization of individual components. However, these are the promising candidates for advanced applications in future. Therefore, a short overview on the TEM studies of these new composites is presented.

The idea behind the self-reinforcement strategy is to make high-strength composites starting from the highly oriented anisotropic fibers or tapes of a single polymer. The technology used to produce single-polymer composites is called "hot compaction" [98, 99].

As in many other aspects of polymer science, EM has provided not only the morphological information of the monocomposites but also the structural basis for the mechanisms of self-reinforcement.

Figure 11.13 shows TEM micrograph of a hot compacted PP composite. The specimens for the EM were prepared by ultramicrotomy of the compact sample that was previously treated with $RuO_4$ in order to stain the loosely bound amorphous part of the sample. The lower part of the TEM micrograph shows the structure of typical oriented PP, whereby the

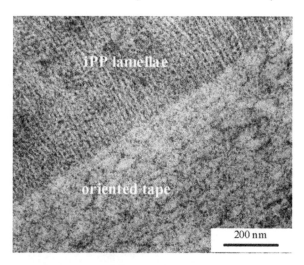

**Figure 11.13**   TEM micrograph showing the morphology of hot compacted PP tape; iPP lamellae epitaxially grown on the surface of oriented PP tape are visible.

crystallites appear as domains. The dark part in the TEM results from the staining by $RuO_4$ staining. Onto the surface of the oriented fiber of the PP are visible the recrystallized PP lamellae that are characterized by typical cross-hatched structure of $\alpha$-form of iPP.

Extensive works have been carried out to develop the technology of the "hot compaction" of the polyolefin fibers and tapes by the researchers in Leeds University and DaimlerChrysler Company [93–95]. However, the technique is extended to other polymeric systems as well [99–102]. Self-reinforced PP is even commercially available under the name Curv®, a trademark of BP AMOCO, whose applications include the construction of automotive undershield, hood for a passenger car, mid range loudspeaker cone, sports goods etc.

## 11.3.4.  Nanofibers and Hybrid Nanofibers

It has been emphasized in the preceding sections that 1D nanostructures have been the subject of intensive research due to their unique properties and intriguing applications in many specific areas. Electrospining, a drawing process based on electrostatic interactions, provides a simple and versatile approach to produce exceptionally long nanofibers (i.e., 1D nanostructures with extremely large surface area per unit volume) having uniform thickness and varying compositions [51, 103–111]. The electrospun nanofibers find applications in diverse fields such as filter applications, tissue engineering, and drug release systems, as templates for manufacturing nanotubes etc.

It has been common observation that the PNCs suffer frequently from the problem of agglomeration under common processing conditions such as melt mixing, solution blending etc., which lead finally to deterioration in some of the important physical properties such as toughness. To combine the properties of nanofiller and that of electrospun fibers in a synergic way, many researchers have succeeded to incorporate nanofillers having variable dimensions (1D, 2D, and 3D) into electrospun nanofibers [108, 112–114]. EM has contributed significantly not only in the structural characterization of the nanofibers

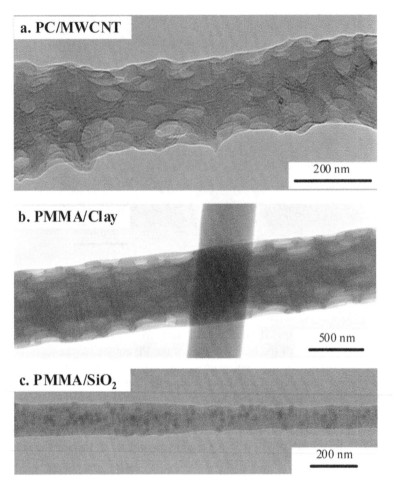

**Figure 11.14**    TEM micrographs showing the morphology of different nanocomposite fibers prepared by electrospinning; (a) PC/4 wt% MWNT, (b) PMMA/5 wt% clay, and (c) PMMA/10 wt% $SiO_2$ (reproduced with permission from [66]).

but also in the comprehensive understanding of various mechanisms responsible for the mentioned applications.

Figure 11.14 shows the TEM micrographs of three different nanocomposite fibers comprising 1D (MWNTs), 2D (organophillic modified layered silicate), and 3D ($SiO_2$ nanoparticles) nanofillers. The nanocomposite fibers, prepared via solution electrospinning, were collected directly onto the TEM grids [66, 108].

TEM analysis of the nanocomposite fibers revealed that fibers containing 1D and 2D fillers had porous surface with the alignment of the filler along the fiber direction. The most interesting phenomenon noted in these fibers was their surprisingly high toughness, which is characterized by necking and yielding of the nanofibers under the application of external load (see Figure 11.20, for example). Thus, electron microscopic examination of the nanocomposite fibers (especially their deformation behavior) demonstrated that the properties of nanofiller and electrospun nanofibers can be synergistically combined.

## 11.3.5.  Biosynthesized Nanocomposites

To search for a new generation of advanced materials, the conventional solutions available to humankind have nearly been exhausted. So the material scientists have turned to biomimetic (or biology-inspired) structures with the goal of developing new synthetic products. For instance, inspired by the composite structures produced by nature, polymer scientists are trying to reconstruct artificially the biomimetic morphologies [115, 116]. Through the course of evolution of living beings on earth, the nature has synthesized countless number of composite materials that are characterized by incredible structural perfection and wonderful matches of properties suited for specific physiological functions of individual organism. For instance, there are layered composites everywhere in nature – both in animal and plant kingdom; the walls of living cell, the rings of a tree, eye lens of mammals are a few examples. This section introduces two composite materials biosynthesized by nature: nacre and human bone (details in [9]).

Although nacre from the shell of abalone snail (commonly known as "see ear") is the most popular example cited in the literature, the material itself seems to follow the same general construction plan also in other nacre products, in mussel shells and pearls.

To keep it protected from the physical injuries and from the attack of the predators, the body of an abalone snail is covered by a specially designed apparatus having excellent mechanical properties. The outstanding mechanical stability of the shell results from the specially designed architecture, that is, a composite arrangement consisting of thousands of alternating hard/soft layers. Figure 11.15(a) depicts the SEM micrograph of nacre from the shell of the Blue Mussel. The specimen for the SEM study was prepared by simply hammering the shell. The fracture surface thus formed was studied without any treatment.

As shown in Figure 11.15(a), the stiff layers approximately 1.5 μm in thickness are made up mainly of calcium carbonate (microcomposite) [9]. One can notice that the layers are organized in such a perfect manner that over a macroscopic dimension of the sample, no defects can be identified.

The abalone and mussel shells demonstrate that the living organisms are capable of synthesizing high-strength and high-toughness mineralized tissues via biomolecular control of mineral crystals type combined with an organic adhesive layer. In the shell, the flat platelets of $CaCO_3$ (the orthorhombic aragonite polymorph) are stacked like bricks, glued together by a thin film of proteins and polysaccharides [117–119]. The biocomposite consists of 95% inorganic mineral and 5% organic polymer, the two being held together by means of electrostatic force of attraction. The organic glue is strong enough to hold layers of tiles firmly together, but weak enough to permit the layers to slip apart absorbing the energy of a heavy blow in the process. In fact, the abalone shell is much stronger than the conventional man-made ceramic materials. The biomineralization processes can turn the brittle materials such as calcium carbonates into a high-durability organic/inorganic laminates [117]. The layering of alternating hard/soft elements makes it a really tough composite material.

With respect to the promising mechanical properties, the bones of vertebrates offer another excellent example of bionanocomposite. Human cortical bone, for instance, comprises hard–soft composite arrangement on the nanoscopic scale giving bone material the excellent balance of stiffness and toughness that is needed to maintain structural integrity of the skeleton (see TEM image in Figures 11.15(b) and (c) [9]). Bone tissue can be regarded as a perfect composite material developed through the course of evolution.

Nanoscopic platelets of hydroxyl-apatite mineral are embedded in a fibrous matrix of organic polymer, the collagen (see Figure 11.15(c)). In other words, bone tissue itself could

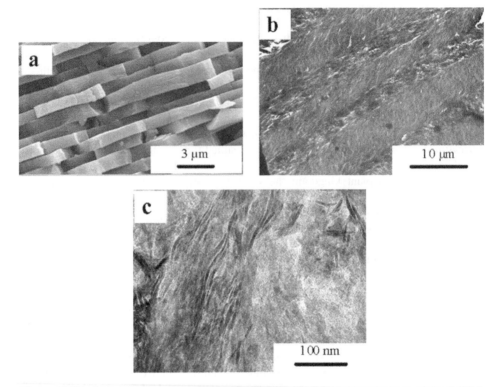

**Figure 11.15**   Structure of nature made nanocomposites; (a) SEM image of nacre and (b, c) TEM image of human cortical bone (reproduced with permission from [9]).

be described as a material consisting of composite nanofibers of different orientations. Bundles of such fibers with identical orientation are alternated with the bundles having different orientation,;orming lamellar structures very similar to plywood (see Figure 11.15(b)). The typical thickness of the lamellae is some microns. It is this hierarchical organization of nano- and microsized structures that gives bone its unique mechanical properties. These conclusions have been drawn on the basis of electron microscopic examination of the bone tissues.

## 11.4.  SFM of Polymer Nanocomposites

So far, applications of different electron microscopic techniques for the study of polymeric nanocomposites were described. Scanning force microscope (also referred to as atomic force microscope) is obviously not a technique belonging to the family of electron microscopy. However, it has been common practice in the last few years to supplement EM with the SFM as the latter enables straightforward surface characterization of polymers and provides additional insight into the structure and properties of polymers, blends, and composites. Therefore, a brief survey of the relevant applications of this technique in nanocomposite research is presented.

SFM not only allows the direct and straightforward analysis of PNC morphology but also enables the easy assessment of morphology development under annealing conditions

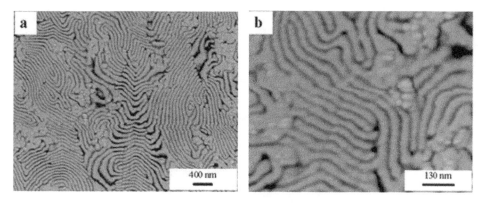

**Figure 11.16** Lower (a) and higher (b) magnifications of SFM phase micrographs showing the structure of SiO$_2$ filled PMMA–PBA copolymer.

and on loading [30, 120–129]. The most popular SFM mode of operation is the so-called tapping mode or intermittent contact mode, in which the sample is scanned with an oscillating probe. The contrast mechanism in the SFM operation is based on the local mechanical properties of the PNCs.

The specimens for the SFM examinations presented in this section were prepared by means of ultramicrotomy operated at cryogenic temperatures. A diamond knife was used to slice about 70-nm-thick sections from a specimen surface approximately 100 μm × 100 μm in area. The thin sections thus obtained were used for TEM examinations, whereas the clean microtomed surface was studied by means of SFM.

The capability of the SFM to image the nanocomposite materials is illustrated by SFM phase images presented in Figure 11.16. The tapping mode SFM was used to image simultaneously the nanostructures of the matrix PMMA–polybutyl-acrylate (PBA) copolymer, a block copolymer of PMMA and PBA as well as that of the embedded 5 wt% silica (SiO$_2$) nanoparticles. One can identify the lamellar structure of the block copolymer and the nanoparticles (bright particles) preferentially located in the PMMA lamellae (brighter lamellae). That the nanoparticles are present both as individual particles and in the form of agglomerates can be clearly demonstrated by the SFM micrographs.

One of the novel advantages of the SFM techniques is that the heterogeneous polymers can be directly imaged without any chemical modifications, thus allowing direct and straightforward evaluation of the nanocomposite structures (and their impact on the morphology of the matrix polymer). Figure 11.17 shows tapping mode SFM images of a thermoplastic elastomer vulcanizate based on 35% PP and 65% nitrile-butadiene rubber filled with carbon black nanoparticles. The filler was introduced into the thermoplastic (i.e., PP phase) after the vulcanization of the rubber was complete. Owing to its higher affinity to the rubbery phase, the filler tends to penetrate into the rubber phase. However, the high concentration of filler is restricted toward the interfacial region due to cross-linking of the rubber component [130].

The height image (Figure 11.17(a)) shows the topography of the cryo-ultramicrotomed surface of the sample, whereas the phase image (Figure 11.17(c)) depicts the material contrast in the SFM image. Figure 11.17(b) was generated via high pass filter of Figure 11.17(a), and reveals dispersed filler as bright dots in the PP matrix. The filler appears as rough surface structures in the PP matrix in the height image (Figure 11.17(a)). The SFM

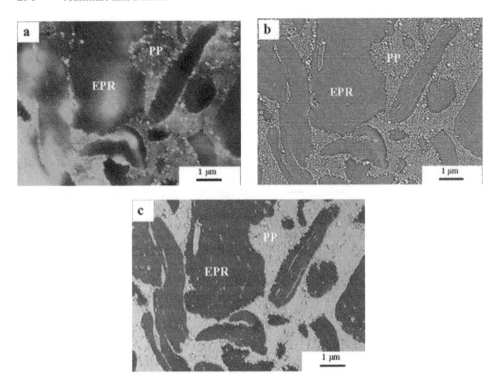

**Figure 11.17** SFM images of a composite comprising PP/nitrile-butadiene rubber (35/65 weight ratio) thermoplastic elastomer vulcanizate containing 10 wt% carbon black nanofiller; recorded in tapping mode; (a) unfiltered height image, (b) height image after high-pass filter, and (c) corresponding phase image.

imaging, on one hand, shows clearly the phase morphology of the polymer blend and the localization of the nanofiller in the restricted locations (here the PP phase) chosen by the processing engineer on the other.

An excellent example of SFM capability in imaging the morphology of heterophase polymeric material is further provided by imaging of multilayered polymer based on PET and PC, see SFM phase images in Figure 11.18. The multilayered composite was annealed at 150°C for 30 min before preparing the specimen by ultramicrotomy for microscopic analyses. Crystalline lamellae that are formed in the PET layers due to cold crystallization can be easily identified. These lamellae, which grow perpendicular to the interface between the layers, are only a few nanometers in thickness.

Similar composites were studied also by TEM, whereby it was too difficult to stain the amorphous phase of the PET in order to image the crystalline lamellae formed in it. In contrast, as one can notice, SFM phase imaging (see Figure 11.18) made excellently the internal structures of the PET layers visible in a straightforward manner.

Optical micrographs and TEM pictures of the similar composites were presented earlier (see Figures 11.3 and 11.12). By means of OM, it was possible to investigate the polymer morphology on spherulitic scale. However, obviously, neither the nanoscale morphology of the layered composite nor the lamellar structure inside the PET layers was visible in Figure 11.3. In contrast, the nanocomposite morphology was easily imaged by the TEM

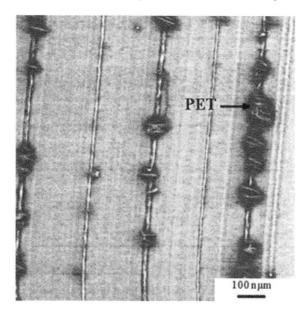

PET ⟶

100 nμm

**Figure 11.18**  Tapping mode SFM phase image showing the lamellae structure of PET phase as in PET/PC (composition 10/90 by volume) multilayered composite.

(Figure 11.12), but resolution of the lamellar nanocrystals was not straightforward (due to difficulty in staining). Those lamellar nanocrystals could be easily imaged by the SFM (see Figure 11.18). These examples illustrate that a combination of several microscopic techniques would generate complementary results enabling deeper insights into the structure and properties of nanocomposites.

A successful SFM imaging requires that the samples are properly prepared with flat, clean, and artifact-free surfaces. Equally important are the choice of imaging modes, selection of the probes, and other experimental parameters.

The ideal samples for SFM operations are the thin films prepared by casting procedure. The films can be annealed to make the structures reaching close to the equilibrium state. However, it should be remembered that the thin films and especially the surfaces of the thin films can have entirely different morphologies than the bulk polymer samples. Therefore, the surface characterization of the thin films does not provide much information on the polymer structures that can be related to the bulk polymer properties of interest. For the study of the bulk polymer structures, one may use the technique of ultramicrotomy and cryo-ultramicrotomy to expose the polymer internal structures on the freshly prepared surface. However, every sectioning step during ultramicrotomy may introduce the sample deformation that will be a reason behind the erroneous interpretation of SFM data. Hence, proper care should be taken in preparing a surface for SFM investigation.

## 11.5. Study of Deformation Micromechanisms

For the application of polymers, mechanical stability is an important aspect to be taken into account. In fact, besides introducing some other functional properties, the researches on nanofiller-based polymer composites are directed toward the simultaneous enhancement of toughness, stiffness, and strength of polymers [15, 66]. Owing to the enormous increase in surface area, the polymers reinforced with nanoparticles should show vastly improved mechanical properties because the energy-dissipating flow processes (i.e., local yielding phenomena) can be initiated at much more locations than in conventional composites.

Mechanical performance of PNCs is determined by their morphology on different length scales, the later being a function of molecular properties of the polymer, filler type, and dimensionality and the nature of interface between the filler and the matrix. The mechanical properties of polymers depend much on how their morphological details on different length scales respond toward the applied load. Therefore, the knowledge of those responses (i.e., that of deformation mechanisms) plays a crucial role in developing polymers with designed property profile [131]. As in homopolymers, blends, and other heterophase systems, EM provides direct information on the deformation mechanisms occurring in PNCs. There are a growing number of scientific literatures dedicated to the structure and micromechanical properties of PNCs (e.g., [24, 53, 66, 132–140]). In particular, the microscopes equipped with *in-situ* stretching device can provide direct information on the deformation micromechanisms in heterogeneous polymers and nanocomposites [128, 132, 135]. It has been, in general, found that the micromechanisms operating in nanocomposites systems are entirely different from those observed in classical filled polymer systems. Besides the large interfacial area resulting from enormous surface area of nanofiller, the mobility of the nanofiller is found to be a key to the unusual performance of the nanocomposites [131]. This section presents an overview of deformation studies on some model nanocomposite systems.

Figure 11.19 shows a series of TEM micrographs of thin film of nanocomposite comprising PS homopolymer and MWNTs recorded during an *in-situ* deformation test [133]. When a tensile stress was applied to the nanocomposite film, a crack was found to nucleate around the notch during mechanical loading. Then the nanotubes were aligned along the tensile direction bridging the crack faces. The bridging nanotubes provided closure stresses across the crack faces, thereby reducing the stress concentration around the crack tip [133].

With increasing strain, the nanotubes undergo debonding and even fracture (note the deformation of nanotubes indicated by A and B in Figure 11.19). Once the nanotubes break or pull out of the matrix, they spring back to their original straight conformation. For instance, the nanotubes indicated by letters A and B in Figure 11.19 are significantly distorted from their axial direction, but after either breaking or pulling out of the matrix, they return to their original straight configuration confirming the high flexibility of the nanotubes. The high flexibility of MWNTs ensures the high toughness of the composite because the nanotubes contribute to crack bridging.

It has been already mentioned that the problem of agglomeration of nanofiller in the composite materials is a critical problem for the frequently reported deterioration of mechanical properties. This particular problem could be eliminated by subjecting the polymer composite solution high-speed electrostatic spinning [104–108, 112]. As a consequence, dramatic effects on the micromechanical properties of the composites, as revealed by the TEM, were observed. The observations were in line with the expectation that the nanofillers could initiate local flow processes quite intensively owing to their exceptionally large surface area. The electrospun nanofibers of PNCs deformed via nanonecking and shear yielding unlike their bulk counterparts, which showed crazing and thence brittle behavior [66].

Without application of EM, it would have been impossible to monitor *in situ* the structural changes taking place in nanocomposite fibers on mechanical loading and thereby collecting information about the yielding process. Figure 11.20 shows, for example, series

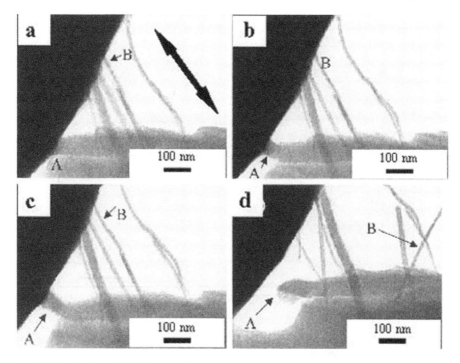

**Figure 11.19**  Series of TEM images showing the deformation of MWNT reinforced PS nanocomposite; note that very flexible MWNTs (denoted by a and b) are bent, then break or pull out of the PS matrix and spring back to their initial orientations under tensile loading (reproduced with permission from [133]).

of TEM micrographs of PC reinforced with 4 wt% MWNT after stepwise tensile deformation [66]. Whereas the bulk composite deformed via formation of craze-like deformation zones, the nanofibers of the same materials deformed via necking and drawing (Figures 11.20(b) and (c)). No local deformation zones were produced in the nanofibers. Thus, a brittle-to-ductile transition in the electrospun nanocomposite fibers, a process of great practical significance, was directly elucidated by means of the EM.

Finally, we present the deformation behavior of a nanocomposite comprising organophillic modified boehmite nanoparticles and PS, a brittle polymer at room temperature as studied by means of SEM, see Figure 11.21 [138]. This is the same sample as that presented in Figure 11.8 but with higher filler content (30 wt%). Sample films were prepared via ultrasonication of the mixture of PS and boehmite nanoparticles dissolved in chloroform followed by solution casting. The SEM imaging was performed on the fracture surface using SEs mode.

The dark oval-shaped particles embedded inside the network of polymer are the aggregates of boehmite nanoparticles. As discussed previously, the particles appear larger due to increased tendency of agglomeration at higher filler content. The most interesting feature revealed by the SEM imaging is that the matrix polymer has been converted to highly drawn filaments, an astonishing observation for an otherwise highly brittle polymer. This effect revealed by SEM is a clear evidence of the fact that the nanofiller paves way for the formation of extremely intense local yielding zones in the PNC, provided the filler is optimally distributed in polymer matrix.

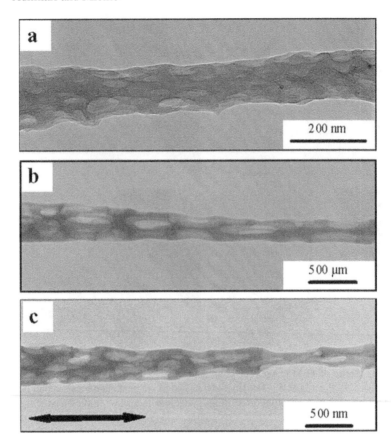

**Figure 11.20** TEM images showing the sequence of mechanical deformation processes in a PC/MWNT composite; (a) undeformed state and (b, c) deformed below and beyond the critical strain; loading direction shown by an arrow (reproduced with permission from [93]).

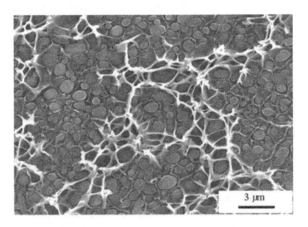

**Figure 11.21** SEM image showing the fracture surface morphology of a composite comprising polystyrene and 30 wt% boehmite nanoparticles; the fibrils pointing out of the fracture surface represent the high ductility polymer (from [138]).

In summary, electron microscopic techniques have provided firm basis for the elucidation of structure–property correlation of the PNCs. With the help of very direct information about various plastic deformation mechanisms, it has been possible to understand and tailor the mechanical properties of nanocomposites.

## 11.6. Concluding Remarks

Polymers are filled with nanometric filers in order to improve mechanical barrier and other functional properties. Generally, on optimum processing of these nanocomposites, significant improvement in their properties is obtained at a very low weight fraction of nanofiller. Different microscopic techniques, especially EM and SFM, provide very direct structural basis for the modification in the properties of the filled polymers. In this chapter, we have highlighted the application of different microscopic techniques in the elucidation of structure–property correlations in nanofilled polymers. The examples presented are the nanocomposites filled with nanofillers with varying dimensionalities, and some special artificial as well as natural composites. In general, the selection of particular microscopic technique depends on the nature of problem at hand.

Different microscopic techniques, especially EM (such as TEM and SEM) and SFM, have been employed in the investigation of PNCs. Besides the routine techniques of morphological analysis, the *in situ* investigations can deliver unparalleled local information about filler–matrix interaction, thereby revealing the mechanisms of property enhancement in the nanocomposites.

Electron microscopic techniques are used for the structural studies of the matrix polymer, the filler, and the composites comprising them. Microscopy, complemented by other integral techniques, has been, and will continue to be, the principal technique for the characterization of polymeric nanocomposites. These techniques reveal the morphology of specific sample location and are not restricted, unlike other nonmicroscopic methods, to the average values. Therefore, very accurate information on structure–property correlations of the nanocomposites can be assessed. EM offers straightforward information on such very fundamental questions of nanocomposite science and technologies as whether the filler has really been dispersed on nanoscale and how the filler has modified the matrix's structures and properties. Thus, electron microscopic techniques have become indispensable tool for the researches on science and technology of PNCs.

## Acknowledgement

We thank Alexander von Humboldt Foundation for generously supporting the research stay of RA at Martin Luther University Halle-Wittenberg (February-July 2009) through Georg-Forster Fellowship.

## References

1. Bhushan, B. (Ed.), Handbook of Nanotechnology, Springer, Berlin, 2003.
2. Bucknall, C. B. Toughened Plastics, Applied Science Publishers, London, 1977.
3. Kausch, H. H. Polymer Fracture, 2nd edition, Springer, Berlin, 1987.
4. Kinloch, A. J.; R. J. Young, Fracture Behaviour of Polymers, Applied Science, New York, 1983.
5. Vaia, R. A.; Ishii, H.; Giannelis, E. P. Chem. Mater. 1993, 5, 1694.

6. Baltá Calleja, F. J.; Giri, L.; Asano T.; Yoshida, T. Rep. Fac. Sci. Shizuoka Univ., 1994, 28, 53.
7. Ash, B. J.; Stone, J.; Rogers, D. F.; Schadler, L. S.; Siegel, R. W.; Benicewicz, B. C.; Apple, T. Mater. Res. Soc. Symp. Proc. Vol. 661 © 2001, Materials Research Society, Warrendale, PA, USA.
8. Ash, B. J.; Siegel R. W.; Schadler, L. S. J. Polym. Sci.: B: Polym. Phys. 2004, 42, 4371.
9. Adhikari, R.; Henning S.; Michler, G. H. Macromol. Symp. 2006, 233, 26.
10. Fu, B. X.; Lee A.; Haddad, T. S. Macromolecules 2004, 37, 5211.
11. Koerner, H.; Liu, W.; Alexander, M.; Mirau, P.; Dowty H.; Vaia, R. A. Polymer 2005, 46, 4405.
12. Ray, S. S.; Okamoto, M. Prog. Polym. Sci. 2003, 28, 1539.
13. Szleifera I.; Yerushalmi-Rozen, R. Polymer 2005, 46, 7803.
14. Wang, C.; Guo, Z.-X.; Fu, S.; Wu, W.; Zhu, D. Prog. Polym. Sci. 2004, 29, 1079.
15. Vaia, R. A., Wagner, H. D. Materials Today, Nov. 2004, 32.
16. Wagner, H. D., Vaia, R. A. Materials Today Nov. 2004, 38.
17. Breuer, O.; Sundararaj, U. Polym. Comp. 2004, 25, 630.
18. Jordan J.; Jacob, K. I.; Tannenbaum, R.; Sharaf, M. A.; Jsiuk, I. Mater. Sci. Eng. A 2005, 393, 1.
19. Alexandre, M. ; Dubois, P. Mater.Sci. Eng. Rep. 2000, 28, 1.
20. Moniruzzaman, M.; Winey, K. I. Macromolecules 2006, 39, 5194.
21. Mert, M.; Yilmazer, U. J. Appl. Polym. Sci. 2008, 3890.
22. Bhimaraj, P.; Yang, H.; Siegel, R. W.; Schadler, L. S. J. Appl. Polym. Sci. 2007, 106, 4233.
23. Gerber, C.; Lang, H. P. Nature Nanotech. 2006, 1(10).
24. Ma, J.; Simon, G. P.; Edward, G. H. Macromolecules 2008, 41, 409.
25. Kurian, M.; Dasgupta, A.; Galvin, M. E.; Ziegler, C. R.; Beyer, F. L. Macromolecules 2006, 39, 1864.
26. Shah, R. K.; Paul, D. R. Macromolecules 2006, 39, 3327.
27. Chung, H.-J.; Ohno, K.; Fukuda, T. Russell; Composto J. Macromolecules 2007, 40,384.
28. Wang, D. H.; Arlen, M. J.; Baek, J.-B.; Vaia, R. A.; Tan, L.-S. Macromolecules 2007, 40, 6100.
29. Lee, D. H.; Chang, J. A.; Kim, J. K. J. Mater. Chem. 2006, 16, 4575.
30. Ginzburg, V. V.; Myers, K.; Malowinski, S. ; Cieslinski, R.; Elwell, M.; Bernius, M. Macromolecules 2006, 39, 3901.
31. Usuki, A.; Hasegawa, N.; Kato, M. Adv. Polym. Sci. 2005, 179, 135.
32. Wang, Z. L. (Ed.), Characterization of Nanophase Materials, Wiley-VCH, Weinheim, 2000.
33. Michler, G. H.; Balta Callea; F. J. (Eds.), Mechanical Properties of Polymers Based on Nanostructure and Morphology, Taylor & Francis Group, Boca Raton, FL, 2005.
34. Ajayan, P. M.; Schadler, L. S.; Braun, P. V. (Eds.), Nanocomposite Science and Technlogy, Wiley-VCH, Weinheim, 2003.
35. Klabunde, C. J. (Ed.), Nanoscale Materials, John Wiley and Sons, New York, 2001.
36. Michler, G. H. (Ed.), Electron Microscopy of in Polymers, Springer, Heidelberg, 2008.
37. Michler, G. H.; Lebek, W. (Eds.), Ultramicrotomie in der Materialforschung, Carl Hanser Verlag, Munich, 2004.
38. Sawyer, L. C.; Sawyer, L. C. Polymer Microscopy, 2nd edition, Chapman and Hall, London, 1996.
39. Chen, S.; Cao, T.; Jin, Y. Polym. Commun. 1987, 28, 314.
40. Kato, K. Polym. Lett. 1966, 4, 35.
41. Vitali, R.; Montani, E. Polymer 1980, 21, 1220.
42. Trent, J. S.; Scheinbeim J. I.; Couchman, P. R. Polym. Sci. Technol. 1983, 2, 589.
43. Lammerlink, R. G. H.; Hempenius, M. A.; Van den Enk, J. E.; Chan, V. Z.-H.; Thomas, E. L.; Vancso, G. J. Adv. Mater. 2000, 12, 98.
44. Park, M.; Harrison, C.; Chaikin, P. M.; Register, R. A.; Adamson, D. H. Science 1997, 276, 1401.
45. Collins, S.; Hamley, I.W.; Mykhaylyk, T. Polymer 2003, 44, 2403.
46. Olley, R. H.; Bassett, D. C. Polymer 1982, 23, 1707.
47. Olley, R. H.; Bassett, D. C.; Hine, P. J.; Ward, I. M. J. Mater. Sci. 1993, 28, 1107.
48. Yuan, Q.; Mishra, R. D. K. Polymer 2006, 47, 4421.
49. Chan, C.-M.; Wu, J.; Li, J.-X ; Cheung, Y.-K. Polymer 2002, 43, 2981.
50. Nowacki, R.; Monasse, B.; Piorkowska, E.; Galeski, A.; Haudin, J. M. Polymer 2004, 45, 487.
51. Trujillo, M.; Arnal, M. L.; Müller, A. J.; Laredo; Bredeau, E. St.; Bonduel, D.; Dubois, Ph. Macromolecules 2007, 40, 6268–6276.
52. Zhang, Q., Lippits, D. R.; Rastogi, S. Macromolecules 2006, 39, 658.
53. Shen, L.; Gao, X.; Tong, Y.; Yeh, A.; Li, R.; Wu, D. J. Appl. Polym. Sci. 2008, 108, 2865.
54. Bassett, D. C. Mechanical Properties of Polymer Based on Nanostructure and Morphology, Michler, G. H.; Balta Calleja, F. J. (Eds.), Taylor & Francis, Bacon Raton, FL, 2005.

55. Bernal-Lara, T.; Ranade, A.; Hiltner, A.; Baer, E. Mechanical Properties of Polymer Based on Nanostructure and Morphology, Michler, G. H.; Balta Calleja, F. J. (Eds.), Taylor & Francis, Bacon Raton, FL, 2005.

56. Adhikari, R; Lebek, W.; Godehardt, R.; Henning, S.; Michler, G. H.; Baer, E.; Hiltner, A. Polym. Adv. Technol. 2005, 16, 95.

57. Baer, E.; Hiltner, A.; Keith, H. D. Science 1987, 235, 1015.

58. Puente Orench, I.; Ania, F.; Baer, E.; Hiltner, A.; Bernal, T.; Balta Calleja, F. J.; Phil. Mag. 2004, 84, 1841.

59. Adhikari, R.; Damm, C.; Michler, G.H.; Münstedt, H.; Balta Callea, F.J. Comp. Interfaces 2008, 15, 453.

60. Adhikari, R.; Henning, S.; Lebek, W.; Godehardt, R.; Illish, S.; Michler, G. H. Macromol. Symp. 2006, 231, 116.

61. Schulte, K.; Nolte, M. C.; Mechanical Properties of Polymer Based on Nanostructure and Morphology', Michler, G. H.; Balta Calleja, F. J. (Eds.), Taylor & Francis, Bacon Raton, FL, 2005.

62. Ania, F.; Broza, G.; Mina, M. F.; Schulte, K.; Roslaniec, Z.; Balta-Calleja, F. J. Comp. Interfaces 2006, 13, 33.

63. Musumeci, A. W.; Silva, G. G.; Liu, J.-W.; Martens, W. N.; Waclawik, E. R. Polymer 2007, 48, 1667.

64. Lo, C.-T.; Lee, B.; Pol, V. G.; Dietz Rago, N. L.; Seifert, S.; Winans, R. E.; Thiyagarajan, P. Macromolecules 2007, 40, 8302.

65. Iijima, S. Nature 1991, 354, 56.

66. Kim, G. M.; Lach, R.; Michler, G. H.; Pötschke, P.; Albrecht, K. Nanotechnology 2006, 17, 963.

67. McNally, T.; Pötschke, P.; Halley, P.; Murphy, M.; Martin, D.; Bell, S. E. J.; Brenman, G. P.; Bein, D.; Lemoine, P.; Quinn, J. P. Polymer 2005, 46, 8222.

68. Zou, Y.; Feng, Y.; Wang, L.; Liu, X. Carbon 2004, 42, 271.

69. Pötschke, P.; Bhattacharyya, A. R.; Janke, A. Polymer 2003, 44, 8061.

70. Ciprari, D.; Jacob, K.; Tannenbaum, R. Macromolecules 2006, 39, 6565–6573.

71. Tong, X.; Liu, C.; Cheng, H.-M.; Zhao, H.; Yang, F.; Zhang, X. J. Appl. Polym. Sci. 2004, 92, 3697.

72. Kaminsky, W.; Funck, A.; Wiemann, K. Macromol. Symp. 2006, 239, 1.

73. Sohn, B. H.; Seo, B. H. Chem. Mater. 2001, 13, 1752.

74. Spatz, J. P.; Moessmer, S.; Hartmann, C.; Moeller, M. Langmuir 2000, 16, 407.

75. Hasegawa, H.; Hashimoto, H. in Holden G, Legge, N. R., Quirk R. P., Schroeder H. E. (Eds.) Thermoplastic Elastomers, 2nd Edition. Hanser Publishers, Munich 1998.

76. Davis, K. A.; Matyjaszewski, K. Adv. Polym. Sci., 2002, 159, 107.

77. Hamley, I. W. Developments in Block Copolymer Science and technology. Wiley InterScience, New York, 2004.

78. Ruzette, A.-V.; Leibler, L. Nature Mater. 2005, 4, 19.

79. Abetz, V.; Simon, P. F. W. Adv. Polym. Sci. 2005, 189, 125.

80. Simon, P.; Huhle, R.; Lehmann, M.; Lichte, H.; Mönter, D.; Bieber, T.; Reschetilowski, W.; Adhikari, R.; Michler G. H. Chem. Mater. 2002, 14, 1505.

81. Kloninger, C.; Knecht, D.; Rehahn, M. Polymer 2004, 45, 8323.

82. Jain, A.; Gutmann, J. S.; Garcia, C. B. W., Zhang, Y.; Tate, M. W.; Gruner, S. M.; Wiesner, U. Macromolecules 2002, 35, 4862.

83. Simon, P. F. W.; Ulrich, R.; Spiess, H. W.; Wiesner, U. Chem. Mater. 2001, 13, 3464.

84. Lopes, W. A.; Jaeger, H. M. Nature 2001, 414, 735.

85. Hodjipanayis, G. C.; Siegel, R. W. (Eds.), Nanophase Materials: Synthesis, Properties and Applications, Kluwer, Dordrecht, 1994.

86. Hillmyer, M. A. Adv. Polym. Sci. 2005, 190, 137.

87. Liu, G.; Ding, J.; Guo, A.; Herfort, M.; Bazzett-Jones, D. Macromolecules 1997, 30, 1851.

88. Carastan, D. J. ; Demarquette, N. R. Macromol. Symp. 2006, 233, 152.

89. Ha, Y.-H.; Kwon, Y.; Breiner, T.; Chan, E. P.; Tzianetopoulou, T.; Cohen, R. E.; Boyce, M. C.; Thomas, E. L. Macromolecules 2005, 38, 5170

90. Baer, E.; Kerns, J.; Hiltner, A. in: 'Structure Development During Polymer Processing', Cunha, A. M.; Fakirov, S. (Eds.), Kluwer, Dordrecht, 2000.

91. Ivankova, E.; Krumova, M.; Michler, G. H.; Koets, P. Colloid Polym. Sci. 2004, 282, 203.

92. Liu, R. Y. F.; Bernal-Lara, T. E.; Hiltner, A.; Baer, E. Macromolecules 2004, 37, 6972.

93. Ward, I.; Hine, P. J. Polymer 2004, 45, 1423.

94. Jordan, N. D.; Bassett, D. C.; Olley, R. H.; Hine, P. J.; Ward, I. M. Polymer 2003, 44, 1133.

95. Bjekovic, R. E. Monocomposite Schichtwerkstoffe auf Basis von Polypropylen, VDI Verlag GmbH, Düsseldorf, 2003.

96. Vincent, M.; Agassant, J. F. in 'Two-Phase Polymer Systems', L. A. Utracki (Ed.), Hanser Publishers, Munich, 1991.

97. Ivankova, E.; Michler, G. H.; Hiltner, A.; Baer, E. Macromol. Mater. Eng. 2004, 289, 787

98. Ward, I. M. Mater. World 1998, 6, 608.

99. Ward, I. M.; Hine P. J. Mechanical Properties of Polymers based on Nanostructure and Morphology, Michler, G. H.; Balta Calleja, F. J. (Eds.), Taylor & Francis, Bacon Raton, FL, 2005.

100. Hine, P. J.; Ward, I. M.; Jordan, N. D.; Olley, R. H.; Bassett, D. C. Polymer 2003, 44, 111.

101. Olley, R. H.; Bassett, D. C.; Hine, P. J.; Ward, I. M. J. Mater. Sci. 1993, 28, 1107.

102. Rassburn, J.; Hine, P. J.; Ward, I. M.; Olley, R. H.; Bassett, D. C.; Kabeel, M. A. J. Mater. Sci. 1995, 30, 615.

103. Li, D.; Xia, Y. Adv. Mater, 2004, 16, 1151.

104. Chronakis, I. S. J. Mater. Process. Technol. 2005, 167, 283.

105. Tan, S. T.; Wendorff, J. H.; Pietzonka, C.; Jia, Z. H.; Wang, G. Q. ChemPhysChem 2005, 6, 1461.

106. Hou, H.; Reneker, D. H. Adv. Mater. 2004, 16, 69.

107. Sun, Z.; Zussman, E.; Yarin, S. A. L.; Wendorff, J. H.; Greiner, A. Adv. Mater. 2003, 15, 1929.

108. Kim, G. M.; Lach, R.; Michler, G. H.; Chang, Y. W. Macromol. Rapid Commun. 2005, 26 728.

109. McCullen, S. D. Stevens, D. R. ; Roberts, W. A.; Ojha, S. S.; Clarke, L. I.; Gorga R. E. Macromolecules 2007, 40, 997.

110. Bashouti, M.; Salalha, W.; Brumer, M.; Zussman, E.; Lifshitz, E. ChemPhysChem 2006, 7, 102.

111. Wang, T.; Kumar, S. J. Appl. Polym. Sci. 2006, 102, 1023.

112. Fong, H.; Liu, W.; Wang, C. S.; Vaia, R. A. Polymer 2002, 43, 775.

113. Hou, H.; Ge, J. J.; Zheng, J.; Li, Q.; Reneker, D. H.; Greiner, A.; Cheung, S. Z. D. Chem. Mater. 2005, 17, 967.

114. Wang, M.; Singh, H.; Hatton, T. A.; Rutledge, G. C. Polymer 2004, 45, 5505.

115. Fritz, M.; Belcher, A. M.; Radmacher, M.; Walters, D. A.; Hansma, P. K.; Stucky, G. D.; Morse, D. E.; Mann, S. Nature 1994, 371, 49.

116. Weiner, S.; Wagner, H. D. Annu. Rev. Mater. Sci. 1998, 28, 271.

117. Nalla, R. K.; Kinney, J. H.; Ritchie, R. O. Nat. Mater. 2003, 2, 164

118. DiMasi, E.; Sarikaya, M. J. Mater. Res. 2004, 19, 1471.

119. Lin, A.; Meyers, M. A. Mater. Sci. Eng. A 2004, 390, 27.

120. Hobbs, J. K.; Humphris, A. D. L.; Miles, M. J. Macromolecules 2001, 34, 5508.

121. Pearce, R.; Vancso, G. J. Polymer 1998, 39, 1237.

122. Nishino, T.; Nozawa, A.; Kotera, M.; Nakamae, K. Rev. Sci. Instrum. 2000, 71, 2093.

123. Godehardt, R.; Rudolf, S.; Lebek, W.; Goerlitz, S.; Adhikari, R.; Allert, E.; Giesemann, J.; Michler, G. H. J. Macromol. Sci. Phys. 1999, 38, 817.

124. Magonov, S. N. in: 'Encyclopedia of Analytical Chemistry', Mayers, R. A. (Ed.), John Wiley & Sons, Chichester, 2000.

125. Godehardt, R.; Lebek, W.; Adhikari, R.; Rosenthal, M.; Michler, G. H. Eur. Polym. J. 2004, 40, 917.

126. Tranchida, D.; Piccarolo, S. M. Soliman, Macromolecules, 2006, 39, 4547.

127. Magonov, S. N.; Reneker, D.H. Annu. Rev. Mater. Sci. 1997, 27, 175.

128. Bamberg, E.; Grippo, C. P.; Wanakamol, P.; Slocum, A. H.; Boyce, M. C.; Thomas, E. L. Prec. Eng. 2006, 30, 71.

129. Fujita, M.; Iwata, T.; Doi, Y. Polym. Degrad. Stab. 2003, 81, 131.

130. Radusch, H. J. Conference Papers of 12th Problem Seminar Polymer Blends: Nanoblends-Bad Lauchstädt, March 28–29, 2007, ISBN 978-3-86010-897-0.

131. Michler, G. H. Polym. Adv. Technol. 1998, 9, 812.

132. Qian, D.; Dickey, E. C. J. Microscopy, 2001, 204, 39.

133. Gersappe, D. Phys. Rev. Lett. 2002, 89, 058301 (1–4).

134. Kim, G. M.; Lee, D. H.; Hofmann, B.; Kressler, J.; Stöpelmann, G. Polymer 2001, 42, 1095.

135. Ding, W.; Eitan, A.; Fisher, F. T.; Chen, X.; Dikin, D. A.; Andrews, R.; Brinson, L. C.; Schadler, L. S.; Ruoff, R. S. Nanolett. 2003, 3, 1593.

136. Jancer, J.; Kucera, J. Polym. Eng. Sci. 1990, 30, 707.

137. Qian, D.; Dickey, E. C.; Andrews, R.; Rantell, T. Appl. Phys. Lett. 2000, 76, 2868.

138. Michler, G. H.; Kausch, H. H. J. Appl. Polym. Sci. 2008, 105, 2577.

139. He, C.; Liu, T.; Tjiu,W. C.; Sue, H.-J.; Yee, A. F. Macromolecules 2008, 41, 193.

140. Boo, W.-J.; Sun, L.; Warren, G. L.; Moghbelli, E.; Pham, H.; Clearfield, A.; Sue, H.-J.; Polymer 2007, 48, 1075.

# 12 Characterization of Polymer Nanocomposites by Free-Volume Measurements

*C. Ranganathaiah*

Department of Studies in Physics, University of Mysore, Manasagangotri, Mysore 570 006, India
cr@physics.uni-mysore.ac.in

**Abstract**

This chapter is designed to introduce the concept of free volume in polymer-based nanocomposites and how this internal material parameter influences the properties of the nanocomposites. The nanometer-sized free-volume holes or cells can be directly and accurately measured by the versatile experimental technique namely positron annihilation spectroscopy (PAS). Briefly the three methods of PAS measurements are described with an emphasis on positron lifetime measurement. Although only few investigations using PAS have been reported in this new exciting field so far, a review of these works is made and a focus on the future work that could be carried out employing the techniques of PAS to nanocomposites to understand their unusual and exciting properties is provided. Hence, this review offers a comprehensive discussion on the new characterization tool namely PAS in the field of polymer nanocomposites.

**Keywords:** nanocomposites, *ortho*-positronium lifetime, free volume, gas transport, permeability, filler size, surface area

## 12.1. Introduction

Nanostructured materials research has been very exciting in the last two decades or so. The impact of these researches has been tremendous and still growing at a fast rate in fundamental science and potential scientific and industrial applications. One can quote many exciting examples of nanomaterials such as colloidal nanocrystal, fullerenes like C60, C70, carbon nanotube, semiconductor nanowire, etc. The field is rapidly evolving and is now intricately interfacing many different scientific disciplines, from chemists to physicists, to materials scientists to engineers and to biologists [1]. The amount of literature published on this subject has been tremendous and still increasing significantly each year. The research on nanostructured materials is highly interdisciplinary because of different synthetic methodologies involved as well as many different physical characterization methods used. Undoubtedly, chemists are playing a vital role since the synthesis of these materials is certainly about how to assemble atoms or molecules into nanostructures of desired coordination environment, size, and shape. Nevertheless, physicists and engineers are also not lagging behind and they are also into development of these materials.

Nanocomposites comprising polymers and isotropic inorganic particles with diameter up to 100 nm are another set of nanostructured systems with material properties, which substantially differ from those of respective composites with larger particles [2]. Polymers containing nanoparticles can exhibit exciting light scattering behavior. It is to be considered that the physical properties of nanocomposites depend on the size of the incorporated particle, like the electrical conductivity, the color, the refractive index of metallic or semi conducting nanoparticles, which can vary with particle size and markedly differ from those values of the bulk materials [3–5].

The high-surface free energy of the inorganic particles renders them possible to fuse thermodynamically to favorable state of a single sphere surrounded by polymer. Therefore, the formation of nanocomposites results in randomly dispersed inorganic particles. Secondly, nanoparticles have a large surface area to volume ratio and their properties depend on the structure on the surface as well. An improvement in a property usually arises when the length scale of the morphology (i.e., nano) and fundamental physics associated with a property coincide. As stated earlier, two principal factors that influence the properties of nanomaterials significantly different from other materials are increased relative surface area and the quantum effects [6]. In the case of particles and fibers, the surface area per unit volume is inversely proportional to the material's diameter; thus, the smaller the diameter, greater is the surface area per unit volume [7]. Common particle geometries and their respective surface area to volume ratios are shown in Figure 12.1.

For the fiber and layered material, the surface area/volume is dominated, especially for nanomaterials, by the first term in the equations given in Figure 12.1. The second term (2/l and 4/l) has a very small influence (and is often neglected) compared to the first term. Depending on the nature of the components used (layered silicate or nanofiber, cation exchange capacity, and polymer matrix) and the method of preparation, significant differences in composite properties may be obtained [8]. As an example, Figure 12.2

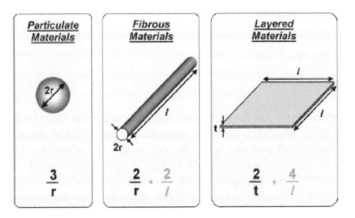

**Figure 12.1** Particle geometries and their respective surface area to volume ratios [7] (reprinted with permission from Elsevier).

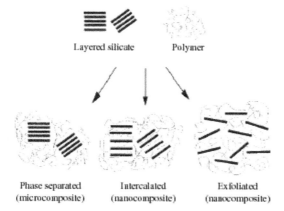

Layered silicate    Polymer

Phase separated    Intercalated    Exfoliated
(microcomposite)    (nanocomposite)    (nanocomposite)

**Figure 12.2**    Scheme of three main types of layered silicates in polymer matrix [9] (reprinted with permission from Elsevier).

shows three main types of composites for layered silicate materials. When the polymer is unable to intercalate (or penetrate) between the silicate sheets, a phase-separated composite is obtained, and the properties stay in the same range as those for traditional micro composites [9]. In an intercalated structure, where a single extended polymer chain can penetrate between the silicate layers, a well-ordered multilayer morphology results with alternating polymeric and inorganic layers. An exfoliated or delaminated structure results when silicate layers are uniformly and completely dispersed in a continuous polymer matrix [9].

Properties that could be controlled in this way include optical absorption, ferromagnetism, hydrogen storage capacity, and mechanical stretching. Some nanocomposites may show properties dominated by the interfacial interactions and others may exhibit the quantum effects associated with nano-dimensional structures [10]. However, the mobility of the particles is usually negligible once the nanocomposites are formed and, therefore, phase separation is generally suppressed in the final products over extended periods.

## 12.2.  Polymeric Nanocomposites

### 12.2.1.  Preparation of Nanocomposites

As mentioned above, the specific surface energies of inorganic materials are high, thus rendering the primary particles to interact strongly with each other in isolated powders. The resulting agglomerates usually do not break into individual primary particles during nanocomposite fabrication. As such, rarely bare nanoparticles have been successfully used for the preparation of nanocomposites with primary particles that are well dispersed in the polymer matrix. We can cite an example of silicon particles of 20–40-nm diameter prepared by high-energy milling of silicon powders, which were subsequently used in the fabrication of nanocomposites by spin coating [11]. Another method considered favorable for the employment of nanoparticles is the use of colloids which are coated with a layer of organic molecules strongly bound to the particle surfaces. Such particles can often be isolated as viscous substances or solids, which readily disperse as individual primary particles in water or some organic solvents. In these cases, agglomeration is suppressed by the surface layer diminishing markedly the specific surface energy and thus decreasing the attraction tendency between the particles. In the past, a number of surface-modified colloids have been isolated, for example, gold or silver nanoparticles with phosphines or thiols attached to their surfaces [12–15] or CdS nanoparticles covered with aromatic thiols [16–18]. Corresponding colloids can be mixed with polymers in solution or in the melt and from this state further processed to nanocomposites. However, nanocomposites have been prepared most commonly by synthesis of the inorganic particles *in situ*, for instance in solution including micellar systems [17]. (a) Up to now, *in-situ* formation of particles in a liquid medium is probably the most widely used method for the preparation of polymer nanocomposites containing isotropic inorganic particles. (b) Formation of the particles in a solid polymer matrix in which inorganic particles can be prepared *in situ* in solid polymer matrices, e.g., by thermal decomposition of incorporated precursors, reaction of incorporated compounds with gaseous species, or when polymer films containing an incorporated precursor are immersed in liquids containing the reactive species required for the formation of the desired colloids. Other chapters in this book describe various other methods employed in the preparation of polymer nanocomposites in greater details and we will not cover much of that here.

## 12.3.  Characterization Techniques for Nanocomposites

Characterization techniques are crucial to comprehend the basic physical and chemical properties of polymer nanocomposites. For applications, it facilitates the study of emerging materials by providing information on some intrinsic properties [19]. Various techniques have been used for characterization in polymer nanocomposites research [19]. The commonly used powerful techniques are wide-angle X-ray diffraction (WAXD), small-angle X-ray scattering (SAXS), scanning electron microscopy (SEM), and transmission electron microscopy (TEM) [9, 20]. The SEM provides images of surface features associated with a sample. However, there are two other microscopies, namely scanning probe microscopy (SPM) and scanning tunneling microscopy (STM), which are indispensable in nanotube research [19]. The SPM uses the interaction between a sharp tip and a surface to obtain an image. In STM, a sharp conducting tip is held sufficiently close to a surface (typically about 0.5 nm) such that electrons can "tunnel" across the gap [6]. These methods provide surface structure and electronic information at atomic level. The invention of the STM inspired the development of other "scanning-probe" microscopes, such as the atomic force microscope (AFM) [6]. The AFM uses a sharp tip to scan across the sample. Raman spectroscopy has also been employed and proved to be a useful probe of carbon-based material properties [19]. However, positron annihilation spectroscopy (PAS), which has been established as a powerful defect spectroscopy tool in metals, ionic materials, polymers, blends, and zeolites because of its sensitivity and nondestructive nature, has not been fully utilized in this field of research and only recently gaining importance in nanomaterials characterization. For the benefit of the scientific community in nanocomposites, an overview of PAS and its applications to condensed matter studies and materials research is presented. The overview is aimed as an introduction for potential users of PAS in these areas. Detailed treatment of PAS methods can be found in several books, review articles, and in conference proceedings [21–26].

The positron annihilation techniques have been extensively used as probes to study defect structure in metals and ceramics and also been used to characterize the physical properties of polymers, polymer blends, and recently nanocomposites. Early investigations using positron annihilation focused on macromolecular systems as ideal amorphous materials and included polymer systems like poly(ethylene), poly(tetrafluoroethylene), poly(styrene), poly(methyl-methacrylate). Polymers are far from being simple or ideal amorphous materials and positron annihilation measurements were found to be sensitive to the physical properties that make polymers unique materials with respect to glass transition and subglass transitions. In the last 28 years or so, studies on polymers and polymer-related materials using positron methods have dramatically increased in number and quality, and our knowledge of the annihilation processes and the way in which they are influenced by the nature of the surrounding molecular system has advanced significantly. It is, therefore, now felt desirable to introduce this technique to the new exciting field of nanocoposites and ask the question: "What do positron annihilation studies tell us about the physical properties of polymer based nanocoposites?"

## 12.4. Positron Annihilation Spectroscopy

In this section, we present the basic principles of positron annihilation and describe the experimental techniques of PAS.

### 12.4.1. Positron Annihilation Process

The positron $(e^+)$ is the antiparticle of electron $(e^-)$, and hence a positron–electron pair is unstable and will annihilate. The annihilation of positrons in collisions with electrons is accompanied by the emission of one, two, or more photons (gamma-quanta). Single-photon emission by electron–positron annihilation is possible only in the presence of a third body (a nucleus or electrons) that carries away the recoil momentum whose probability is highly negligible. When a free positron and a free electron annihilate, at least two photons are created. The positron annihilation cross-section rapidly decreases with an increase in the number of emitted photons. When the number of created photons increases by one, the annihilation cross-section is multiplied by the fine-structure constant $\alpha$ (1/137), i.e., the probability of the annihilation process decreases more than by two orders of magnitude [23, 27–32].

The cross-section of two-photon annihilation of a free positron and a free electron was calculated by P Dirac. In the nonrelativistic approximation, this cross-section increases with decreasing relative velocity $v$ of the colliding particles:

$$\sigma = \frac{\pi r_o^2}{v/c}, \quad v \ll c, \tag{12.1}$$

where $r_0$ is the classical electron radius and $c$ the velocity of light in vacuum. Consequently, for a positron embedded in a sea of "cold" electrons with a density $n_e$, one obtains the decay rate

$$\lambda = \pi r_0^2 n_e c. \tag{12.2}$$

Here $n_e$ is the electron density at the site of positron annihilation. By measuring the annihilation rate $\lambda$, the inverse of which is the mean lifetime $\tau$, one directly obtains the electron density $n_e$ encountered by the positron. Thus a positron can serve as a test particle for the electron density of the medium. However, because of the opposite charges,

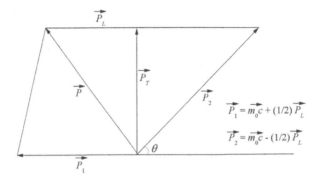

**Figure 12.3**  The vector diagram of the momentum conservation in the 2$\gamma$-annihilation process. The momentum of the annihilation pair is denoted by $P$, subscripts $L$ and $T$ refer to longitudinal and transverse components, respectively.

a strong Coulomb attraction exists between the positron and electrons of the medium. Consequently, the electron density $n_e$ is enhanced from the equilibrium value in matter due to the Coulomb screening of the positron. Measurement of positron lifetime in a given medium constitutes the lifetime spectroscopy (PALS).

The kinetic energy of the annihilating pair is typically a few electron volts. In their center-of-mass frame, each photon energy is exactly $m_0c^2 = 511$ keV and the two photons go strictly into the opposite directions, i.e., emission direction is collinear. Because of the nonzero momentum of the pair mainly due to the finite momentum of the electron, the photons deviate from collinearity in the laboratory frame. As shown in Figure 12.3, the momentum conservation yields a result

$$\theta \cong \frac{P_T}{m_0c}, \tag{12.3}$$

where $180° - \theta$ is the angle between the two photons in the laboratory frame and $P_T$ is the momentum component of the electron–positron pair transverse to the photon emission direction. Usually $\theta$ is very small ($\theta < 1°$), and hence the validity of Eq. (12.3). Since the momentum of the thermalized positrons is almost zero, the measured angular correlation curve describes the momentum distribution of the electrons of the medium with which the positrons annihilated. This constitutes the angular correlation of the annihilated photons (ACAR) experiment.

The motion of the electron–positron pair also causes a Doppler shift in the energy of the annihilation photons measured in the laboratory system. The frequency shift is $\Delta v/v = v_L/c$, where $v_L$ is velocity of the pair and hence the change in energy of the photon is given by

$$\Delta E = \frac{v_L}{c}; \quad E = \frac{cP_L}{2}, \tag{12.4}$$

where $P_L$ is longitudinal momentum component of the pair. Thus, in the case of two-photon annihilation, measuring the deviation angle $\theta$ of the photons from $180°$ or the Doppler shift $\Delta E$ of the annihilation photon (0.511 MeV) makes it possible, in principle, to determine the momentum of the electron–positron pair in the laboratory frame of reference. In the laboratory, this change in energy is very small and is usually described in terms of a line-shape parameter of the annihilation radiation. This experiment constitutes the third method of positron annihilation namely Doppler broadening of annihilation photons (DBAR) measurement.

## 12.4.2. Positronium

The positron–electron pair can also form a quasistationary state called positronium (Ps) atom. Apart from the effects due to annihilation, Ps is an analog of the hydrogen atom. Positronium, therefore, is found either in the *para* state ($S = 0$, spins antiparallel) or in the triply degenerate *ortho* state ($S = 1$, spins parallel). The cross-section and the nature of annihilation depend on the mutual orientation of the spins of the particles participating in the annihilation process. The probabilities of the spontaneous annihilation of $p$-Ps and $o$-Ps atoms are also different. Owing to conservation of spin angular momentum, the *para* positronium ($p$-Ps) will decay by two photon emission with a rate of 1/(125 ps), and the *ortho* positronium ($o$-Ps) will decay by three photon emission at a rate of 1/(140 ns) in free space. Positronium is generally not formed in metals due to the fact that positron sees a sea of free electrons and the probability of it forming Ps is very small, but it is formed in metal oxides, molecular solids, liquids, and gases.

The most commonly used experimental methods for observing the positron annihilation in matter are based on measuring the positron mean lifetime (PLS), observing the angular correlation of the annihilation photons (ACAR), and determining the line-shape parameter of the annihilation photon (DBAR). Other methods also exist, such as the combination of measurements of the positron lifetime with that of the angular correlation of the annihilation photons or DB (the combined method), etc. Here we shall describe the basic methods of positron annihilation [32] namely the three methods mentioned above.

When energetic positrons from a radioactive source such as $^{22}$Na born with a kinetic energy of 540 keV are injected into a medium, they first slow down to thermal energies in a very short time of the order 1 ps ($1 \times 10^{-12}$ s). The mean implantation range usually varies from 10 to 1000 μm depending upon the type of the medium that guarantees [23] that positrons reach the bulk of the sample material very quickly. Finally, after living in thermal equilibrium, the positron annihilates with an electron from the surrounding medium dominantly into two 511-keV gamma quanta. The average lifetime of positrons is the characteristic of each material and varies from 100 to 500 ps.

Figure 12.4 shows schematically the positron annihilation experiment, where the most commonly used radioisotope $^{22}$Na is employed. Within a few picoseconds after the

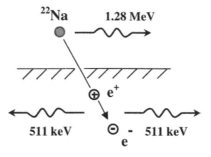

**Figure 12.4** Schematic representation of the annihilation of the positrons from $^{22}$Na source resulting into two annihilation photons of energy 511 keV in a medium.

positron emission, the nucleus emits an energetic 1.28 MeV photon, which serves as a birth signal for the positron. The lifetime of the positron can thus be measured as the time delay between the birth and annihilation gamma photons. As described above, the momentum of the annihilating electron–positron pair is transferred to the annihilation quanta and it can be detected as small-angle deviation from collinearity between the 511 keV annihilation photons. The motion of the pair also produces a Doppler shift in the energy of the annihilation photon and this is measured accurately using a high-resolution solid-state detector like Ge(Li) or HpGe.

### 12.4.3. Positrons in Defects and Positronium in Free Volume

In metals containing defects, the positron may localize in the defect. If the positron is trapped in a defect, it exhibits a characteristic lifetime that is different from annihilation with a free electron. Positron annihilation has been intensively studied in metals [24, 26, 33–38], and the reason for this is that the positron annihilation method makes it possible to determine such important characteristics of metals as the electron momentum distribution, the Fermi energy $\varepsilon_F$ (usually measured in eV), the number of free electrons $Z_C$ per metal atom, and the concentration $n_E$ (in cm$^3$) of such electrons in the conduction band. These characteristics, as is well known, largely determine the mechanical, electrical, and magnetic properties of metals [32]. Thus, it is proved that the positron methods are defect-sensitive and nondestructive in nature as the information is carried away by the annihilation photons and material after the test can be reused.

The Ps atom, like the positron, also tends to localize in larger defects or cavities called free-volume holes. In vacuum, $o$-Ps decays through three-photon emission. In condensed matter, however, the Ps atom undergoes many collisions during its lifetime. Surrounding electrons of the condensed media compete with the original electron for the positronium. This process, referred to as pick-off, has a rate $\sim 1/(125$ ps$)$, which is substantially faster than the vacuum decay rate of $o$-Ps (1/140 ns). The pick-off process results predominantly in the emission of two photons, as does the annihilation of the $p$-Ps. The annihilation of $o$-Ps reflects the electron momenta as in the case of the positron, and the annihilation of the $p$-Ps mainly reflects the momentum of the Ps atom itself. The positron lifetime in matter depends on the number density of electrons in the positron annihilation region. In the presence of free volume (voids), the positronium lifetime is determined by the void size [34] in which it localizes before annihilation. Therefore, the application of positron annihilation to the study of defects in solids is based both on the ability of the positron to become trapped and on the electron density in the traps [24, 26, 33–38].

### 12.4.4. Ore Model of Positronium Formation

The probability of formation of Ps in a given medium depends on the energy of the electron lying within an energy gap where no other electronic energy transfer process is possible. To capture an electron from a molecule with ionization energy $E_i$, the kinetic energy $E$ of the $e^+$ must be greater than $E_i - E_{Ps}$, where $E_{Ps}$ is the binding energy of Ps. In vacuum, $E_{Ps}$ is 6.8 eV but may be smaller in the medium. When $E > E_i$, the Ps atom is formed with a kinetic energy greater than its binding energy and will immediately breakup. Inelastic collisions will compete with Ps formation until the $e^+$ kinetic energy is less than $E_{ex}$, the lowest electronic excitation energy. Thus the probability of Ps formation is highest in the range $E_i - E_{Ps} < E < E_{ex}$, called the Ore gap [39] and the Ps yield can be calculated from the size of this gap. Mogensen [39] proposed the spur model of Ps formation. The basic premise of this model is that when positron loses its last few hundred electron volts of kinetic energy, it creates a track, or the so-called spur, in which it resides along with the atoms, molecules, ions, and electrons. The electrostatic attraction between positron and electrons in the spur can result in Ps formation. The Ps formation will compete with other processes such as ion–electron recombination, diffusion of electrons out of the spur, and annihilation of electrons with positron. This model indicates a correlation between the Ps formation probability and the properties of electron spurs studied in radiation chemistry. The main difference between these two theories is that in the spur model, Ps is formed only when the $e^+$ is thermalized, while in the ore model, Ps is formed during thermalization.

## 12.4.5. Free-Volume Model and Positronium Formation

The free volume in an amorphous material may be considered to be a cluster of defects, which are pinned in space but are of sufficient size to allow motion into the defect of neighboring molecular entities [4]. In a small molecule gas, the species diffusing into the defect may be a molecule or in the case of macromolecular system the moving entity may be a small number of backbone units. A simplified quantitative treatment of the free-volume model [40, 41], which assumes that Ps is always localized in a low-density region that is free-volume hole or cavity in the material, relates the change of the $o$-Ps lifetime in such a "hole" or "cavity" to the change of the total volume of the material [42–51] and takes into account the factors influencing the Ps yield. An alternative approach has been proposed based on the ore gap model [39], which assumes that Ps is formed only if a free volume larger than a certain critical value is present in the matrix. This model has been refined to relate both the $o$-Ps yield and pick-off lifetime to the free volume in molecular materials and in this way one obtains a theoretical relationship between these two quantities [42, 43]. A slightly different approach [44] has been developed to explain the lifetime variation with pressure, the free energy–surface tension of the cavity being used to connect the void size with the Ps annihilation parameters. In high electron density materials, such as metals, Ps cannot be formed unless defects of size greater than the critical size are present. In polymers, generally the Ps lifetimes varies in the range 1 to 5 ns.

Many theories have been proposed to describe the properties of microvoids. Values of microvoids or free volume have been calculated using the group contribution method [45] or estimation of the occupied volume (van der Waals) [46]. Also, the size of the free-volume holes can be calculated using the semiempirical relation (12.7).

## 12.5. Experimental Techniques of PAS

### 12.5.1. Positron Annihilation Lifetime Spectroscopy (PALS)

The principle of the positron lifetime measurement with $^{22}$Na as positron source is given in Figure 12.5. The positron is implanted into the sample almost simultaneously with the birth gamma ray of 1278 keV. Lifetimes of individual positrons, $t$, can be measured as time differences between emission of the birth gamma quantum and one of the annihilation photons. Positron lifetime measurement then consists of measuring the spectrum of delayed coincidences between gamma (1278 keV photon) and annihilation gamma (511 keV photon). In Figure 12.5, the simplest configuration of a positron lifetime spectrometer, referred to as the fast–fast coincidence system, is shown. Detectors D1 and D2 (BaF$_2$ or plastic

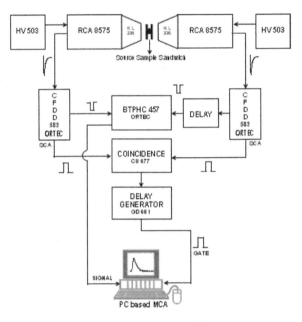

**Figure 12.5**   Block diagram of positron lifetime spectrometer.

scintillators) detect the birth and annihilation photons, respectively. Fast signals related to time of emission of corresponding gamma quanta are generated by constant-fraction differential discriminators (CFDDs). These signals are fed to the start and stop inputs of a time-to-pulse height converter (TPHC). Energy selection is also provided by the CFDDs. The fast coincidence circuit (CU677) produces the gate signal for the TPHC provided that the coincidence event of photons with proper energies occurred simultaneously. The TPHC output signal amplitude, which is proportional to time delay $t$ between the birth and the annihilation photons, is digitalized by an analog-to-digital converter (ADC) and the event is then stored at corresponding address of a histogram memory in a multichannel analyzer [52]. Thus the spectrum of positron lifetimes is obtained as the histogram of counts $N(t)$. Spectrum of positron lifetimes is a sum of exponentials of positron lifetimes in a medium. In practice only up to three lifetime components can typically be resolved in molecular substances like polymers and blends, whereas in zeolites it can be up to five. Standard computer programs are available for this, like PATFIT [53–56] used by many positron workers world over. Experimentally obtained spectra differ from the analytical description given in Eq. (12.5) by convolution with time resolution function, which is the response of the spectrometer to prompt coincidences. The resolution function resembles Gaussian-like shape and is characterized by the full width at half maximum (FWHM). The FWHM and count rate are crucial characteristics of the spectrometer, which determine the quality of experimental data. The time resolution may be improved using state of the art electronics and fast timing detectors with precise energy selection of coincidence events.

The decay curve of positrons obtained experimentally can be mathematically represented as

$$y(t) = N_s(t) \sum_{i=1}^{n} I(t)\, \alpha_i \lambda_i \exp[-\lambda_i t], \qquad (12.5)$$

where $N_s(t)$ is the total number of counts, $\alpha_i$ is the fraction of positrons annihilating with annihilation rates $\lambda_i$, and $I(t)$ is the instrumental resolution, which can be obtained using standard computer procedures [53–56]. Number of lifetime components $n$ is equal to the number of different states from which positrons annihilate. In practice, the positron lifetime spectrum of Eq. (12.5) is convoluted with a Gaussian resolution function (sometimes a combination of two or three Gaussians) whose FWHM ranges from 180 to 280 ps depending of the system configuration. About 8–10% of positrons annihilate in the source material ($^{22}$Na and the source backing) and proper "source corrections" shall be made. Constant background counts have to be corrected before the spectrum is convoluted [57, 58].

In homogeneous crystalline materials, all the positrons see the same environment and have the same annihilation probability per unit time. Positrons have a deBroglie wavelength of the order of 10 Å at room temperature, which is considerably greater than the lattice spacing in crystalline materials. Consequently, only one mode of annihilation is involved in such cases, and the resulting lifetime distribution is usually a pure exponential decay curve of the type

$$N(t) = N_0 + e^{-\lambda_1 t}. \tag{12.6}$$

However, in molecular media like polymers, blends, nanocomposites, etc., each positron or Ps will see a different environment and have a different lifetime. As a result, the measured lifetime spectra will represent many decaying exponentials which will produce a curve showing a more complex spectrum. Figure 12.6 shows a typical lifetime spectrum in a polymer.

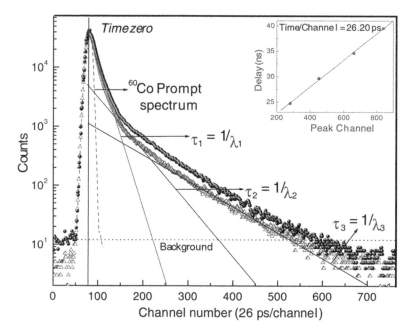

**Figure 12.6**  Typical spectrum of positrons in a polymer. The instruments resolution is also shown along with the various lifetime components resolved.

In polymers, three lifetime components are typically resolved. The attribution of the three lifetime components is as follows. The shortest lifetime component $\tau_1$ with intensity $I_1$ is attributed to contribution from *para*-positronium (*p*-Ps) and free positron annihilations. The intermediate lifetime component $\tau_2$ with intensity $I_2$ is considered to be due to the annihilation of positrons trapped at defects present in the crystalline regions or trapped at the crystalline–amorphous interface boundaries. The longest-lived component $\tau_3$ with intensity $I_3$ is due to pick-off annihilation of the *ortho*-positronium (*o*-Ps) in the free-volume sites present mainly in the amorphous regions of the system under investigation [57, 58]. It is generally accepted that *o*-Ps can serve as a measure of the free-volume hole size seen by it. The simple model of Ps in a spherical potential well of radius $R$ leads to a correlation between *o*-Ps lifetime $\tau_3$ and $R$ [30, 59, 60, 61–63],

$$\frac{1}{\tau_3} = 2\left[1 - \left(\frac{R}{R_o}\right) + \frac{1}{2\pi}\sin\left(\frac{2\pi R}{R_o}\right)\right] \text{ ns}^{-1}, \tag{12.7}$$

where $R$ is the radius of the free-volume hole, and $R_o$ equals $R + \Delta R$, where $\Delta R$ is an empirical parameter that stands for the electron layer thickness around the free-volume cavity. A value of 1.656 Å has been determined for $\Delta R$ empirically making use of the known hole sizes in molecular media like zeolites [34]. It follows that the average hole volume, $V_f$, can be calculated as $(4/3)\,\pi R^3$. The total fractional free volume, $F_V$, is then a constant $(C)$ times the product of $\tau_3$ and $I_3$, where $I_3$ is the *o*-Ps intensity. Despite the simplicity of the model assumptions, Eq. (12.7) seems to hold surprisingly well in the region of $R$ up to 10 nm and constitutes a base for numerous PAS applications to studies of the free volume and its change in polymers. To mention only a few, glass transitions [24, 34], pressure dependencies [24, 34], gas transport [64], diffusion, [65–68], structural relaxation [69–71], and blending [57, 58, 72, 73]. On the other hand, interpretation of the *o*-Ps intensities as a measure of the hole concentration must be taken with certain caution because of the possible influence of various factors on Ps formation, e.g., irradiation effects [32, 33], influence of chemical species influencing Ps formation probability, which have not yet been sufficiently understood. However, the fractional-free volume $F_V$ derived from PAS is well accepted and understood.

### 12.5.2. Free-Volume Distribution-Lifetime Analysis by Laplace Transform Method

The difficulties encountered in the analysis of positron decay curves limit the determination of free-volume sizes only to average values. In reality, any molecular media will not consist of free-volume holes of the same size but of different sizes. So to obtain the distribution of free-volume sizes close to the actual situation, advances have been made in the analysis of positron annihilation lifetime data using integral transform methods from which it is possible to extract continuous distributions of annihilation rates. One such integral transform method to derive free-volume and pore-size distributions from positron lifetime data is CONTIN program [74].

In many porous media like polymers, composites, and proteins, the heterogeneity of the local molecular environment in which positron annihilation takes place is expected to generate distributions of lifetimes. In such cases, it is necessary to replace the sum in Eq. (12.5) by an integral

$$y(t) = N_s \int_0^\infty \lambda \alpha(\lambda) I(t) \exp(-\lambda_i t) d\lambda,$$

where $\alpha(\lambda)$ is the annihilation rate probability density function (PDF). Reliable algorithms have been developed for the solution of the above equation and related Fredholm integral equations based either on eigenfunction expansion of the Laplace integral [75–77] or on the method of regularization [78–83]. A constrained, regularized least-squares method for the solution of Fredholm integral equations, which is available as a FORTRAN program called CONTIN, has been developed by Provencher [81–83]. Gregory has modified this program to solve integral equations with convoluted exponentials as kernels (CONTINPALS-2) [84–86]. This approach avoids direct determination of $I(t)$ by employing the decay curve of a reference material with a well-known single lifetime, $\lambda_r^{-1}$. CONTIN PALS-2 also corrects for source components in the sample and reference decay curves without the need for direct knowledge of $I(t)$ [85]. The method can define discrete components in the annihilation rate PDF if their lifetimes are well separated but have the added flexibility to describe continuous distributions of annihilation rates if it is demanded by the data.

The fraction of positrons annihilating with rates between in $\lambda$ and $\lambda + d \ln \lambda$ is represented as $\lambda \alpha(\lambda) d \ln \lambda$. This is the representation of the annihilation rate PDF employed by CONTIN-PALS-2. The corresponding annihilation lifetime PDF is given as $\lambda^2 \alpha(\lambda)$. The fraction of positrons annihilating with lifetimes between $\tau + d\tau$ is represented as $\lambda^2 \alpha(\lambda) d\tau$.

The transformation of the annihilation rate PDF, i.e., $\alpha(\lambda)$, to the corresponding radius PDF for the free-volume regions in which $o$-Ps annihilates is possible. Values of the radius $R$ are most conveniently determined using Newton's method. The radius PDF is given by

$$f(R) = 2 \Delta R \{\cos[2\pi R/(R + \Delta R)] - 1\} \alpha(\lambda)/(R + \Delta R)^2, \tag{12.8}$$

where $\Delta R$ means the same as described earlier and takes the value 1.656 Å. The fraction of positrons annihilating in cavities with radii between $R$ and $R + \Delta R$ is given as $f(R)dR$.

Assuming a spherical cavity, the free-volume PDF is given as $G(V) = f(R)/4\pi R^2$. Therefore, the fraction of positrons annihilating in cavities with volumes between $V$ and $V + dV$ is then represented as $g(V)dV$. Theoretical treatment using molecular dynamics and kinetic theory [87, 88] has predicted that the radii and the hole volumes of the free volume in a polymer obey the distribution functions $f(R)$ and $g(V)$, respectively as described above.

## 12.5.3. Free-Volume Distributions in Polymers

For several polymers, like amorphous PTFE (polytetrafluoethylene) [86], epoxy polymer DGEBA/DDH/DAB [89], positron annihilation rate distributions have been determined and reported. Few such studies on the free-volume properties in polymers report variations in the free volume with temperature [90], pressure [91, 92], physical aging and structural relaxation[93], stress-induced structural deformation [94], etc. We give here a typical distribution showing the cavity radius PDFs and free-volume PDFs for PTFE at 5.3°C taken from Ref. [86] in Figures 12.7 and 12.8.

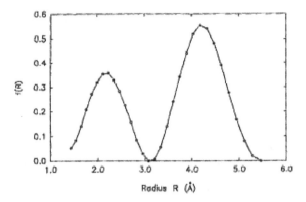

**Figure 12.7**  Cavity radius PDF for amorphous PTFE at 5.3°C derived from *o*-Ps and q-Ps components of the annihilation rate PDF for the four-peak solution.

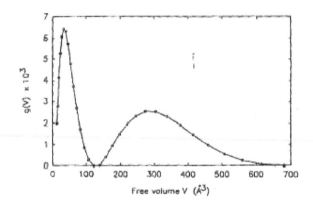

**Figure 12.8**  Free-volume PDF for amorphous PTFE at 5.3°C derived from the radius PDF shown in Figure 12.7.

The radius PDF for free-volume regions in PTFE preserves the bimodal character of the parent annihilation rate PDF and consists of two fairly symmetric peaks with mean radii of about 2.2 and 4.2 Å corresponding to mean free volumes of 45 and 310 Å$^3$, respectively. The free-volume PDF consists of two highly skewed peaks. The distribution of free-volume cavities sampled by *o*-Ps extends from 140 to 650 Å$^3$, the most probable free-volume size being about 275 Å$^3$. The most probable free-volume size associated with quasi-Ps (q-Ps) is 33 Å$^3$, while the density associated with this species extends to free volumes of about 120 Å$^3$. From this, it is very clear that this kind of information is not forthcoming from the conventional analysis of positron lifetime spectra, which provides the average or mean *o*-Ps lifetime.

An interesting study by Singru et al. [95] using annihilation rate PDFs reports physical properties of the solution-cast films of the proton-conducting polymer polyethylene oxide (PEO) complexed with ammonium perchlorate ($NH_4ClO_4$) studied in the temperature range 300–370 K. One can see the beautiful changes in free-volume distribution as a function of temperature as shown in Figures 12.9 and 12.10. The authors observe that the total free volume in PEO:$NH_4ClO_4$ increases with temperature, and the rate of increase

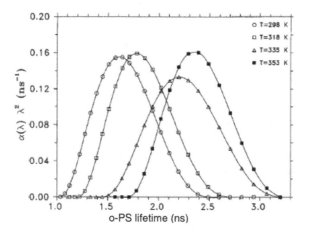

**Figure 12.9**  *o*-Ps lifetime distribution functions in PEO:NH$_4$ClO$_4$ (80:20 wt%) at different temperatures.

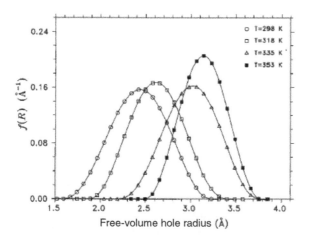

**Figure 12.10**  Free-volume distribution functions in PEO:NH$_4$ClO$_4$ (80:20 wt%) at different temperatures.

changes its slope at $T_m$. It appears that the increase in total free volume is determined mainly by the enhancement in the hole size rather than the increase in the number of holes. The pattern of increase in free volume with temperature was found to be independent of mobility of the ions with temperature.

Another example of the annihilation rate PDFs and of much significance comes from Mills et al. [96, 97]. In this work, the authors extract information on the nanoparticles surfaces embedded in MgO. A combined study of variable-energy positron lifetime spectroscopy PDFs and measurements of Doppler broadening annihilation radiation was used to find the formation of vacancy clusters on the surfaces of Au nanoparticles embedded in MgO. The Au implants were not detectable until they form nanoparticles which are associated with vacancy defects. This study clearly demonstrates that the positron probe can be used to characterize defect structures of nanoparticle surfaces, which is an important parameter as nanoscale fabrication becomes more important at present.

### 12.5.4. Angular Correlation of Annihilation Radiation (ACAR)

The origin of angular correlation of annihilation radiation has been explained earlier and can be seen clearly in Figure 12.3. The two annihilation photons are emitted collinearly if the annihilating pair has zero momentum. Owing to the finite momentum of the pair, there is a deviation from this collinearity. Thus, by measuring the coincidence counting rate for the photons in two-photon positron annihilation as a function of the angle $\theta$ (the deviation of the photons' flying apart from 180°), it is possible to determine the momentum of $e^{-}-e^{+}$ pairs (which is electrons momentum as positron momentum is small and negligible). In the angular correlation experiment, the momentum resolution is very good compared to DBAR experiment as it can be adjusted in the range of 0.2–5 mrad [27, 35]. This corresponds to the energy resolution of DBAR measurement in the range of 0.05–1.3 keV. Thus ACAR technique provides essentially the same kind of information as DBAR; however, the momentum resolution of the method is much better. On the other hand, DBAR is a simple experiment and involves little time in getting the information. ACAR studies provide very accurate estimate of Fermi energy of the system under study and even the activation energy/vacancy migration energy can be readily obtained.

The angular correlation device can resolve neither the angular deviation in the $x$-direction nor the Doppler shift in the $y$-direction and consequently, the counting rate experimentally one observes is given by

$$N\left(\theta_z\right) = c \int\limits_{-\infty}^{+\infty} \int\limits_{-\infty}^{+\infty} dp_x dp_y \rho\left(p_x, p_y, \theta_z m_o c\right), \tag{12.9}$$

where $\rho(p_x, p_y, p_z)$ is the momentum distribution of the annihilating positron–electron pairs in the medium. When the positron is free and in thermal equilibrium with the medium, its momentum is negligible and then the angular correlation curve represents the $p_z$ – the distribution of the momentum of the annihilated electrons. As an example, we show the ACAR curve obtained in the case of Cu and Al and it is evident that the Fermi energy can be obtained readily from ACAR measurement.

From Figure 12.11, the contribution of free electrons and the core electrons to the annihilation process is very clearly seen. The inverted parabolas represent the free electrons participating in the annihilation and the Gaussian parts stand for the core electrons.

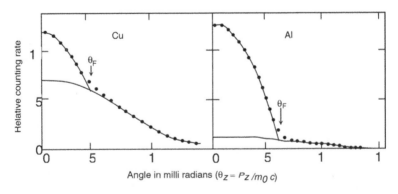

**Figure 12.11** Typical angular correlation curves in metals (aluminum and copper). The inverted parabolas are due to free electrons and the Gaussian parts correspond to core electrons.

### 12.5.5. Doppler Broadening of the Annihilation Radiation (DBAR)

Doppler broadening of the annihilation radiation (DBAR) can be measured with a standard gamma-ray spectrometer equipped with the HPGe detector. Energy resolution of such devices will be around $\Delta E_\gamma = 1.1$ keV at 511 keV. The line shapes are usually characterized with shape parameters $S$ and $W$. The $S$ parameter is experimentally determined as the ratio of area of the central part to the total area under the curve and $W$ shape parameter expresses the relative contributions of tails to the total peak area. Thus the $S$-parameter is higher if relative contribution of lower momentum electrons to positron annihilation is more, while the $W$-parameter becomes greater if contribution of the core electrons with higher momenta is more [98, 99]. A typical DBAR spectrum after background counts subtracted is shown in Figure 12.12 along with the $S$- and $W$-parameters.

For instance, relative increase in the fraction of positrons trapped at the open-volume defects can be markedly reflected by an increase in the observed $S$-values. A useful approach is to present experimental data in terms of $S$–$W$ plot, which allow to draw qualitative conclusions on the evolution of defects participating in positron trapping.

A substantial problem in the DBAR experiment is high background counts [28, 99–102]. This is especially a problem for measurements of the high momentum parts of the annihilation peak. The background can be drastically reduced by the use of the coincidence technique, which registers both annihilation photons in coincidence and the background is drastically reduced. Since the time required for DB measurement is very much smaller than for ACAR, DBAR is often the preferred second method of PAS for any investigation [27, 28]. If one employs coincidence between the two annihilation photons, again the time required for acquisition of a statistically good spectrum is more.

### 12.5.6. Surface Area Measurement from PAS

The $o$-Ps intensity, which is a measure of relative number of free-volume cavities in polymers and the like also, represents the probability of Ps formation in the medium. Jean

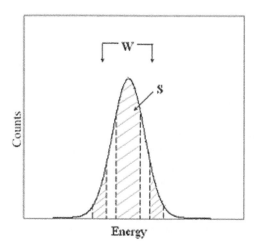

**Figure 12.12**  Typical annihilation gamma photon spectrum obtained with an HpGe detector.

et al. [102] have derived an empirical relation to calculate the surface area $S$ encountered in the annihilation of $o$-Ps from the intensity $I_3$, which is

$$I_3 = 3.0 + 0.0333S. \tag{12.10}$$

Although $o$-Ps lifetime parameter is unquestionably the true parameter related to free voulme, $o$-Ps intensity ($I_3$) could also serve as a useful parameter in measuring the surface area surrounding the nanoparticles in nanocomposites. Few attempts have been made to ascertain this aspect in explaining the structural relaxation in polymers under isochronal treatment [69, 70]. However, not much has been done in this direction. There are arguments that the change in $I_3$ may be due to several other factors like charging effects of the continuous irradiation of the polymer by positrons and gamma rays of the positron source, or if the polymer is nonpolar in character, etc. The influence of these effects can be conveniently ruled out if the strength of the source used is very weak and by knowing the lifetime of positrons in the medium of study by careful scrutiny [69, 70]. Normally BET is the standard method of measuring the surface area of mesopores in a system, and the positron method could well be used as a new probe to measure qualitatively the surface area of nanopores. Few measurements will certainly set the ball rolling and efficacy of the method tested.

## 12.6. PAS Results in Polymer Nanocomposites

Metal-to-semiconductor like transitions in polymer nanocomposites could be detected clearly by free-volume measurements similar to the electrical resistivity measurements, although the later is not a very sensitive method. It is expected that the transition temperature depends on the average particle size of the nanoparticle. Since the properties of nanocomposites depend on the surface area to volume ratio, the change in surface tension in the composite could be possibly detected by free-volume parameters. Gas permeability and selectivity, which depends largely on the porosity of the material, is another important area where PAS can certainly provide valuable information. In the following sections, we review the work carried out on polymer nanocomposites so far employing positron methods. Unfortunately, the reported literature tells us that only lifetime technique has been used often compared to the other two methods namely ACAR and DBAR.

### 12.6.1. Polymer Nanocomposites: Silica Fillers, Ag, and Carbon Nanoparticles

The physical dispersion of nonporous, nanoscale, and fumed silica particles in glassy amorphous poly(4-methyl-2-pentyne) (PMP) simultaneously and surprisingly enhances both membrane permeability and selectivity for large organic molecules over small permanent gases. These highly unusual property enhancements have been discovered in a recent study [103], in contrast to results obtained in conventional filled polymer systems. This reflects fumed silica-induced disruption of polymer chain packing and an accompanying subtle increase in the size of free-volume elements through which molecular transport occurs, as discerned by PALS. Such nanoscale hybridization represents an innovative means to tune the separation properties of glassy polymeric media through systematic manipulation of molecular packing. It is known that large free-volume cavities have been described as interconnected pathways through which most penetrant molecule transport occurs [104]. In this study, the authors observe that addition of fumed silica to PMP systematically increases the average size of the large free-volume elements. The total increase in free-volume radius at high fumed silica loading is less than 0.1 nm, emphasizing the true molecular-scale alteration of PMP chain packing engineered by nanoscale-fumed silica. Material modification on this scale may be necessary to manipulate, in controllable fashion, the transport characteristics because penetrant size of difference of less than 0.02 nm can yield substantial differences in flux through polymer membranes. The importance of using small filler particles to achieve the desired effect on transport in PMP has been well demonstrated in this work. Additionally the authors conducted experiments with several other silica of varying sizes, as well as carbon black and a-alumina powder, which were added to PMP. From this study, it is concluded that for a fixed volume fraction, smaller particles yield more polymer/particle interfacial area and provide more opportunity to disrupt polymer chain packing and effect molecular transport, which is true to what has been emphasized in the introduction. The free-volume distribution studies would have further strengthened the above observation.

The observed fumed silica dispersion is consistent with the corresponding transport data, because highly networked fumed silica would be expected to promote Knudsen diffusion and thereby reduce reverse selectivities. In comparison with the earlier study [105] to harness the selective pores of inorganic zeolite particles dispersed within a polymeric matrix to enhance membrane performance, these authors have demonstrated that nonporous particles, having dimensions comparable to those of individual polymer chains, can be used to regulate the manner in which these chains pack and thereby favorably manipulate molecular transport and selectivity.

Petra Winberg et al. [106] conducted experiments on polydimethylsiloxane (PDMS)/fumed silicon dioxide ($SiO_2$) nanocomposites using PALS to study the effect of filler content and filler particle size on the free-volume properties in the temperature range $-185°C$ and $100°C$. An interesting transition temperature at $-35°C$ was observed depicting dependence of the $o$-Ps lifetime in PDMS on temperature (Figure 12.13). Above this temperature, the authors find a relationship between $o$-Ps lifetime $\tau_3$ and the square

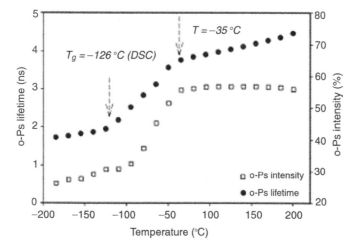

**Figure 12.13** *Ortho*-positronium lifetime and intensity in PDMS between $-185°C$ and $100°C$. Arrows indicate glass transition as measured by DSC and the temperature where the leveling off effect is observed (reprinted with permission from Elsevier).

root of the temperature-dependent surface tension of PDMS, which is in close agreement with that observed for low molecular liquids indicating $o$-Ps bubble formation. In this temperature region, the lifetime of $o$-Ps reflects the equilibrium between the repulsive forces of the $o$-Ps acting on the bubble wall and the surface tension of the polymer. In the temperature region above $-35°C$, they found a systematic decrease in $\tau_3$ when nanosized silica was added suggesting the apparent surface tension increase.

The DSC results indicated reduced mobility in the polymer due to the presence of nanosized silica. This reduced mobility in the polymer is also likely due to increase in the apparent surface tension and thereby reduce $o$-Ps lifetime according to the observed relationship between $o$-Ps lifetime and surface tension. In contrast to the $o$-Ps lifetime behavior at temperatures above $-35°C$, at temperatures below $-35°C$ the nanocomposite containing 44 wt% silica showed an increased $o$-Ps lifetime due to increased free-volume cavity sizes. The increased free-volume cavity sizes can be due to inability of polymer segments in close contact with the filler surface to change conformation when the temperature is decreased. The $o$-Ps yield is strongly reduced by crystallization and by the addition of $SiO_2$. The reduction due to filler addition is, however, in the case of nanosized $SiO_2$ not following a linear weight averaging relationship between filler content and $o$-Ps intensity observed for micron-sized fillers. When nanosized silica is added to PDMS, the $o$-Ps yield displays a positive deviation. The authors ascribe this to the small size of the particles in which $o$-Ps diffuses out from the filler particles to the surface of the particles and results in increased $o$-Ps intensity. This particular aspect needs to be further investigated to support the authors' claim.

The same group [107] found that the interstitial cavities have a significant positive effect on the permeability of gases in nanocomposite membranes, which outweigh the generally observed effect of reduced diffusivity when impenetrable fillers are incorporated into a polymeric membrane. The authors have investigated the free-volume sizes and interstitial mesopore sizes in poly(1-trimethylsilyl-1-propyne) (PTMSP)/silica nanocomposites and observed a correlation between nitrogen permeability and cavity sizes through PALS for various filler concentrations between 0 and 50 wt%. Interstitial mesopores were observed in PTMSP/fumed silica nanocomposites at filler concentrations between 10 and 50 wt% and in as-received fumed silica. Since polymer segments partially fill these interstitial cavities in the PTMSP/$SiO_2$ nanocomposites, the mean radius of the cavities showed decrease with the decrease in filler concentration as evident from Figure 12.14.

The mean size of the larger free-volume cavities in the bimodal free-volume size distribution in PTMSP increased significantly with increasing filler concentration as shown in Figure 12.15. Here the annihilation PDFs helped in showing clearly the bimodal size distribution, which would not have been possible with average size estimation. The significant increase observed even at relatively low filler concentrations say 20 wt% suggests that the filler particles affect the free-volume sizes at longer length scales like what is generally observed in polymers. The larger free-volume cavity size distribution in PTMSP was broadened with increasing filler concentration, indicating that the filler particles affect only a fraction of the polymer, presumably polymer segments close to the filler surface. This study reveals that a good correlation exists between nitrogen gas permeability and interstitial cavity volume that is as the filler content ($>10$ wt%) increases, the permeability increases.

Mukherjee et al. [108] investigated composites of polyacrylamide in which silver nanoparticles of diameters in the range 16.4–33.3 nm were embedded by a chemical method. Positron lifetime and Doppler broadening measurements of these samples were carried out. It was found that positrons get trapped into diffuse vacancy clusters on the

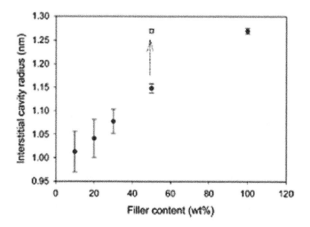

**Figure 12.14** Interstitial cavity radius in PTMSP/hydrophobic fumed silica nanocomposite membranes (●) at room temperature as a function of silica content and interstitial cavity radius in silica recovered from PTMSP50 membranes (o). Arrows indicate shift of the interstitial cavity radius upon removal of polymer matrix in PTMSP50 (reprinted with permission from American Chemical Society 2005).

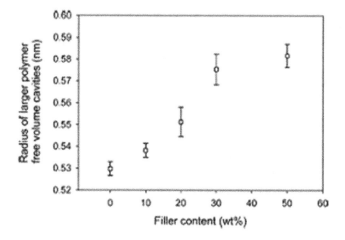

**Figure 12.15** Large free-volume cavity radius in PTMSP/hydrophobic fumed silica nanocomposite membranes at room temperature as a function of silica content (reprinted with permission from American Chemical Society 2005).

grain surface and the nature and size of these clusters could be quantitatively understood from the measured positron lifetime. The results predicted the vacancy clusters of sizes of about – three to four mono vacancies on the grain surfaces. It was further observed that the open spaces between the grain surface and the surrounding innermost polymer layer could also trap positrons, resulting in a characteristically long lifetime. The authors propose that molecular-dynamic simulation studies possibly throw more light on this concept, which is at least qualitatively identical to a solid bubble in a polymer matrix.

The $o$-Ps lifetime $\tau_3$ in this study essentially reflected the properties of the free-volume holes in the polymer matrix. The isochronal annealing of the composites exhibited

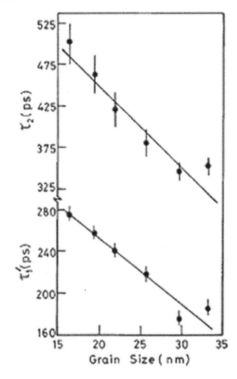

**Figure 12.16** Estimated lifetimes of positrons annihilating at the diffuse vacancy clusters on the grain surface ($\tau_1'$) and at the grain–polymer interface ($\tau_2$) vs the average grain size (reprinted with permission from American Physical Society 1998).

shrinking of the defects, which was evident from the resultant decrease in $o$-Ps lifetime. Another interesting finding of this investigation is that the isothermal annealing results at 473 K showed an inter diffusion of the surface atoms on to the nanoparticles and the polymer. Further the authors find a metal-to-semiconductor like phase transition around 80K as indicated by the $S$-parameter predicted for the case of nanosized metal particles, which has been recently seen in the electrical resistivity measurements [109]. The authors observe that the transition temperature seems to be rather sensitive to the average particle size, as expected from the established theoretical predictions. This study has clearly shown the strength of PAS techniques in the study of nanocomposites.

A study of the effects of plasticizer and stretching strain on the percolation transition in polyisoprene–carbon nanocomposites (PCNC) is reported [110]. The free-volume sites accessible to *ortho*-positronium ($o$-Ps) have been measured by PALS under two conditions of the sample – one in relaxed state and the other stretched state of PCNC containing different amounts of plasticizer (Figure 12.17). The positron data showed that the number of free-volume cavities decreased during stretching regardless of the content of either carbon nanoparticles (CNP) or plasticizer namely STRUKTOL. The free-volume cavity size reached its maximum value in the region of percolation transition measured from electrical conductivity measurement. The relative number of free-volume cavities represented by $I_3$ also decreases at CNP concentrations exceeding the percolation threshold.

The authors infer that plasticizer molecules apparently go into the interspace formed by side-branches of the elastomer molecules as well as the nanospace between the CNPs.

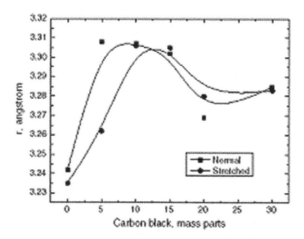

**Figure 12.17**  Mean radius of free-volume cavities as a function of CNP mass parts in normal and stretched ($\Delta l = 15$ mm) states in PCNC measured in PALS. Basis length of the sample in normal state, $l_0 = 50$ mm (reprinted with permission from Elsevier).

The effect of STRUKTOL on free-volume cavities was that the polymer chains were more closely packed upon addition of STRUKTOL in both the normal and stretched states. The results support the notion that CNPs are better dispersed by STRUKTOL but can still form a conductive network at lower loadings. The same group conducted experiments on polyimide–silica hybrid composite membranes to understand the inorganic–organic interactions or what we call interfacial phenomena, which is generally neglected in composites theory. The type of alkoxide sol employed in the sol–gel process was found to have appreciable effect on the free-volume distribution and gas transport properties in the hybrids. Crosslinking of pure polyimide with APTEOS-coupling agent causes a reduction in free-volume element size and concentration, which affects the gas transport properties. They concluded that the interaction between the inorganic nanodomains and the polymer chains contributes significantly to the free volume and gas transport properties of hybrid membranes.

PAS as depth-sensitive probe was applied to study diamond-like nanocomposite (DLN) of films [111]. DLN films were deposited from a plasma discharge of polyphenilmetil-siloxane on rf-biased Si substrates. The film properties show systematic changes with variations in the voltage applied to the substrate. The film deposited at an rf-bias voltage of 1000 V was stable under a 450°C annealing. The positron study shows that in spite of an amorphous structure, these films were found to be very stable against structural distortion and films with highest breakdown voltage have the highest open-volume regions suggesting the influence of open pores on the electrical properties of these films.

### 12.6.2. Polymer/Layered Nanocomposites

The free-volume study in polymer/layered nanocomposites could throw light on the interlayer spacing and how it influences the properties of the composites like gas barrier property, selectivity, etc. Olson et al. [112] reported positron investigation on a layered organosilicate–polystyrene nanocomposite to compare with that reported already namely an organically modified fluorohectorite (C18FH), an atactic polystyrene specimen, and a

C18FH/polystyrene nanocomposite. The neat C18FH material was produced by a cation exchange reaction between the mica-like silicate and an alkyl ammonium salt with a length of 18 carbon atoms (C18), and has a structure consisting of weakly cohesive silicate layers connected by galleries with an average spacing of 2.253 nm. The presence of the complexed alkylammonium cations in the galleries between the silicate layers renders the material hydrophobic. On forming the nanocomposite, the interlayer spacing was found to increase to 3.031 nm. This study found that polystyrene possesses a slightly larger free volume than in the bulk. Theoretically, the authors found that $o$-Ps pickoff lifetime for the nanocomposite shows essentially a linear increase with temperature. This indicates that the glass transition in the nanoconfined polystyrene occurs at the same temperature as the bulk, but is masked by the transition in C18FH, which occurs at about the same temperature.

Wanga et al. [113] measured the free-volume hole property by PALS on polymer–clay nanocomposite of styrene–butadiene rubber (SBR) and layered silicate clay of rectorite and conventional composite materials N326 (carbon black)/SBR. The PALS and differential scanning calorimeter (DSC) results showed that layered rectorite has a stronger effect on restraining polymer chain mobility, which results in the decrease of fractional free volume. The gas barrier of rectorite/SBR exhibits a 68.8% reduction in permeability compared to pure SBR. Incorporation of nanolayers of rectorite effectively improves gas barrier property attributed to the tortuous diffusion path and lower fractional free volume. The dispersion of nanoscale rectorite clay in SBR largely enhances gas barrier property in contrast to results obtained in N326/SBR system. The observed results are attributed to the clay platelet like morphology mainly influencing gas permeability in rectorite/SBR nanocomposite.

A recent PALS study on intercalated polyamide/clay nanocomposite showed that high concentrations of clay (>19 wt%) caused an increase in the mean free-volume cavity size (Figure 12.18), suggesting a lower chain packing efficiency in the intercalated PA6/clay nanocomposites [114]. Interesting observation is that lower packing efficiency was found both above and below the glass transition of the polymer. In the opinion of the authors, the

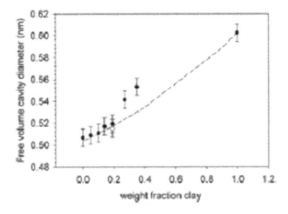

**Figure 12.18** Mean free-volume cavity diameter in PA6/clay nanocomposites as a function of clay content (•) and the mean free-volume cavity diameter in PA6 with 7 wt% organic modifier (o). Simulated free-volume cavity diameter in a physical mixture of PA6 and modified clay is depicted as a dashed line (reprinted with permission from Elsevier).

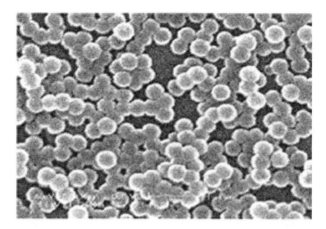

**Figure 12.19** Scanning electron micrograph of monodisperse PS particles used as seed (solid content = 30 wt%, Dpz = 190 nm) (reproduced with permission from Springer-Verlag 2001).

presence of clay affects not only the polymer but also the behavior of the organic modifier, which constituted a significant fraction of these nanocomposites.

The order to disorder transition of the organic matter, when exchanged with cations between the clay layers, was shifted toward lower temperature by about 12°C, as compared to the free organic modifier. The transition was shifted to even lower temperatures when the organically modified clay was dispersed in the PA6-forming intercalated polymer nanocomposites. It is, therefore, likely that the measured increase in free-volume sizes in the nanocomposites originates from structural changes both in the polymer as well as the organic modifier. The changes found in the behavior of the organic modifier in combination with the large effects that the clay was shown to have on the crystalline structure of the polymer makes the process of deducing a quantitative measure of the effect of organically modified clay on the free-volume sizes in PA6 is a challenging one. Further investigations are necessary in this area.

## 12.6.3. Epoxy/Elastomer Latex Particle Nanocomposites

Lizama et al. [115] have prepared a series of thermoplastic/elastomer composite (latex particles) by two-step emulsion polymerization techniques. In both cases, PS particles constituted the first step polymer. PBuA or PmeA was employed as the second phase. A tendency to form core–shell structures at low second polymer contents was observed because of the polar nature of acrylics, which favors segregation from the polystyrene nucleus and the apparition of a thin layer of the second polymer near aqueous medium in the emulsion. The morphology of the composite particles as a function of system composition has been studied by TEM. SEM photograph for 30% monodisperse PS particles is shown in Figure 12.19. The authors obtained a correlation between the structural behavior of the composites and the $o$-Ps lifetime and formation probability ($o$-Ps intensity $I_3$). The changes in free volume are attributed to the phase transitions associated with the particles.

The covering of the hard initial nodules with soft polymer containing a reduced number of large free-volume holes produces a diminution of the $o$-Ps formation probability and an augmentation of their average lifetime. This behavior is more evident when isolated particles rich in the second-step polymer are formed. All these transitions produce drastic changes in the *o*-Ps-related parameters. Therefore, the authors conclude that *ortho*-positronium annihilation data correlate with the composite particles morphology. More studies of this nature using PAS are very much needed.

SiO$_2$ nanoparticles embedded in epoxy to produce nanocomposites have been investigated. Ultrasonic and mechanical methods were used to disperse the nanoparticles in the epoxy resin. The nanocomposites were characterized by tensile and impact testing as well as TEM studies. Specially to study the effects of nanometer-sized SiO$_2$ particles on free volume of nanocomposites, positron annihilation lifetime measurements have been carried out [116]. The studies on the morphology and mechanical properties show that the introduction of SiO$_2$ nanoparticles in the CYD-128 matrix polymer has dramatic effects on nanocomposites. Uniform dispersion of nanoparticles is critical to the morphological structure of the nanocomposites, which in turn affects the impact strength of the SiO$_2$/CYD-128. They opine that if more interfacial surfaces are generated between polymer and nanoparticles, this will assist in absorbing the stress. This study also supports that the free-volume parameters of nanocomposites change with the addition of nanoparticles.

In another positron lifetime study in epoxy/rectorite nanocomposites, the fractional free volume showed a dramatic decrease with increasing epoxide equivalent and free-volume hole size was affected slightly by epoxide equivalent. It was observed that lower the epoxide equivalent, higher the crosslink density in epoxy and they suppose strong interaction between layered rectorite and epoxy molecules, which in turn result in the decrease of free-volume concentration and formation of interfacial layer in nanocomposites [117]. The XRD patterns of the nanocomposites indicated that intercalated structure exists with epoxide equivalent 188 and 222, and rectorite platelets have been exfoliation in nanocomposite while epoxide equivalent increases from 263 to 1110. The *S*-parameter determined from DB measurement indicates the rectorite structure changes with the entry of rectorite into epoxy demonstrated by high sensitivity of positron annihilation studies.

### 12.6.4. Polymer/Calcium Carbonate

Zhang et al. have performed an interesting PALS study in HDPE/CaCO$_3$ nanocomposites. The positron lifetime spectra were measured as a function of weight fraction of CaCO$_3$ nanoparticles [118]. It is inferred that, first, *o*-Ps is formed and annihilates in the free-volume hole of HDPE and the free-volume concentration decreases with increasing fraction of the CaCO$_3$ nanoparticles in the HDPE/CaCO$_3$. The impact strength of HDPE/CaCO$_3$ nanocomposites was also measured and found that from 0% to 7% of CaCO$_3$ content, the strength increases and then decreases while *o*-Ps intensity showed continuous decrease. The results are indicative of the fact that the interfacial layers between CaCO$_3$ nanoparticles and HDPE matrix play an important role for the increase in the impact strength, while CaCO$_3$ content is small. The authors opine that there is strong interaction between CaCO$_3$ nanoparticles and HDPE matrix, which can restrict the main-chain segmental motion and reduce the mobilization of PE chains. These results further confirm that positron annihilation is an effective tool to probe the nanocomposite materials.

### 12.6.5. Inorganic–Organic Hybrid System

Last in this series, is the following: triblock, diblock, and random copolymers of poly(ethylene oxide) and poly(propylene oxide) were used as molecular templates in poly(methyl silsesquioxane) (MSQ) matrices to fabricate ultra low-*k* dielectric materials ($k \leq 2.0$). In this work, the authors complement the positron results with solid-state NMR results and conclude that polymer architecture plays an important role in the polymer domain size and the polymer–matrix interface in the nanocomposites. PALS reveals an important aspect about the percolation that porous MSQ templated by triblock copolymers have smallest pores and highest percolation threshold compared to those templated by diblock and random copolymers [119]. This study further supports the results of earlier studies on percolation.

## 12.7.  Conclusions

From the review of the work described above, it is clear that not much work has been done using positron annihilation techniques in the study of polymer nanocomposites in particular free-volume measurements, although this exciting field of nanocomposites is progressing at a rapid rate. It is observed that wherever this new technique is used, the obtained information is very useful in understanding certain aspects of nanocomposites properties. Metal to semiconductor like phase transition observed clearly showed that the nanoparticle size played a crucial role. Another important fact that is evident from the study on the influence of intestinal cavities and their effect on gas transport and permeability. The effect of plasticizer in nanocomposites under normal and stretched condition is better understood in terms of free volume of the system. Another noticeable result is that the increase in surface tension, with the addition of silica nanoparticles, is well revealed by PAS, pointing the ability of PAS to probe the surface area that we frequently talk about in nanocomposites. The results of Ag nanoparticles in polyacrylamide predicted the vacancy clusters of sizes of about three to four mono vacancies on the grain surfaces. The authors of this study would have further explored the possibility of evaluating the surface area through positron data and check whether one can get more information on this aspect. The studies have shown that small filler particles can be used to achieve the desired effect of transport properties. The importance of higher open volume to get higher electrical breakdown properties in DNL films is an eye opener to begin more positron annihilation research in this area. In polymer/layered nanocomposites, crosslinking and strong interactions could be qualitatively understood through free-volume monitoring, which equally holds good in inorganic–organic hybrid systems. Here also more investigations are needed to test the efficacy of the technique to establish the prowess of the positron probe to understand how the surface area volume ratio influences properties of the polymer nanocomposite materials. As already said earlier, polymers containing nanoparticles can exhibit exciting behavior and the properties depend on the size of the incorporated particle, like the electrical conductivity, color, refractive index of the system, etc. So far, no investigation with an aim to understand the optical properties of the nanocomposites through free-volume measurements have been made and this problem needs immediate attention as there is a great need for materials whose optical properties can be tailored. Even the semiconducting properties monitored in terms of free volume have to be addressed in future.

## 12.8.  Future Outlook

The potential applications of nanocomposites in various commercial and technological fields and the related research is promising and attracting increasing investment from all over the world [6]. Although there are few applications where nanotechnology has penetrated the current market, the major impact is still to be felt which may be at least another decade away. Since only a moderate success has been made over the last 20 years or so, research should continue to find strategies to optimize the fabrication of nanomaterials to achieve both improved mechanical, optical and transport properties [120]. There is vast list of commercial applications for polymer nanocomposites like inkjet markets, nanoparticles in cosmetics, and automotive applications such as body moldings, engine covers and catalytic converts, batteries, computer chips [6]. Additionally, memory devices, biosensors for diagnostics, advances in lighting in particular are all possible through nanocomposites. These could be explored by the application of novel techniques of positron annihilation. It is suggested that investigations motivated to understand the optical properties of nanocomposites through free-volume measurements be immediately undertaken so that useful information can be obtained, as there is a great need for materials for optical waveguides, etc., whose properties to be tailored. As the size of the nanoparticles plays an important role, it is envisaged that positron annihilation techniques namely the lifetime measurements will help to quantify the surface area and hence the resulting properties.

# References

1. *The Chemistry of Nanostructured Materials*; Peidong Yang, University of California, Berkeley, USA
2. R. Roy, Purpose design of nanocomposites: Entire class of new materials. in *Ceramic Microstructures'86*; Pask, J. A.; Evans, E. G. (Eds.); Plenum Press: New York, 1987, p. 25.
3. A. J. Kinloch, *Adhesion and Adhesives*; Chapman and Hall: London, 1987.
4. W. Mahler, Polymer-trapped semiconductor particles. *Inorg. Chem.* **27** (1988), 435.
5. Godovski D. Yu, Electron behavior and magnetic properties of polymer nanocomposites, *Adv. Polym. Sci.* **119** (1995), 79.
6. *Nanoscience and Nanotechnologies*; The Royal Society & the Royal Academy of Engineering: July 2004.
7. E. Thostenson, C. Li, T. Chou, *Compos. Sci. Technol.* **65** (2005), 491.
8. C. Park, O. Park, J. Lim, H. Kim, *Polymer* **42** (2001), 7465.
9. M. Alexandre, P. Dubois, *Mater. Sci. Eng. Rep.* **28** (2000), 1.
10. O.O. Christopher, M. Lerner, Nanocomposites and Intercalation Compound, Encyclopedia of Physical Science and Technology; Elsevier: 3rd edn, 2001, p. 10.
11. F. Papadimitrakopoulos, P. Wisniecki, D. E. Bhagwagar, *Chem. Mater.* **9** (1997), 2928
12. G. Schmid, *Chem. Rev.* **92** (1992), 1709.
13. G. Schmid, S. Peschel, T. Sawitowski, *Z. Anorg. Allg. Chem.* **623** (1997), 719.
14. M. Brust, M. Walker, D. Bethell, D. J. Schiffrin, R. Whyman, *J. Chem. Soc., Chem. Commun.* (1994), 801.
15. Y. Dirix, C. Bastiaansen, W. Caseri, P. Smith, *J. Mater. Sci.* **34** (1999), 3859.
16. N. Herron, Y. Wang, H. Eckert, *J. Am. Chem. Soc.* **112** (1990), 1322.
17. T. Hirai, M. Miyamoto, T. Watanabe, S. Shiojiri, I. Komasawa, *J. Chem. Eng. Jpn.* **31** (1998), 1003.
18. Y. Wang, N. Herron, *Chem. Phys. Lett.* **200** (1992), 71.
19. M. Meyyappan, *Carbon Nanotubes, Science and Application*, CRC Press: Boca Raton, 2004.
20. E.P. Giannelis, *Adv. Mater.* **8** (1996), 29.
21. Positrons Solid-State Physics: W. Brandt, A. Dupasquier, *Proc. International School of Physics Enrico Fermi*, Course LXXXIII, Varenna (1981); North-Holland: Amsterdam, 1983.
22. Positron Spectroscopy of Solids: A. Dupasquier, A.P. Mills Jr., Proc. International School of Physics Enrico Fermi, Course CXXV, Varenna (1993): IOS Press: Amsterdam, 1995.
23. P. Hautojarvi, *Positrons in Solids*, Springer: Berlin, Heidelberg, New York, 1979.
24. *Positron and Positronium Chemistry*, D.M. Schrader, Y.C. Jean, Elsevier Science: Amsterdam, 1988. Positron Solid-State Physics; W. Brandt, A. Dupasquier, North-Holland: Amsterdam, 1984.
25. Positron Annihilation in Fluids: S.C. Sharma, World Scientific: Singapore, 1987.
26. R.N. West, Positron studies of condensed matter. *Adv. Phys.* **22** (1973), 263.
27. K. Saarinen, P. Hautojarvi, C. Corbel, Identification of defects in semiconductors; in M. Stavola (Ed.), *Semiconductors and Semimetals*, Academic Press: San Diego, Vol. 51A, 1998.
28. R. Krause-Rehberg, H.S. Leipner, *Positron Annihilation in Semiconductors*; Springer: Berlin, 1999.
29. S. Berko, Positrons Solid-State Physics: W. Brandt, A. Dupasquier, *Proc. International School of Physics: Enrico Fermi*, Course LXXXIII, Varenna (1981); North-Holland, Amsterdam, 1983, 64.

30. Y. Nakanishi, S.J. Wang, Y.C. Jean, in *Positron Annihilation Studies of Fluids*; S.C. Sharma (Ed.), World Scientific: Singapore, 1988, 292.
31. Y.C. Jean, F. Deng, *J. Polym. Sci. Part B. Phys.* **30** (1992), 1359
32. V.I. Grafutin, E.P. Prokop'ev, Positron annihilation spectroscopy in materials structure studies: Instruments and methods of investigation: Physics-Uspekhi Uspekhi Fizicheskikh Nauk, Russian Academy of Sciences, **45(1)** (2002), 59-74
33. M.J. Puska, R.M. Nieminen, *Rev. Mod. Phy.* **66** (1994), 841.
34. Y.C. Jean, *Microchem. J.* **42** (1990), 72.
35. V.I. Goldanskii, in *Positron Annihilation*, A.T. Stewart, L.O. Roellig (Eds.), Academic Press: New York, 1967.
36. V.I. Goldanskii, *At. Energy Rev.* **6** (1968), 3.
37. I.K. Mackenzie, *Positron Solid State Physics*, North Holland: Amsterdam, 1983, p. 196.
38. V. I.Grafutin et al. Fiz. Tverd. Tela (St.-Petersburg) **41** (1999), 929 [Phys. Solid State **41** (1999), 843]
39. A. Ore, J.L. Powell, *Phys. Rev.* **75** (1949), 1969; O.E. Mogensen, *J. Chem. Phys.* **60** (1974), 998.
40. W. Brandt, S. Berko, W.W. Walker, *Phys. Rev.* **120** (1960), 1289.
41. W. Brandt, S. Berko, W.W. Walker, *Phys. Rev.* **121** (1961), 1864E.
42. B.V. Tosar, V.G. Kulkarni, R.G. Lagu, G. Chandra, *Phys. Lett.* **28A** (1969), 760.
43. B.V. Tosar, V.G. Kulkarni, R.G. Lagu, G. Chandra, *Phys. Status. Solidi.* **B55** (1973), 415.
44. R.K. Wilson, P.O. Johnson, R. Stump, *Phys. Rev.* **129** (1963), 2091.
45. D.W. Van Krevlan, P.J. Hoftzer, *Properties of Polymers*; 2nd edn.; Elsevier: New York, 1976, p. 51.
46. W.M. Lee, *Polym. Eng. Sci.* **20** (1980), 65
47. W. Brandt, J. Wilkenfeld, *Phys. Rev.* **B12** (1975), 2579; **B13** (1976), 2243E.
48. W. Brandt, I. Spim, *Phys. Rev.* **142** (1966), 231.
49. W. Brandt, H. Feibus, *Phys. Rev.* **174** (1968), 454.
50. W. Brandt, H. Feibus, *Phys. Rev.* **184** (1969), 277.
51. W. Brandt, J.H. Fahs, *Phys. Rev.* **B2** (1970), 1425; **B3** (1971), 2370E
52. M.C. Thimmegowda, PhD Thesis, Studies on the influence of free volume controlled water diffusion in contact lens polymers and their tolerance to UV radiation using positron lifetime technique, University of Mysore, 2004.
53. P. Kirkegaard, N.J. Pedersen, M. Eldrup, *Riso Nat. Lab. Reports*; Denmark **M-2740** (1989)
54. D.W. Marquardt, *J. Soc. Ind. Appl. Math.* **11** (1963), 431.
55. P. Kirkegaard, Some aspects of the general least squares problem for data fitting; Riso, **M-1399** (1971), 16.
56. P. Kirkegaard, M. Eldrup, *The Least Squares Fitting Program POSITRONFIT: Principles and Formulas*; Riso **M-1400** (1971)
57. H.B. Ravikumar, G.N. Kumaraswamy, S. Thomas, C. Ranganathaiah, Polymer. **46** (2005), 2372.
58. H.B. Ravikumar, C. Ranganathaiah, *Polym. Int.* **54** (2005), 1288.
59. S.J. Tao, *J. Chem. Phys.* **56** (1972), 5499.
60. M. Eldrup, D. Lightbody, J.N. Sherwood, *Chem. Phys.,* **63** (1981), 51.
61. C. Wastlund, H. Berndtsson, F.H.J. Maurer, *Macromolecules.* **31** (1998), 3322
62. A.J. Hill, M.D. Zipper, M.R. Tant, G.M. Stack, T.C. Jordan, A.R. Shultz, *Phys. Condens. Matter* **8** (1996), 3811.
63. C. Wastlund, F.H.J. Maurer, *Macromolecules.* **30** (1997), 5870.
64. Ranimol Stephen; C. Ranganathaiah, S. Varghese, K. Joseph, S. Thomas, *Polymer.* **47** (2006), 858.
65. R. Ramani, C. Ranganathaiah, *Polym. Int.* **50** (2001), 237.
66. R. Ramani, P. Ramachandra, G. Ramgopal, C. Ranganathaiah, *Eur. Polym. J.* **33** (1997), 1753.
67. Gani Shariff, P.M. Sathyanarayana, M.C. Thimmegowda, M.B. Ashalatha, R. Ramani, D.K. Avasthi, C. Ranganathaiah, *Polymer* **43** (2002), 2819.
68. M.V. Deepa Urs, C. Ranganathaiah, R. Ramani, Sarfaraz Alam, *J. Appl. Polym. Sci.* **102** (2006), 2784.
69. M.B. Ashalatha, P.M. Sathyanarayana, Gani Shariff, M.C. Thimmegowda, R. Ramani, C. Ranganathaiah, *Appl. Phys. A.* **76** (2003), 1.
70. P.M. Sathyanarayana, Gani Shariff, M.C. Thimmegowda, M.B. Ashalatha, R. Ramani, C. Ranganathaiah, *Polym. Int.* **50** (2002), 765.
71. P. Ramachandra, R. Ramani, G. Ramgopal, C. Ranganathaiah, *Eur. Polym. J.* **33** (1997), 1707.
72. G.N. Kumaraswamy, C. Ranganathaiah, M.V. Deepa Urs, H.B. Ravikumar, *Eur. Polym. J.* **42** (2006), 2655.
73. G.N. Kumaraswamy, C. Ranganathaiah, *Polym. Eng. Sci.* **46** (2006), 1231.
74. R.B. Gregory, *J. Appl. Phys.* **70** (1991), 4665.
75. L.M. Delves, J.L. Mohamed, *Computational Methods for Integral Equations*, Cambridge University Press: Cambridge, 1985.

76. J.G. McWhirter, E.R. Pike, *J. Phys. A.* **11** (1978), 1729.
77. D.M. Schrader, S.G. Usmar, in *Positron Annihilation Studies of Fluids*, S. C. Sharma (Ed.), World Scientific: Singapore, 1988.
78. A.N. Tikhonov, V.Y. Aisenin, *Solutions of Ill-Posed Problems*, Winston: Washington, DC, 1977.
79. C.W. Groetsch, *The Theory of Tikhonov Regularization for Fredholm Equations of the First Kind*, Pitman: Boston, 1984.
80. S.W. Provencher, *Comput. Phys. Commun.* **27** (1982), 229.
81. A.K. Livesey, Skilling, *J. Acta. Cryst. A* **41** (1985), 113.
82. A.K. Livesey, J.C. Brochon, *Biophys. J.* **52** (1987), 693.
83. S.W. Provencher, *Comput. Phys. Commun.* **27** (1982), 213.
84. R.B. Gregory, A. Procyk, in *Positron Annihilation in Fluids*, S. C. Sharma (Ed.), World Scientific: Singapore, 1988.
85. R.B. Gregory, Y. Zhu, *Nucl. Instrum. Methods A.* **290** (1990), 172.
86. R.B. Gregory, *Nucl. Instrum. Methods A.* **302** (1991), 496.
87. D. Rigby, R.J. Roe, *Macromolecules.* **23** (1990), 5312.
88. R.E. Robertson, R. Simha, J.G. Curro, *Macromolecules.* **18** (1985), 2239.
89. Y.C. Jean, in *Positron and Positronium Chemistry*, Y.C. Jean (Ed.), World Scientific, Singapore, 1990.
90. Q. Deng, F. Zandiehnadem, Y.C. Jean, *Macromolecules.* **25** (1992), 1090.
91. Q. Deng, Y.C. Jean, *Macromolecules.* **26** (1993), 30.
92. Y.C. Jean, Q. Deng, *J. Polym. Sci. Polym. Phys. Ed.* **30** (1992), 1359.
93. K.J. Heater, P.L. Jones, *Nucl. Instrum. Methods B* **56/57** (1991), 610.
94. C.L. Wang, B. Wang, S.Q. Li, S.J. Wang, *J. Phys.: Condens. Matter.* **5** (1993), 7515.
95. B. Haldar, R.M. Singru, K.K. Maurya, S. Chandra, *Phys. Rev. B.* **54** (1996), 7143.
96. Jun Xu, A.P. Mills Jr., A. Ueda, D.O. Henderson, R. Suzuki, S. Ishibashi, *Phys. Rev. Letts.* **83** (1999), 4586.
97. Jun Xu, J. Moxom, S.H. Overbury, C.W. White, A.P. Mills Jr., R. Suzuki, *Phys. Rev. Letts.* **88** (2002), 175502-1
98. R.A. Pethrick, *Prog. Polym. Sci.* **22** (1997), 1.
99. W.J. Davis, R.A. Pethrick, *Polym. Int.* **45** (1998), 395.
100. W.J. Davis, R.A. Pethrick, *Eur. Polym. J.* **34** (1998), 1747.
101. K. Venkateswaran, K. Cheng, Y.C. Jean, *J. Phys. Chem.* **88** (1984), 2465.
102. D.M. Schrader, Y.C. Jean, *Positron and Positronium Chemistry*: Elsevier: The Netherlands, 1988.
103. T.C. Merkel, B.D. Freeman, R.J. Spontak, Z. He, I. Pinnau, P. Meakin, A.J. Hill, *Science.* **296** (2002), 519.
104. I. Pinnau, L.G. Toy, *J. Membr. Sci.* **116** (1996), 199.
105. R. Mahajan, C.M. Zimmerman, W.J. Koros, in *Polymer Membranes for Gas and Vapor Separation: Chemistry and Materials Science*; B.D. Freeman, I. Pinnau (Eds.), American Chemical Society: Washington DC, 1999, p. 277.
106. P. Winberg, M. Eldrup, H.J.M. Frans, *Polymer.* **45** (2004), 8253.
107. P. Winberg, K. DeSitter, C. Dotremont, S. Mullens, F.J.V. Ivo, H.J.M. Frans, *Macromolecules.* **38** (2005), 3776.
108. M. Mukherjee, D. Chakravorty, *Phys. Rev. B.* **57** (1998), 848.
109. M. Wood, N.W. Ashcroft, *Phys. Rev. B.* **25** (1982), 6255.
110. M. Knite, A.J. Hill, S.J. Pas, V. Teteris, Zavickis, *J. Mater. Sci. Eng.* **C26** (2006), 771.
111. P. Asoka-Kumar, B.F. Dorfman, M.G. Abraizov, D. Yan, Fred H. Pollak. *J. Vac. Sci. Technol.* **A13(3)** (1995), 1044
112. B.G. Olson, Z.L. Peng, J.D. McGervey, A.M. Jamieson, E. Manias, E.P. Giannelis, *Mater. Sci. Forum.* **255–257** (1997), 336.
113. Z.F. Wanga, B. Wanga, N. Qia, H.F. Zhangb, L.Q. Zhangb, *Polymer.* **46** (2005), 719.
114. P. Winberg, M. Eldrup, Niels Jørgen Pedersenb, Martin A. van Esc, H.J.M. Frans, *Polymer.* **46** (2005), 8239.
115. B. Lizama, R. López-Castañares, V. Vilchis F. Vázquez *Mater. Res. Innovat.* **5** (2001), 63.
116. Yaping Zhenga, Ying Zhengb, Rongchang Ninga *Mater. Lett.* **57** (2003), 2940.
117. L.M. Liu, P.F. Fang, S.P. Zhang, S.J. Wang, *Mater. Chem. Phys.* **92** (2005), 361.
118. M. Zhang, P.F. Fang, S.P. Zhang, B. Wang, S.J. Wang, *Radiat. Phys. Chem.* **68** (2003), 565.
119. Shu Yanga, Peter Miraua, Jianing Sunb, David W. Gidley, *Radiat. Phys. Chem.* **68** (2003), 351.
120. M.J. Schulz, A.D. Kelker, M.J. Sunderasen, Enhancement of the Mechanical Strength of Polymer Based Composites Using Carbon Nanotubes in Nanoengineering of Structural, Functional, and Smart Materials, 1st edn.: CRC Taylor & Francis: USA (2006)

# 13 Thermal Characterization of Nanocomposites

*Robert A. Shanks*

School of Applied Sciences, RMIT University, GPO Box 2476 Melbourne, 3001, Australia
robert.shanks@rmit.edu.au

## Abstract

Nanocomposites provide enhanced properties at low filler fractions and often-unique properties conferred by the special size, surface, or shape of the nanoparticles. The most useful assessment of these properties comes from the thermal analysis. The main thermal analysis techniques used are differential scanning calorimetry (DSC), thermogravimetry (TGA), thermomechanical analysis (TMA), and dynamic mechanical spectroscopy (DMS). The application of each of these techniques to nanocomposites is reviewed with attention given to the approaches suitable for probing the structure of the nanocomposites. Nanofillers can nucleate crystallization, adsorb chain segments retarding motions, and limit phase separations – these can be revealed by DSC. Increased thermal stability has been detected by TGA. TMA and DMS are used in many programmed and geometrical modes to provide a measure of many mechanical properties as well as interpretation of structural features and equilibration. Nanocomposites are of increasing interest contemporaneously with new thermal analysis techniques that can provide increased information about nanocomposites and, indeed, all polymeric materials.

**Keywords:** differential scanning calorimetry; thermogravimetry; thermomechanical analysis; dynamic mechanical analysis; modulated temperature; modulated force; molecular relaxation; crystallization; melting temperature; glass transition temperature

## Abbreviations

| | |
|---|---|
| ABS | poly(acrylonitrile-*co*-butadiene-*co*-styrene) |
| AFM | atomic force microscopy |
| ASTM | American Standard Test Method |
| DDSC | dynamic differential scanning calorimetry |
| DMA | dynamic mechanical analysis |
| DMS | dynamic mechanical spectroscopy |
| DSC | differential scanning calorimetry |
| FTIR | Fourier transform infrared |
| HN | Havriliak–Negami |
| KWW | Kohlrausch–Williams–Watts |
| MDI | methanediphenyl diisocyanate |

| MDSC | modulated differential scanning calorimetry |
| MS | mass spectroscopy |
| MT-DSC | modulated temperature differential scanning calorimetry |
| MT-TGA | modulated temperature thermogravimetry |
| MWCT | multiwall carbon nanotube |
| PCL | polycaprolactone |
| PET | poly(ethylene terephthalate) |
| PLLA | Poly(L-lactide) |
| PMMA | poly(methyl methacrylate) |
| PP | polypropylene |
| PP-$g$-MA | polypropylene grafted with maleic anhydride |
| QI-MT-DSC | quasiisothermal modulated temperature differential scanning calorimetry |
| SBR | poly(styrene-$co$-butadiene) |
| $T_c$ | crystallization temperature |
| $T_d$ | decomposition temperature |
| TEOS | tetraethylorthosilicate |
| $T_g$ | glass transition temperature |
| TGA | thermogravimetry |
| $T_m$ | melting temperature |
| TMA | thermomechanical analysis |
| TTS | time-temperature superposition |
| WLF | Williams–Landel–Ferry |

## 13.1. Introduction

Nanocomposites are filled polymeric systems where the filler particles have a least one dimension in the nanometer scale. Consequently the fillers may be particulate (three dimensions of nanometer scale), platelet (two dimensions of nanometer scale) or fibrous (one dimension of nanometer scale). The polymer may be either thermoplastic or of network structure. They are characterized by the extremely high surface area to volume ratio. A small filler volume fraction will provide significant enhancements to structure, stability, and properties. Nanocomposites have become increasingly important as new materials are developed and as elements in many new high-performance materials.

Thermal analysis is a group of techniques where the temperature of a material is controlled, either by heating, cooling, isothermal, or modulated conditions while a property of the material is measured [1]. There are many specialized thermal analysis techniques; however, this chapter is restricted to the most widely available instruments. Properties of interest include the following:

- Exothermic or endothermic heat exchange by the sample in differential scanning calorimetry (DSC);
- Mass in thermogravimetry (TGA);
- Specimen dimensions with or without an applied force in thermomechanical analysis (TMA); and
- the complex modulus and its components of the storage modulus and loss modulus with the imposition of an oscillating force at a defined frequency in dynamic mechanical

spectroscopy (DMS). In this chapter, the term DMS is used to emphasize the dependence of properties on frequency, instead of the more commonly used dynamic mechanical analysis (DMA). The use of DMS emphasizes the measurement of properties rather than the technique providing an analysis of the material.

Other instruments are able to measure similar properties, but they are not classified as thermal analysis instruments. An inherent feature of thermal analysis instruments is the small sample/specimen size that provides rapid response of the specimen to changes in temperature and consequent imparting of other stimuli and measurements to be made rapidly. Thermal analysis instruments mostly make measurements when the material is not at thermodynamic equilibration though equilibrium conditions can be approached. A feature of the technique is that while the instrument is imparting its conditions upon the specimen, the specimen is providing a perturbation to the instrument. For example during an exothermic event, the specimen will add excess heat to the system so that the temperature may change from the specified program rate. In TGA, oxidative degradation is often most exothermic and the heat may cause localized heating of the sample in excess of the specified program. In TMA and DMS, the specimen may distort or otherwise physically change creating a stress that is combined with the instrument stress being imparted upon the specimen. Since the specimen must change its dimensions during the thermomechanical tests, these tests are the most difficult to conduct and isolate from extraneous events.

## 13.2. Differential Scanning Calorimetry and Temperature Modulation

### 13.2.1. Differential Scanning Calorimetry

DSC is used to measure polymer transitions, relaxations, crystallization, and melting associated with the polymeric components of nanocomposites. The filler is normally not detected except when it causes a change in the behavior of the polymer. DSC measures the heat flow associated with the material being tested as the material is subjected to a controlled temperature program. Many temperature programs are available, from isothermal and linear cooling and heating to specialized, complex, and innovative programs. The temperature program interacts with the polymer so that the results obtained are not purely a measure of the isolated properties of the polymer. The DSC graph is usually shown as heat flow versus temperature. In the examples shown and for our use of DSC, we convert the heat flow axis into heat capacity or specific heat, more correctly described as apparent heat capacity since the thermodynamic data are not measured under equilibrium conditions due to finite scanning rate and because the polymer is usually not at thermal equilibrium. Conversion of heat flow data to heat capacity requires that an independent baseline is recorded using matched empty sample pans and the calculation takes into consideration the sample mass and heating rate. The resulting apparent heat capacity curves enable DSC curve to be quantitatively compared, added, or subtracted to account for mixtures of components of a polymer composition. Figure 13.1 shows a DSC curve for crystallization on cooling of a poly(propylene-*ran*-ethylene) where the exotherm is shown as positive by heat capacity convention. Figure 13.2 shows a DSC curve for melting of a metallocene polypropylene.

A typical DSC curve of a polymer may reveal an inflection due to a glass transition, an exotherm due to cold crystallization, and one or more melting endotherms that may involve complex premelting endotherms and melting-recrystallization-melting phenomena

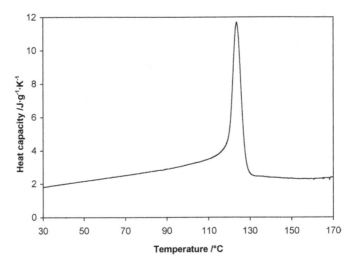

**Figure 13.1**   DSC heat capacity curve for crystallization of poly(propylene-*ran*-ethylene) during cooling at 2 K min$^{-1}$.

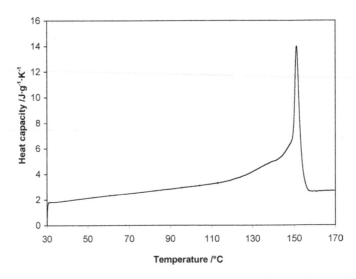

**Figure 13.2**   DSC heat capacity curve for melting of metallocene polypropylene at a heating rate of 2 K min$^{-1}$.

due to equilibrations during the scan [2]. Other components, such as filler, may affect these transitions. A typical DSC scan will consist of a heating scan to characterize the original as-received polymer followed by application of a cooling program to allow the polymer to crystallize and vitrify under the conditions of the scan, from the melted state after several minutes delay to allow the melt to equilibrate. Then a second heating scan will be applied to reveal the properties of the polymer that have been formed during the cooling scan. The difference between the two heating scans will give information on the thermal history of the original material compared with the material whose morphology has been generated during the cooling scan. The difference between the pure polymer and a composite can be attributed to the influence of the filler on the polymer.

The filler effects on the glass transition are often due to the adsorbed polymer segments having reduced degrees of freedom, although free-volume effects of the filler add complexity. The glass transition has been observed to increase due to restriction of degrees of freedom, or polymer segment mobility, but this is not always the case. Relaxation or physical aging effects upon annealing below the glass transition will generally be reduced.

## 13.2.2. Crystallization and Nucleation

Crystallization is often first modified by nucleation by the filler. This is particularly the case with nanofillers due to their large specific surface area. Nucleation will cause the onset of crystallization on cooling to occur at a higher temperature. In the case of isothermal crystallization, the onset will occur after a shorter incubation time [3]. The rate of crystallization can be quickly calculated by measuring the difference between the onset and peak temperatures for crystallization or the width at half height of the crystallization exotherm. The rate of crystallization may be increased due to the enhanced nucleation [4], or it may be decreased due to adsorption and immobilization of chain segments. A broader exotherm will signify a slower rate of crystallization. Quantitative analysis of crystallization kinetics requires isothermal DSC measurements and analysis of the data using the Avrami equation. Alternative temperature scanning methods for analysis of crystallization kinetics have been developed [5].

Melting of crystals occurs during a heating scan and often complex endotherms occur that reveal aspects of the thermal history created during processing, cooling, isothermal crystallization, and annealing. The melting endotherm peak of crystals formed during crystallization is the first peak to identify. The peak temperature will correlate with the isothermal crystallization temperature or the cooling rate when crystallization has occurred under controlled conditions in the DSC instrument. A lower temperature premelting peak is often observed. The premelting peak is associated with secondary crystallization or thermal annealing of the crystalline polymer. The premelting peak temperature and enthalpy vary considerably with thermal history. A double melting peak may be associated with the main melting peak. The higher temperature peak is usually due to recrystallization of the already formed crystals during the DSC scan. It is an artifact of the DSC scanning conditions. If scanning were very fast, then the higher temperature peak of the double melting peak would be absent. If scanning is very slow, then the higher temperature peak may dominate to the extent that it is the only peak observed. In this case, information about the original crystals will be absent. The process by which this occurs is called melting-recrystallization-melting.

If the polymer is crystallized under isothermal conditions, then the associated melting temperature of the original crystals is determined by the crystallization temperature. A plot of melting temperature ($T_m$) versus crystallization temperature ($T_c$) can be extrapolated to a line where $T_m = T_c$ to obtain the equilibrium or thermodynamic melting temperature ($T_m°$) for the polymer. For a copolymer, $T_m°$ is decreased by the fraction of comonomer or non-crystallizable unit, so extrapolation of a $T_m$ versus $T_c$ graph will give the copolymer melting temperature ($T_m^c$). Fillers and, particularly, nanofillers contribute through nucleation and interference with growth during crystallization. This consequentially can cause change in the melting temperature and the melting endotherms observed, such as premelting, main melting peak, and melting-recrystallization-melting peak.

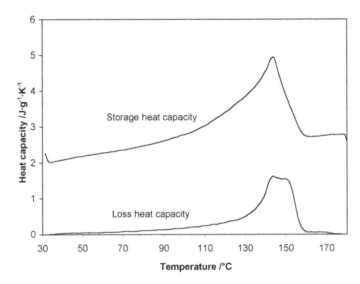

**Figure 13.3**  MT-DSC melting of poly(propylene-*ran*-ethylene) revealed by storage and loss heat capacities measured using an underlying heating rate of 2 K min$^{-1}$.

### 13.2.3. Thermodynamic and Kinetic Events During Melting

The complex processes occurring during melting can be considered as thermodynamic or reversible events, coupled with kinetic or nonreversing events. These events can be resolved using modulated temperature DSC (MT-DSC) that is available in several forms [6]. MT-DSC involves imparting an oscillating temperature change upon the underlying temperature program. The total heat flow signal, which is equivalent to a conventional DSC scan under the same underlying program rate, is resolved into a rapidly reversing heat flow and a nonreversing heat flow. The reversing heat flow is reversible under the scanning conditions, but named differently to distinguish from thermodynamic reversibility. The nonreversing heat flow is the difference between the total and reversing heat flows. Another method involves calculation of a complex heat capacity. The complex heat capacity is resolved into an in-phase component, the storage heat capacity, and an out-of-phase component, the loss heat capacity. Separately a total heat capacity is obtained from this method as the average heat capacity over one temperature modulation cycle. The storage heat capacity is interpreted similarly to the reversing heat capacity. Figure 13.3 shows an MT-DSC curve for melting of poly(propylene-*ran*-ethylene) where storage and loss heat capacity provide separate information on the processes. Figures 13.4 and 13.5 show the MT-DSC cures for melting of Ziegler–Natta polypropylene and metallocene polypropylene for comparison with Figure 13.3.

The storage-specific heat is equivalent to the thermodynamic-specific heat for the materials when no transitions or phase changes are occurring. The glass transition is observed in the storage or reversing signal [7]. Melting due to poorly formed crystals or where molecular segments become partially detached from a crystal are reversing since they can recrystallize during a cooling part of the modulated cycle [8]. When molecules are detached from a crystal, or crystals completely melt, the melting is nonreversing since recrystallization would require a nucleation step. The loss specific heat is more difficult to

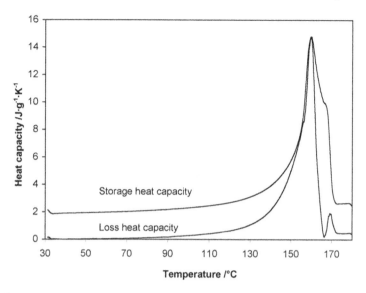

**Figure 13.4** MT-DSC melting of polypropylene revealed by storage and loss heat capacities measured using an underlying heating rate of 2 K min$^{-1}$.

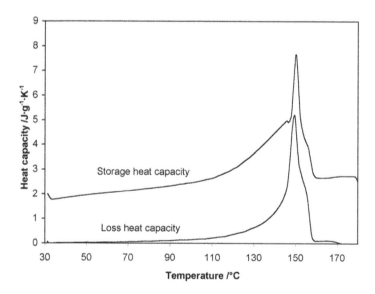

**Figure 13.5** MT-DSC melting of metallocene polypropylene revealed by storage and loss heat capacities measured using an underlying heating rate of 2 K min$^{-1}$.

ascribe to transition phenomena. The lost specific heat is zero except when a phase change is taking place. The lost specific heat has been interpreted as due to entropy changes as molecules gain degrees of freedom due to the associated phase change [9].

MT-DSC has been applied to the crystallization and melting of poly(L-lactide) (PLLA) composites of multiwall carbon nanotubes (MWCT), where the filled and polymer are grafted to form MWCT-$g$-PLLA. Crystal perfection, secondary crystallization, and double

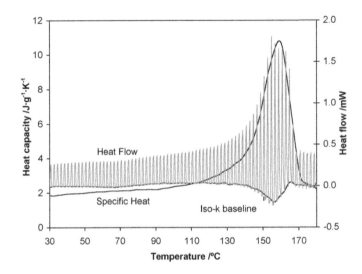

**Figure 13.6** MT-DSC melting of polypropylene revealed by the step-wise heat flow data, thermodynamic-specific heat and the iso-k baseline, or kinetic signal, measured using the step-scan method.

melting endotherms were observed and distinguished in the reversing and nonreversing heat flow curves [10].

Quasi-isothermal MT-DSC (QI-MT-DSC) involves a modulated temperature superimposed on an underlying isothermal condition. Data for each temperature must be obtained separately as the isothermal temperature is incremented. The temperature modulations are performed with small amplitude, and data collection is delayed at each isothermal step so that equilibrium is approached. The resulting QI-MT-DSC curve approaches thermodynamic equilibrium conditions and represents most closely the true equilibrium transition phenomenon for the polymer. Crystallization and melting are typically studied using QI-MT-DSC [11]. Step-scan DSC provides an alternative method for measuring an apparent thermodynamic-specific heat using a series of temperature steps, followed by equilibration as the heat flow approaches a constant value before moving to the next step. Figure 13.6 shows step-scan curves for the melting of polypropylene (PP).

## 13.3. Thermogravimetry

TGA is the most useful initial way to study all materials [12]. The methods range from simple comparative scans to innovative temperature and purge gas combinations. TGA has been combined with other instruments such as Fourier transform infrared (FTIR) spectroscopy and mass spectroscopy (MS). These hybrid or hyphenated techniques will not be considered in this chapter as here the objective is to review the many methods that use TGA alone. In addition to the TGA method, there are ways, such as differential thermogravimetry (DTG) and curve subtractions, to present the data to emphasize comparisons between materials. The data are often used to calculate kinetic parameters using a range of kinetic models and data conversion techniques such as isoconversion, and transposition of data measured at different scan rates into isothermal curves. TGA interpretation can lead to the knowledge of the reactions taking place at various temperatures and their mechanisms.

Consider a typical nanocomposite that will be analyzed by TGA. The first method is typically a temperature scan at a heating rate of 10 or 20 K min$^{-1}$ using an inert purge gas such as nitrogen. The gas can be changed to air at 700–800°C for a scan to 900–1000°C. The air causes any carbonaceous residue to oxidize to carbon dioxide so that the residue will only consist of the filler. The dependent variable is normally mass loss as a fraction (percent) of the starting mass. The starting mass should be small (2–5 mg) to avoid thermal lag, though the starting mass must be chosen to provide a representative and uniform sample. The mass loss curve will show a series of steps that may be due to volatiles such as water or residual solvent, higher boiling liquids such as plasticizers (> 200°C), decomposition starting with intercalating agents of layered clay nanocomposites (> 300°C), and polymer degradation (300–500°C) that can occur in several steps depending on the characteristic mechanism. Polymer degradation may occur in a single step when depropagation or random scissions are involved, whereas loss of a small molecule such as dehydration or dehydrohalogenation will be followed by decomposition leading to aromatization and carbonization with loss of hydrocarbon fragments. When the purge gas is switched to air, any carbonaceous residue will be oxidized to carbon dioxide leaving inorganic residue. Superimposed on the polymer degradation will be filler decomposition, typically through loss of water of hydration, carbon dioxide from carbonates, and sulfur dioxide from sulfates.

While a general degradation sequence is described above, TGA is often used to compare any effect of fillers on the stability of the polymer. Some nanofillers increase the onset degradation temperature of the polymer, the peak temperature of the DTG curve, and/or the end-set temperature of degradation. Shifting of polymer degradation to higher temperatures may be due to adsorption of polymer onto the filler, adsorption of pyrolysate molecules onto filler, or the tortuous path created by the filler to impede diffusion of volatiles to the surface. The effect of filler may be reversed in that fillers may possess catalytic properties that will enhance polymer degradation, thus moving it to lower temperatures. Increase in degradation temperatures by fillers is particularly strong with nanofillers due to their large surface areas – this has been associated with fire retardant properties.

Several techniques are available to aid comparison of filler effects. TGA curves can be recorded for each of the components of a blend or composite, or groups of components. The individual TGA curves are then combined in the same ratio as the constituents of the composition. This is a synthetic or calculated TGA curve that can be overlaid with the experimental curve for comparison. If the curves are the same, then there has been no interaction revealed in the composition. If the curves are different, then the effect must be due to interactions such as stabilization or catalysis of degradation. Similarities or differences can be revealed more clearly by subtraction of the synthetic curve from the experimental curve. Positive differences will mean that the stability has been increased, while negative differences will mean that the composition is less stable than the components.

Sample controlled temperature programs enhance TGA resolution. Instead of the operator specifying a temperature program rate, the rate is determined by the differential of the mass loss, that is, the mass loss rate. When there is no mass loss, the scan rate is increased by the instrument, usually to an upper limit set by the operator. When mass is lost, the scan rate is decreased in proportion to the mass loss rate by an amount or sensitivity set by the operator. At highest mass loss rate, a minimum scan rate or an isothermal condition will apply as set by the operator. In this way, the lowest scan rates and hence highest resolution will occur during rapid mass loss. The overall time for the complete scan can still occur in a convenient time for the productive use of the instrument. Sample controlled TGA is still better than conducting the whole scan over a slow scan rate since the high scan rates prevent the kinetics of mass loss from prevailing at lower temperatures. Any premature mass loss is prevented by the rapid scan rate so that a high onset temperature is maintained. The temperature is then increased slowly until the mass loss rate declines thus keeping the mass loss within a narrow temperature range, and then the scan rate increases again until another mass loss is detected. The result is narrow mass loss steps or DTG peaks that have maximum separation or resolution.

Modulated temperature TGA (MT-TGA) has been used similar to MT-DSC. The linear temperature program is superimposed with a sinusoidal, or otherwise modulated, temperature fluctuation. Rapid or near instantaneous reactions will change with the modulation, that is, they will be in-phase with the instantaneous temperature. Slower or kinetically controlled reactions will lag the instantaneous temperature changes and be out-of-phase. The mass loss can thus be resolved into reversing or apparent thermodynamic mass loss, and nonreversing or kinetic mass loss. This interpretation of the resolved mass losses with respect to reactions occurring within the analyte can contribute to knowledge of the reaction mechanism and diffusion processes as the volatiles are evolved.

## 13.4. Thermomechanical Properties

### 13.4.1. Coefficient of Thermal Expansion

The starting test for TMA is a measurement of thermal expansion coefficient and thermal transitions revealed by inflections. This is a zero load test, though a small load is required to keep the probe in contact with the specimen, and the area of the probe in contact with the specimen should be large so that the penetration is prevented. The test can be performed with heating or cooling programs. In addition to thermal transitions, thermal aging can be detected as a sudden increase in position upon reaching the glass transition. Thermal aging can be assessed by measuring slow isothermal contraction of the specimen at temperatures below but near to the glass transition. Upon cooling, the contraction will decrease from the equilibrium curve at the glass transition. Subtle changes will be resolved by a derivative probe position curve. This test monitors overall changes occurring within the specimen. Filler will limit expansion according to its volume fraction. Filler will perhaps increase transition temperatures by actively adsorbing polymer segments. Nanofillers are relatively active in this role due to their high relative surface area. Graphite nanocomposites have shown changes in conductivity during elastic deformations that have been related to the disruption and formation of expanded nanographite agglomerates [13].

### 13.4.2. Static Force Penetration Tests

A penetration test requires a compressive force and an indentation probe [14]. The probe may be of hemispherical shape with the diameter determined from those available from the instrument's manufacturer. The load is selected according to the specimen properties, so preliminary tests are often required for materials unfamiliar to the operator. The indentation shows an increase inflection at the glass transition. The inflections may be resolved by a derivative curve, and more detail is often shown than with expansion mode since penetration is exploring the internal regions of the specimen. Penetration mode in a three-point bend configuration is used to determine the standard Vicat softening temperature. A small dynamic stress superimposed on the static stress can be used for these measurements, but the test is not considered a dynamic test since static properties are being determined. Fillers will limit penetration by physically impeding the movement of the probe. Penetration will be dependent on the orientation for fillers with a nonspherical shape. Fillers adsorb polymer segments and restrict polymer molecular mobility, thereby further impeding the probe penetration. Complex morphology-dependent data may be obtained from penetration measurements, so resolving the distinction between amorphous polymer constrained by crystals or filler, the rigid amorphous phase, and the isolated amorphous polymer regions. The thermal behavior of aliphatic–aromatic polyesters in the region of the glass transition demonstrates the application and distinction of thermal expansion and penetration methods in detecting the mobile and rigid amorphous fractions [15].

### 13.4.3. Stress–Strain Analysis

A stress–strain analysis can be preformed with the TMA-DMS instrument with a limitation that the elongation is restricted to small values by the instrument; moderate strain can be obtained by using small specimen size. The instruments are typically stress controlled so a rate of increase of force or stress is applied, and the resulting deformation or strain is measured. Strain can be controlled via a feedback loop in the instrument software so that strain rates may be able to be specified. The stress–strain curve is used to obtain the static modulus and other parameters typical of such tests on larger scale test instruments. The stress–strain data serve another purpose with TMA-DMA measurements in that it is used to establish the linear viscoelastic regions for DMS measurements and suitable stress for creep measurements. These preliminary studies should be used prior to performing complex TMA-DMS measurements to establish and confirm experimental parameters. The stress–strain curves reveal changes in properties contributed by fillers, such as increased modulus, orientation effects, increased yield stress, and decreased ultimate elongation if the strain can be great enough. Figure 13.7 shows a stress–strain curve for natural rubber where the stress has been increased and then decreased to give a hysteresis curve.

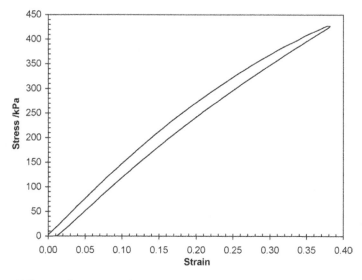

**Figure 13.7**  Tensile stress-strain hysteresis of natural rubber measured in TMA mode.

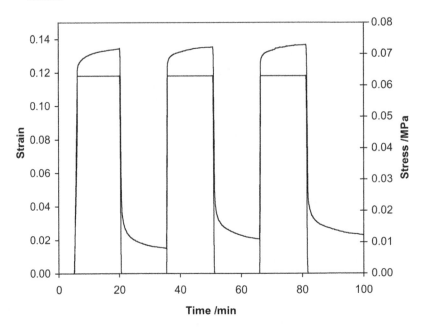

**Figure 13.8**  Successive creep and recovery cycles for poly(ethylene-*co*-propylene) foamed elastomer in the static penetration mode at ambient temperature.

The TMA instrument is ideal for creep and stress relaxation measurements since precise control of conditions, complex programs, and temperature control are available. The creep test involves immediate application of force or stress for a set time. The strain is measured throughout. The immediate strain response is the elastic response. The viscoelastic and viscous flow elements increase the strain over time. Upon removal of the stress, the elastic recovery is immediate, followed by the viscoelastic recovery, leaving the viscous flow or permanent set. The recovery time is recommended by ASTM to be four times the time under stress. Further cycles of stress and recovery can be applied. Increasing stress steps can be used to calculate an isochronous stress–strain curve. Measurements at different temperatures can be used to construct a creep master curve.

Stress relaxation is a complementary test where the strain is maintained and the stress measured over time. Various strain steps can be applied. Creep and stress relaxation tests allow all facets of the elastic, viscoelastic, and viscous response of the specimens to be studied. Examples of creep and recovery for a poly(ethylene-*co*-propylene) foamed elastomer in compression mode and expanded stress relaxation of natural rubber are shown in Figure 13.8. Figure 13.9 shows a stress relaxation curve for natural rubber. Stress relaxation of shape memory polyurethane nanocomposites was measured by a DMA instrument in tensile mode after applying 100% strain at 60°C. Stress relaxation ratio was measured relative to the initial stress over 7 min of recovery [16]. The polyurethane was based on polycaprolactone (PCL), butanediol, and methanediphenyl diisocyanate (MDI) formed in the presence of Chlosite 30B treated clay where the quaternaryammonium ions contain an alkyl with hydroxyl functionality that can be tethered to the polyurethane structure.

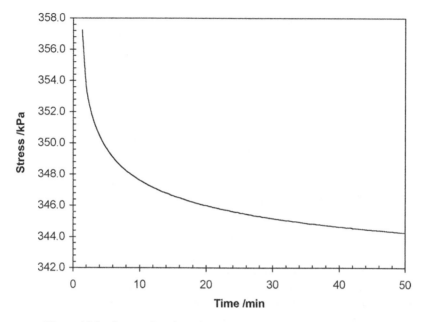

**Figure 13.9**   Stress relaxation of natural rubber at ambient temperature.

The universal viscoelastic model is used to analyze the creep parameters [17]. Uniaxial creep, stress relaxation, and constant strain-rate data have been analyzed using a viscoelastic model [18]. The model has been applied to two poly(acrylonitrile-*co*-butadiene-*co*-styrene) (ABS) with the application of different stresses [19]. Fillers are expected to increase the modulus and so reduce the immediate elastic response. The creep and subsequent permanent set will be reduced by fillers acting as physical crosslinks. The viscoelastic response will be slowed by polymer adsorption onto filler giving higher relaxation times, higher creep modulus, and higher creep viscosity element. Nanocomposites will display similar characteristics with lower filler content. The response will be dependent on orientation of the material and the filler fibers or platelets within the composite. Nonlinear tensile creep of high-density polyethylene–cycloolefin copolymer blends has been revealed using time-strain superposition, and the segmental motions were controlled by available fractional free volume [20]. Creep of glass-reinforced aromatic and aliphatic thermoplastics with 30% filler has shown viscoelastic response that can be extended to construct isochronous stress–strain diagrams, creep modulus charts, and the Maxwell linear viscoelastic constants [21].

The viscoelastic creep, recovery, and stress relaxation model takes into account the elastic, viscoelastic, and viscous elements. Data have typically been fitted using an exponential response; however this has been shown to be better fitted using a stretched exponential function, such as the Kohlrausch–Williams–Watts (KWW) function. The spring-dashpot model has been extended to provide a mechanical analogy of a latch to illustrate the KWW concept. During creep and recovery, not all molecules can respond due to constraints from nearby molecules. When the relaxation of a constrained molecule or molecular segment becomes active, a latch is opened in the mechanical analogy. The latch is described by the nonexponentiality parameter [22]. The stretched exponential function of the KWW model is known as the Weibull distribution, a model that is widespread in the analysis of material failures [23]. The total strain ($\varepsilon_{ctot}$) under an applied stress is given by

$$\varepsilon_{ctot}(t) = \varepsilon_i + \varepsilon_c \left[ 1 - \exp \left( - \left\{ \frac{t}{\eta_c} \right\}^{\beta} \right) \right], \tag{13.1}$$

where the subscripts refer to: $i$ = instantaneous, $c$ = creep, $\eta$ = viscosity, and $\beta$ = non-exponality or shape parameter. Upon removal of the stress, there may be elastic recovery (subscript=e) and contributions from viscoelasticity ($f$) and viscous flow ($f$):

$$\varepsilon_{rvis}(t) = \varepsilon_r \left[ 1 - \exp \left( - \left\{ \frac{t}{\eta_r} \right\}^{\beta_r} \right) \right] + \varepsilon_f. \tag{13.2}$$

Each equation is a four-element model that has been found to correlate with experimental data.

Viscoelastic behavior from creep and dynamic mechanical measurements on layered silicate nanocomposites have been modeled using the Havriliak–Negami (HN) equation and cole–cole functions for poly(ethylene terephthalate) PET copolyesters [24]. The time-domain KWW and frequency-domain HN relaxation functions equivalence has been compared and proven for the $\alpha$-relaxation of glass forming poly(hydroxy ethers of bisphenol-A) [25].

## 13.5. Dynamic Mechanical Spectroscopy

### 13.5.1. Elasticity and Damping Characteristics

The normal DMS test will be a temperature scan with a constant sinusoidal stress applied and the strain and phase delay measured. The resultant parameters are the storage modulus ($E'$ or $G'$, elasticity), loss modulus ($E''$ or $G''$, viscoelasticity), and the damping factor ($\tan(\delta) = E''/E'$). The purpose of the measurement is to identify various thermal transitions due to segmental, branching, and substituent motions of the polymer. Another purpose is to measure each parameter at various temperatures of interest. Since the stress applied is being repeated each cycle, the material must be within its linear viscoelastic region for the results to be valid. This will involve some preliminary static measurements to identify this region. The results will show the contribution of filler to the dynamic mechanical parameters, such as changes in the transition temperatures and the magnitude of the parameters at significant temperatures. The measurement is preferably performed using a constant strain within the linear viscoelastic region, and the force is modified by the instrument to maintain the strain. Scan rates must be kept low relative to other thermal analytical techniques to avoid thermal lag [26] and the thermal environment inside the furnace [27].

A frequency scan is performed by increasing the frequency while maintaining the temperature and the constant sinusoidal stress. Alternatively, the experiment can be performed at constant strain, within the linear viscoelastic region, by the instrument progressively changing the applied stress. A frequency scan at a single temperature is useful in choosing a frequency for a temperature scale. The most common frequency used is 1 Hz, although this is arbitrarily chosen. Frequencies in the range 1–10 Hz are often selected after preliminary frequency scans. It is important to not be near the resonance frequency, though these are typically higher in the 100 Hz range. The resonance frequency can be modified by changing the specimen dimensions. The process for the measurement of polymers in three-point bending mode has been evaluated and improved using a Taguchi experiment design, and the results demonstrate that different experiments should be conducted to obtain the best results for all properties [28].

Storage and loss modulii and damping factor are obtained with temperature. Various frequencies can be applied and the consequent data used to calculate activation energies, master-curves, and analyze the uniformity of distribution of nanoparticles. Similar data are provided by dielectric relaxation spectroscopy. Multiple frequency–temperature scans can be performed by a step-wise temperature program. A soak time is set at each temperature to allow the specimen to equilibrate and then each of the frequencies is applied sequentially. The temperature is then changed to the next step and the process repeated. A limited range of frequencies can be scanned simultaneously using a synthetic combined frequency. The component frequency data are resolved using a Fourier transform. A review of the application of Fourier transform to rheology provides comprehensive theory and exper-imental information that can be applied to DMS [29]. During synthetic frequency scans, the temperature can be changed linearly as for a single frequency, except that the combined frequency profile is used as shown in the DMS graph in Figure 13.10 for very low-density polyethylene. A DMS analysis for poly(styrene-co-butadiene) rubber (SBR) shown in Figure 13.11 was measured using a synthetic frequency of five component frequencies to obtain the multifrequency data. Nanosilica generated from tetraethylorthosilicate (TEOS) was generated in situ in poly(perfluoroethylene) ion exchanger (Nafion, Du Pont). The nanosilica composites were mechanically characterized using DMA in tensile mode at 1 Hz from $-10°C$ to $260°C$ at $4 \text{ K min}^{-1}$ and the thermomechanical stability was found to be increased relative to the unfilled acid form [30]. Biopolymer nanocomposites based on PLLA and bentonite or microcrystalline cellulose were investigated and compared using DMA in tensile mode at 1 Hz, 0.05% strain from $15°C$ to $100°C$ at $3 \text{ K min}^{-1}$ and the damping peak moved to higher temperatures in the composites compared with PLLA [31]. PP and PP grafted with maleic anhydride (PP-g-MA) were used to prepare nanocomposites with layered silicate (Closite 20A, Southern Clay Products), and DMA was studied in tensile mode at 10 Hz from $-40°C$ to $160°C$ at $2 \text{ K min}^{-1}$ [32].

## 13.5.2.  Frequency Dependence of Viscoelasticity and Time-Temperature-Superposition

The multifrequency data for storage modulus, loss modulus, and tan($\delta$) can be analyzed in several ways. First using the Arrhenius equation, log(frequency) is plotted against 1/temperature ($T$ in K) of the peak of tan($\delta$) and the slope is $-E/RT$, where $E$ is the activation energy for the transition and $R$ is the gas constant. The activation energy ($E$) is the characteristic of the transition and the composition of the material. Nanocomposites are expected to show increased activation energies due to the retardation of segmental motions by adsorptions and steric hindrances of the filler particles, platelets, or fibers. The nonlinear behavior of polyethylene foams is influenced by their cell-edge bending, cell-face stretching, and gas inside the cells providing complex behavior resulting from small-scale structural features [33]. DMA was performed using single cantilever bending mode on sheets of poly(trimethylene terephthalate) using a slow scan rate of $1 \text{ K min}^{-1}$ at frequencies of 0.1–10 Hz to reveal the constraining influence of the crystal–amorphous interface region [34].

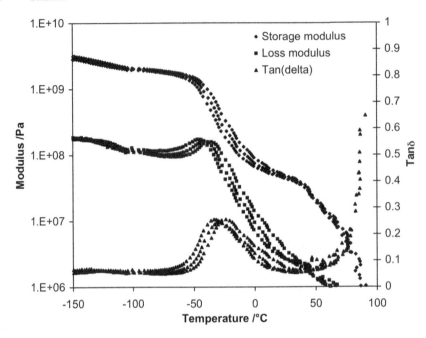

**Figure 13.10**  Tensile DMS of metallocene very low-density polyethylene at three simultaneous frequencies (0.1, 1, 10 Hz).

The modulus and damping factor versus temperature curves measured at a single frequency (typically between 1 and 10 Hz) provide detailed and sensitive information about thermal transitions. Often the thermal transition temperatures are increased when the polymer motions are retarded by adsorption onto filler. Alternatively, decreases have been observed due to increased free volume in a composite or to change phase separation in polymer blends with added filler. After allowing for the filler dilution, the magnitude of the damping peak is usually decreased in filler systems. Crystal–crystal transitions are observed at temperatures above the glass transition, and filler effects on the initial nucleation and crystallization or during the crystal–crystal transition range often influence these transitions. The data can be measured in tensile, compression, three-point bend, dual cantilever bend, and shear sandwich modes. There are newer modes becoming available, and fixtures where the specimens can be exposed to various liquids or humid environments.

The second and most common method of analysis is time-temperature superposition (TTS) using the Williams–Landel–Ferry (WLF) equation to construct a master-curve. The master-curve is constructed by shifting the modulus-frequency data, measured at different temperatures, along the frequency axis until curves measured at different temperatures from a chosen reference temperature are superpositioned, or overlapped in regions of common data. The amount of shift along the frequency axis is the shift factor ($\log(a_T)$). The values of the shift factor are correlated with the difference in temperature from the reference temperature according to the WLF equation. Superposition of data can be improved by applying a simultaneous vertical shift to allow for changes in free volume, or density, with temperature. There are many ways to perform the TTS, the simplest being a

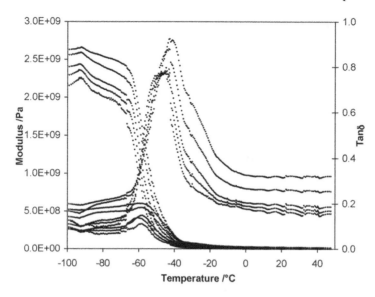

**Figure 13.11**  Tensile DMS of SBR using a synthetic frequency of five component frequencies (1, 2, 4, 10, 20 Hz).

visual superposition of the DMA curves. Alternatively, the theoretical values for the WLF coefficients ($C_1$ = 17.44 and $C_2$ = 51.6) can be used:

$$\log(a_T) = \frac{-C_1[T - T_0]}{C_2 + [T - T_0]}.$$  (13.3)

The reference temperature is often chosen as ambient temperature though $T_g$, or other application temperature, is sometimes chosen. Another approach is to fit each temperature data set to the WLF equation and select the pair of $C_1$ and $C_2$ with the best fit and use with the overall data. An example of TTS master-curve for an acrylate ester copolymer is shown in Figure 13.12. Other computational methods are available that are not discussed here.

DMA has been used to measure the adhesive characteristics of polyacrylate pressure sensitive adhesive using multilayer lap shear test geometry. The first approach was to apply a frequency sweep, isothermal test over nine frequencies, 100 to 0.01 Hz. A second approach was a frequency temperature sweep at 100, 10, 1, and 0.1 Hz at temperatures from −50°C to 60°C in 5°C intervals with 5 min soak time at each temperature. The latter data were used to construct TTS master-curves at 23°C according to the WLF equation [35].

The master-curves allow extremes of frequency to be considered – these are equivalent to short and long times for prediction of properties at times inconvenient or impossible to measure. Data for the terminal and plateau regions can be obtained to compare properties at very low frequency where viscoelastic flow may dominate, and very high frequencies where only elastic response can occur. The master-curve data for storage modulus ($E'$) can be plotted against loss modulus ($E'$) to compare the Cole–Cole relationship. The Cole–Cole graph should be a semicircle with its diameter along the $E''$ horizontal axis for a homogeneous material. A Cole–Cole graph that is distorted to the upper right represents an increasing heterogeneous material. This may correspond to filler content and uneven

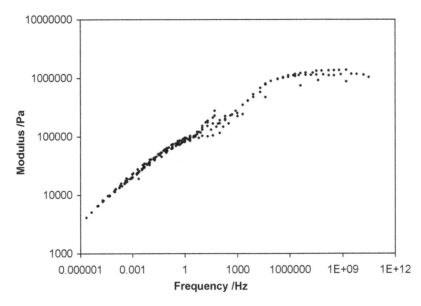

**Figure 13.12**  Storage modulus master-curve with reference temperature 25°C, measured in shear sandwich mode, for an acrylate ester copolymer.

filler dispersion or agglomeration. Many aspects of nanocomposite are revealed in master-curves and Cole–Cole analysis.

DMS master-curves have been constructed for two- and three-component polyurethanes using parallel plate mode with cooling and heating scans. The dynamic properties characterize phase separation and the presence of mesophases in the complex chemistry of the polyurethanes [36]. Relaxation phenomena have been interpreted according to conformational changes, vibrational modes, and phase changes of crystalline polymer blends with emphasis on interfaces and orientation [37]. The concept of dynamic fragility has been applied to the dynamic, creep behavior, and thermally stimulated recovery of poly(methyl methacrylate) (PMMA). Master-curves were established for PMMA, and the nonexponentiality of the structural relaxation function was related to dynamic fragility. A change from Vogel to Arrhenius behavior was observed [38].

## 13.6.  Future Trends

Advanced thermal analysis techniques will further probe nanocomposite properties. Further insight into the interpretation of MT-DSC will enable reversible events to be distinguished from entropic and kinetic events, since equilibrations and thermodynamic stability of morphologies is important. New MT-DSC programs and methods of data analysis are being developed such as the step-scan method. High heating rate DSC will be important for measuring nonequilibrium properties by preventing equilibrations during slower than normal DSC scanning times.

TGA has been enhanced through sample controlled heating programs to achieve higher resolution. The interpretation of data through kinetics and by correlation with burning tests, particularly cone calorimetry, means that increasingly TGA is the prime analytical technique for first assessment of flammability of polymer compositions and assessment of fire retardants. Coupling of TGA with FTIR spectroscopy and gas chromatography–MS is providing new insights into the mechanism of degradation and burning.

The combination of TMA and DMS, often in the same instrument, is exceptional in its ability to measure many mechanical properties. This has been increased by better and more stable instrument design and Fourier transform of frequency data to isolate analytical frequencies from noise. The Fourier transform of the raw data now makes available multiplexed frequency DMS where typically five frequencies are imparted simultaneously as a compound stress wave on the specimen and deconvoluted for the presentation as a final data set. Temperature-modulated DMA has been developed [39]. The TMA and DMS instruments are distinguished from traditional larger instruments by small sample size, low total force, and small strain reading capability. Larger dynamic instruments are closing this gap. The TMA and DMS instruments provide more capabilities in a single instrument with versatility of heating, atmosphere/environmental control, and variety of stress, strain, temperature, frequency, and time programs. Submicron [40] or even nanoscale TMA and DMS experiments are now used including application of atomic force microscopy (AFM).

## 13.7.  Conclusion

Nanocomposites provide enhanced nucleation of crystallization or encourage growth of particular polymorphs, though retardation of segmental motions may decrease the crystal growth rate. The amorphous phase may have constrained motions because of segmental adsorption to a nanofiller, thus increasing the glass transition temperature, or increasing the fraction of rigid amorphous phase. In some cases, the glass transition temperature is decreased since the nanoparticles may increase the free volume within the composite. In many cases, the thermal stability of a polymer is increased by a nanofiller and this is detected by TGA. Nanofillers have been found to be effective fire retardants. Some intercalating agents decompose at low temperature so that the onset of degradation occurs at lower temperature than normally exhibited by the polymer. Some nanofillers may be catalytically active toward degradation thereby decreasing the degradation temperature of the composite. DSC reveals much about the morphology, transitions, and relaxations of nanocomposites, while TGA quantifies the stability and composition.

Measurement of mechanical properties is complex since many specific modes and parameters need to be considered. Overall the modulus will be increased, whether tensile, shear, or compressive. The strength will tend to increase. Compared with micron-sized fillers that increase the strength and decrease the toughness, nanofillers often increase both strength and toughness due to crack arrest mechanisms following impacts. Adsorption of polymer onto the large nanoparticle surfaces will provide effective physical crosslinks that will decrease creep and enhance recovery though with an increase in relaxation time; the complementary stress relaxation will be similarly changed. Thermal expansion coefficient is likely to be decreased, as is volume relaxation on cooling, or with time. The dynamic properties will typically show decreased damping, increased transition temperatures, increased elastic storage modulus, the modification of plateau and terminal regions of the frequency-dependant complex modulii, the phase distribution available from the relationship between loss and storage modulii, and the activation energy of thermal transitions and relaxation. Thermomechanical and dynamic mechanical techniques revealed much about nanocomposite structure as well as specific properties.

# References

1.  J.O. Hill, Thermal analysis techniques: An overview; in Lee S (Ed.), Encyclopedia of Chemical Processing, Tayor & Francis, New York, (2006) 2965–2973.
2.  Z. Ziaee, P. Supaphol, Non-isothermal melt- and cold-crystallization kinetics of poly(3-hydroxybutyrate), Polym. Test., **25** (2006) 807–818.
3.  B. Osowiecka, A. Bukowski, J. Zielinski, W. Ciesinska, T. Zielinski, Investigations on nucleated polypropylene with using thermal analysis method, J. Therm. Anal. Calorim., **74** (2003) 673–679.
4.  S. Nagasawa, A. Fujimori, T. Masuko, M. Iguchi, Crystallisation of polypropylene containing nucleators, Polymer, **46** (2005) 5241–5250.
5.  E. Piorkowska, A. Galeski, J.M. Haudin, Critical assessment of overall crystallisation kinetics theories and predictions, Prog. Polym. Sci., **31** (2006) 549–575.
6.  J. Schawe, A comparison of different evaluation methods in modulated temperature DSC, Thermochimica Acta, **260** (1995) 1–16.
7.  J. Schawe, Principles for the interpretation of modulated temperature DSC measurements Part 1: Glass transition, Thermochimica Acta **261** (1995) 183–194.
8.  J. Schawe, Principles for the interpretation of temperature-modulated DSC measurements Part 1: A thermodynamic approach, Thermochimica Act **304/305** (1997) 111–119.
9.  G.W.H. Hohne, Remark on the interpretation of the imaginary part C" of the complex heat capacity, Thermochimica Acta **304/305** (1997) 121–123.
10. Y.T. Shieh, G.L. Liu, Effects of carbon nanotubes on crystallization and melting behaviour of poly(L-lactide) via DSC and TMDSC studies, J. Polym. Sci. B. Polym. Phys. **45** (2007) 1870–1881.
11. C. Schick, M. Merzlyakov, A. Minakov, A. Wuurm, Crystallization of polymers studied by temperature modulated calorimetrc measurements at different frequencies, J. Therm. Anal. Calorim., **59** (2000) 279–288.
12. J.O. Hill, Thermogravimetric Analysis (TGA); in S. Lee (Ed.), Encyclopedia of Chemical Processing, Tayor & Francis, New York, (2006) 3009–3021.
13. M. Krzesinska, A. Celzard, B. Grzyb, J.F. Mareche, Elastic properties and electrical conductivity of mica/expanded graphite nanocomposites, Mater. Chem. Phys. **97** (2006) 173–181.
14. G.M. Spinks, H.R. Brown, Z. Liu, Indentation testing of polystyrene through the glass transition, Polym. Test. **25** (2006) 868–872.
15. G. Karayannidis, E. Kirikou, C. Roupakias, G. Papageorgiou, Study of the behaviour of alipharomatic polyesters around the glass–rubber transition region by thermomechanical analysis: the mobile and rigid amorphous fraction, Polym. Int. **56** (2007) 158–166.
16. C. Feina, S.C. Jana, Nanoclay-tethered shape memory polyurethane nanocomposites, Polymer **48** (2007) 3790–3800.
17. R.D. Sudduth, Evaluation of the characteristics of a viscoelastic material from creep analysis using the universal viscoelastic model I. Isolation of the elastic component designated as the "projected elastic limit", J. Appl. Polym. Sci. **89** (2003) 2923–2936.
18. R.D. Sudduth, Development of a simplified relationship between uniaxial creep, stress relaxation, and constant strain-rate results for viscoelastic polymeric materials, J. Appl. Polym. Sci. **82** (2001) 527–540.
19. .R.D. Sudduth, Comparison of the viscous and elastic components of two ABS materials with creep, stress relaxation and constant strain rate measurements using the universal viscoelastic model, J. Appl. Polym. Sci. **90** (2003) 1298–1318.
20. J. Kolarik, A. Pegoretti, L. Fambri, A. Penati, High-density polyethylene/cycloolefin copolymer blends, part 2: Nonlinear tensile creep, Polym. Eng. Sci. **46** (2006) 1363–1373.
21. J.S. Lyons, Linear viscoelastic analysis of the room-temperature creep behavior of glass-reinforced aromatic and aliphatic thermoplastics, Polym. Test. **22** (2003) 545–551.
22. K.S. Fancey, A mechanical model for creep, recovery and stress relaxation in polymeric materials, J. Mater. Sci. **40** (2005) 4827–4831.
23. K.S. Fancey, A latch-based Weibull model for polymeric creep and recovery, J. Polym. Eng. **21** (2001) 489–509.
24. R.A. Kalgaonkar, S. Nandi, S. Tambe, J.P. Jog, Analysis of viscoelastic behavior and dynamic mechanical relaxation of copolyester based layered silicate nanocomposites using the Havriliak–Negami model, J. Polym. Sci. Part B: Polym. Phys. **42** (2004) 2657–2666.
25. F. Alvarez, A. Alegria, J. Colmenero, Relationship between the time-domain Kohlrausch–Williams–Watts and frequency-domain Havriliak–Negami relaxation functions, Phys. Rev. B **44** (1991) 7306–7312.

26. I. Lacik, I. Krupa, M. Stach, A. Kucma, J. Jurciova, I. Chodak, Thermal lag and its practical consequence in the dynamic mechanical analysis of polymers, Polym. Test. **19** (2000) 755–771.
27. N.M. Alves, J.F. Mano, J.L Gomez-Ribelles, Analysis of the thermal environment inside the furnace of a dynamic mechanical analyser, Polym. Test. **22** (2003) 471–481.
28. M.A. Rodriguez-Perez, L.O. Acros y Rabago, A. Gonzalez, J.A. de Saja, Improvement of the measurement process used for the dynamic mechanical characterization of semicrystalline polymers in three point bending, Polym. Test. **22** (2003) 63–76.
29. M. Wilhelm, Macromol. Fourier-transform rheology, Mater. Eng. **287** (2002) 83–105.
30. S.K. Young, K.A. Mauritz, Dynamic mechanical analyses of Nafion/organically modified silicate nanocomposites, J. Polym. Sci. B. Polym. Phys. **39** (2001) 1282–1295.
31. L. Petersson, K. Oksman, Biopolymer based nanocomposites: Comparing layered silicates and microcrystalline cellulose as nanoreinforcement, Compos. Sci. Technol. **66** (2006) 2187–2196.
32. I.L. Dubnikova, S.M. Berezina, Y.M. Korolev, G.M. Kim, S.M. Lomakin, Morphology, deformation behaviour and thermomechanical properties of polypropylene/maleic anhydride grafted polypropylene/layered silicate nanocomposites, J. Appl. Polym. Sci. **105** (2007) 3834–3850.
33. M.A. Roodriiguez-Perez, J.A. de Saja, Dynamic mechanical analysis applied to the characterisation of closed cell polyolefin foams, Polym. Test. **19** (2000) 831–848.
34. S. Kalallunnath, D.S. Kalika, Dynamic mechanical and dielectric relaxation characteristics of poly(trimethylene terephthalate), Polymer **47** (2006) 7085–7094.
35. H. Yang, W. Zhang, R.D. Moffitt, T.C. Ward, D.A. Dillard, Multi-layer *in-situ* evaluation of dynamic mechanical properties of pressure sensitive adhesives, Int. J. Adhes. Adhesives **27** (2007) 536–546.
36. H. Valentova, J. Nedbal, M. Ilavsky, P. Pissis, DSC, dielectric and dynamic mechanical behavior of two- and three-component ordered polyurethanes, Polymer **46** (2005) 4175–4182.
37. A. Galeski, Dynamic mechanical properties of crystalline polymer blends. The influence of interface and orientation,e-Polymers**26**(2002)1–29.
38. N.M. Alves, J.F. Mano, J.L Gomez-Ribelles, J.A. Gomez-Tejedor, Departure from the Vogel behaviour in the glass transition – thermally stimulated recovery, creep and dynamic mechanical analysis studies, Polymer **45** (2004) 1007–1017.
39. A. Wurm, M. Merlyakov, C. Schick, Temperature modulated dynamic mechanical analysis, Thermochimica Acta, **330** (1999) 121–130.
40. F. Oulevey, N.A. Burnham, G. Gremaud, A.J. Kulik, H.M. Pollock, A. Hammiche, M. Reading, M. Song, D.J. Hourston, Dynamic mechanical analysis at the submicron scale; Polymer **41** (2000) 3087–3092.

# 14 Wear, Fatigue, Creep, and Stress-Relaxation Behavior of Polymer Nanocomposites

*Chong-gui Li\* and You Wang*

Department of Materials Science, School of Materials Science and Engineering, Harbin Institute of Technology, Harbin 150001, People's Republic of China

**Abstract**

Polymer nanocomposite materials have got increasing applications. Polymer nanocomposites are polymers filled with some kind of dispersed fillers, which exhibit at least one dimension in the nanometer range. Dispersing nanosize fillers in a polymer matrix induces superior properties compared to conventional fillers. The nanolevel reinforcement affects the performance of the matrix and needs to be clearly understood before using in practical applications. Extensive research has been conducted concerning the development of polymer nanocomposites. In this chapter, wear, fatigue, creep, and stress-relaxation behavior of polymer nanocomposites are discussed. Nanomaterials such as nanofibers, nanoparticles, or nanolayers can offer significant advantages over conventional polymer materials. The wear resistance of most of the polymer materials reinforced by various nanofibers, nanoparticles, and nanolayers can be significantly improved compared to pure polymer materials. Most of the polymer materials with special nanofillers can provide low frictional coefficient and low specific wear rate. The fatigue resistance of the polymer nanocomposites was improved and the fatigue mechanism was quite different from the pure polymer materials due to the addition of various nanofillers. The creep resistance of most of the polymer nanocomposites was increased compared to pure polymer materials. The influence of the nanofillers on the stress-relaxation behavior of the polymer nanocomposites is also described.

## 14.1. Wear of Polymer Nanocomposites

### 14.1.1. Fiber-Reinforced Polymer Nanocomposites

There are more and more technical applications in which friction and wear are critical issues. Polymer composites containing different fillers and/or reinforcements are frequently used for these purposes. Nowadays, polymer composites have been increasingly used as engineering components in various industries as lower weight alternatives to metallic materials.

---

\*Correspondence should be addressed to e-mail: chongguili@gmail.com

In particular, they are now being used as sliding elements, which were formerly composed of metallic materials only. Nevertheless, new developments are still under way to explore other fields of application for these materials and to tailor their properties for more extreme loading and environmental temperature conditions [1].

In order to improve the friction and wear behavior of polymeric materials, one typical concept is to reduce their adhesion to the counterpart material and to enhance their hardness, stiffness, and compressive strength. This can be achieved quite successfully by using special nanofillers, which are expected to be able to strongly influence the wear performance of polymers and composites. Short fiber-reinforced polymer composites are nowadays used in numerous tribological applications.

Normally the matrix should possess a high temperature resistance and have a high cohesive strength. However, sometimes it is also advantageous to have a PTFE-based matrix in which a stiffer and more wear-resistant polymer phase along with other fillers provide more optimum conditions for the tribological situation under particular consideration, e.g., its use at cryogenic temperatures [2]. Additional fillers that enhance the thermal conductivity are often of great advantage, especially if effects of temperature enhancement in the contact area must be avoided in order to prevent an increase in the specific wear rate. It should also be noted that not all the fillers are of benefit to the wear performance of composites. The wear resistance is increased when fillers decompose and generate reaction products, which enhance the bonding between the transfer film and the counterface [3], whereas other fillers decrease the wear resistance because they generate more discontinuities in the material. It is thus important to understand the growth, bonding, and loss of transfer films, which are strongly related to the wear mechanisms. It should also be noted that chemical and mechanical interactions of transfer films are very complicated; therefore, further efforts to understand these relationships in more detail are still a subject of current and future studies [4, 5].

Table 14.1 summarizes the specific wear rate of various filler modified and/or short fiber-reinforced thermoplastic composite systems tested against steel counterparts on a block-on-ring test configuration. They have been developed for special applications in which low friction, high wear resistance, and good thermal conductivity under sliding

**Table 14.1** Example of various newly developed, filled polymer systems with excellent tribological properties under various loading conditions [1][a].

| Material compositions (vol%) | | | | | | | Specific wear rate ($10^{-6}$ mm$^3$/N m) |
|---|---|---|---|---|---|---|---|
| PTFE | PPS | PEEK | Graphite | CF | Bronze | Al$_2$O$_3$ | |
| 51.6 | 31.8 | – | 4.8 | 11.8 | – | – | 1.53 |
| 51 | 31.6 | – | 3.9 | 13.5 | – | – | 1.40 |
| 51.9 | 32.4 | – | 2.6 | 13.1 | – | – | 1.20 |
| 12.4 | – | 61.8 | 11.7 | 14.1 | – | – | 4.31 |
| 9.7 | – | 49.7 | 12.5 | 28.1 | – | – | 6.33 |
| 76.8 | – | – | 19.8 | – | – | 3.4 | 22.40 |
| 84.1 | – | – | – | 12.6 | – | 3.3 | 1.25 |
| 52.5 | 28 | – | – | 19.5 | – | – | 1.69 |
| 78.6 | – | – | – | 21.4 | – | – | 1.75 |
| 80 | – | – | – | 10 | 10 | – | 0.565 |

[a]Block-on-ring tests. Testing conditions: $p = 2$ MPa; $v = 1$ m/s; $T = RT$; $t = 8$ h; counterpart: steel.

wear conditions against smooth steel counterparts were of great importance. It can be found that PTFE +10 vol%Bronze +10 vol%CF (carbon fiber) exhibits an excellent wear resistance. In particular, the addition of bronze improves significantly the tribological properties of the composite because of its outstanding thermal conductivity. This is, in fact, a good example to prove the effectiveness of this kind of filler for the design of a wear-resistant polymer composite. It should be mentioned that there are certainly many more compositions available on the commercial market, which, due to lack of space, cannot all be listed here.

Thermoplastic polymer matrix composites are used as coating materials for the bore of downhole tubulars used as water injectors in the oil industry. These coatings are primarily employed for corrosion resistance but must also resist mechanical damage from the inspection tools lowered at speed down the tubing. This mechanical damage is produced by the wearing action of the supporting wire against the coating and by direct impact of the tool against the coating. Filler materials are added to these polymeric coatings and these additions are known to affect the wear resistance of the coating.

Three types of thermoplastic polymeric coatings were subjected to wear tests according to a study by Xu et al. [6]. Both abrasive wear tests, using silicon carbide papers as the abrasive, and wireline wear tests, utilizing a true tribocouple consisting of the coating and a length of "slickline" wire on a modified pin-on-disk apparatus, were carried out to study the wear resistance of these three coatings. Detailed scanning electron microscopy was performed on the wear tracks produced to elucidate the wear mechanism and in particular the role of fillers. In abrasive wear, a polymer with a brittle filler has a higher wear rate than an unfilled polymer due to the fact that the brittle fillers can be easily fractured and detached from the polymer matrix. In general, the weak bond between the filler and a thermoplastic polymer matrix leads to the filler particles detaching from the matrix causing enhanced wear. In wireline wear, the presence of voids and unmelted particles is particularly deleterious.

Samples from each type of coating were prepared using standard metallographic methods and studied using SEM. For ease of reference, the three different coatings were designated as A, B, and C. The substrate material for all coatings was a 0.15C mild steel. Coating A was formulated from a thermoplastic nylon resin with no filler. This was powder coated on to mild steel by means of electrostatic spraying. Coating B was a $500 \pm 10$ μm thick thermoplastic fusion bond nylon powder coating filled with 23% by volume of fused silica and dolomite fillers with a particle size of 5–10 μm. Coating C was a $650 \pm 10$ μm thick semicrystalline thermoplastic PVDF powder coating, free from any additives. The thermoplastic powder coating C was manufactured by electrostatic spraying on substrate plates. During all abrasive wear tests, no significant edge loading on the pin samples was detected.

Figures 14.1 and 14.2 show the specific wear rate ($m^3$/N m) of polymeric coatings A, B, and C as a function of the sliding distance of the pin in abrasive wear tests without and with pin rotation, respectively. For tests carried out with pin rotation, it leads to an additional distance slid. However, the magnitude of this additional distance varies across the pin, being greatest at the perimeter of the pin. Calculations showed that the additional distance slid by the perimeter of the pin was less than 1.5% of the total distance slid. Hence, this additional distance was not taken into account in calculating the sliding distances given in Figures 14.1 and 14.2. Error bars are included for the data in Figures 14.1 and 14.2, where duplicate tests were carried out. Based on numerous experiments, the error in each

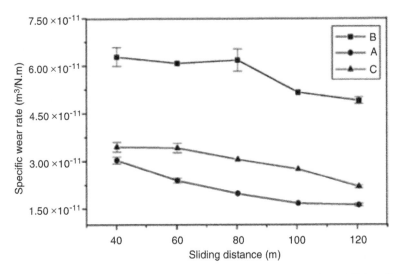

**Figure 14.1** Specific wear rate of the three polymeric coatings as a function of sliding distance in abrasive wear tests without pin rotation.

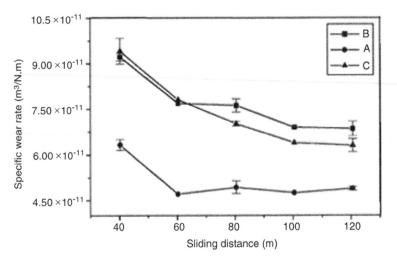

**Figure 14.2** Specific wear rate of the three polymeric coatings as a function of sliding distance in abrasive wear tests with pin rotation.

individual determination is estimated as ±5%. In the tests without pin rotation, the filled polymeric coating B had a much higher specific wear rate than coatings A and C.

Long ploughed furrows can be found on the micrograph of the surface of the unfilled polymeric coating A after sliding 120 s without pin rotation in an abrasive wear test. The wear scars indicate that the coating is behaving in a predominantly ductile manner and most material is being displaced with only some being cut and detached from the surface, resulting in low wear rates determined from weight loss measurements.

Short furrows and tendril production can be found on the micrograph of the surface of the filled polymeric coating B after sliding 120 s without pin rotation in an abrasive wear test. Instead of smooth and ductile furrows, short and fragile furrows and swarf-like tendrils were found on the surface of coating B. The fillers present in the coating can be seen at the position where fracture of the furrow has occurred. Thus, the filler is aiding the detachment of material. In addition as the bond between the fillers and the matrix is poor, fillers are easily detached and contribute to abrade the surface.

Furthermore, the voids left on the surface of the coating due to filler detachment act as stress raisers and produce cracks, so aiding the detachment of large sections of the coating. All the above features will exacerbate the wear rate of coating B.

Ploughed furrows and tendril production can be found on the micrograph of the surface of the polymeric coating C after sliding 150 s without pin rotation in an abrasive wear test. Its surface exhibits features seen in both coatings A and B. The ploughed furrows were smooth. However, their fracture can be clearly seen. It was caused by the voids and the unmelted powders in the coating, disrupting the furrows and causing their detachment. This effect was more severe in coating B due to the greater volume fraction of filler and to the abrasive nature of the fillers themselves. Hence, coating C has a wear rate intermediate between that of coatings A and B.

In the abrasive wear tests with pin rotation, the wear mechanism of these three coatings changed considerably, although the ranking of the specific wear rate of these three coatings was almost the same as in the tests without pin rotation. Comparing the results in Figures 14.1 and 14.2, it can be seen that the specific wear rate of filled coating B with pin rotation was only 1.2–1.5 times greater than that without pin rotation, while coatings A and C showed a 2–3 times increase.

It can be shown that the unfilled nylon coating A had the best wear resistance in the study. The silica- and dolomite-filled polymeric coating B was found to have the worst abrasive wear resistance. This was because when the coated pin was slid on the abrasive paper, the fillers, rather than supporting the load, were pulled out due to the poor bond between the fillers and the matrix material. The poor bond also caused cracks associated with the fillers. The cracks reduced the wear resistance of the coating. The unfilled PVDF coating C had a worse abrasive wear resistance than coating A because the presence of many voids and unmelted powder particles reduced the cohesive strength and aided fracture of the ploughed furrows. It was found that unfilled coatings A and C had worse wear resistance when the pin was rotated rather than not rotated because the deep ploughed furrows were cut into short intersecting ones, which were easily detached from the coating. Rotating the pin produced a less significant increase in the wear rate of coating B than for coatings A and C.

Poly(ether ether ketone) (PEEK) is a semicrystalline high-performance thermoplastic with excellent mechanical properties such as elastic modulus, strength, and toughness, even at elevated temperatures. PEEK is an attractive bearing material as it is comparatively fatigue resistant and exhibits a low creep rate up to about 250°C, especially when fiber reinforced [7]. Furthermore, fibrous reinforcements can improve the wear behavior of PEEK.

It has been shown that vapor-grown carbon nanofibers (CNF) can be homogeneously incorporated and dispersed in PEEK composites by twin-screw extrusion and injection molding in recent study [8]. The resulting nanocomposites showed a linear increase in stiffness and strength with increasing nanofiber content. Taking into account the small size and crystal structure of the CNFs, these PEEK nanocomposites were expected to display improved wear behavior.

A study by Werner et al. [9] investigated the influence of vapor-grown CNFs, of average diameter 150 nm, on the wear behavior of semicrystalline PEEK. Unidirectional sliding tests against two different counterpart materials (100Cr6 martensitic bearing steel, and X5CrNi18-10 austenitic stainless steel) were performed on injection-molded PEEK–CNF nanocomposites. The specific wear rates of the nanocomposites were measured as a function of filler loading fraction and the results are compared to a variety of commercial PEEK grades. Analysis of the wear behavior, and examination of the wear surfaces, revealed a variety of effects compared to the control materials. Most importantly, the CNFs were found to reduce the wear rate of PEEK significantly. In the light of these promising results, combinations of nanofibers with the conventional fillers, polytetrafluoroethylene and/or carbon fibers, were explored; it was found that such optimized compounds can be used to tailor the wear properties of the PEEK–CNF compounds.

The PEEK matrix used in this study was Victrex PEEK powder grade 450P. The starting vapor-grown CNF material consists of loosely aggregated hollow nanofibers with a mean diameter of $155 \pm 30$ nm.

Two different commercially available steels were used as counterparts for the wear tests: 100Cr6 (chromium content: 1.35–1.6 wt%) steel balls were used for comparison to data obtained from the literature, while X5CrNi18-10 (CrNi) stainless steel (chromium content: 17.0–19.5 wt%) was investigated as it is frequently used as a relatively inexpensive counterpart for composite materials under corrosive conditions. Prior to testing, the balls were cleaned in acetone.

The wear tests were performed according to ISO 7148-2 with a ball-on-prism test system. A metallic prism with an opening angle of 90° containing the specimens was pressed against a steel bearing ball. A normal load component of $F_N = 21.2$ N at 45° with respect to the load axis was applied to the specimen surface by dead weights via a lever. Due to the spherical contact geometry, the pressure–velocity values ($p–v$) decreased during the tests and cannot be quoted. However, the wear marks had a diameter of about 1–3 mm leading to a contact pressure of 3–27 MPa and a $p–v$-value of less than 0.76 MPa m/s. The bearing balls had a diameter of $d = 12.7$ mm and rotated uniformly at a frequency of $f = 1$ s$^{-1}$, which results in a continuous sliding speed of $v = \pi d \cos 45°$, $f = 28.2$ mm/s.

Figure 14.3(a) shows the results of the wear tests for the pure PEEK and PEEK–CNF nanocomposites with CNF contents of 5–15 wt%, tested against the 100Cr6 steel. Both the raw data, shown as light gray scatter, and the fitting to the steady state part of the wear test

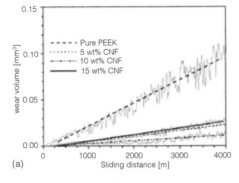
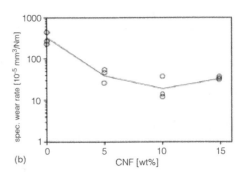

**Figure 14.3**    (a) Wear volume as a function of sliding distance and (b) semilog plot of specific wear rate of PEEK–CNF compounds versus 100Cr6 steel, as a function of CNF content.

(solid line) are included. There is a clear reduction in wear volume for the PEEK–CNF nanocomposites compared to the pure matrix, although it appears only weakly dependent on filler weight content.

The specific wear rates for these materials show similar trends to the total wear volume, as shown in Figure 14.3(b). The CNFs significantly reduce the specific wear rate of PEEK; however, given the error bars, the effect is not strongly dependent on the CNF content within the range of 5–15 wt%. Nevertheless, there is a small optimum in the specific wear rate for a loading of 10 wt% of nanofibers.

Light microscopy images of the polymer nanocomposite wear surfaces after testing against 100Cr6 steel was obtained. The image of the pure PEEK reveals depositions of polymer debris forming large plaques on the wear surface. The CNFs tend to reduce the accumulation of the PEEK wear debris, such that smaller chips of matrix were found on the nanocomposite wear surfaces. Furthermore, the sliding surfaces become smoother with increasing CNF content due to the lubricating effect of the nanofibers. The sharpness of the sliding tracks seems to show a minimum for the 10 wt% sample mirroring the trend for the specific wear rate. The 15 wt% composite reveals a greater number of larger dimples on the worn surface which, in turn, corresponds to an increased specific wear rate compared to the 10 wt% sample.

In this study, the wear behavior of carbon-fiber-reinforced PEEK was found to be controlled by the counterpart material. Tests against the 100Cr6 steel showed a strong time dependence of the specific wear rate under the chosen wear conditions. After 600 m sliding distance, a change in the slope of the wear curve can be observed in Figure 14.4(a), accompanied by the onset of rust formation. Tests performed against the more corrosion-resistant CrNi steel did not show this unfavorable time dependence, as shown in Figure 14.4(b).

The influence of PTFE and carbon fibers on the wear behavior of the 10 wt% PEEK–CNF nanocomposite was explored. The results show that the addition of 10 wt% CF to the PEEK–CNF composite has no influence on the specific wear rate but does affect the initial wear behavior for the 100Cr6 steel. While the PEEK–CNF nanocomposite shows a rather constant wear rate throughout the entire test, a pronounced running-in phase can be observed for the CNF–CF composite until a preliminary steady state is reached. Furthermore, after about 4500 m of sliding distance, a wear transition is observed. This transition is attributed to a tribochemical reaction even though it is not as pronounced as for the PEEK–CF composite without CNFs and occurs later. In contrast, the addition of the carbon fibers significantly reduces the specific wear rate of the optimized composite against the CrNi steel.

It can be found that vapor-grown CNFs are an interesting additive for polymers in tribological applications. In homogeneously dispersed systems, they can significantly reduce the wear rate of PEEK and demonstrate a superior wear behavior to all the commercial PEEK composites in the test. It could also be concluded that such optimized combinations of nanofibers with the conventional fillers, polytetrafluoroethylene and/or carbon fibers, can be used to tailor the wear properties of the PEEK–CNF compounds.

Poly(methyl methacrylate) (PMMA) is a very important thermoplastic because of its favorable mechanical properties, good solvent resistance, and outstanding climate resistance. Now PMMAs are widely used in architecture, automotive, air, and railway transport systems for tribological applications. However, the most difficult in the tribological applications is the relatively poorer wear resistance for PMMA. Due to the effects of

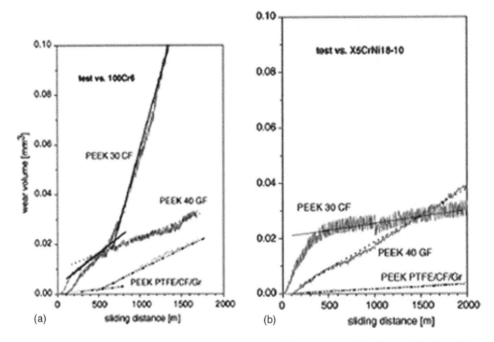

**Figure 14.4** Comparative wear volume versus sliding distance curves for glass and carbon fiber modified PEEK compared to an optimized commercial PEEK blend containing PTFE, carbon fibers, and graphite when tested against 100Cr6 (a) and X5CrNi18-10 (b) steels.

the reinforcement, CNTs can be used to fabricate the nanocomposites with excellent tribological properties. Therefore, CNTs would be expected to significantly improve the tribological properties of PMMA-based nanocomposites.

Poly(methyl methacrylate)-carbon nanotubes (PMMA–CNTs) nanocomposites with different contents have been prepared successfully by means of *in-situ* polymerization process according to a study conducted by Yang et al. [10]. CNTs discussed in the paper were multiwalled nanotubes, which were synthesized by a catalytic chemical vapor deposition (CVD) method. The microhardness of the PMMA–CNTs nanocomposites was measured. It was found that CNTs effectively increased the microhardness of the nanocomposites. The tribological behaviors of the nanocomposites were investigated by a friction and wear tester under dry conditions. Comparing with pure PMMA, PMMA–CNTs nanocomposites showed not only higher wear resistance but also smaller friction coefficient. CNTs could dramatically reduce the friction and improve the wear resistance behaviors of the nanocomposites. The mechanisms of the significant improvements on the tribological properties of the PMMA–CNTs nanocomposites were also discussed.

CNTs are centrally hollow tubes, which are typical of multiwalled carbon nanotubes. The outer diameters of most CNTs range from 10 to 20 nm, and their lengths are several micrometers. After subsequent purification treatment, as-synthesized CNTs with a purity exceeding 98% were obtained.

The influence of CNTs content on the microhardness of the PMMA–CNTs nanocomposites was investigated. The results show that the microhardness of the nanocomposites increases sharply when the CNTs content is below 1.0 wt%. The microhardness values

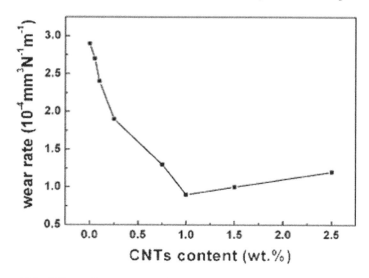

**Figure 14.5**    Effects of CNTs content on wear rate of PMMA–CNTs nanocomposites.

decrease slightly when the CNTs content is above 1.0 wt%. It is attributed to the conglomeration of CNTs in the nanocomposites matrix. It is thus inferred that the incorporation of CNTs as a reinforcing agent helps to increase the load-carrying capacity and mechanical properties of PMMA and improve the tribological behavior as well.

The influence of CNTs content on the friction coefficient of PMMA–CNTs nanocomposites for steady-state sliding against the stainless steel ring under dry conditions was also investigated. The results show that the friction coefficient of PMMA–CNTs nanocomposites decreases with increasing CNTs content. The friction coefficient values of nanocomposites sharply decrease when the CNTs content is below 1.0 wt%. As the content of CNTs in nanocomposites is higher, the friction coefficient becomes lower. Moreover, the variation of the friction coefficient reaches a relatively stable value when the CNTs content surpasses 1.0 wt%.

Figure 14.5 indicates the effects of CNTs content on the wear rate of PMMA–CNTs nanocomposites. It can be clearly seen that the incorporation of CNTs significantly decreases the wear rate of PMMA. The wear rate of PMMA–CNTs nanocomposites decreases sharply from $2.9 \times 10^{-4}$ to $9 \times 10^{-5}$ mm$^3$ N$^{-1}$m$^{-1}$ with the content of CNTs from 0 to 1.0 wt%. It is found that PMMA-1.0 wt% CNTs nanocomposites exhibit the smallest wear rate. When the content of CNTs in the nanocomposites exceeds 1.0 wt%, the wear rate of PMMA–CNTs nanocomposites increases slightly with increasing CNTs content. Similar results were also observed on the microhardness of the nanocomposites.

In order to explain the effect of CNTs on tribological behavior of PMMA–CNTs nanocomposites, the morphologies of the worn surfaces of the nanocomposites blocks were observed using scanning electron microscope. The SEM images of the worn surfaces of PMMA and PMMA-1.0 wt% CNTs nanocomposites under the same testing conditions are shown in Figures 14.6(a) and (b), respectively. The worn surface of pure PMMA shows signs of adhesion and abrasive wear (Figure 14.6(a)). The corresponding surface is very rough, displaying plucked and ploughed marks indicative of adhesive wear and plowing. This phenomenon corresponds to the relatively poorer wear resistance of the pure PMMA in sliding against the steel. It can be seen that more obvious plowed furrows appear on

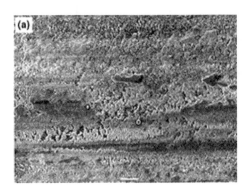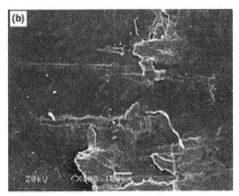

**Figure 14.6**    SEM images of the typical worn surfaces of PMMA (a) and PMMA–CNTs nanocomposites (b).

the worn surface of the PMMA block specimen. By contrast, the scuffing and adhesion on the worn surface of the PMMA/1.0 wt% CNTs nanocomposites are considerably reduced (Figure 14.6(b)). We can see a relatively smooth, uniform, and compact worn surface, which is in good agreement with the considerably increased wear resistance of the PMMA–CNTs nanocomposites. Therefore, it can be deduced that the incorporation of CNTs contributes to restrain the scuffing and adhesion of the PMMA matrix in sliding against the steel counter face. As a result, the PMMA–CNTs nanocomposites show much better wear resistance than the pure PMMA.

It was found that CNTs significantly increased the wear resistance of the nanocomposites and decreased their friction coefficient. The improvements on the tribological properties of PMMA–CNTs nanocomposites were attributed to the excellent mechanical properties and unmatched topological tubular structure of CNTs. As a result of high stiff and toughness of CNTs, the wear resistance of the nanocomposites was also improved. During the course of wear and friction, CNTs which were dispersed uniformly in the nanocomposites could serve as medium, preventing the close touch of the two surfaces between the applied loading and the nanocomposites. Moreover, the self-lubricate properties of CNTs resulted in reduction of the wear rate and the friction coefficient.

## 14.1.2. Particle-Reinforced Polymer Nanocomposites

Inorganic particles are well known to enhance the mechanical properties of polymers, which has been widely investigated in the past decades. It has been found that the size of the particles plays an important role to improve, in particular, stiffness and toughness simultaneously. Reducing the particle size to a nanoscale level is assumed to reach a significant efficiency. Nanoparticle-filled polymers, so-called polymer nanocomposites, are very promising materials for various applications. They are expected to replace polymers, polymer blends, and their traditional composites in products produced by melt-processing techniques. This prediction is justified by the improvements in properties without sacrificing the melt rheological properties [1].

Nanoparticles are entities with diameters in the range of 1–100 nm. This new field of nanoparticles is lying between the traditional fields of chemistry and solid-state physics. Therefore, a significant gap exists between these regimes with unique characteristics that neither obeys the law of physics nor quantum chemistry. The smaller is the particle, the higher is the surface-to-volume ratio. Thus more atoms tend to reside on the surface than inside the particle itself. Particle chemical/mechanical properties that are once determined by the molecular structures are now influenced by the defects on the surface [11].

Polymers are widely used in aeronautics, automobiles, constructions, oil and gas industries, and so on. However, they are susceptible to damage by scratching and abrasive wear. Such processes impair the appearance and also reduce the mechanical strength by the introduction of flaws [12]. Polymers scratch and abrasive properties are of practical importance, and the use of reinforced polymer composites is becoming more common. When nanoparticles are embedded in polymer, the resulted composite material is known as polymer nanocomposite. Hard particulate fillers made of ceramic particles such as $Al_2O_3$ have been used with a PTFE matrix to dramatically improve the wear resistance, even up to three orders of magnitude [13]. Recent investigations on polymer nanocomposites reveal their significant potential in producing materials with low friction and/or high wear resistance [14–18]. Besides, polymer coatings are one of the most effective and convenient methods to protect carbon steel from corrosion.

At present, functional polymer coatings are used in increasingly demanding applications that require good adhesion and specific resistance to damage. High-performance polymer composite materials [19] are used increasingly for engineering applications under hard-working conditions. The materials must provide unique mechanical and tribological properties combined with a low specific weight and a high resistance to degradation in order to ensure safety and economic efficiency. This led to an increasing interest in improving the mechanical properties of polymer coatings, especially their abrasion, scratch, and wear resistance. An understanding of abrasion resistance and the associated surface deformation mechanisms is of primary importance in materials engineering and design.

For an improvement of the wear resistance, various kinds of micrometer-sized particles [1], e.g., $TiO_2$, $ZrO_2$, SiC, copper compounds (CuO, CuS, $CuF_2$), were incorporated into different polymer matrices, e.g., PEEK, polyamide (PA), polyphenylene sulfide (PPS), polyoxymethylene (POM), and polytetrafluoroethylene (PTFE). Based on the experience from various studies, fine particles seem to contribute better to the property improvement under sliding wear conditions than larger particles. However, opposite size effects were also found.

In comparison with the widely used micrometer-sized particles, nanoparticles are believed to have the following advantages [15, 16]: (i) the abrasiveness of the hard microparticles decreases remarkably as a result of a reduction in their angularity while the mechanical behavior of the bulk materials remain competitive; (ii) the transferred film can be strengthened because the nanoparticles would have the capability of blending well with wear debris; (iii) the material removal of nanoparticulate composites would be much milder than that of conventional composites because the fillers have the same size as the segments of the surrounding polymer chains.

If the particle size is reduced down to the nanoscale level (<100 nm), the wear performance of these nanocomposites is significantly different compared to that of micron particle filled systems. Polymers filled with nanoparticles are recently under discussion because of some excellent properties they have shown under various testing conditions. Some results were achieved in various studies which give hints that this method is also promising for new processing routes of wear-resistant materials. For example, Xue et al. [20] found that various kinds of SiC particles, i.e., nano, micron, and whisker, could reduce the friction and wear when incorporated into a PEEK matrix at a constant filler content,

**Table 14.2**  Wear of nanoparticle-filled polymer composites [1].

| Matrix/nanoparticle | Nanoparticle size (nm) | Lowest wear rate achieved ($10^{-6}$ mm$^3$/N m) | Optimum particle content (vol%) |
|---|---|---|---|
| EEK/Si$_3$N$_4$ | <50 | 1.3 | 2.8 |
| PEEK/SiO$_2$ | <100 | 1.4 | 3.4 |
| PEEK/SiC | 80 | 3.4 | 1–3 |
| PEEK/ZrO$_2$ | 10 | 3.9 | 1.5 |
| PPS/Al$_2$O$_3$ | 33 | 12 | 2 |
| PPS/TiO$_2$ | 30–50 | 8 | 2 |
| PPS/CuO | 30–50 | 4.6 | 2 |
| Epoxy/SiO$_2$ | 9 | 45 | 2.2 |
| Epoxy/SiO$_2$-$g$-PAAM | 9 | 11 | 2.2 |
| Epoxy/Si$_3$N$_4$ | <20 | 2.0 | 0.8 |
| Epoxy/TiO$_2$ | 300 | 14 | 4 |
| Epoxy/Al$_2$O$_3$ | 13 | 3.9 | 2 |
| Epoxy/SiO$_2$ | 13 | 22 | 3 |
| PTFE/ZnO | 50 | 13 | 15 |
| PTFE/Al$_2$O$_3$ | 40 | 1.2 | 12 |

e.g., 10 wt% (~4 vol%). However, nanoparticles resulted in the most effective reduction. Nanoparticles were observed to be of help to the formation of a thin, uniform, and tenacious transfer film, which led to this improvement. The variation of ZrO$_2$ nanoparticles from 10 to 100 nm was conducted by Wang et al. [21]. The results showed a similar trend like most of the micron particles, i.e., the smaller the particles were applied, the better was the wear resistance of the composites.

Up to now, various inorganic nanoparticles, e.g., Si$_3$N$_4$, SiO$_2$, SiC, ZrO$_2$, Al$_2$O$_3$, TiO$_2$, ZnO, CuO, and CaCO$_3$, were incorporated into PEEK [20–25], PPS [15, 26], PMMA [27], epoxy [16, 18, 19, 28–31], and PTFE [13, 17] matrices, in order to improve their wear performance. In most of these cases, optimum nanoparticle filler contents could be acquired at which the wear resistance of these polymers was the best. The results have been summarized in Table 14.2. It can be seen that the optimum filler content of the small particles was always in a range between 1 and 4 vol%, except for PTFE matrix composites. Although in most of these works the morphologies of nanoparticle dispersion were not provided in detail, it should be clear that high filler contents lead to deterioration in the wear properties, which may be due to a tendency of particle agglomeration. It is of high importance that the nanoparticles are uniformly dispersed rather than being agglomerated in order to yield a good property profile, in general.

In the case of PEEK matrix nanocomposites [20–25], a small amount of inorganic nanoparticles, such as Si$_3$N$_4$, SiO$_2$, SiC, and ZrO$_2$, contributed to a reduction in frictional coefficient from about 0.4 for the neat matrix down to about 0.2 at high nanofiller content. The dominant wear mechanisms were modified from adhesive and fatigue wear in case of the neat PEEK to a mild abrasive wear of the nanocomposites. However, with a further increase in nanofiller content, particle agglomeration occurred and resulted in a severe abrasive wear, as observed on the worn surfaces.

Bahadur et al. [15, 26] demonstrated a similar tendency for PPS, filled with various kinds and amounts of nanoparticles. Here 2 vol% Al$_2$O$_3$ exhibited an optimum reduction in the wear rate of the composites at two different surface roughness of the steel counterpart, i.e., $Ra$ = 60 and 100 nm. However, with the roughness of $Ra$ = 27 nm, which was smaller than the particle size (being 33 nm in average), any amount of nanoparticles increased the wear rate. Very recently, nanosize $TiO_2$ and CuO particles were found to be able to reduce wear in PPS matrix composites, whereas ZnO and SiC exhibited an opposite effect. The optimum wear resistance was obtained with 2 vol% of CuO or $TiO_2$. An interesting experimental approach, which was used to observe the transfer film counterpart bond strength by a peeling study, found that the bond strength strongly correlated to the wear resistance.

Sliding wear studies using inorganic nanoparticle filled epoxy were also carried out recently [16, 18, 19, 28–31]. It was found that well-dispersed nanoparticles could significantly improve both the mechanical and the tribological properties of the thermosetting matrix composites. The incorporation of $SiO_2$ nanoparticles (9 nm), grafted with another polymer (polyacrylamide, PAAM) to enhance the adhesion of the particle agglomerates with the surrounding epoxy resin matrix, was conducted by Zhang et al. The grafting polymerization technique increased the interfacial interaction between the nanoparticles and the matrix through chemical bonding, which in turn led to an improved tribological performance of these nanocomposites. Nano-$Si_3N_4$ particles (<20 nm) seem to be very effective in reducing friction and wear when being incorporated into an epoxy matrix at a very low filler content (<1 vol%). Here 300 nm $TiO_2$ contributed to an improved wear resistance by changing from a severe abrasive wear of the neat epoxy to a mild abrasive wear caused by the formation of a compacted transfer film due to the filler particles.

Wang et al. [32] investigated the tribological properties of $Al_2O_3$/polymer nanocomposite coatings using microhardness test, single-pass scratch test, and abrasive wear test.

Optical micrographs of polymer coating and polymer coating mixed with 10 wt% $Al_2O_3$ nanoparticles were obtained. In the polymer coating, the surface of polymer is even and smooth. Some lines formed on the surface could be due to the inherent property of the polymer itself or caused by the surface roughness of the substrate, which may obstruct the polymer layer set perfectly during curing process. As for the polymer coating mixed with 10% $Al_2O_3$ nanoparticles, 10 wt% $Al_2O_3$ nanoparticles seem evenly distributed in the polymer matrix. However, those should not be the nanoparticles but small aggregates or clustered $Al_2O_3$ nanoparticles exposed on the coating surface due to the resolution of light optical micrograph.

Figure 14.7 shows the average microhardness results for the polymer coatings with and without embedded $Al_2O_3$ nanoparticles. The difference between the measured maximum and minimum hardness values is also shown in Figure 14.7. The 5 and 10 wt% $Al_2O_3$/polymer nanocomposite coatings displayed similar microhardness values as the polymer coating which is 55 $HV_{500\,mN}$. But, as $Al_2O_3$ content increases to 20 wt%, there is a slight increase in the microhardness of the nanocomposites coating.

Figure 14.8 displays the wear rate in unit volume loss ($cm^3$) of the samples per rotating cycle versus the content of particles in the polymer. This figure shows the wear rate of $Al_2O_3$/polymer nanocomposite coatings. From the graph, it is evident that the wear rate of 10 wt% $Al_2O_3$/polymer nanocomposite coating is lower than the polymer coating. A minimum wear rate occurs at a content of 20 wt% $Al_2O_3$ particles. However, further increasing the amount of $Al_2O_3$ particles in the polymer could result in an increased abrasive wear. At an $Al_2O_3$ particle content of 30 wt%, for example, the nanocomposite coating displays a significant increase in wear rate. The wear rate in this case is higher than that for the polymer coatings. The sudden increase in wear rate of 30 wt% $Al_2O_3$/polymer nanocomposite coating may be attributed to the excessively high amount of particles in the polymer. When the amount of particles is too high, there is insufficient polymer to bond the

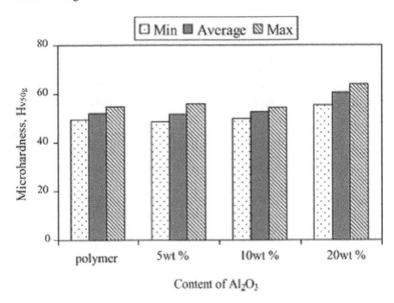

**Figure 14.7**  Microhardness results of different polymer coatings with and/or without $Al_2O_3$ nanoparticles.

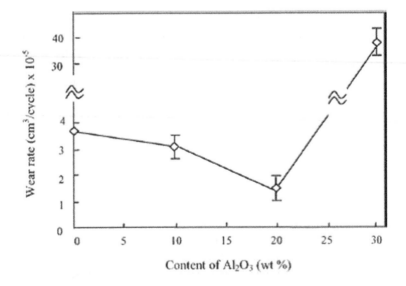

**Figure 14.8**  Wear rates of $Al_2O_3$ nanocomposite coatings.

particles together in the polymer matrix. As a result, a weak adhesion between the polymer with particles and polymer with substrate is anticipated.

A series of wear tracks for the polymer and $Al_2O_3$-containing polymer composite coatings were observed. The results show that the depth of the wear tracks decreases from high to low wear loss. The visibility or depth of the wear tracks is an indication to the wear level. A deeper wear track means that the coating is less resistant to the abrasion wear and vice versa.

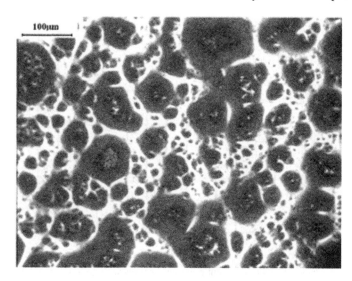

**Figure 14.9**  Typical micrograph of coating containing 10 wt% nanostructured WC particles. Dark areas represent WC and light areas polymer.

It can be found that the coatings containing $Al_2O_3$ nanoparticles showed improvement in scratch and abrasive resistance compared with that of polymer coating. The improvement in scratch and abrasive resistance is attributed to the dispersion hardening of $Al_2O_3$ nanoparticles in polymer coatings.

Tungsten carbide (WC) is a technologically important material and has been widely used as cutting tools, rock drills, punches, and wear-resistant coating materials. Bulk WC is known to exhibit excellent friction characteristics and improved wear resistance. Since the resistance to mechanical and chemical damage of the composite coatings is highly desirable, nanosized WC particulates in polymer composite coatings should also provide the polymers with improved tribological properties.

Wang et al. [33] investigated the tribological behavior of nanostructured WC particles/polymer composite coatings using microscratch technology.

The nanostructured tungsten carbide particles, about 40 nm grain size, were obtained from Inframat Corporation, USA. Particles size was about 200 nm. The polymer incorporate was obtained from Solomon Coatings, Edmonton, Alberta. It is known as Xylan 1810/D1864 commercially. The base metal used in this experiment was a 1018 grade carbon steel cut into small pieces with a working surface area of 1.95 $cm^2$.

Figure 14.9 shows an optical micrograph of a polymer coating mixed with 10 wt% nanostructured WC particles. It can be seen that WC particles were not evenly dispersed in the polymer coating as isolated particles, but often remained as aggregates. We are searching a method to disperse the particles well in the polymer coatings.

Although material properties determining resistance to wear and erosion are not well defined and hard to predict, materials with high hardness, rapid work hardening rate, and good oxidation and corrosion resistance, in general, exhibit an enhanced wear and erosion resistance [34, 35]. In polymeric materials, a certain degree of rigidity and hardness is essential to improve resistance to mechanically induced surface damage.

Figure 14.10 shows the microhardness results for the polymer coatings with and without embedded nanostructured WC particles. The average hardness is seen to increase with

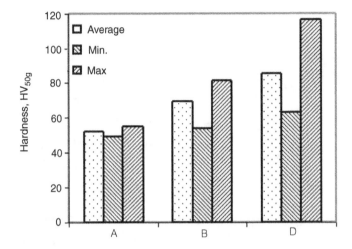

**Figure 14.10**    Hardness results of different polymer coatings with and/or without n-WC particles.

increasing amount of nanostructured WC particles in polymer coatings. The difference between the measured maximum and minimum hardness values also increases with increasing amount of nanostructured WC particles in polymer coatings.

Morphologies of scratch tracks on coating samples were examined using an optical microscope. The obtained micrograph of the pure polymer coating shows a scratch track on a pure polymer coating created during scratch test. For nanostructured WC particles/polymer composite coating, the scratch track is not clearly visible. Also shown are the fragments at the edge and at the bottom of the scratch track on the pure polymer coatings, while such fragments were not observed on the nanostructured WC particles/polymer composite coatings.

It could be concluded that the coatings containing nanostructured WC particles showed a significant increase in hardness and scratch resistance compared to that of pure polymer coating. The improvement in hardness and scratch resistance is attributed to the dispersion hardening of nanostructured WC particles in polymer coatings.

Small ceramic particles are known to enhance the mechanical and tribological properties of polymers. Introduced into an epoxy resin, the filler morphology, size, particle amount, and the dispersion homogeneity influence extensively the composite's performance.

Various amounts of micro- and nanoscale particles (calcium silicate $CaSiO_3$, 4–15 mm, alumina $Al_2O_3$, 13 nm) were systematically introduced into an epoxy polymer matrix for reinforcement purposes according to a study conducted by Wetzel et al. [19]. This study focused on the development of nanocomposites with properties superior to the neat matrix. The influence of these particles on the block-on-ring wear behavior was investigated. The addition of alumina nanoparticles into epoxy resin demonstrates their ability to improve the wear resistance slightly at 2 vol% $Al_2O_3$. And the introduction of calcium silicate microparticles into a nanocomposite matrix, which contains an optimum amount of nanoparticles (2 vol%), results in further increase in the wear resistance.

Xian et al. [36] applied 5 vol% nano-$TiO_2$ or micro-$CaSiO_3$ to a polyetherimide (PEI) matrix composite, which was filled additionally with short carbon fibers (SCF) and graphite flakes. The influence of these inorganic particles on the sliding behavior was investigated

**Table 14.3**  The fillers and their basic properties.

| Fillers | Particle size ($\mu$m) | Density (g/cm$^3$) |
|---|---|---|
| TiO$_2$ | 0.3 | 4.05 |
| CaSiO$_3$ | 2.5 | 2.85 |
| CaSiO$_3$ | 3.5 | 2.85 |
| Graphite | $\sim$ 20–14.5 (diameter) | 2.25 |
| SCF | $\sim$ 90 (length) | 1.6 |

**Table 14.4**  Composites studied in the paper.

| Designation | SCF (vol%) | Graphite (vol%) | TiO$_2$ (vol%) | CaSiO$_3$ (vol%) |
|---|---|---|---|---|
| 0-C | 10 | 15 | 0 | 0 |
| Nano-C | 10 | 15 | 5 | 0 |
| Micro 1-C | 10 | 15 | 0 | 5 (2.5 $\mu$m) |
| Micro 2-C | 10 | 15 | 0 | 5 (3.5 $\mu$m) |

with a pin-on-disk testing rig at room temperature and 150°C. Experimental results showed that both particles could reduce the wear rate and the frictional coefficient of the PEI composites under the applied testing conditions. At room temperature, the microparticles-filled composites exhibited a lower wear rate and the frictional coefficient, while the nano-TiO$_2$-filled composites possessed the lowest wear rate and the frictional coefficient at elevated temperature. Enhancement in tribological properties with the addition of the nanoparticles was attributed to the formation of transfer layers on both sliding surfaces together with the reinforcing effect.

In this study, a PEI was used as the matrix material. Three kinds of inorganic particles, 300 nm TiO$_2$, 2.5 and 3.5 $\mu$m CaSiO$_3$, were used, as presented in Table 14.3. The detailed compositions and designations of the materials are given in Table 14.4. The basic composition (denoted as 0-C) was filled with only 10 vol% SCF and 15 vol% graphite flakes, and the others were filled additionally with 5 vol% inorganic particles of different sizes.

The specific wear rate of PEI composites was carried out at room temperature and 150°C, respectively, and the results are summarized in Figure 14.11. It was found that the addition of inorganic particles to the basic PEI composite (0-C) could significantly enhance the wear resistance at both temperatures. At room temperature, the bigger particles brought in the better wear resistance, which coincides with the results of neat polymer filled only with particles. In this case, micro 2-C exhibited the lowest wear rate. The increase in the sliding temperature led to a dramatic increase in the wear rate for all studied composites because of the degradation of mechanical properties of composites. However, the particle size exhibited a different effect compared to that at room temperature. The nanoparticles proved more effective to the wear resistance than the microparticles at elevated temperature. Moreover, the sizes of these two microparticles exhibited negligible effect on the wear rate at 150°C, which was more obvious at room temperature.

Figure 14.12 gives the recorded frictional coefficient ($\mu$) as a function of sliding time at room temperature. It was found that all three kinds of inorganic particles reduced the $\mu$ in the steady state from 0.35 of the basic composite (0-C) to around 0.2. No obvious difference between nano- and microparticles was found.

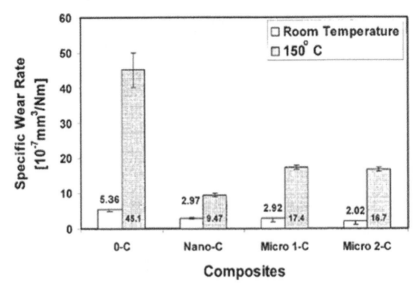

**Figure 14.11**  The specific wear rate of various PEI composites at room temperature or 150°C. Sliding conditions: 1 m/s and 2 MPa.

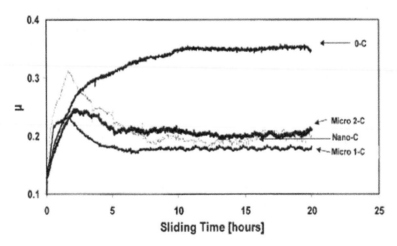

**Figure 14.12**  Frictional coefficient ($\mu$) versus sliding time at room temperature under 1 m/s and 2 MPa.

From Figure 14.12, the running-in time during which $\mu$ increases with gradually increasing sliding time can also be determined. For 0-C composite, the running-in time exceeds 10 h, while the inorganic particles-filled PEI composites show a much shorter running-in time, around 2 h for all three materials. The running-in period is related to smoothing counterface and forming transfer film. The existence of a peak $\mu$ at the end of the running-in period suggested the formation of an effective transfer film. The shorter running-in time for inorganic particles-filled composites indicates that the particles accelerate the formation of the transfer film or accelerate smoothing the surfaces of worn pairs.

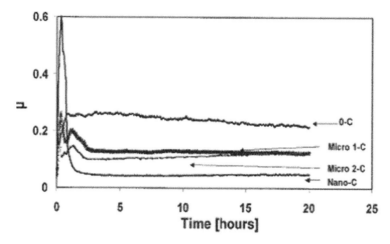

**Figure 14.13**   Frictional coefficient ($\mu$) versus sliding time at 150°C under 1 m/s and 2 MPa.

The variation of $\mu$ as a function of sliding time at elevated temperature is presented in Figure 14.13. All composites showed a lower $\mu$ in the steady state than that at room temperature. The composite without inorganic particles (0-C) possessed the maximum value, and the nanocomposite shows the minimum, and the microparticles-filled composites have intermediates values.

At 150°C, the running-in time is substantially reduced to less than 1 h for all composites (Figure 14.13), much shorter than that at room temperature (Figure 14.12). This is because the transfer film can be easily formed because of the improved PEI molecular chain mobility at elevated temperature. It is interesting to note that the nanocomposite shows a peak value of $\mu$ at around 0.6, which is one time higher than that at room temperature. After that, it reduces sharply to a stable value. This relatively high $\mu$ peak is supposed to be related with the high real contact area realized in the running-in period. On the other hand, the micrometer-sized particles exhibited a fluctuation in $\mu$ during the running period, which reflects the inhomogeneity of the materials with bigger hard particles.

It can be found that the introduction of either micro- or nano-inorganic particles can significantly reduce both the wear rate and the coefficient of friction of the PEI composites (filled additionally with both SCF and graphite) at both room temperature and 150°C. Once the sliding temperature was increased up to 150°C, nanoparticle-filled composite possessed the lowest wear rate and the coefficient of friction.

Chang et al. [37] investigated the tribological properties of polyamide 66 (PA66) composites, filled with $TiO_2$ nanoparticles, SCFs, and graphite flakes. Sliding tests were performed on a pin-on-disk apparatus under different contact pressures, $p$, and sliding velocities, $v$. It was found that nano-$TiO_2$ could effectively reduce the frictional coefficient and wear rate, especially under higher $pv$ (the product of $p$ and $v$) conditions.

Tribological properties of PA66 and PA66 matrix composites under various wear conditions were shown in Table 14.5.

It was found that conventional fillers, e.g., SCF and graphite flakes, could effectively reduce the frictional coefficient and wear rate of PA66 at lower $pv$ condition, e.g., 1 MPa and 1 m/s. The serious fiber removal and rapid increase in wear rate were occurred at the contact temperature increased up to the $T_g$ of matrix. With the addition of nano-$TiO_2$,

**Table 14.5**  Tribological properties of PA66 and PA66 matrix composites under various wear conditions [37].

| Composition | $pv$ factors | Frictional coefficient | Contact temperature (°C) | Specific wear rate ($10^{-6}$ mm$^3$/N m) |
|---|---|---|---|---|
| Neat PA66 | 1 MPa, 1 m/s | 1.16 | 45.09 | 6.70 |
| Graphite + SCF/PA66 | 1 MPa, 1 m/s | 0.60 | 30.57 | 0.53 |
|  | 2 MPa, 1 m/s | 0.57 | 46.07 | 0.80 |
|  | 4 MPa, 1 m/s | 0.69 | 94.98 | 2.92 |
|  | 8 MPa, 1 m/s | 0.35 | 97.37 | 16.93 |
|  | 2 MPa, 2 m/s | 0.78 | 97.67 | 4.87 |
|  | 2 MPa, 3 m/s | 0.62 | 115.80 | 5.73 |
| Nano-TiO$_2$ + graphite + SCF/PA66 | 1 MPa, 1 m/s | 0.44 | 26.66 | 0.50 |
|  | 2 MPa, 1 m/s | 0.34 | 33.21 | 0.72 |
|  | 4 MPa, 1 m/s | 0.26 | 45.55 | 0.80 |
|  | 8 MPa, 1 m/s | 0.22 | 64.69 | 3.26 |
|  | 2 MPa, 2 m/s | 0.34 | 45.94 | 0.63 |
|  | 2 MPa, 3 m/s | 0.38 | 62.36 | 3.35 |

the frictional coefficient and contact temperature of the composite were further decreased, especially with very high pv products.

The tribological behavior of nano-TiO$_2$ particle filled polyetherimide (PEI) composites, reinforced additionally with SCF and lubricated internally with graphite flakes was investigated according to another study conducted by Chang et al. [38]. The wear tests were conducted on a pin-on-disk apparatus, using composite pins against polished steel counterparts under dry sliding conditions, different contact pressures, and various sliding velocities.

Figure 14.14 shows the typical variations of the frictional coefficient and the contact temperature against sliding time for the neat PEI compared to other two composites. It is obvious that the two parameters are strongly correlated. Besides a high coefficient of friction, the neat PEI exhibited also a very poor wear resistance at 1 MPa and 1 m/s. The wear test had to be stopped only after 2.5 h due to an enormous wear loss. However, filled with conventional reinforcement fillers, i.e., SCF and graphite flakes, the wear resistance of material was remarkably enhanced. It can be seen that the frictional coefficient of the composite without nanoparticles was initially increased due to the increase in the real contact area during running-in stage. Thereafter, it stabilized at a mean value of about 0.55, and with a specific wear rate even less than $1 \times 10^{-6}$ mm$^3$/N m. With the addition of nano-TiO$_2$, the duration of the running-in stage and the stable coefficient of friction were further reduced, and the lowest wear rate was achieved.

The worn surface of the neat PEI was observed. It can be concluded that an abrasive mechanism dominated the wear process. The worn surface of the PEI composite without nanoparticles was also observed. Modified with SCF and graphite flakes, a polymer film layer can be transferred to the steel counterpart, which results in a new counter surface producing primarily an adhesive wear mechanism. It was found that the worn surface appears smooth, i.e., without obvious grooves. However, microcracks in the region of the fiber/matrix interface occurred, caused by a brittle fracture of the PEI matrix. With the propagation of these interfacial cracks, the fibers exposed to the asperities of the counterpart are finally removed, leaving voids on the worn surface.

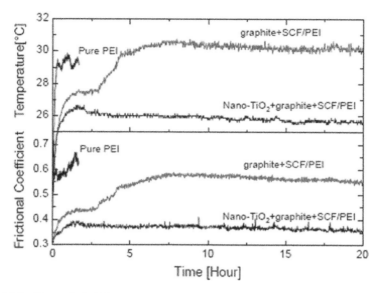

**Figure 14.14**  The typical sliding process curves of frictional coefficient and contact temperature against sliding time of neat PEI and the composites without and with nano-TiO$_2$ under a standard wear condition of 1 MPa and 1 m/s.

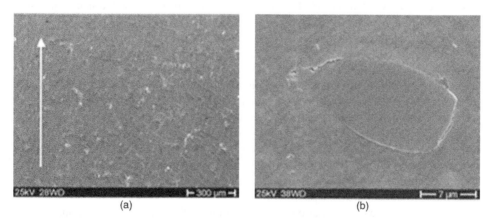

**Figure 14.15**  (a) SEM micrographs of the worn surfaces of the composite nano-TiO$_2$+ graphite+SCF/PEI at 1 MPa and 1 m/s, and (b) the magnified view of the worn fibers. Arrow line represents the sliding direction.

The worn surface of the PEI-based nanocomposite is presented in Figure 14.15. It is obvious that the amount of fiber removal was greatly restricted, and the surface is much smoother in comparison to the samples without nanoparticles.

It was found that the conventional fillers, i.e., SCF and graphite flakes, could remarkably improve both the wear resistance and the load-carrying capacity. With the addition of nano-TiO$_2$, the frictional coefficient and the contact temperature of the composite were further reduced, especially under high $pv$ (the product of the normal pressure, $p$, and the sliding velocity, $v$) conditions.

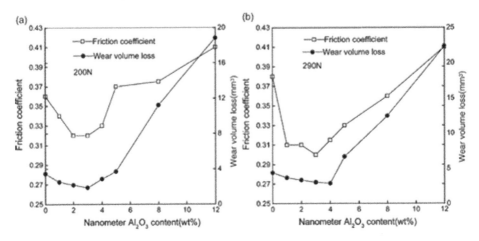

**Figure 14.16**  Effect of the nanometer $Al_2O_3$ content on the friction coefficient and wear volume loss of the $PI/Al_2O_3$ nanocomposite at (a) 200 N and (b) 290 N (sliding speed: 0.431 m/s).

A study by Cai et al. [39] investigated the tribological properties of polyimide (PI) nanocomposites with different proportions of nanoparticle $Al_2O_3$ made by compression molding at elevated temperature. The bending strength and microhardness of the nanocomposite specimens were determined, and the tribological behavior of the nanocomposite blocks in dry sliding against a plain carbon steel ring was evaluated on an M-2000 friction and wear tester. The morphologies of the worn nanocomposite surfaces and transfer films on the counterpart steel ring were observed on a scanning electron microscope. Results indicated that the PI-based nanocomposites with appropriate proportions of nanometer $Al_2O_3$ exhibited lower friction coefficient and wear volume loss than PI under the same testing conditions. The nanocomposite containing 3.0–4.0 wt% nanometer $Al_2O_3$ registered the lowest wear volume loss under a relatively high load. The differences in the friction and wear behaviors of PI and PI–$Al_2O_3$ nanocomposites were attributed to the differences in their worn surface morphologies, transfer film characteristics, and wear debris features. The agglomerated abrasives on the worn composite and transfer film surfaces contributed to increase the wear volume loss of the nanocomposites of higher mass fractions of nanometer $Al_2O_3$.

Figure 14.16 shows the variations of the friction coefficient and wear volume loss of $PI/Al_2O_3$ nanocomposites with the mass fraction of nanometer $Al_2O_3$. The lowest friction coefficient at a load of 200 N is recorded when the mass fraction of the nanometer $Al_2O_3$ in the nanocomposite reaches 3%, and so is the smallest wear volume loss, Figure 14.16(a). In combination with the friction coefficient and wear volume loss in this case, it is therefore rational to recommend the optimal $Al_2O_3$ content in the $PI/Al_2O_3$ nanocomposite as 3%. Similarly, the friction coefficient and wear volume loss of $PI/Al_2O_3$ nanocomposite at a load of 290 N show the same tendencies with increasing nanometer $Al_2O_3$ content, except that the lowest friction coefficient is recorded at a nanometer $Al_2O_3$ mass fraction of 3%, while the smallest wear volume loss registered at a nanometer $Al_2O_3$ mass fraction of 4%, Figure 14.16(b).

Figure 14.17 shows the SEM pictures of the worn surfaces of pure PI and nanocomposites blocks. The worn surface of pure PI, Figure 14.17(a), shows signs of scuffing and

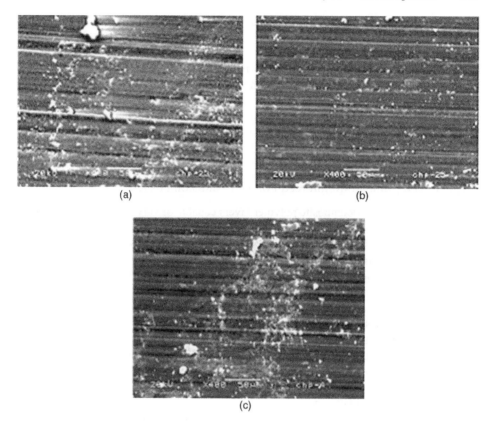

**Figure 14.17**  SEM pictures of (a) worn surface of pure PI, (b) worn surface of PI/3.0% Al₂O₃, (c) worn surface of PI-12.0% Al₂O₃.

adhesion, which is obviously abated on the worn surface of PI/3% $Al_2O_3$ nanocomposite, Figure 14.17(b), while that of PI/12% $Al_2O_3$ shows signs of agglomerated abrasives, Figure 14.17(c), which accounts for the increased wear of the nanocomposite with higher mass fraction of nanometer $Al_2O_3$.

It can be concluded that the incorporation of appropriate content of nanometer $Al_2O_3$ into PI improved the tribological behavior significantly. This is especially so at an extended load and sliding speed. The differences in the friction and wear behaviors of PI and PI–$Al_2O_3$ nanocomposites were attributed to the differences in their worn surface morphologies, transfer films characteristics, and wear debris features. The increased wear volume losses of the nanocomposites of higher mass fractions of nanometer $Al_2O_3$ were attributed to the abrasion action of the agglomerated abrasives on the worn composite and transfer film surfaces.

Inorganic fullerene-like tungsten disulfide (IF-WS2) nanoparticles were incorporated into PEEK coatings by Hou et al. [40] with the aim of reducing the coefficient of friction and improving the wear resistance of the coatings. Tribological tests had also been carried out to evaluate the friction and wear behaviors of IF-WS2/PEEK nanocomposite coatings. The results showed that significant improvement can be achieved in the tribological properties of the nanocomposite coatings by incorporating IF-WS2 nanoparticles.

Bhimaraj et al. [41] studied the friction and wear properties of poly(ethylene) terephthalate (PET) filled with alumina nanoparticles. The test matrix varied particle size, loading, and crystallinity to study the coupled effects on the tribological properties of PET-based nanocomposites. Crystallinity was found to be a function of the processing conditions as well as the particle size and loading, while tribological properties were affected by crystallinity, filler size, and loading. Wear rate and friction coefficient were lowest at optimal loadings that ranged from 0.1 to 10 depending on the crystallinity and particle size. Wear rate decreased monotonically with decreasing particle size and decreasing crystallinity at any loading in the range tested.

### 14.1.3. Layer/Laminate-Reinforced Polymer Nanocomposites

It was proposed that nanocomposite polymer layers capable of very large elastic deformation can exhibit superior nano- and microtribological properties [42].

To reduce the adhesion, internal lubricants such as polytetrafluoroethylene (PTFE), and graphite flakes are frequently incorporated. One of the mechanisms of the corresponding reduction in the coefficient of friction is the formation of a PTFE-transfer film on the surface of the counterpart [1].

Dasari et al. [43] examined the role of nanoclay on the wear characteristics of nylon 6 nanocomposites processed via different routes.

In the study, four types of specimens were prepared: (a) neat nylon 6 (designated E1); (b) nylon 6/as-received pristine clay composite (E2) where no water was injected in compounding; (c) nylon 6/organoclay nanocomposite (E3) in which clay was pretreated with the organic alkyl ammonium surfactant; and (d) nylon 6/pristine clay nanocomposite (E4) prepared with the aid of water.

Wear tests and transmission electron microscopy were used to determine the friction coefficient and wear morphology of the composites, respectively.

Figure 14.18 shows the friction coefficient versus sliding distance at a constant applied load of 5 N and different sliding speeds for all the studied materials. At a low sliding speed of 0.04 m/s, the friction coefficients for all materials are similar but small in the range 0.10–0.14. The presence of clay, either exfoliated (E3, E4) or as-received (E2), and indeed absence of clay (E1) do not seem to induce any significant changes in the friction data. No transfer film can be found on the counterface of the silicon nitride ball. With an increase in sliding speed (0.08 m/s), even though the friction coefficients have increased for all tested materials, the increase is not large in absolute value for E1 (from ~0.14 at 0.04 m/s to ~0.25 at 0.08 m/s).

Figure 14.19 shows wear features of nylon 6 nanocomposite when clay was pretreated with the organic surfactant. At a sliding speed of 0.08 m/s, no obvious damage is observed (Figure 14.19(a)) except wear grooves/furrows. With increasing sliding speed (0.12 m/s, Figure 14.19(b)), the wear grooves become more prominent and occasional damage due to pitting is observed (Figure 14.19(c)). Therefore, interlaminar delamination of fine clay layers in E3, as opposed to that in the large clay aggregates in E2, does not increase wear of nylon due to the hard asperities of the counterface. At the highest sliding speed of 0.20 m/s, there is more surface damage at the microscale (Figures 14.19(d) and (e)). Plowed grooves/furrows are formed similar to E2. However, the severity of wear damage is much less than E2 and E1. In addition, the organoclay layers exert a high constraint on the surrounding nylon 6 matrix material making it difficult to be stretched plastically and torn, thus increasing the wear resistance of E3.

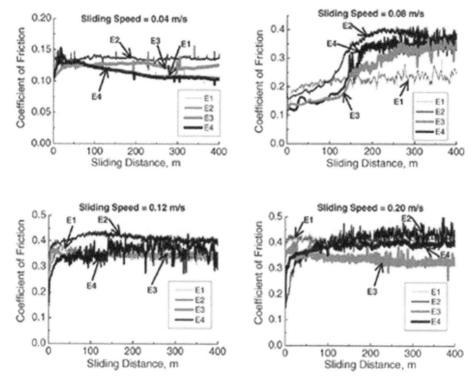

**Figure 14.18**  Plots of friction coefficient versus sliding distance at constant normal load of 5 N and at sliding speeds of 0.04, 0.08, 0.12, and 0.20 m/s for all tested materials (E1–E4).

The results show that nylon 6/pristine clay composite had the worst wear resistance due to the large aggregated clay particles. Nylon 6/organoclay nanocomposite (E3) exhibited higher resistance to wear in contrast to E2 and E4.

It was revealed that the interfacial adhesion of clay to matrix, and not the exfoliated morphology of clay, played a critical role in wear. However, exfoliated clay morphology is preferred to aggregate morphology. Therefore, the superior wear performance of nylon 6/organoclay nanocomposite is brought about by a combined effect of fine dispersion of clay platelets in nylon 6, high interfacial interaction between nylon 6 and clay layers, and effective constraint on surrounding nylon 6 material exerted by the clay platelets.

A polymer trilayer (sandwiched) film with a thickness of 20–30 nm has been designed to serve as a wear-resistant nanoscale coating for silicon surfaces according to a study conducted by Sidorenko et al. [42]. The surface structures are formed by a multiple grafting technique applied to self-assembled monolayers (SAM) and functionalized triblock copolymer, followed by the photopolymerization of a topmost polymer layer.

Highly polished single-crystal silicon wafers of {1 0 0} orientation were used as substrates. (3-Glycidoxypropyl)trimethoxysilane (Aldrich) was used to prepare epoxy-terminated SAM on the silicon surface. The primary compliant layer was formed from the triblock copolymer, poly[styrene-*b*-(ethylene-*co*-butylene)-*b*-styrene] (SEBS) (Kraton 1901, Shell), with glassy polystyrene (PS) domains embedded in rubber polyethyleneb-uthylene (PEB) matrix. This layer was grafted to the epoxy-terminated SAM. The thickness

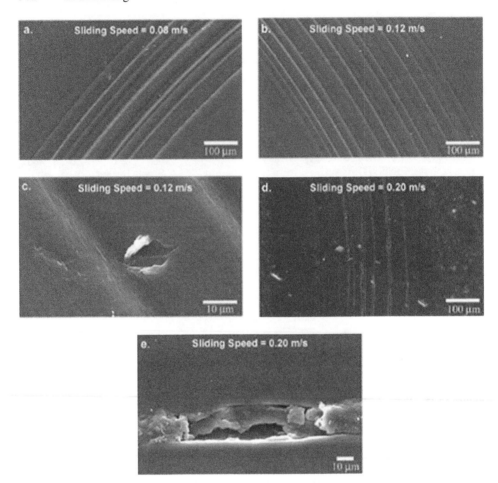

**Figure 14.19**   SEM micrographs showing beneficial effect of adding organoclay to nylon 6 (E3) on wear-resistance at different sliding speeds (0.08, 0.12, and 0.20 m/s) and a normal load of 5 N.

of the grafted SEBS layer was 8.4 ± 0.4 nm. Both the compounds HDM and photoinitiator 4-(dimethylamino) benzophenone (PI) (Aldrich) were used as received. The solution of monomers was deposited directly on to the rubber layer, covered with a glass plate, and exposed to UV light to initialize photopolymerization and the formation of PMA. After polymerization, the glass plate was removed, the sample was rinsed thoroughly with toluene, and treated in an ultrasonic bath to remove residual monomer and ungrafted polymer.

It can be demonstrated that a polymer trilayer film with a nanoscale thickness of 20–30 nm can be designed to serve as a wear-resistant protective nanoscale coating for silicon surfaces. This surface structure is formed by a multiple grafting technique involving chemical grafting of SAM and functionalized triblock copolymer, followed by the photopolymerization of the topmost polymer layer. The unique design of this coating includes a sandwiched, triplex structure with a compliant rubber interlayer mediating local stresses transferred through the topmost hard layer. This architecture provides nonlinear mechanical response

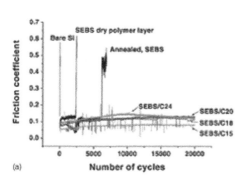
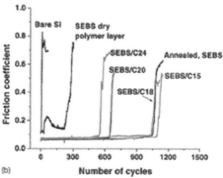

**Figure 14.20**   The coefficient of friction as a function of a number of sliding cycles for the bare silicon, dry polymer layer, annealed polymer layer, and polymer gel layers at the normal load of 0.3 N (a) and 1.8 N (b).

under compression stresses and allows additional dissipation of mechanical energy via the reversibly compressed rubber interlayer. At modest loads, this coating shows friction coefficient against hard steel below 0.06, which is lower than that for a classic molecular lubricant, alkylsilane SAM. At the highest pressure tested in this work, 1.2 GPa, and at the relatively high sliding velocity of 4.4 mm/s, the sandwiched coating designed here were four times more wear resistant than the SAM coating. The predominant wear mechanism was stress- and temperature-induced oxidation in the contact area followed by severe plowing wear.

Ahn et al. [44] reported results on microtribological studies of chemically grafted, nanoscale polymer layers with enhanced wear stability. A 8–10 nm thick polymer gel layer composed of an elastomer was chemically grafted to a solid substrate and saturated with paraffinic molecules with different lengths of alkyl chains (15–24 carbon atoms, molecular weight $M = 212$–338). A microtribometer and a friction force microscope were used to accumulate the frictional characteristics and to study wear stability of the polymer layers.

The grafted polymer films possess uniform, smooth, and homogeneous surface with only a few aggregates observed over surface areas of 10 μm across.

Figure 14.20 shows the coefficient of friction calculated from microtribological data as a ratio of the lateral forces to the normal load as a function of the number of sliding cycles. At a low normal load of 0.3 N, the grafted polymer gel layers exposed to oils showed a performance much better than the uncoated silicon and the dry polymer film. The polymer gel layers did not show a failure or significant deterioration up to 20,000 cycles (the maximum number of cycles tested here), whereas the bare silicon and the dry polymer layer failed within 200 and 2700 cycles, respectively (both values were averaged over three independent measurements).

The polymer gel layers with short-chain (C15) paraffinic oil showed the lowest friction coefficient among the layers (~0.05). The friction coefficient increased with the increasing molecular weight of paraffinic oils to 0.11–0.12 for the polymer gel layer with C24. The friction coefficient value reached the highest value of 0.13–0.17 for the dry polymer layer and a bare silicon substrate (Figure 14.20). The effect of paraffinic oil on the frictional behavior of the polymer layer became more significant for longer runs when the dry polymer layer finally started to deteriorate. At a higher normal load of 1.8 N, the dry

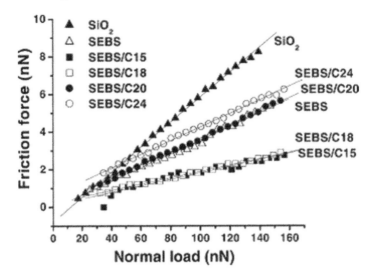

**Figure 14.21** Friction forces versus normal loads as measured for the bare silicon, dry polymer layer, annealed polymer layer, and polymer gel layers with paraffinic oil (C15–C24).

polymer failed only after 250 cycles, while the polymer gel layers were stable up to 500–700 cycles (C20–C24) and 900–1000 cycles (C15 and C18).

It is worth noting that additional annealing of the polymer layer at 60°C for 24 h also resulted in significant improvement of wear resistance, especially under high normal load (Figure 14.20). Obviously that additional annealing of the polymer layer allowed MA functional groups of the rubbery block poly(ethylene-*co*-butylene) to react with the epoxy groups of the supporting surface, thus resulting in a stronger adhered layer. However, the presence of alkyl-chain molecules is still more important factor in enhanced wear resistance at low normal loads (Figure 14.20(a)).

Finally, the nanotribological properties were characterized with friction force microscopy (Figure 14.21). Loading curves were obtained for a bare silicon substrate, annealed polymer layer, and the oil-exposed polymer layers under identical conditions (identical probe, scanning velocity, scanning size, and the range of normal loads). The friction coefficients calculated as a slope of a linear approximation were determined to be the lowest for polymer layers with C15 and C18 oils. The friction coefficient for their layers decreased to 0.02, which was much lower than that for the silicon substrate (0.07–0.1). Despite the fact that the absolute values of the friction coefficient calculated from SPM data were systematically lower than those determined with the microtribotester, general trends were very consistent. Therefore, we can conclude that the polymer gel layer treated with C15 and C18 paraffinic molecules showed the best tribological performance.

It could be concluded that the presence of shorter chain paraffinic oil ($C_{15}H_{32}$ and $C_{18}H_{38}$) resulted in a lower value of the friction coefficient and higher wear resistance as compared to a dry polymer layer and a polymer gel layer with longer chain paraffinic oil ($C_{20}H_{42}$ and $C_{24}H_{50}$). The approach of trapping mobile lubricants within a compliant nanoscale surface layer could lead to exceptionally robust molecular lubrication coatings for complex surface topography with developed nanosized features.

The microtribological performance of molecularly thick ($\sim$10 nm) thermoplastic elastomeric films grafted to a silicon surface was enhanced by adding a minute amount

**Table 14.6**  Polymer composite film's thickness and experiments used in the study.

| Samples | Films thickness | Experiments performed |
|---|---|---|
| Free oil surface | NA | Weight measurements: oil evaporation |
| Bulk polymer | 1.5 mm | Weight measurement with balance: oil diffusion |
| Polymer gel spin-coated thick films | 500 nm | FTIR studies: oil evaporation |
| Polymer gel spin-coated thin films | 8–20 nm | AFM, ellipsometry: oil evaporation |
| Polymer gel grafted films | 8–20 nm | AFM, ellipsometry, microtribometers: surface morphology, tribological properties |

of paraffinic oil, which was adsorbed from vapor phase and held by the rubber matrix according to a study by Julthongpiput et al. [45].

Poly[styrene-*b*-(ethylene-*co*-butylene)-*b*-styrene] (SEBS) with $M_n$ = 41,000 g/mol, $M_w/M_n$ = 1.16 SEBS copolymer were used in this study. The paraffinic oil, $C_{15}H_{32}$, and its deuterated analog, $C_{15}D_{32}$ (Aldrich), were used as lubrication additives. Two different SEBS polymer gel layers were fabricated on the {100} silicon wafer, cut into pieces of 1.5×2 cm$^2$ (Table 14.6).

The presence of a minute amount of the paraffinic oil after evaporation (estimated to be below 5%) affected dramatically the tribological characteristics of the polymer gel layer. The friction coefficient between the silicon tip and surface (estimated radius of the contact area is below 5 nm) as measured by SPM decreased by 40% as compared with the dry polymer layer. Wear resistance of the polymer gel layer increased dramatically.

Figure 14.22 shows the variation of the coefficient of friction, calculated as a ratio of the lateral forces measured to the normal load applied, as a function of a number of sliding

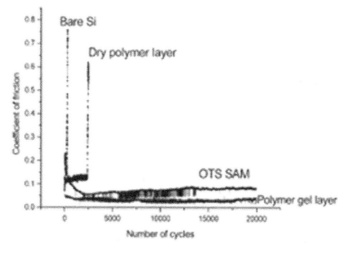

**Figure 14.22**  The coefficient of friction as a function of a number of sliding cycles for the bare silicon (1.2 nm thickness of silicon oxide layer), OTS SAM (2.4 nm), dry grafted polymer layer (8 nm), and polymer gel layer (8 nm) at the normal load of 0.3 N. A sharp increase in friction forces indicates ultimate surface damage.

cycles for the polymer gel layer in comparison with similar data for a bare silicon, the alkylsilane SAM, and the dry SEBS polymer layer. It revealed that at a normal load of 0.3 N, the grafted polymer gel and OTS coating showed a microtribological performance much better than the bare silicon and the dry polymer film. The polymer gel layer had a significantly lower friction coefficient (~0.03) than the SAM coating (0.07).

It could be concluded that these polymer gel layers exhibited a very steady friction response and a small value of the coefficient of friction as well as greater wear-resistance as compared to the initial polymer coating. The performance of polymer gel coatings was much better than the performance of a classic "boundary lubricant" for silicon surfaces, an alkylsilane SAM. The proposed approach demonstrated a new efficient route toward enhanced tribological performance of ultrathin polymer coatings.

CNF-reinforced ultrahigh molecular weight polyethylene (UHMWPE) nanocomposites containing up to 10 wt% of nanofibers were prepared by Galetz et al. [46]. A detailed investigation of the resulting nanocomposite microstructure and of the static mechanical properties revealed that the CNFs lead to improved mechanical properties of the UHMWPE related to the wear performance of such systems. Unidirectional sliding tests against a 100Cr6 steel under dry conditions verified the significant potential of dispersed CNFs to reduce the wear rate of this polymer.

In order to reveal the wear mechanism of polyamide 66 (PA 66) and its composites consisting of styrene–(ethylene/butylene)–styrene triblock rubber grafted with maleic anhydride (SEBS-*g*-MA) particles and organoclay nanolayers, the tribological behavior and transfer films of polyamide 66 and its composites were investigated under dry sliding by Yu et al. [47]. The results show that when SEBS-*g*-MA rubber particles and organoclay nanolayers are added simultaneously to PA 66, the wear resistance of PA 66 can be improved markedly. The main reason is that PA 66/SEBS-*g*-MA/organoclay ternary nanocomposite can form a thin and uniform transfer film on the steel ring surface.

### 14.1.4. Effect of Matrix and Fillers on the Tribological Behavior of Polymer Nanocomposites

When dealing with polymer nanocomposites for tribological applications, particle size, aspect ratio, hardness, concentration, orientation, and the nature of interface between the polymer matrix and particle are all important factors [43].

The tribology of nanoparticle-filled polymers is of significant interest because of the ability of nanoparticles to alter the properties of the matrix and the surfaces involved while remaining as small defects in the matrix [48].

The friction and wear properties of PET filled with alumina nanoparticles were studied according to a study conducted by Bhimaraj et al. [48].

The nanoparticle loading was varied from 1 to 10 wt%. The nanocomposite samples were tested in dry sliding against a steel counterface.

The coefficient of friction data are plotted versus sliding distance for the five different filler contents; this is shown in Figure 14.23. The 1 and 2 wt% composites have slightly lower average coefficients of friction than the unfilled material, with the minimum coefficient of friction obtained at 2 wt%. On further increase in filler content, the coefficient of friction increased, with the 10 wt% sample displaying the highest value. The time-averaged coefficient of friction versus filler content and the standard deviation of the fluctuations in the coefficient of friction during the tests is shown in Figure 14.24. The wear rate for the quenched and annealed samples is also shown in Figure 14.24. At low filler concentrations,

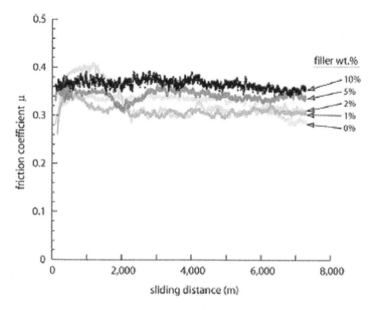

**Figure 14.23** Friction coefficient versus sliding distance for the PET composites (normal load: 340 N, sliding speed: 25 mm/s).

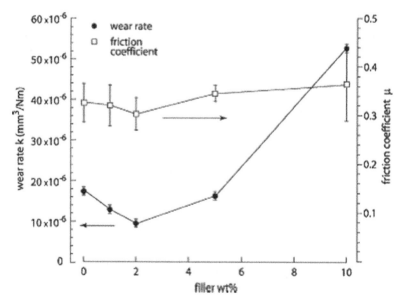

**Figure 14.24** Variation in wear rate of the PET and average coefficient of friction plotted against loading of 38 nm alumina. Wear rate decreases for low loadings with optimum filler content at 2 wt% (normal load: 340 N, sliding speed: 25 mm/s). The error bars on the wear rate are the experimental uncertainty, while for those for the coefficient of friction represent the standard deviation of the measurement during the experiment.

the wear rate decreased with increasing filler concentration. The minimum wear rate is obtained at 2 wt% and is half the wear rate of the unfilled sample. Thereafter, the wear rate increases with increasing concentration and at 10 wt% it is higher than the wear rate of unfilled polymer.

The results show that the addition of nanoparticles can increase the wear resistance by nearly 2× over the unfilled polymer. The average coefficient of friction also decreased in many cases. The nanocomposites form a more adherent transfer film that protects the sample from the steel counterface, although the presence of an optimum filler content may be due to the development of abrasive agglomerates within the transfer films in the higher wt% samples. This study varied both crystallinity and weight percent of filler in a PET matrix in an attempt to separate the effects of nanofillers and crystallinity on the tribology.

Luo et al. [49] investigated the effect of controlled nanoporosity on the wear resistance of polymeric composites reinforced with silica gel powders and determined the mechanisms controlling the abrasive wear properties of these unique nanostructured materials.

Silica gels were prepared by hydrolysis, and condensation of tetraethylorthosilicate (TEOS) using four different catalysts to modify the porous structure of the resulting polysilicate silanation, an organic monomer (TEGDMA) containing various initiators, was introduced into the gel powders to form a paste. The various pastes were then polymerized inside a glass mold. A pin-on-disk apparatus was then used to record the specimen length and number of revolutions. Abrasive wear rates were determined by regression analysis, and statistical differences were determined by analysis of variance and multiple comparisons. BET was used to characterize the filler pore structure, and scanning electron microscopy was used to visually examine the abraded surfaces.

Significant differences in the wear rates of the experimental composites were noted. Within the range of filler porosities examined, wear resistance was found to be linearly dependent ($R = 0.983$) on filler pore volume, as is shown in Figure 14.25. The wear rates

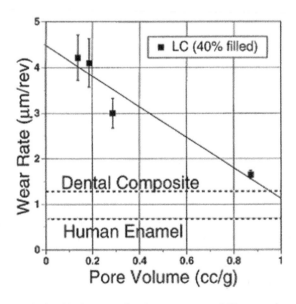

**Figure 14.25** Linear relationship between abrasive wear rate and filler porosity of comparably filled (40 wt%) experimental composites ($R = 0.983$).

decreased with increasing filler porosity. HCl-catalyzed gels having low porosity produced composites having relatively limited abrasion resistance. In contrast, high-porosity HF-catalyzed gels produced more wear-resistant composites. The abrasive wear resistance of these nanocomposites was not significantly affected by the level of silane coupling used in these experiments. SEM evaluation suggested that better wear resistance was associated with fine-scale plastic deformation of the wear surface and the absence of filler particle pullout.

It can be found that porous particles prepared via sol–gel show some promise as fillers that improve the wear resistance of photopolymerized resins. The wear resistance of the fillers appears to be directly related to nanoporous structure of the gel particles. These materials rely primarily on nanomechanical coupling for improved wear resistance.

To overcome the disadvantages generated by the loosened nanoparticle agglomerates dispersed in polymer composites, an irradiation grafting method was applied to modify nanosilica by covalently bonding polyacrylamide (PAAM) onto the particles according to a study by Zhang et al. [28].

When the grafted nanosilica was added to epoxy, the curing kinetics of the matrix was accelerated. Moreover, the grafting PAAM can take part in the curing of epoxy so that chemical bonding was established between the nanometer fillers and the matrix. Sliding wear tests of the materials demonstrated that the frictional coefficient and the specific wear rate of nanosilica/epoxy composites are lower than those of the unfilled epoxy. With a rise in nominal load, both frictional coefficient and wear rate of the composites decrease, suggesting a wear mechanism different from that involved in wearing of epoxy. Grafted nanosilica reinforced composites have the lowest frictional property and the highest wear resistance of the examined composites. Compared with the cases of microsized silica and untreated nanosilica, the employment of grafted nanosilica provided the composites with much higher tribological performance enhancement efficiency.

It can be concluded that unlike micrometer silica, nanosilica can simultaneously provide epoxy with friction- and wear-reducing functions at low filler content (∼2 vol%). The coefficient of friction and specific wear rate of nanosilica-filled epoxy composites decrease with increasing nominal pressure from 3 to 5 MPa. Grafting of PAAM onto nanosilica increases the interfacial interaction between the particles and the matrix through chemical bonding. It proves to be an effective way to further enhance the nanoeffect of the nanoparticles on the improvements of the tribological performance.

## 14.2. Fatigue of Polymer Nanocomposites

Polymers and composites are more and more involved as structural parts in industry, and so specific data such as long-term mechanical properties (creep, relaxation, fatigue) is needed to help the designer. Besides impact, fatigue is the most critical loading mechanism for a material. The consideration of the fatigue behavior of higher load bearing components is of great importance. The failure process in fatigue can be always depicted as a sequence of two steps [50, 51]: the first step is the initiation of microcracks at defects or other inhomogeneities, which may occur at load levels far below the yield strength or the tensile strength of the material. The second step is the propagation of these cracks, which leads to total failure of the component.

The molecular aspects influencing the fatigue behavior of a polymeric material, which are currently the subject of much materials research, should be considered. Examples of

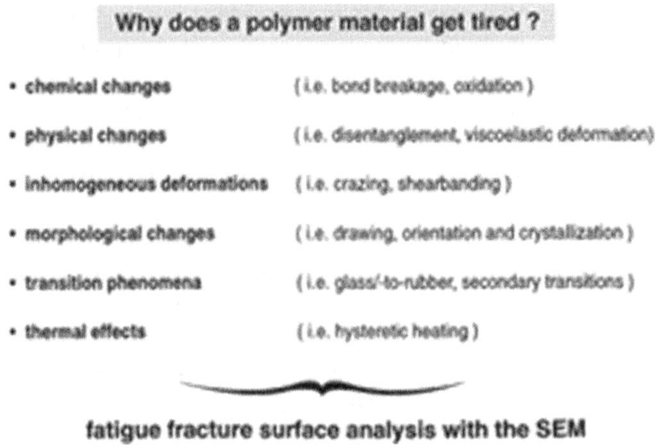

**Why does a polymer material get tired ?**

- chemical changes    ( i.e. bond breakage, oxidation )

- physical changes    ( i.e. disentanglement, viscoelastic deformation)

- inhomogeneous deformations    ( i.e. crazing, shearbanding )

- morphological changes    ( i.e. drawing, orientation and crystallization )

- transition phenomena    ( i.e. glass/-to-rubber, secondary transitions )

- thermal effects    ( i.e. hysteretic heating )

**fatigue fracture surface analysis with the SEM**

**Figure 14.26**    Molecular aspects inluencing fatigue of polymers.

causes of fatigue behavior could be chemical changes such as the breaking of chemical bonds or an oxidative attack. Further examples could be physical changes such as disentanglement and viscoelastic deformation (Figure 14.26). Unlike metals, the cyclic loading of polymeric materials often leads to non-negligible heating due to high material damping and low thermal conductivity. Thus, for the evaluation of the fatigue properties, care should be taken to distinguish mechanical fatigue from fatigue as a result of hysteretic heating. The latter can be suppressed by the reduction of the test frequency [51].

It is well known that the specific fatigue strength of long fiber reinforced polymer matrix composites is excellent compared to other materials. Considering short fiber or particulate-reinforced polymer composites, the characterization of fatigue strength is more difficult.

Elastomeric matrix composites are usually reinforced by mineral particles such as carbon black and sometimes by long metallic or organic fibers. In the absence of fiber, rubbers can be considered as nanocomposites. In service conditions, the fatigue damage of rubbers is a combination of (a) mechanical damage, (b) chemical damage, and (c) thermal damage. Experience shows that in cyclic loading, rubbers are damaged to the point of formation of one or several cracks which then propagate. As for metal, it is recommended to study separately initiation of cracks and then their propagation. Generally speaking, the fatigue resistance is affected by chemical transformation such as crystallization.

A study conducted by Legorju et al. [52] investigated fatigue initiation and propagation in natural and synthetic rubbers. The study of fatigue crack growth and damage mechanisms in rubbers has led to some conclusions. It can be found that the fatigue damage depends on three basic mechanisms: chemical (composition, crystallization), environmental (oxygen), and mechanical (stretching, triaxial stresses). A mean stress in tension improves the fatigue behavior by crystallization of the stretched bonds for exclusively tension cycles. On the other hand, a minimum stress in compression seriously damages the material. In addition to the mechanical damage, an important chemical damage due to gaseous oxygen in the air is present. This chemical damage is more extensive at higher temperatures because of the acceleration of the oxidation reaction. Furthermore, the location of the initiation of damage is strongly related to the stress distribution, shear stress, and hydrostatic pressure.

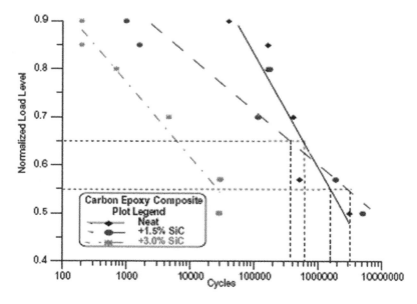

**Figure 14.27**    S–N diagram (flexural fatigue).

Another study carried out by Chisholm et al. [53] investigated matrix properties by introducing micro- and nanosized SiC fillers into an epoxy matrix. The study has revealed that with equal amount of loading, nanoparticle infusion brings about superior thermal and mechanical properties to the matrix than what is usually given by the microfillers infusion. The nanophased matrix is then utilized in a vacuum-assisted resin transfer molding set up with satin weave carbon preforms to fabricate laminated composites. The fillers were nano- and micronsize silicon carbide particles which were mixed with the SC-15 epoxy resin using an ultrasonic processor. Amount of particle loading varied from 1.5% to 3.0% by weight of the resin. Ultrasonic mixing utilized high energy sonic waves to force an intrinsic mixing of particles with the matrix via sonic cavitations.

Fatigue tests were performed under flexural loading, and the performance of the nanoinfused system was seen to be superior to that of the neat system.

In the study, the flexural fatigue tests were conducted at a stress ratio of 0.1 and frequency of 3 Hz. S–N diagrams were then generated for the three systems and are shown in Figure 14.27. It was observed in general that above 60% load level, the neat system is performing better than any of the nanoinfused systems. The 3% system is demonstrating a very poor fatigue performance over the entire range of loading. During fatigue, once the load level goes below 60%, the fatigue performances of the neat and 1.5% system begin to reverse. For example, at 65% load level, the ratio of cycle numbers to failure between the neat and the 1.5% system is around 1.75. Whereas this ratio changes to 0.56 at 55% load level. The reversal of the fatigue phenomenon is not yet fully understood. However, it is seen in the S–N diagram that the slope of the neat system is much steeper than that of the 1.5% system suggesting a lesser sensitivity of the neat system with respect to applied load. This indicates that for a change in the load level, the corresponding change in cycle numbers will be much smaller in case of the neat system. Since this continued up to 50% load level, an intersection point at around 60% marker, which was defined as a threshold load level, can be found. Below this threshold stress level, the fatigue failure

**Table 14.7** Flexural fatigue response (frequency = 3 Hz, $R = 0.1$).

| | Load level | | | | | |
|---|---|---|---|---|---|---|
| Type | 50% | 57% | 70% | 80% | 85% | 90% |
| Neat | 3,109,800 | 496,090 | 621,000 | 66,260 | 69,260 | 38,824 |
| | 3,187,786 | 509,136 | 500,000 | 155,577 | 165,577 | 40,989 |
| | | 560,050 | 244,900 | 37,503 | 37,503 | 41,860 |
| | | | 278,589 | 386,579 | 386,579 | |
| Average | 3,148,793 | 521,759 | 411,122 | 161,480 | 164,730 | 40,558 |
| 1.5% SiC | 5,090,088 | 2,945,712 | 200,000 | 416,285 | 1284 | 1604 |
| | 5,146,479 | 120,281 | 39,350 | 55,327 | 1576 | 538 |
| | | 2,607,053 | 166,393 | 63,490 | 608 | 1066 |
| | | | 105,525 | 52,744 | 3575 | |
| | | | | 300,087 | 1400 | |
| Average | 5,118,284 | 1891,015 | 127,817 | 177,587 | 1689 | 1069 |
| 3.0% SiC | 29,897 | 4688 | 200,000 | 643 | 286 | 121 |
| | 27,765 | 5890 | 39,350 | 534 | 101 | 249 |
| | 28,578 | 3420 | 166,393 | 764 | | |
| | 29,657 | | 105,525 | 1065 | | |
| Average | 28,974 | 4666 | 127,817 | 752 | 194 | 185 |

mechanisms such as matrix cracks, filament splitting, and delamination are significantly slowed down with 1.5% system. And that is quite possible since nanoinfusion (1.5%) certainly improves the matrix properties and matrix-dominated failure modes. Data shown in Table 14.7 represent all failed samples. It is also noted in Figure 14.27 that the slopes of the two nanophased systems (1.5% and 3%) are similar indicating that they are more or less equally sensitive to applied load with the exception that 1.5 wt% system is outperforming the 3 wt% system by a large margin.

It could be concluded that during flexural fatigue, a threshold load level around 60% of the ultimate flexural strength seems to exist. Below this threshold load level, the 1.5% system superceded the neat system whereas the situation is reversed if stress level is above this level. Throughout the entire loading range, the 3% system is seen to be inferior to the other two systems.

Jen et al. [54] investigated the fatigue behavior by adding nanoparticles into PEEK APC-2 composite laminates. The nanocomposite laminates were made by sol–gel method and modified diaphragm method. The constant stress fatigue testing was performed at room temperature; it was found a little improvement in fatigue behavior in nanoparticle laminates apparently.

The received S–N curves of normalized stress versus cycles in semi-log coordinates were plotted in Figures 14.28 and 14.29 for crossply and quasi-isotropic specimens, respectively. The polynomial curve fitting method was adopted; the solid curve represents the data without nanoparticles, while, the dotted curve represents data with nanoparticles in both figures. Apparently, both curves are very close from the beginning to $10^4$ cycles in Figure 14.28; after $10^5$ cycles the dotted curve bends down significantly. However, it must be borne in mind that the ultimate strength of the laminate with nanoparticles was significantly higher than that of original APC-2 laminate. Therefore, at the same normalized stress level, the laminate with nanoparticles is subjected to a higher absolute applied stress level. Similarly, we can explain the phenomenon in Figure 14.29. Therefore,

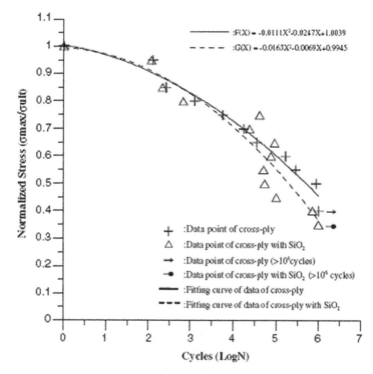

**Figure 14.28**    Normalized stress versus cycles for both with and without SiO₂ in crossply laminates.

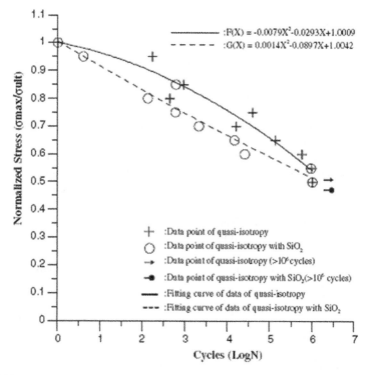

**Figure 14.29**    Normalized stress versus cycles for both with and without SiO₂ in quasi-isotropic laminates.

the laminate of optimal weight percent of nanoparticles can obviously improve the fatigue behavior.

The mechanical response of a polyamide-6 montmorillonite clay nanocomposite and of a polyamide-6 was monitored during axial fatigue tests performed at $R$-ratios of 0.1 and $-1$ by Bellemare et al. [55]. For both materials, two transitions were usually observed in the evolution of all the stress–strain–time parameters studied after similar numbers of loading cycles, suggesting interrelationships between the mechanisms of molecular reorganization. Fatigue test monitoring indicated an initial decrease in the storage modulus and a subsequent trend for this modulus to increase, especially in polyamide-6. During all tests, a partially recoverable strain was accumulated because of viscoelastic deformation. Nanoparticles reduced this strain in the initial cyclic straining regime but not in the last regime, probably because such particles cannot inhibit viscoelastic events constrained in a volume larger than their interaction volume within the matrix. Based on the accumulated volume variation measured, the nucleation and growth of microvoids can be expected to occur in the last cyclic straining regime.

Juwono et al. [56] investigated the fatigue behavior and the mechanism of fatigue failure of an epoxy resin with a dispersion of modified layered silicates in the polymer matrix. Clay–epoxy nanocomposites were successfully synthesized with a commercially available 1-methylimidazole curing agent. The fatigue performance and fatigue failure mechanism of the clay–epoxy materials were studied under repetitive bending loads. The results showed that the fatigue life of filled epoxy improved significantly at strain amplitudes below a threshold value. The E-SEM observations of the epoxy and the clay–epoxy fracture surfaces showed different patterns. In conclusion, the addition of silicate strongly determines the fracture mechanism and enhances the fatigue performance.

## 14.3. Creep of Polymer Nanocomposites

Creep is the time-dependent deformation of materials subjected to a continuous stress. This deformation can be both elastic and plastic and, therefore, it can be nonrecoverable when the load is removed. Creep can lead to unacceptable deformation and eventually even to structural failure and is, therefore, a very important property for engineering plastics and composites. The creep response is quantified by the creep compliance, which is the strain divided by the applied stress [57]. In polymers, the deformation process under load is strongly dependent on the mobility of the chains. The chain mobility is not only temperature dependent but also time dependent. The time-dependent change of the properties in amorphous polymers below the glass transition temperature ($T_g$) is known as physical aging [58].

Two major concerns in using advanced polymer composites in structural applications are their long-term dimensional stability and long-term strength, as a consequence of the viscoelasticity of the polymer matrix [59]. Consequently, the creep behavior concerning the lone-term deformation and strength of the polymer nanocomposites has attracted much interest, e.g., [60–65].

Vlasveld et al. [57] described the creep and physical aging behavior of various types of PA6 nanocomposites and unfilled PA6. The nanocomposites with MEE were prepared by mixing the layered silicate with PA6 (K222D) in a corotating twin-screw extruder. The nanocomposites with ME-100 were made by feeding a mixture of cryogenically milled PA6 (K222D) and ME-100 silicate powder in a corotating twin-screw extruder.

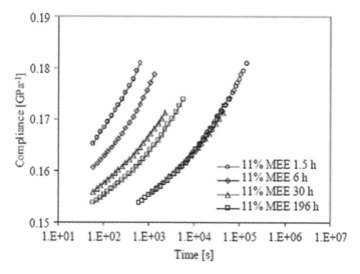

**Figure 14.30**   Creep compliance for 11% MEE nanocomposite as a function of ageing time. The overlapped curves on the right are the curves shifted over the time axis onto one master-curve.

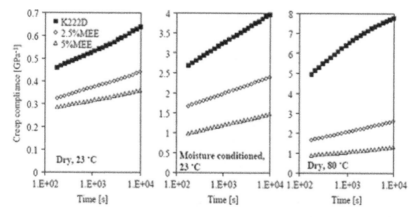

**Figure 14.31**   Effect of absorbed moisture and temperatures above $T_g$ on the creep behavior of PA6 and PA6 nanocomposites.

The samples were tested on a Zwick 1445 tensile tester with a 10 kN force cell. The test machine was equipped with a force feedback loop to be able to test at a constant stress. A constant stress of 16 MPa was used in all ageing experiments, which is approximately 1/5 of the yield stress.

The compliance values in Figure 14.31 for dry samples at 23°C differ slightly from the values in Figure 14.30, because the applied stress was higher and the ageing was performed for several weeks at 80°C. The stress in the creep tests in Figure 14.31 was approximately two-third of the yield stress under the tested conditions; the decrease of the yield stress with increasing temperature and moisture content was taken into account. It can be seen that under these conditions the creep compliance of PA6 is approximately six times higher in moisture-conditioned samples (2.5% water) and 12 times higher in dry samples at 80°C. However, the addition of nanofillers is very effective in reducing the creep

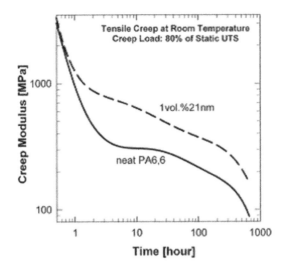

**Figure 14.32** Tensile creep modulus versus test duration curves under 80% of the static UTS at room temperature.

compliance under these conditions, even much more effective than in the dry samples at room temperature. These results clearly show the large benefits of PA6 nanocomposites compared to unfilled PA6 under these conditions.

Pegoretti et al. [61] studied creep properties of PET nanocomposites in solid state (dumb-bell shape specimens). They found an increase in creep-resistance properties in PET nanocomposites.

21-nm $TiO_2$/PA6,6 nanocomposites were compounded using a twin-screw-extruder according to a study conducted by Zhang et al. [62]. The final specimens were formed using an injection-molding machine. Static tension and tensile creep tests were carried out at room and an elevated temperature (50°C).

In the study, creep measurements were carried out at room temperature, in which the creep stress was selected as 80% of the static ultimate tensile strength (UTS). In general, a creep strain versus time curve can be considered as four stages: (i) initial rapid elongation, (ii) primary creep, (iii) secondary creep, and (iv) tertiary creep. Neat PA6,6 exhibited a relatively long creep life (more than 600 h), but at a high creep strain under this load level. On the other hand, 1 vol% 21 nm $TiO_2$ particles significantly reduced the creep strain of PA6,6 over all the three creep stages, although the final creep life was not very much different from that of the neat polymer under the present testing conditions, and was much smaller for the 21-nm particle-filled PA6,6 system.

It could be also found that enhancing the creep load to 90% UTS definitely accelerated the creep process. A very high initial creep rate occurred for neat PA6,6 in the primary creep stage, which was followed by a relatively short secondary creep stage with a steady-state creep rate, being clearly higher than that at the lower load level.

Creep curves of the neat PA6,6 and the nanocomposites, measured at 50°C, are plotted in Figure 14.33. The unfilled matrix was measured at two constant stresses. The higher loading level resulted in a very high initial rapid elongation and a primary creep stage up to about 100 h. The steady-state secondary creep stage lasted more than 1200 h; thereafter,

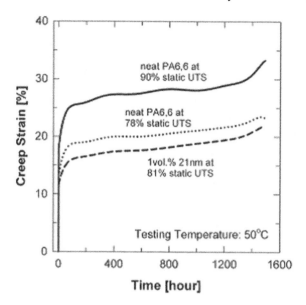

**Figure 14.33** Tensile creep strain versus test duration curves under various load levels at 50°C. Constant creep stresses: 50 MPa (90% of UTS at 50°C) or 43 MPa (78% of UTS at 50°C) for neat PA6,6, and 46 MPa (81% of UTS at 50°C) for nanocomposite.

a rapid increase in the creep rate occurred in the tertiary creep stage, until the polymer fell into creep failure.

It can be found that with a very low volume fraction of inorganic nanoparticles, the creep resistance of thermoplastic could be significantly improved. It is expected to be able to promote the engineering applications of these materials.

In order to comprehensively and deeply understand the effect of nanoparticles, Yang et al. [63] provided a further systematic investigation on various kinds of nanofillers, i.e., spherical particles and nanoclay, modified polyamide 66 under different stress levels (20, 30, and 40 MPa) at room and elevated temperatures (23, 50, and 80°C), respectively. Static tensile tests are also performed at corresponding temperatures. Creep was characterized by considering the isochronous stress–strain curves, creep rate, and creep compliance with the influence of temperature and stress, respectively. It was found that the creep resistance of nanocomposites was significantly enhanced by nanoparticles without sacrificing the tensile properties.

Ranade et al. [64] investigated nonlinear time-dependent creep of polyethylene (PE) montmorillonite layered silicate (MLS) nanocomposites. In the study, PE-grafted maleic anhydride (PE-*g*-MA) was used, as a coupling agent to improve the miscibility between PE and organically modified MLS. The creep and tensile response of maleated and non-maleated PE nanocomposites were determined. The results show that the tensile properties of maleated PE nanocomposites were higher than the nonmaleated nanocomposites. And the increase in physical and creep properties of the maleated PE nanocomposites was due to synergistic contributions from PE-*g*-MA and MLS. Maleated PE nanocomposites showed not only better creep recovery but also lower creep strains than neat PE and nonmaleated nanocomposites.

Nanoindentation technique has been used to investigate the mechanical properties of exfoliated nylon-66 (PA66)/clay nanocomposites according to a study conducted by Shen et al. [65]. In the study, the creep behavior of the nanocomposites has been evaluated. The creep behavior of the nanocomposites shows an unexpected increasing trend as the clay loading increases (up to 5 wt%). It is believed that the lowered creep resistance with increasing clay content is mainly due to the decrease in crystal size and degree of crystallinity as a result of clay addition into PA66 matrix.

The long-term tensile creep of polyamide 66 and its nanocomposites filled with 1 vol% $TiO_2$ nanoparticles 21 and 300 nm in diameter is studied by Starkova et al. [66]. It is assumed that the dominant mechanisms of creep deformation are of viscoelastic nature, while the contribution of plastic strains is not essential in the stress (< 0.6 of the ultimate stress) and time (about 100 h) ranges considered. The creep isochrones obtained show that the materials exhibit a nonlinear viscoelastic behavior and the degree of nonlinearity is reduced significantly by incorporation of the nanoparticles. The evolution of viscoelastic strains is less pronounced for the nanocomposite filled with smaller nanoparticles. Smooth master curves are constructed by applying the time–stress superposition (TSS). The Boltzmann–Volterra hereditary theory is used for the creep modeling. The nonlinearity of viscoelastic behavior is taken into account by using the TSS principles and introducing a stress reduction function into an exponential creep kernel. The master curves are employed to predict the creep for time periods more than 60 times exceeding the test time. A comparison of relaxation spectra of the polymers shows that the incorporation of nanoparticles restricts the mobility of polymer chains. The smaller the nanoparticles, the greater the enhancement in the creep resistance. An empirical approach and a three-parameter law are also used for creep approximation.

In a recent work conducted by Zhou et al. [67], $SiO_2$ nanoparticles were pretreated by silane-coupling agent to introduce reactive C=C double bonds onto their surfaces. Afterwards, in the melt mixing with matrix polypropylene (PP), butyl acrylate monomers reacted with the silane-coupling agent on the nanoparticles and then crosslinked to interconnect the particles with each other, while the PP chains penetrated into the networks, forming semi-interpenetrating polymer network (semi-IPN) structure. Taking the advantage of the specific microstructure, the nanoparticles were well distributed in the matrix with enhanced interfacial interaction. As a result, mobility of the matrix molecules was restricted to certain extent, and creep resistance of the nanocomposites became much higher than those of untreated nano-$SiO_2$/PP system and unfilled PP as well.

## 14.4. Stress-Relaxation Behavior of Polymer Nanocomposites

Polymer materials are often operating under conditions where factors such as fatigue, creep, stress relaxation, and chemical attack are significant. Such materials can fail by progressive degradation resulting in hardening, softening, set, crack growth, or fracture.

The stress-relaxation effect comes from a combination of physical and chemical processes. The physical process involves the motion of chains toward new configurations in equilibrium at the new perturbed (strained) state such as movement of entanglements and the relaxation of chain ends [68]. The main difference between physical and chemical relaxation processes is that certain physical processes such as entanglements, diffusion of chain ends, or cell wall buckling have the potential to recover with time, whereas all chemical relaxation processes are nonrecoverable [68, 69].

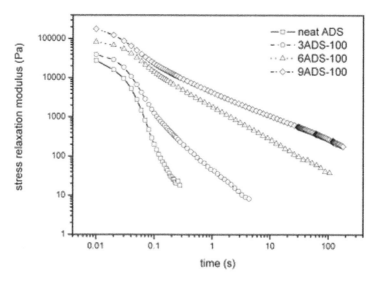

**Figure 14.34**    Linear stress relaxation modulus $G(t)$ for ADS-based nanocomposites with different clay content, extruded at 100 rpm.

The prediction and understanding of the stress-relaxation behavior is of particular importance, because it provides information about the molecular mechanisms affecting the macroscopic properties of the material [70]. Creep (constant stress) and stress-relaxation (constant strain) tests are perhaps the simplest techniques for analyzing the relaxation of polymeric structures [71]. The stress-relaxation behavior of polymer nanocomposites has been investigated in several studies [72–77].

Incarnato et al. [72] investigated the stress-relaxation behavior of the copolyamide nanocomposites. It is believed that the formation of an extended structural network across the polymer matrix is due to strong polymer–silicate interactions that slow the relaxation times of the macromolecules.

Transient stress-relaxation measurements, obtained at low strain in order to assure linear regime, are reported in Figure 14.34 for ADS (the selected copolyamide) based nanocomposites. It can be observed that for any fixed time after the imposition of strain, the modulus $G(t)$ increases with silicate loading. Furthermore, while at short times the stress-relaxation behavior is qualitatively similar for the hybrids and unfilled polymer, and at long times the unfilled polymer relaxes like a liquid, while the hybrids with high silicate contents behave like a pseudo-solid material. Similar trend was also observed for PA6-based nanocomposites, but in this last case the differences are smaller (Figure 14.35). Both the dynamic oscillatory shear and the stress relaxation moduli indicate that the addition of the layered silicate to PA6 and ADS resins significantly modify the long-time relaxation of the hybrids by increasing their relaxation times due to the formation of a three-dimensional superstructure. This phenomenon is particularly relevant for ADS hybrids with silicate loading higher than 6 wt%.

Baeurle et al. [73] develop a new multiscale modeling method, which combines the self-consistent field theory approach with the kinetic Monte Carlo method to simulate the structural-dynamical evolution taking place in thermoplastic elastomers, where hard glassy and soft rubbery phases alternate. In the study, the investigation provides an explanation

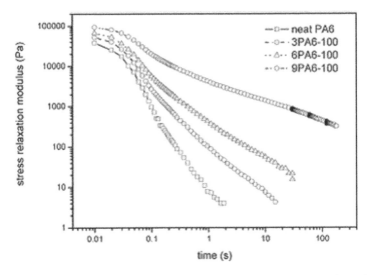

**Figure 14.35** Linear stress relaxation modulus $G(t)$ for PA6-based nanocomposites with different clay content, extruded at 100 rpm.

and confirms the importance of the chain-pullout mechanism, as a result of fluctuational melting, in the viscoelastic and stress-relaxation behavior of polymer-based multiphase nanomaterials.

Nathani et al. [74] studied the stress whitening behavior of melt-intercalated polybutene–clay nanocomposites during tensile straining. It has been found that in polybutene–5 wt% clay-reinforced nanocomposites, the early stage of plastic deformation involved nucleation and growth of large size voids followed by stress relaxation of the locally stressed region and fine cracks in the relaxed zones.

Wienecke et al. [75] produced large surface area of the polymer membrane electrode by vacuum thermal evaporation coating with metals. The study shows that the morphology of the sintered PTFE membranes has been changed due to the thermal stress relaxation during the cooling of the metallic film.

Privalko et al. [76] investigated the thermoelasticity and stress-relaxation behavior of polychloroprene/organoclay nanocomposites (ENC). The results show that the stress-relaxation behavior of the ENC is qualitatively consistent with the original assumption that after initial stretching to the highest elongation ($\lambda_{\lim}$) the preexisting infinite clusters of filler particles are broken into the isolated clusters, which remain structurally similar, whatever the subsequent stretching to $\lambda_f$ (predetermined fixed extension) $< \lambda_{\lim}$.

A so-called three-dimensional filler network structure will be constructed in the polymer/layered silicate nanocomposites when the content of layered clay reaches a threshold value, at which the silicate sheets are incapable of freely rotating, due to physical jamming and connecting of the nanodispersed layered silicate. Wang et al. [77] investigated the effect of such clay network on the mobility and relaxation of macromolecular chains in isotactic polypropylene (iPP)/organoclay nanocomposites. The results show that the turning point of macroscopic properties appeared at 1 wt% organoclay content. After this point, a reduced mobility of chains and a retarded chain relaxation were observed and attributed to the formation of a mesoscopic filler network.

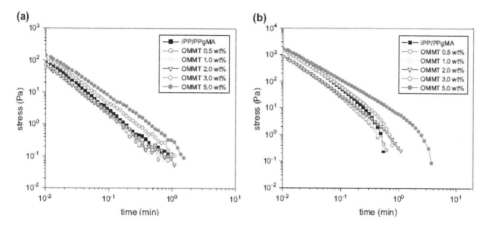

**Figure 14.36** Stress-relaxation spectra of compatiblized iPP/OMMT hybrids under (a) low-amplitude strain (10%) and (b) large-amplitude strain (200%).

The stress-relaxation behaviors of unfilled iPP and iPP/OMMT (organically modified montmorillonite) composites in the linear viscoelastic regime (strain $= 10\%$) and nonlinear viscoelastic regime (strain $= 200\%$) are presented in Figure 14.36. In the stress-relaxation experiments, a single-step strain was imposed on the melt at time $= 0$, and the shear stress $\sigma(t)$ was recorded immediately as a function of time. Within the testing strain range from 0.05% to 300%, the melt stress is linearly increased with strain amplitude; therefore the magnitude of stress at large strain (nonlinear regime) is considerably larger than that at small strain (linear regime). In the linear strain regime (Figure 14.36(a)), the stress relaxation of the unfilled polymer and iPP/OMMT composites is independent of the OMMT content. When the stress-relaxation measurements were conducted in the nonlinear regime (Figure 14.36(b)), a slow stress-relaxation phenomenon is observed in the iPP/OMMT composite with 5 wt% OMMT, in which the highest degree of mesoscopic filler network is expected. It is believed that the organoclay network confines the chain motion and relaxation of iPP.

Fluorescence spectroscopy was used by Priestley et al. [78] to characterize the rate of physical ageing at room temperature in nanocomposites of silica (10–15 nm diameter) nanoparticles in PMMA. The physical ageing rate was reduced by more than a factor of 20 in 0.4 vol% silica–PMMA nanocomposites relative to neat PMMA. The molecular-scale origin of this nearly complete arresting of physical ageing was investigated with dielectric spectroscopy. The strength of the beta-relaxation process was reduced by nearly 50% in the nanocomposite relative to neat PMMA. This reduced strength of the beta process results from dipoles (ester groups) having hindered motions or being virtually immobile on the timescale being probed at a frequency of 100 Hz. This hindered mobility results from hydrogen bonding between PMMA ester side groups and hydroxyl units on the surface of the silica nanoparticles. In contrast, no reduction in physical ageing rate was observed upon addition of silica to PS, which cannot form hydrogen bonds with the silica surfaces. Thus, the molecular origin of the suppressed physical ageing in silica PMMA nanocomposites is the interfacial hydrogen bonding, which leads to a major reduction in the strength of the beta process.

## Acknowledgments

The authors would like to thank Elsevier Science Publishers Ltd. and John Wiley & Sons Ltd. for their permission to reproduce the figures and tables in this chapter from copyright journals. Thanks are also due to Ph.D. Y. Yang and to F. J. Wang from Harbin Institute of Technology for helpful suggestions.

## References

1. Friedrich, K.; Zhang, Z.; Schlar, A. K. Compos. Sci. Technol. 2005, 65, 2329–2343.
2. Hübner, W.; Gradt, T.; Schneider, T.; Borner, H. Wear. 1998, 216, 150–159.
3. Briscoe, B. J. In Advances in Composite Tribology; Friedrich, K.; Ed.; Composite Materials Series; Elsevier: Amsterdam, the Netherlands, 1993; Vol. 8, pp 3–15.
4. Bahadur, S. Wear. 2000, 245, 92–99.
5. Gao, J. Wear. 2000, 245, 100–106.
6. Xu, Y. M.; Mellor, B. G. Wear. 2001, 251, 1522–1531.
7. Lu, Z. P.; Friedrich, K. Wear. 1995, 181–183, 624–631.
8. Sandler, J.; Werner, P.; Shaffer, M. S. P.; Demchuk, V.; Altstädt, V.; Windle, A. H. Compos. Part A. 2002, 33, 1033–1039.
9. Werner, P.; Altstädt, V.; Jaskulka, R.; Jacobs, O.; Sandler, J. K. W.; Shaffer, M. S. P.; Windle, A. H. Wear. 2004, 257, 1006–1014.
10. Yang, Z.; Dong, B.; Huang, Y.; Liu, L.; Yan, F. Y.; Li, H. L. Mater. Lett. 2005, 59, 2128– 2132.
11. Klabunde, K. J. et al. J. Phys. Chem. 1996, 100, 12142–12153.
12. Adams, M. J. et al. Wear. 2001, 251, 1579–1583.
13. Sawyer, G. W.; Freudenberg, K. D.; Bhimaraj, P.; Schadler, L. S. Wear. 2003, 254, 573–580.
14. Ng, C. B.; Schadler, L. S.; Siegel, R. W. Nanostruct. Mater. 1999, 12, 507–510.
15. Schwartz, C. J.; Bahadur, S. Wear. 2000, 237, 261–273.
16. Rong, M. Z. et al. Ind. Lubric. Tribol. 2001, 53, 72–77.
17. Li, F.; Hu, K.; Li, J.; Zhao, B. Wear. 2002, 249, 877–882.
18. Zhang, M. Q. et al. Macromol. Mater. Eng. 2002, 287, 111–115.
19. Wetzel, B.; Haupert, F.; Zhang, M. Q. Compos. Sci. Technol., 2003, 63, 2055–2067.
20. Xue, Q.; Wang, Q. Wear, 1997, 213, 54–58.
21. Wang, Q.; Xue, Q.; Liu, H.; Shen, W.; Xu, J. Wear. 1996, 198, 216–219.
22. Wang, Q.; Xu, J.; Shen, Q.; Liu, W. Wear. 1996, 196, 82–86.
23. Wang, Q.; Xue, Q.; Shen, W. Tribol Int. 1997, 30, 193–197.
24. Wang, Q.; Xu, J.; Shen, W.; Xue, Q. Wear. 1997, 209, 316–321.
25. Wang, Q.; Xue, Q.; Shen, W.; Zhang, J. J. Appl. Polym. Sci. 1998, 69, 135–141.
26. Bahadur, S.; Sunkara, C. Wear. 2005, 258, 1411–1421.
27. Avella, M.; Errica, M. E.; Martuscelli, E. Nano Lett. 2001, 1, 213–217.
28. Zhang, M. Q.; Rong, M. Z.; Yu, S. L.; Wetzel, B.; Friedrich, K. Wear. 2002, 253, 1086–1093.
29. Shi, G.; Zhang, M. Q.; Rong, M. Z.; Wetzel, B.; Friedrich, K. Wear. 2003, 254, 784–796.
30. Wetzel, B.; Haupert, F.; Friedrich, K.; Zhang, M. Q.; Rong, M. Z. Polym. Eng. Sci. 2002, 42, 1919–1927.
31. Sreekala, M. S.; Eger, C. In Polymer Composites – from Nano- to Macro-Scale; Friedrich, K.; Fakirov, S.; Zhang, Z.; Ed.; Polymer Sciences Book, XXII, 367; Springer: New York, NY, 2005; pp 91–105.
32. Wang, Y.; Lim, S.; Luo, J. L.; Xu, Z. H. Wear. 2006, 260, 976–983.
33. Wang, Y.; Lim, S. Wear. 2007, 262, 1097–1101.
34. Wang, Y.; Chen, W. J. Mater. Sci. Lett. 2003, 22, 845–848.
35. Johnson, M.; Mikkola, D. E.; March, P. A.; Wright, R. N. Wear. 1990, 140, 279–289.
36. Xian, G. J.; Zhang, Z.; Friedrich, K. J. Appl. Polym. Sci. 2006, 101, 1678–1686.
37. Chang, L.; Zhang, Z.; Zhang, H.; Schlar, A. K. Compos. Sci. Technol. 2006, 66, 3188–3198.
38. Chang, L.; Zhang, Z.; Zhang, H.; Friedrich, K. Tribol. Int. 2005, 38, 966–973.
39. Cai, H.; Yan, F. Y.; Xue, Q. J.; Liu, W. M. Polym. Test. 2003, 22, 875–882.
40. Hou, X. H.; Shan, C. X.; Choy, K. L. Surface & Coatings Technol. 2008, 202, 2287–2291.
41. Bhimaraj, P.; Burris, D.; Sawyer, W. G.; Toney, C. G.; Siegel, R. W.; Schadler, L. S. Wear. 2008, 264, 632– 637.

42. Sidorenko, A.; Ahn, H. S.; Kim, D. I.; Yang, H.; Tsukruk, V. V. Wear. 2002, 252, 946–955.
43. Dasari, A.; Yu, Z. Z.; Mai, Y. W.; Hu, G. H.; Varlet, J. Compos. Sci. Technol. 2005, 65, 2314–2328.
44. Ahn, H. S.; Julthongpiput, D.; Kim, D. I.; Tsukruk, V. V. Wear. 2003, 255, 801–807.
45. Julthongpiput, D.; Ahn, H. S.; Sidorenko, A.; Kim, D. I.; Tsukruk, V. V. Tribol. Int. 2002, 35, 829–836.
46. Galetz, M. C.; Blass, T.; Ruckdaeschel, H.; Sandler, J. K. W.; Altstaedt, V.; Glatzel, U. J. Appl. Polym. Sci. 2007, 104, 4173–4181.
47. Yu, S. R.; Hu, H. X.; Zhang, Y. B.; Liu, Y. H. Polym. Int. 2008, 57, 454–462.
48. Bhimaraj, P.; Burris, D. L.; Action, J.; Sawyer, W. G.; Toney, C. G.; Siegel, R. W.; Schadler, L. S. Wear. 2005, 258, 1437–1443.
49. Luo, J. Z.; Lannutti, J. J.; Seghi, R. R. Dent Mater. 1998, 14, 29–36.
50. Trotignon, J. P. Polym. Test. 1995, 14, 129–147.
51. Kallrath, J.; Altstädt, V.; Schlöder, J. P.; Bock, H. G. Polym. Test. 1999, 18, 11–35.
52. Legorju-jago, K.; Bathias, C. Internat. J. Fatigue. 2002, 24, 85–92.
53. Chisholm, N.; Mahfuz, H.; Rangari, V. K.; Ashfaq, A.; Jeelani, S. Compos. Struct. 2005, 67, 115–124.
54. Jen, M. H. R.; Tseng, Y. C.; Wu, C. H. Compos. Sci. Technol. 2005, 65, 775–779.
55. Bellemare, S. C.; Dickson, J. I.; Bureau, M. N.; Denault, J. 2005, 26, 636–646.
56. Juwono, A.; Edward, G. J. Nanosci. Nanotechnol. 2006, 6, 3943–3946.
57. Vlasveld, D. P. N.; Bersee, H. E. N.; Picken, S. J. Polymer. 2005, 46, 12539–12545.
58. Struik, L. C. E. Polymer. 1987, 28, 1521–1533.
59. Raghavan, J.; Meshii, M. Compos. Sci. Technol. 1997, 57, 1673–1688.
60. Galgali, G.; Ramesh, C.; Lele, A. Macromolecules. 2001, 34, 852.
61. Pegoretti, A.; Kolarik, J.; Peronia, C.; Migliaresi, C. Polymer. 2004, 45, 2751–2759.
62. Zhang, Z.; Yang, J. L.; Friedrich, K. Polymer. 2004, 45, 3481–3485.
63. Yang, J. L.; Zhang, Z.; Schlarb, A. K.; Friedrich, K. Polymer. 2006, 47, 2791–2801.
64. Ranade, A.; Nayak, K.; Fairbrother, D.; D'Souza, N. A. Polymer. 2005, 46, 7323–7333.
65. Shen, L.; Phang, I. Y.; Chen, L.; Liu, T. X.; Zeng, K. Y. Polymer. 2004, 45, 3341–3349.
66. Starkova, O.; Yang, J. L.; Zhang, Z. Compos. Sci. Technol. 2007, 67, 2691–2698.
67. Zhou, T. H.; Ruan, W. H.; Yang, J. L.; Rong, M. Z.; Zhang, M. Q.; Zhang, Z. Compos. Sci. Technol. 2007, 67, 2297–2302.
68. Patel, M.; Morrell, P. R.; Murphy, J. J. Polym. Degrad. Stab. 2005, 87, 201–206.
69. Patel, M.; Soames, M.; Skinner, A. R.; Stephens, T. S. Polym. Degrad. Stab. 2004, 83, 111–116.
70. Baeurle, S. A.; Hotta, A.; Gusev, A. A. Polymer. 2005, 46, 4344–4354.
71. Andreassen, E. Polymer. 1999, 40, 3909–3918.
72. Incarnato, L.; Scarfato, P.; Scatteia, L.; Acierno, D. Polymer. 2004, 45, 3487–3496.
73. Baeurle, S. A.; Usami, T.; Gusev, A. A. Polymer. 2006, 47, 8604–8617.
74. Nathani, H.; Dasari, A.; Misra, R. D. K. Acta Materialia. 2004, 52, 3217–3227.
75. Wienecke, M.; Bunescu, M. C.; Pietrzak, M.; Deistung, K.; Fedtke, P. Synth. Met. 2003, 138, 165–171.
76. Privalko, V. P.; Ponomarenko, S. M.; Privalko, E. G.; Schön, F.; Gronski, W. Eur. Polym. J. 2005, 41, 3042–3050.
77. Wang, K.; Liang, S.; Deng, J. N.; Yang, H.; Zhang, Q.; Fu, Q.; Dong, X.; Wang, D. J.; Han, C. C. Polymer. 2006, 47, 7131–7144.
78. Priestley, R. D.; Rittigstein, P.; Broadbelt, L. J.; Fukao, K.; Torkelson, J. M. J. Phys. Conden. Mater. 2007, 19, 205120–205132.
79. Xia, H. S.; Song, M.; Zhang Z. Richardson, M. J. Appl. Polym. Sci. 2007, 103, 2992–3002.

# 15 Thermal Degradation and Combustion Behavior of the Polyethylene/Clay Nanocomposite Prepared by Melt Intercalation

*S. M. Lomakin*[\*,1], *I. L. Dubnikova*[2], *A. N. Shchegolikhin*[1],
*G. E. Zaikov*[1], *R. Kozlowski*[3], *G.-M. Kim*[4], *and G. H. Michler*[4]

[1]NM Emanuel Institute of Biochemical Physics of Russian Academy of Sciences,
119934 Kosygin 4, Moscow, Russia
[2]NN Semenov Institute of Chemical Physics of Russian Academy of Sciences,
119991 Kosygin 4, Moscow, Russia
[3]Institute of Natural Fibres, Poznan, ul. Wojska Polskiego 71 b, Poland
[4]Martin-Luther-Universität Halle-Wittenberg, Geusaer Straße, D-06217 Merseburg, Germany

## Abstract

Studies of thermal and fire-resistant properties of the polyethylene/organically modified montmorillonite nanocomposites prepared by means of melt intercalation are discussed. The sets of the data acquired with the aid of nonisothermal thermogravimetric analysis experiments have been treated by the model kinetic analysis. The extra acceleration of thermal-oxidative degradation of the nanocomposite that has been observed at the first stage of the overall process has been analyzed and is explained by the catalytic effect of the clay nanoparticles. The results of cone calorimetric tests lead to the conclusion that char formation plays a key role in the mechanism of flame retardation for nanocomposites.

**Keywords:** intercalation polymerization, kinetics, layered clay, nanocomposite, oxidation, polyethylened, thermal degradation, combustion

---

*Correspondence should be addressed to e-mail: lomakin@sky.chph.ras.ru

## 15.1. Introduction

The properties and thermal degradation of model polyethylene (PE) are reasonably well known [1–12]. It is generally accepted that pristine PE has relatively low thermal stability and flame resistance. The approach to the enhancement of thermal stability and fire resistance based on the use of polymer nanocomposites has been extensively developed in the last years [13–18].

Intercalation in polymer melts is widely used as a simple method for the synthesis of corresponding materials [18]. In these experimental studies, it was found that PE nanocomposites based on layered silicates exhibit lower flammability as compared with the parent polymer, which can be achieved by introducing a small amount of an inorganic silicate ingredient [13–16].

It is believed that, in the course of high-temperature pyrolysis and/or combustion, clay nanoparticles are capable of promoting formation of protective clay-reinforced carbonaceous char that is responsible for the reduced mass loss rates and hence the lower flammability.

Along with an enhancement of flame resistance of such nanocomposites, the rise of thermal stability was observed during thermogravimetric analysis (TGA) experiments. Generally, TGA cannot be used to elucidate a complex mechanism of polymer thermal degradation. Nevertheless, dynamic TGA has been frequently used to study the overall thermal degradation kinetics of polymers because it gives reliable information on the kinetic parameters, such as preexponential factor, the activation energy, and the overall reaction order [20–22]. The thermal degradation kinetics of PE has been studied by many investigators. Various suggestions for the kinetic parameters for polymer degradation have been reported [8–10, 20–22]. Although some attempts have been made to understand the complex nature of decomposition of a polymer, involving numerous reactions, some authors have found it sufficient to consider a global first-order kinetic expression to represent the overall decomposition rate [8].

In this chapter, we applied the model kinetic analysis of PE and PE/organically modified montmorillonite (MMT) thermal-oxidative degradation in order to predict an increase of flame resistance for PE nanocomposite, that is the reduction of the mass loss rate and the rate of heat release.

## 15.2. Experimental

### 15.2.1. Materials

PE, Basel Lupolen, was used to prepare PE nanocomposites in combinations with polar compatibilizer, maleic anhydride-modified PE oligomer (MAPE) Polybond 3109, provided by Crompton. A Cloisite 20A (purchased from Southern Clay Products, Inc.) has been used as the MMT to prepare PE–MMT and PE–MAPE–MMT nanocomposites throughout this study. The content of an organic cation-exchange modifier, $N^{+}2CH_{3}2HT$ (HT = hydrogenated tallow, C18 $\approx$ 65%; C16 $\approx$ 30%; C14 $\approx$ 5%; anion: Cl$^{-}$), in the MMT was 38% by weight.

### 15.2.2. Preparation of Nanocomposites

PE–MAPE–MMT nanocomposites were prepared by the melt mixing of the components in a double-rotor laboratory (Brabender) mixer in two steps. PE and MAPE (Polybond 3109 by Crompton Co.) were blended/mixed using a laboratory Brabender mixing chamber in two steps. In the first step, the two polymers PE and MAPE were blended in a 4:1 ratio for 2 min; after that, an MMT powder was added in an amount of 3 and 7 wt%. The mixing of components at the second step lasted for 10 min at a temperature of 190°C and rotor speeds of 60 rpm. Samples designed for cone calorimeter testing in the form of $70 \times 70 \times 3$ mm$^3$ plates with a mass of $14.0 \pm 0.1$ g were prepared by molding at 190°.

## 15.2.3.  Investigation Techniques

### 15.2.3.1.  *Transmission Electron Microscopy*

The filler dispersion in the composites was studied by the transmission electron microscopy (TEM) with a Philips EM-301 electron microscope (the Netherlands) at an accelerating voltage of 80 kV. Thin sections of film samples for TEM examination were prepared with an LKB Ultratome III® ultramicrotome.

### 15.2.3.2.  *Small-Angle X-Ray Scattering*

Small-angle X-ray scattering (SAXS) analysis of nanocomposite-layered structure was performed in the reflection mode over the angular range of film samples at the room temperature using a DRON-2 X-ray diffractometer (CuK$_\alpha$ radiation) with modified collimation. Diffraction patterns were collected in reflection-mode geometry from 2° to 10°2θ. Interlayer distances in MMT were determined from the angular positions of base reflections ($d_{001}$) in the diffraction patterns of composites. The degree of polymer intercalation was determined by a change in the interlayer distance in MMT.

### 15.2.3.3.  *Thermogravimetric Analysis*

A Perkin-Elmer TGA-7 instrument calibrated by Curie points of several metal standards has been employed for nonisothermal TGA. The measurements were carried out at a desired heating rate (in the range of 2.5–10 K/min) in air. A kinetic analysis of the thermal degradation of composites was performed with the use of the NETZSCH-Gerätebau Thermokinetics software. The algorithm of the kinetic analysis program was based on the calculation of regression by the fifth-order Runge–Kutta method using the dedicated Prince-Dormand formula for automatic optimization of the number of significant digits [23, 24].

### 15.2.3.4.  *Combustibility Characteristics (Cone Calorimeter)*

Ignitability tests were performed according to the standard procedures ASTM 1354-92 and ISO/DIS 13927 using a cone calorimeter [25].

## 15.3.  Results and Discussion

### 15.3.1.  Investigation of Nanocomposite Structure

SAXS has been used to evaluate the degree of intercalation of the organoclay particles in the polymer matrix. SAXS diffraction patterns for SAXS patterns for the original Cloisite 20A, PE-7% Cloisite 20A, PE–MAPE-7% Cloisite 20A, and MAPE-7% Cloisite 20A are displayed in Figure 15.1.

The Cloisite 20A itself has a single peak at around 3.7° with *d*-space of 2.4 nm (Figure 15.1, curve 1). The shift of the clay basal spacing $d_{001}$ from 3.7° to 2.7° in PE–MAPE-7%

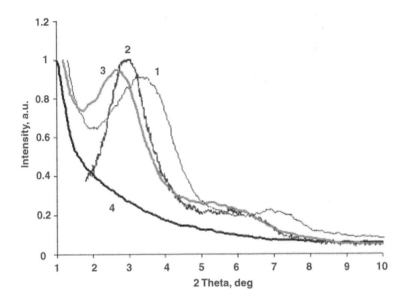

**Figure 15.1**  SAXS patterns for the original Cloisite 20A (1), PE-7% Cloisite 20A (2), PE–MAPE-7% Cloisite 20A (3), and MAPE-7% Cloisite 20A (4).

Cloisite 20A (Figure 15.1, curve 3) sample suggests that the intercalated nanocomposite sample have higher $d$-space (3.3 nm) than that in the original clay (2.4 nm), it may have some exfoliated structures, and considering the smearing of peak in nanocomposite sample, the nanocomposite structure have not been well exfoliated. Figure 15.1 also shows the shift of the base reflection toward smaller angles for PE-7% Cloisite 20A sample from 3.7° to 2.9° (curve 2), which indicates an increase in the interplanar distance in the silicate structure of the PE-7% Cloisite 20A nanocomposite from 2.4 to 3.07 nm by intercalation.

TEM technique has been used to evaluate the degree of exfoliation of the organoclay particles in the polymer matrix. Figures 15.2–15.4 show the influence of polar compatibilizer (MAPE) on the exfoliation degree of the MMT in the nanocomposites.

It is seen that the full exfoliation of the MMT particles to the monolayers takes place under the action of MAPE compatibilizer in the MAPE–MMT nanocomposite (Figure 15.2), whereas in the PE–MAPE–MMT nanocomposite, the hybrid structure consisting of intercalated tactoids and exfoliated MMT monolayers has been formed (Figure 15.3). Finally, the pure PE–MMT composition has the intercalated structure of nanocomposite (Figure 15.4).

### 15.3.2. Study of Thermal-Oxidative Degradation of PE–MAPE–MMT Nanocomposite

It is acknowledged that the thermal stability of polymer nanocomposites is higher than that of pristine polymers, and this gain is explained by the presence of anisotropic clay layers hindering diffusion of volatile products through the nanocomposite material.

The radical mechanism of thermal degradation of PE has been widely discussed in a framework of random scission-type reactions [1–10]. It is known that PE decomposition

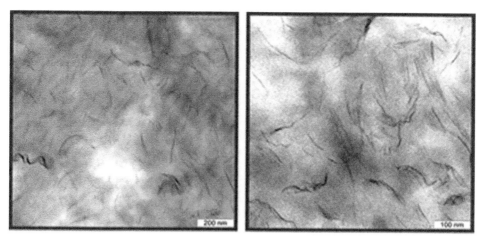

**Figure 15.2** TEM micrographs of MAPE-7% Cloisite 20A – maleic anhydride-modified PE oligomer (Polybond 3109) at different magnification showing mainly exfoliated nanocomposite.

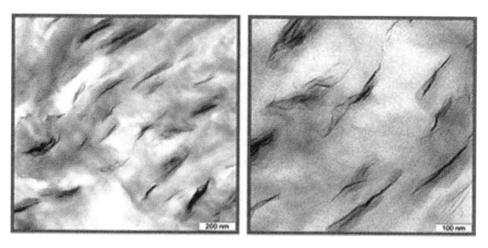

**Figure 15.3** TEM micrographs of PE–MAPE-7% Cloisite 20A at different magnification showing discrete (hybrid) structure of nanocomposite consisting of intercalated tactoids and exfoliated monolayers.

products comprise a wide range of alkanes, alkenes, and dienes. The polymer matrix transformations, usually observed at lower temperatures and involving molecular weight alteration without formation of volatile products, are principally due to the scission of weak links, for example, oxygen bridges, incorporated into the main chain as impurities. The kinetics of thermal degradation of PE is frequently described by a first-order model of mass conversion of the sample [10, 21]. A broad variation in Arrhenius parameters can be found in literature, that is, activation energy ($E$) ranging from 160 to 320 kJ/mol and pre-exponential factor ($A$) variations in the range of $10^{11}$ and $10^{21}$ $s^{-1}$ [7–10] are not unusual. It is believed that the broad range of $E$ values reported may be explained by the polymers molecular mass variations, by the use of various additives, and by different experimental conditions [10] employed by different authors.

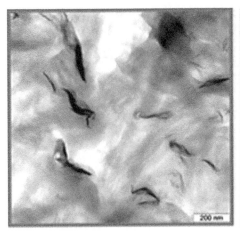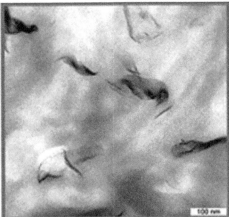

**Figure 15.4**   TEM micrographs of PE-7% Cloisite 20A at different magnifications (for the most part intercalated nanocomposite).

Thermal-oxidative degradation of PE and PE nanocomposites has been extensively studied over the last decades [14–17, 26–28]. It has been reported that the main oxidation products of PE are aldehydes, ketons, carboxylic acids, esters, and lactones [26, 27]. According to Lacoste and Carlsson [26], β-scission plays an important role in thermal oxidation of UHMWPE. Notably, the feasibility of intramolecular hydrogen abstraction by the peroxy radicals for PE has been questioned in frames of a thermal-oxidation mechanism proposed by Gugumus [27, 28]. A mechanism describing oxidation of organic molecules by the virtue of complex chain reactions has been proposed earlier by Benson [29].

In this chapter, the processes of thermal degradation of PE and PE–MAPE–MMT nanocomposite with MMT content of 7 wt% have been investigated by TGA in the air at the heating rates of 2.5, 5, and 10 K/min. According to the dynamic TGA data, the polymer thermal-oxidative degradation starts at about 300°C and then, through a complex radical chain process, the material totally destructs and completely volatilizes in the range of 450–500°C (Figure 15.4).

The diverse behavior of PE and PE–MAPE–MMT (Figures 15.5 and 15.6) shows that the influence of MMT nanoparticles on the thermal-oxidation process resulted in higher thermal-oxidative stability of hybrid PE–MAPE–MMT nanocomposite. It can be seen as a regular increase in the temperature values of the maximum mass loss rates (about 50°C) for the PE–MAPE–MMT as compared with pristine PE (Figure 15. 6).

It seems reasonable to suggest that thermally stable crosslinked carbonized layer on the nanocomposite surface is formed during the thermal-oxidative degradation and starts to hinder the diffusion transport of both the volatile degradation products (out of the polymer melt into the gas phase) and the oxygen (from the gas phase into the polymer). The above set of events results in actual increase of the nanocomposite thermal stability in the temperature range of 350–500°C, where normally a general degradation of the main part of PE takes place.

### 15.3.3.  Kinetic Analysis of PE–MAPE–MMT Thermal-Oxidative Degradation

Kinetic studies of material degradation have long history, and there exists a long list of data analysis techniques employed for the purpose. Often, TGA is the method of choice for

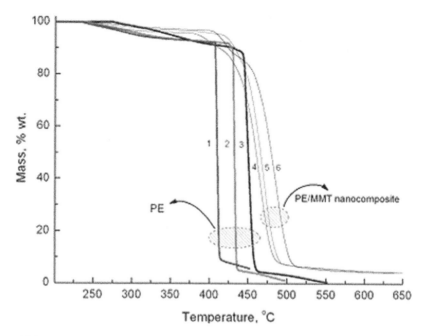

**Figure 15.5**  TGA curves for PE (1–2.5, 2–5, and 1–10 K/min,) and PE–MAPE–MMT (4–2.5, 5–5, and 6–10 K/min).

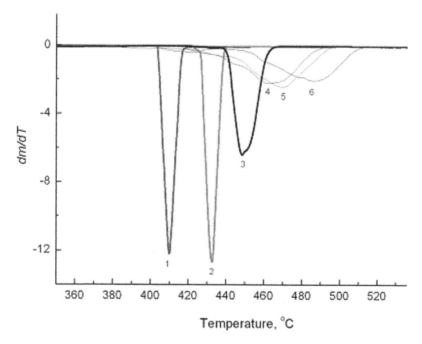

**Figure 15.6**  DTG curves for PE (1–2.5, 2–5, and 1–10 K/min) and PE–MAPE–MMT (4–2.5, 5–5, and 6–10 K/min).

acquiring experimental data for subsequent kinetic calculations, and namely, this technique was employed here.

It is commonly accepted that the degradation of materials follows the base equation (15.1) [23]

$$\frac{dc}{dt} = -F(t, T, c_o, c_f),\tag{15.1}$$

where $t$ is the time, $T$ the temperature, $c_o$ the initial concentration of the reactant, and $c_f$ the concentration of the final product. The right-hand part of the equation $F(t, T, c_o, c_f)$ can be represented by the two separable functions, $k(T)$ and $f(c_o, c_f)$:

$$F(t, T, c_o, c_f) = k[T(t)f(c_o, c_f)],\tag{15.2}$$

Arrhenius equation (15.4) will be assumed to be valid for the following:

$$k(T) = A \exp\left(-\frac{E}{RT}\right).\tag{15.3}$$

Therefore,

$$\frac{dc}{dt} = -A \exp\left(-\frac{E}{RT}\right) f(c_o, c_f).\tag{15.4}$$

All feasible reactions can be subdivided into classic homogeneous reactions and typical solid state reactions, which are listed in Table 15.1 [23]. The analytical output must provide good fit to measurements with different temperature profiles by means of a common kinetic model.

**Table 15.1** Reaction types and corresponding reaction equations, $dc/dt = -A \exp(-E/RT) f(c_o, c_f)$.

| Name | $f(c_o, c_f)$ | Reaction type |
|---|---|---|
| $F_1$ | $c$ | First-order reaction |
| $F_2$ | $c^2$ | Second-order reaction |
| $F_n$ | $c^n$ | $n$th-order reaction |
| $R_2$ | $2 \cdot c^{1/2}$ | Two-dimensional phase boundary reaction |
| $R_3$ | $3 \cdot c^{2/3}$ | Three-dimensional phase boundary reaction |
| $D_1$ | $0.5/(1-c)$ | One-dimensional diffusion |
| $D_2$ | $-1/\ln(c)$ | Two-dimensional diffusion |
| $D_3$ | $1.5 \cdot e^{1/3}(c^{-1/3}-1)$ | Three-dimensional diffusion (Jander's type) |
| $D_4$ | $1.5/(c^{-1/3}-1)$ | Three-dimensional diffusion (Ginstling–Brounstein type) |
| $B_1$ | $c_o \cdot c_f$ | Simple Prout–Tompkins equation |
| $B_{na}$ | $c_o^n \cdot c_f^a$ | Expanded Prout–Tompkins equation ($n_a$) |
| $C_{1-X}$ | $c \cdot (1+K_{cat} \cdot X)$ | First-order reaction with autocatalysis through the reactants, $X \cdot X = c_f$ |
| $C_{n-X}$ | $c^n \cdot (1 + K_{cat} \cdot X)$ | $n$th-order reaction with autocatalysis through the reactants, $X$ |
| $A_2$ | $2 \cdot c \cdot (-\ln(c))^{1/2}$ | Two-dimensional nucleation |
| $A_3$ | $3 \cdot c \cdot (-\ln(c))^{2/3}$ | Three-dimensional nucleation |
| $A_n$ | $N \cdot c \cdot (-\ln(c))^{(n-1)/n}$ | $n$-dimensional nucleation/nucleus growth according to Avrami/Erofeev |

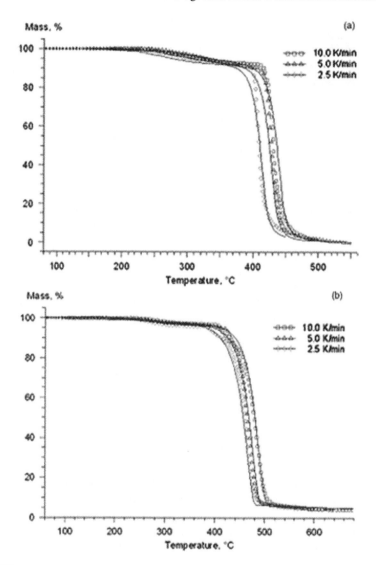

**Figure 15.7** Nonlinear kinetic modeling of PE (a) and PE–MAPE–MMT (b) thermal-oxidative degradation in air. Comparison between experimental TGA data (dots) and the model results (firm lines) at several heating rates.

Kinetic analysis of thermal-oxidative degradation of PE and PE–MAPE–MMT at the heating rates of 2.5, 5, and 10 K/min (Figures 15.7(a) and (b)) has been accomplished by using the interactive model-based nonlinear fitting approach in accordance with a formalism we proposed earlier [30]. In order to assess the activation energy for the development of a reasonable model for kinetic analysis of pristine PE and PE–MAPE–MMT thermal degradation processes, a few evaluations by model-free Friedman analysis have been done as the starting point [31].

Further, nonlinear model fitting procedure for PE and PE-*n*-MMT TGA curves has led to the following triple-stage model scheme of successive reactions, wherein a general

**Table 15.2**  Results of the multiple-curve kinetic analyses for thermal-oxidative degradation of PE and PE–MAPE–MMT in accordance with the reaction model (15.5).

| Material | Parameter | Value | Correlation coefficient |
|---|---|---|---|
| PE | $\log A1\ (\text{s}^{-1})$ | 3.7 | 0.9989 |
|  | $E_1$ (kJ/mol) | 65.5 |  |
|  | $n_1$ | 1.28 |  |
|  | $\log A_2\ (\text{s}^{-1})$ | 15.5 |  |
|  | $E_2$ (kJ/mol) | 238.5 |  |
|  | $n_2$ | 0.59 |  |
|  | $\log A_3\ (\text{s}^{-1})$ | 16.9 |  |
|  | $E_3$ ( kJ/mol) | 250.2 |  |
|  | $n_3$ | 1.79 |  |
| PE–MAPE–MMT | $\log A_1\ (\text{s}^{-1})$ | 5.2 | 0.9992 |
|  | $E_1$ (kJ/mol) | 87.5 |  |
|  | $n_1$ | 1.34 |  |
|  | $\log A_2\ (\text{s}^{-1})$ | 16.5 |  |
|  | $E_2$ (kJ/mol) | 266.2 |  |
|  | $n_2$ 0.78 |  |  |
|  | $\log A_3\ (\text{s}^{-1})$ | 17.2 |  |
|  | $E_3$ (kJ/mol) | 279.6 |  |
|  | $n_3$ | 0.89 |  |

$n$th-order ($F_n$) reaction was used for all steps of the overall process of thermal-oxidative degradation (Table 15.1) (Figures 15.7(a) and (b)):

$$A \xrightarrow{F_n} B \xrightarrow{F_n} C \xrightarrow{F_n} D. \tag{15.5}$$

Data in Table 15.2 for the first stage of thermal-oxidative degradation reaction show the activation energies values for PE and PE–MAPE–MMT amount to 65.5 and 87.5 kJ/mol, respectively, indicating that the degradation of these samples is initiated by the similar oxygen-induced reactions. At the same time, the values of activation energy found at the second and third stages of thermal-oxidative degradation for PE-$n$-MMT (266.2 and 279.6 kJ/mol) are higher than those for PE (238.5 and 250.2 kJ/mol) (Table 15.2). This difference may be attributed to a shift of the PE-$n$-MMT degradation process to a diffusion-limited mode, owing to emergence in the system of a carbonized crosslinked material. It is known [11] that the contribution of radical recombination reactions, which lead to the intermolecular crosslinking, increases under the oxygen-deficient conditions. Such conditions are realized at thermal-oxidative degradation of the PE nanocomposites due to the labyrinth effect of the silicate layers toward the diffusing gas. The formation of chemical crosslinking during the thermal-oxidative degradation of the PE nanocomposites is a necessary condition of PE carbonization process.

This fact infers that the last stage of the PE-$n$-MMT degradation process is governed mainly by random scission of C–C bonds, rather than by an oxygen-catalyzed reactions. On the contrary, these results are also consistent with the barrier model mechanism, which suggests that inorganic clay layers can play a role of barriers retarding the diffusion of oxygen from the gas phase into the nanocomposite.

On the basis of the calculated kinetic parameters of thermal-oxidative degradation of PE and PE–MAPE–MMT, we designed the curves of mass loss rate during the isothermal (600°C) heating of the samples for more than 60 s (Figure 15.8). The choice of the specified temperature 600°C is not occasional, since this temperature corresponds to an incident heat flux of 35 kW/m$^2$ that has been used in working tests on the ignitability of samples with a cone calorimeter [25]. The tests made it possible to evaluate important combustibility characteristics, such as mass loss rate and heat release rate. The calculation of the heat release rate as a fundamental parameter measured by the cone calorimeter was based on the oxygen absorption principle. According to this principle, the heat released during the burning of a material is proportional to the amount of oxygen required for its combustion. For solid materials, the consumption of 1 kg of oxygen for their combustion is basically accompanied by evolution of 13.1 MJ of heat [25]. One of the tasks of this study was the correlation-based evaluation of the heat release rate under the cone calorimetry test conditions and the rate of mass low under the conditions of isothermal-oxidative degradation. In the general form, the basic equation relating the mass loss rate to the heat release rate during combustion is as follows:

$$\dot{Q}_{tot} \, (kW/m^2) = \chi \cdot \Delta H_{comb} \cdot \dot{m}, \tag{15.6}$$

where $\chi$ is the combustion efficiency, $\Delta H_{comb}$ the heat of complete combustion, and $m$ the rate of mass loss per unit surface. If $\Delta H_{comb}$ is a constant value for PE and the PE–MAPE–MMT nanocomposite (i.e., the silicate additive does not inhibit gas-phase processes in the flame and has no effect on the heat of combustion), the heat release rate linearly depends on the mass loss rate. In this case, the coefficient $\chi$ for the linear equation characterizing the combustion efficiency or the completeness of combustion directly depends on the amount and structure of the carbonaceous residue.

It is clearly seen that under conditions of polymer ignition and initial surface combustion, the mass loss rate for PP–MAPE is noticeably lower than adequate values for the neat PE (Figure 15.8). An improvement in flame resistance of PE–MAPE over the neat PE should happen as a result of the char formation providing a transient protective barrier. In the present study, this phenomena was interpreted in terms of kinetic approach.

## 15.3.4. Flammability Characteristics of PE and PE–MAPE–MMT Nanocomposites

The flammability characteristics of pristine PE and PE–MAPE–MMT with 3% and 7% by weight Cloisite 20A nanocomposites were examined with a cone calorimeter at the incident heat flux of 35 kW/cm$^2$ for the samples having a standard surface area of 70 × 70 mm$^2$ and identical masses of 12.0 ± 0.2 g. Figures 15.9–15.11 depict the plots of the basic ignitability characteristics: heat release rate, mass loss rate, and specific heat of combustion, versus time for PE, as well as for the PE–MAPE-7 wt% MMT and PE–MAPE-3 wt% MMT nanocomposites.

From Figure 15.9, it is seen that the maximum heat release rate for pristine PE is 2005 kW/m$^2$, whereas that for the PE–MAPE–MMT-3% wt. nanocomposite and the PE–MAPE–MMT-7% wt. nanocomposite is 789 and 728 kW/m$^2$, respectively; thus, the peak heat release rate decreases by 40%. A similar trend is observed in Figure 15.10 that illustrates the dependence of the mass low rate on the combustion time; Figure 15.11 shows

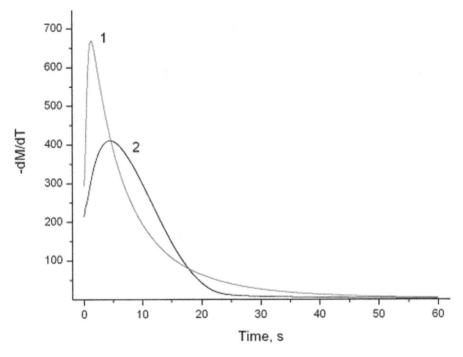

**Figure 15.8** Mass loss rates versus time for PE (1) and PE–MAPE–MMT (2) under the isothermal heating condition at 600°C.

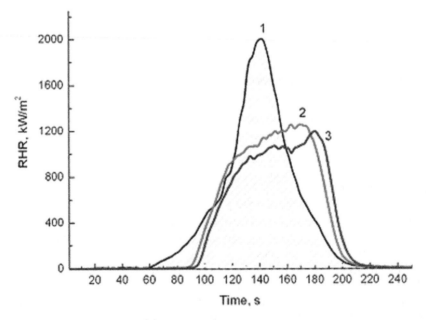

**Figure 15.9** Rate of heat release versus time for PE (1), PE–MAPE–MMT-3% wt. (2), and PE–MAPE–MMT-7% wt. (3) obtained by cone calorimeter at the incident heat flux of 35 kW/cm$^2$.

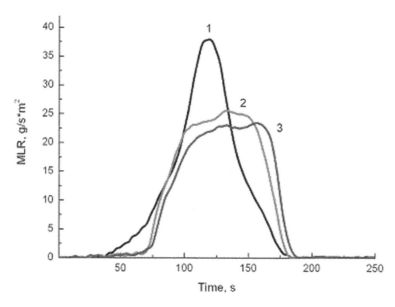

**Figure 15.10**    Mass loss rate versus time for PE (1), PE–MAPE–MMT-3% wt. (2), and PE–MAPE–MMT-7% wt. (3) obtained by cone calorimeter at the incident heat flux of 35 kW/cm$^2$.

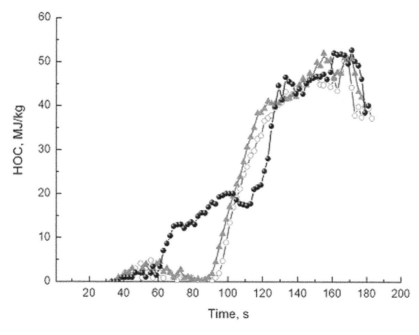

**Figure 15.11**    Heat of combustion versus time for PE ($\bullet$), PE–MAPE–MMT-3% wt. ($\circ$), and PE–MAPE–MMT-7% wt. ($\blacktriangle$) obtained by cone calorimeter at the incident heat flux of 35 kW/cm$^2$.

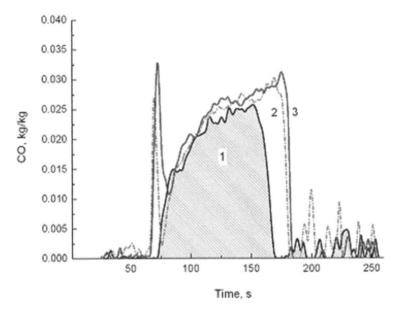

**Figure 15.12**   CO yield versus time for PE (1), PE–MAPE–MMT-3% wt. (2), and PE–MAPE–MMT-7% wt. (3) obtained by cone calorimeter at the incident heat flux of 35 kW/cm$^2$.

the time dependence for the specific heat of combustion, which is practically identical for all of the three samples. This resemblance confirms that the silicate agent does not inhibit the gas-phase combustion process and, thus, does not affect the heat of combustion of the samples; the effect of twofold reduction in the maximum rate of heat release may be explained in terms of formation of a protective char layer on the burning polymer surface. The value characterizing the average amount of released carbon monoxide remains practically unchanged over the entire set of test samples; however, a small increase in the maximum CO yield for nanocomposite samples at the end of combustion time intervals indicates the crossover of active combustion to the oxygen-deficient smoldering phase (Figure 15.12).

It is noteworthy that despite effective charring, the maximum level of smoke formation during the combustion of PE nanocomposites does not exceed the level of pristine PE, and its total yield is practically the same in all cases (Figure 15.12).

The obtained results validate the conclusion that char formation plays a key role in the mechanism of flame retardation for nanocomposites. The sample surface coated with a composition of silicate particles and the heat-resistant organic ingredient of char is a very effective barrier on the way of flame propagation over the surface. The ideal structure of the protective layer containing silicate particles and organic char must be a densely crosslinked network structure and must possess a considerable mechanical strength sufficient for the protective layer to remain intact during the burning up of the polymer from the surface.

## 15.4.  Conclusion

Dynamic TGA experiments show substantial increase in thermal-oxidative stability of the PE–MAPE–MMT nanocomposite as compared with pristine PE.

Collective action of chemical crosslinking and catalytic dehydration promoted by MMT presents a necessary and sufficient condition of solid phase carbonization reactions, which is observed in the process of thermal-oxidative degradation and combustion of PE–MAPE–MMT nanocomposites. Carbonized layer formation leads to appreciable increase of thermal stability of PE–MAPE–MMT nanocomposite, owing to a hindrance of the mass transfer in the nanocomposite. This fact explains an increase of thermal-oxidative stability in the PE–MAPE–MMT nanocomposite as well as its fire-proof properties.

Design data of the isothermal-oxidative degradation at 600°C provided by the model kinetic analysis reveal substantial reduction of the mass loss rate of PE–MAPE–MMT nanocomposites as compared with pristine PE. These results are in close agreement with the experimental cone calorimeter tests performed at an incident heat flux of 35 kW/cm$^2$.

# References

1. Lacey DJ; Dudler VV Polym Degrad Stab 1996, 51, 1011.
2. Paabo M; Levin BC Fire Mater 1987, 11, 55.
3. Lattimer RP J Anal Appl Pyrolysis 1995, 31, 203–226.
4. Kuroki T; Sawaguchi T; Niikuni S; Ikemura T Macromolecules 1982, 15, 1460–1462.
5. Kiran E; Gillham JK J Anal Appl Pyrolysis 1976, 20, 2045–2068.
6. Blazso MJ Anal Appl Pyrolysis 1993, 25, 25–35.
7. Hornung U; Hornung A; Bockhorn H Chem Ing Tech 1998, 70, 145–148.
8. Hornung U; Hornung A; Bockhorn H Chem Eng Tech 1998, 21, 332–337.
9. Bockhorn H; Hornung A; Horung UJ Anal Appl Pyrolysis 1998, 46, 1–13.
10. Bockhorn HA; Hornung A; Hornung U; Schawaller DJ Anal Appl Pyrolysis 1999, 48, 2, 93.
11. Grassie N; Gerald S Polymer Degradation and Stabilization; Cambridge University Press: Cambridge, New York, Melbourne, Sydney, 1988, 222.
12. Breen C; Last PM; Taylor S; Komadel P Thermochim Acta 363 (2000) 93–04.
13. Zanetti M; Lomakin SM; Camino G Macromol Mater Eng 2000, 279, 1–9.
14. Alexandre M; Dubois P Mater Sci Eng R 2000, 28, 1–63.
15. Giannelis EP Adv Mater 1996, 8, 29–35.
16. Gilman JW; Kashiwagi T; Nyden M; Brown JT; Jackson CL; Lomakin SM; Gianellis EP; Manias E In: Chemistry and Technology of Polymer Additives; Al-Maliaka S, Golovoy A, Wilkie CA, Eds.; Blackwell Scientific: London, 1998; 249–265.
17. Zanetti M; Bracco P; Costa L; Polym Degrad Stab 85 (2004) 657–665.
18. Ray SS; Okamoto M Prog Polym Sci 28 (2003) 1539–1641.
19. Kojima Y; Usuki A; Kawasumi M; Okada A; Fukushima Y; Kurauchi T; Kamigaito O J Mater Res. 8 (1993) 1185.
20. Breen C; Last PM; Taylor S; Komadel P Thermochim Acta 363 (2000) 93–04.
21. Gao Z; Amasaki I; Kaneko T; Nakada M Polym Degrad Stab 81(2003) 125–130.
22. Jin Woo Park J; Oh S; Lee H; Kim H; Yoo K, Polym Degrad Stab 67 (2000) 535–540.
23. Opfermann JJ Thermal Anal Cal, 60 (2000) 641.
24. Marquardt D J Appl Math 11 (1963) 431.
25. Babrauskas V Fire Mater 19 (1995) 243.
26. Lacoste L; Carlsson DJ J Polym Sci Part A Polym Chem 1992, 30, 493–500.
27. Gugumus F Polym Degrad Stab 2002, 76, 2, 329.
28. Gugumus F Polym Degrad Stab 2002, 77, 1, 147.
29. Benson SW Thermochemical Kinetics; Wiley, New York, 1976, 114.
30. Lomakin SM; Dubnikova IL; Berezina SM; Zaikov GE Polym Int 54 (2005) 7, 999–1006.
31. Friedman HL J Polym Sci C 6 (1965), 1, 175.

# 16 Image Analysis of Pore Size Distribution in Electrospun Nanofiber Webs: New Trends and Developments

*M. Ziabari, V. Mottaghitalab, and A. K. Haghi*[*]

The University of Guilan, P.O. Box 3756, Rasht, Iran

**Abstract**

Nanofibers produced by electrospinning method are widely used for drug delivery as tissue-scaffolding materials and filtration purposes where specific pore characteristics are required. For continued growth in these areas, it is critical that the nanofibers be properly designed for these applications to prevent failure. Most of the current methods only provide an indirect way of determining pore structure parameters and contain inherent disadvantages. In this study, we developed a novel image analysis method for measuring pore characteristics of electrospun nanofiber webs. Five electrospun webs with different pore characteristics were analyzed by this method. The method is direct and so fast and presents valuable and comprehensive information regarding the pore structure parameters of the webs. Two sets of simulated images were generated to study the effects of web density, fiber diameter, and its variations on pore characteristics. The results indicated that web density and fiber diameter significantly influence the pore characteristics, whereas the effect of fiber diameter variations was insignificant.

## 16.1. Introduction

Fibers with a diameter of around 100 nm are generally classified as *nanofibers*. What makes nanofibers of great interest is their extremely small size. Nanofibers compared with conventional fibers, with higher surface area to volume ratios and smaller pore size, offer an opportunity for use in a wide variety of applications. To date, the most successful method of producing nanofibers is through the process of *electrospinning*. The electrospinning process uses high voltage to create an electric field between a droplet of polymer solution at

---

[*]Correspondence should be addressed to e-mail: Haghi@Guilan.ac.ir

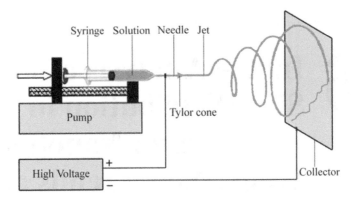

**Figure 16.1** Electrospinning setup.

the tip of a needle and a collector plate. When the electrostatic force overcomes the surface tension of the drop, a charged, continuous jet of polymer solution is ejected. As the solution moves away from the needle and toward the collector, the solvent evaporates and jet rapidly thins and dries. On the surface of the collector, a nonwoven web of randomly oriented solid nanofibers is deposited [1–5]. Figure 16.1 illustrates the electrospinning setup.

Material properties such as melting temperature and glass transition temperature as well as structural characteristics such as fiber diameter distribution, pore size distribution, and fiber orientation distribution determine the physical and mechanical properties of the nanofiber webs. The surface of electrospun fibers is important when considering end-use applications. For example, the ability to introduce porous surface features of a known size is required if nanoparticles need to be deposited on the surface of the fiber and if drug molecules are to be incorporated for controlled release, as tissue-scaffolding materials and for acting as a cradle for enzymes [6]. Besides, filtration performance of nanofibers is strongly related to their pore structure parameters, that is, percent open area (POA) and pore-opening size distribution (PSD). Hence, the control of the pore of electrospun webs is of prime importance for the nanofibers that are being produced for these purposes. There is no literature available about the pore size and its distribution of electrospun fibers and in this work, the pore size and its distribution was measured using an image analysis technique.

Current methods for determining PSD are mostly indirect and contain inherent disadvantages. Recent technological advancements in image analysis offer great potential for a more accurate and direct way of determining the PSD of electrospun webs. Overall, the image analysis method provides a unique and accurate method that can measure pore-opening sizes in electrospun nanofiber webs.

## 16.2. Methodology

The porosity, $\varepsilon_V$, is defined as the percentage of the volume of the voids, $V_v$, to the total volume (voids plus constituent material), $V_t$, and is given by

$$\varepsilon_V = \frac{V_v}{V_t} \times 100.$$

Similarly, the POA, $\varepsilon_A$, that is defined as the percentage of the open area, $A_o$, to the total area $A_t$, is given by

$$\varepsilon_A = \frac{A_o}{A_t} \times 100.$$

Usually, porosity is determined for materials with a three-dimensional structure, for example, relatively thick nonwoven fabrics. Nevertheless, for two-dimensional textiles such as woven fabrics and relatively thin nonwovens, it is often assumed that porosity and POA are equal [7].

The size of an individual opening can be defined as the surface area of the opening, although it is mostly indicated with a diameter called equivalent opening size (EOS). EOS is not a single value, and may differ for each opening. The common used term in this case is the diameter, $O_i$, corresponding with the equivalent circular area, $A_i$, of the opening

$$O_i = \left(\frac{4A_i}{\pi}\right)^{1/2}.$$

This diameter is greater than the side dimension of a square opening. A spherical particle with that diameter will never pass the opening (Figure 16.2(a)) and may therefore not be considered as an equivalent dimension or equivalent diameter. This will only be possible if the diameter corresponds with the side of the square area (Figure 16.2(b)). However, although not all openings are squares, yet the equivalent square area of openings is used to determine their equivalent dimension because this simplified assumption results in one single opening size from the open area. It is the diameter of a spherical particle that can pass the equivalent square opening; hence, the equivalent opening or pore size, $O_i$, results from

$$O_i = (A_i)^{1/2}.$$

From the EOSs, pore size distribution (PSD) and an equivalent diameter for which a certain percentage of the opening have a smaller diameter ($O_x$, pore-opening size that is $x$ percent of pores is smaller than that size) may be measured.

a                                         b

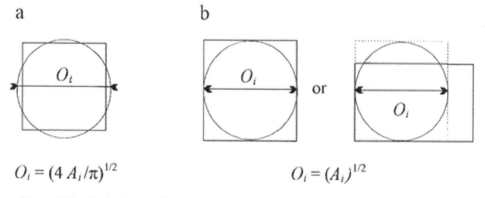

**Figure 16.2**  Equivalent opening size, $O_i$, based on (a) equivalent area and (b) equivalent size.

The PSD curves can be used to determine the uniformity coefficient, $C_u$, of the investigated materials. The uniformity coefficient is a measure for the uniformity of the openings and is given by

$$C_u = \frac{O_{60}}{O_{10}}.$$

The ratio equals 1 for uniform openings and increases with decreasing uniformity of the openings [7].

Pore characteristic is one of the main tools for evaluating the performance of any nonwoven fabric and electrospun webs as well. Understanding the link between processing parameters and pore structure parameters will allow for better control over the properties of electrospun fibers. Therefore, there is a need for the design of nanofibers to meet specific application needs. Various techniques may be used to evaluate pore characteristics of porous materials including sieving techniques (dry, wet, and hydrodynamic sieving), mercury porosimetry, and flow porosimetry (bubble point method) [8, 9]. As one goes about selecting a suitable technique for characterization, the associated virtues and pitfalls of each technique should be examined. The most attractive option is a single technique that is nondestructive, yet capable of providing a comprehensive set of data [10].

### 16.2.1. Sieving Methods

In dry sieving, glass bead fractions (from finer to coarser) are sieved through the porous material. In theory, most of the glass beads from the first glass bead fraction should pass. As larger and larger glass bead fractions are sieved, more and more glass beads should become trapped within and on top of the material. The number of pores of a certain size should be reflected by the percentage of glass beads passing through the porous material during each glass bead fraction sieved; however, electrostatic effects between glass beads and between glass beads and the material can affect the results. Glass beads may stick to fibers making the pores effectively smaller and they may also agglomerate to form one large glass bead that is too large to pass through the any of the pores. Glass beads may also break from hitting each other and the sides of the container, resulting in smaller particles that can pass through smaller openings.

In hydrodynamic sieving, a glass bead mixture is sieved through a porous material under alternating water flow conditions. The use of glass bead mixtures leads to results that reflect the original glass bead mixture used. Therefore, this method is only useful for evaluating the large pore openings such as $O_{95}$. Another problem occurs when particles of many sizes interact that likely results in particle blocking and bridge formation. This is especially a problem in hydrodynamic sieving because the larger glass bead particles will settle first when water is drained during the test. When this occurs, fine glass beads that are smaller than the pores are prevented from passing through by the coarser particles.

In wet sieving, a glass bead mixture is sieved through a porous material aided by a water spray. The same basic mechanisms that occur when using the hydrodynamic sieving method also take place when using the wet sieving method. Bridge formation is not as pronounced in the wet sieving method as in the hydrodynamic sieving method; however, particle blocking and glass bead agglomeration are more pronounced [8, 9].

The sieving tests are very time-consuming. Generally, 2 h are required to perform a test. The sieving tests are far from providing a complete PSD curve because the accuracy of the tests for pore sizes smaller than 90 μm is questionable [12].

## 16.2.2. Mercury Porosimetry

Mercury porosimetry is a well-known method that is often used to study porous materials. This technique is based on the fact that mercury as a nonwetting liquid does not intrude into pore spaces except under applying sufficient pressure. Therefore, a relationship can be found between the size of pores and the pressure applied.

In this method, a porous material is completely surrounded by mercury, and pressure is applied to force the mercury into pores. As mercury pressure increases, the large pores are filled with mercury first. Pore sizes are calculated as the mercury pressure increases. At higher pressures, mercury intrudes into the fine pores and when the pressure reaches a maximum, total open pore volume and porosity are calculated.

The mercury porosimetry thus gives a PSD based on total pore volume and gives no information regarding the number of pores of a porous material. Pore sizes ranging from 0.0018 to 400 μm can be studied using mercury porosimetry. Pore sizes smaller than 0.0018 μm are not intruded with mercury, and this is a source of error for porosity and PSD calculations. Furthermore, mercury porosimetry does not account for closed pores as mercury does not intrude into them. Because of applying high pressures, sample collapse and compression is possible; hence, it is not suitable for fragile compressible materials such as nanofiber sheets. Other concerns would include the fact that it is assumed that the pores are cylindrical, which is not the case in reality. After the mercury intrusion test, sample decontamination at specialized facilities is required as the highly toxic mercury is trapped within the pores. Therefore, this dangerous and destructive test can only be performed in well-equipped labs [6, 8, 9].

## 16.2.3. Flow Porosimetry (Bubble Point Method)

The flow porosimetry is based on the principle that a porous material will only allow a fluid to pass when the pressure applied exceeds the capillary attraction of the fluid in largest pore. In this test, the specimen is saturated with a liquid and continuous air flow is used to remove liquid from the pores. At a critical pressure, the first bubble will come through the largest pore in the wetted specimen. As the pressure increases, the pores are emptied of liquid in order from largest to smallest and the flow rate is measured. PSD, number of pores, and porosity can be derived once the flow rate and the applied pressure are known. Flow porosimetry is capable of measuring pore sizes within the range of 0.013–500 μm.

As the air only passes through the pores, characteristics of these pores are measured, whereas those of closed and blind pores are omitted. Many times, 100% total flow is not reached. This is due to porewick evaporation from the pores when the flow rate is too high. Extreme care is required to ensure that the air flow does not disrupt the pore structure of the specimen. The flow porosimetry method is also based on the assumption that the pores are cylindrical, which is not the case in reality. Finding a liquid with low surface tension that could cover all the pores has no interaction with the material, and causing swelling in material is not easy all the times and sometimes is impossible [6, 8, 9].

## 16.2.4. Image Analysis

Because of its convenience to detect individual pores in a nonwoven image, it seemed to be advantageous to use image analysis techniques for pore measurement. Image analysis was used to measure pore characteristics of woven [11] and nonwoven geotextiles [12]. In the former, successive *erosion* operations with increasing size of *structuring element* were used to count the pore openings larger than a given structuring element. The main purpose of the erosion was to simulate the conditions in the sieving methods. In this method, the voids connected to border of the image that are not complete pores are considered in measurement. Performing opening and then closing operations preceding pore measurement cause the pore sizes and shapes to deviate from the real ones. The method is suitable for measuring pore sizes of woven geotextiles with fairly uniform pore sizes and shapes and is not appropriate for electrospun nanofiber webs of different pore sizes.

In the later case, cross-sectional image of nonwoven geotextile was used to calculate the pore structure parameters. A *slicing* algorithm based on a series of morphological operations for determining the mean fiber thickness and the optimal position of the uniform slicing grid was developed. After recognition of the fibers and pores in the slice, the pore-opening size distribution of the cross-sectional image may be determined. The method is useful for measuring pore characteristics of relatively thick nonwovens and cannot be applied to electrospun nanofiber webs due to extremely small size.

Therefore, there is a need for developing an algorithm suitable for measuring the pore structure parameters in electrospun webs. In response to this need, we have developed a new image analysis-based method and presented in the following.

In this method, a binary image of the web is used as an input. First of all, voids connected to the image border are identified and cleared using *morphological reconstruction* [13], in which mask image is the input image and marker image is zero everywhere except along the border. Total area that is the number of pixels in the image is measured. Then the pores are labeled and each is considered as an object. Here the number of pores may be obtained. In the next step, the number of pixels of each object as the area of that object is measured. Having the area of pores, the porosity and EOS corresponding to each pore may be calculated. The data in pixels may then be converted in nanometers. Finally, PSD curve is plotted and $O_{50}$, $O_{95}$, and $C_u$ are determined.

### 16.2.4.1. *Real Webs*

In order to measure pore characteristics of electrospun nanofibers using image analysis, images of the webs are required. These images called micrographs usually are obtained by scanning electron microscope (SEM), transmission electron microscope (TEM), or atomic force microscope (AFM). The images must be of high quality and taken under appropriate magnifications.

The image analysis method for measuring pore characteristics requires the initial segmentation of the micrographs in order to produce binary images. This is a critical step because the segmentation affects the results dramatically. The typical way of producing a binary image from a gray scale image is by *global thresholding* [13], in which a single constant threshold is applied to segment the image. All pixels up to and equal to the threshold belong to object and the remaining belong to the background. One simple way to choose the threshold is picking different thresholds until one is found that produces a good result as judged by the observer. Global thresholding is very sensitive to any inhomogeneities in the gray-level distributions of object and background pixels. In order

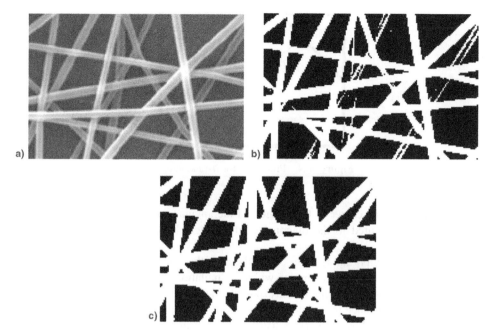

**Figure 16.3**    (a) A real web, (b) global thresholding, and (c) local thresholding.

to eliminate the effect of inhomogeneities, *local thresholding* scheme [13] could be used. In this approach, the image is divided into subimages for which the inhomogeneities are negligible. Then optimal thresholds are found for each subimage. A common practice in this case is to preprocess the image to compensate for the illumination problems and then apply a global thresholding to the preprocessed image. It can be shown that this process is equivalent to segment the image with locally varying thresholds. In order to automatically select the appropriate thresholds, *Otsu's method* [14] is employed. This method chooses the threshold to minimize interaclass variance of the black and white pixels. As it is shown in Figure 16.3, global thresholding resulted in some broken fiber segments. This problem was solved using local thresholding. Note that, since the process is extremely sensitive to noise contained in the image, before the segmentation, a procedure to clean the noise and enhance the contrast of the image is necessary.

### 16.2.4.2. *Simulated Webs*

It is known that the pore characteristics of nonwoven webs are influenced by web properties and so are those of electrospun webs. There are no reliable models available for predicting these characteristics as a function of web properties [15]. In order to explore the effects of some parameters on pore characteristics of electrospun nanofibers, simulated webs are generated. These webs are images simulated by straight lines. There are three widely used methods for generating random network of lines. These are called S-randomness, $\mu$-randomness (suitable for generating a web of continuous filaments), and I-randomness (suitable for generating a web of staple fibers). These methods have been described in details by Abdel-Ghani and Davis [16] and Pourdeyhimi et al. [17]. In this study, we used $\mu$-*randomness* procedure for generating simulated images. Under this scheme, a line with a specified thickness is defined by the perpendicular distance $d$ from a fixed reference

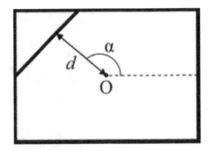

**Figure 16.4**   Procedure for $\mu$-randomness.

point $O$ located in the center of the image and the angular position of the perpendicular $\alpha$. Distance $d$ is limited to the diagonal of the image. Figure 16.4 demonstrates this procedure.

One of the most important features of simulation is that it allows several structural characteristics to be taken into consideration with the simulation parameters. These parameters are web density (controlled as line density), angular density (sampled from a normal or random distribution), distance from the reference point (sampled from a random distribution), line thickness (sampled from a normal distribution), and image size.

## 16.3. Experimental

Nanofiber webs were obtained from electrospinning of PVA with average molecular weight of 72,000 g/mol (MERCK) at different processing parameters for attaining different pore characteristics. Table 16.1 summarizes the electrospinning parameters used for preparing the webs. The micrographs of the webs were obtained using Philips (XL-30) environmental SEM, under magnification of 10,000× after being gold coated. Figure 16.5 shows the micrographs of the electrospun webs.

## 16.4. Results and Discussion

Because of the previously mentioned reasons, sieving methods and mercury porosimetry are not applicable for measuring pore structure parameters in nanoscale. The only method that seems to be practical is flow porosimetry. However, since in this study the nanofibers

**Table 16.1**   Electrospinning parameters used for preparing nanofiber webs.

| No. | Concentration (%) | Spinning distance (cm) | Voltage (kV) | Flow rate (ml/h) |
|-----|-------------------|------------------------|--------------|------------------|
| 1 | 8 | 15 | 20 | 0.4 |
| 2 | 12 | 20 | 15 | 0.2 |
| 3 | 8 | 15 | 20 | 0.2 |
| 4 | 8 | 10 | 15 | 0.3 |
| 5 | 10 | 10 | 15 | 0.2 |

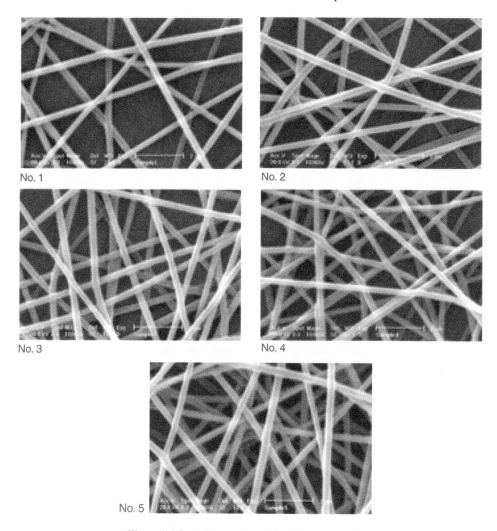

No. 1

No. 2

No. 3

No. 4

No. 5

**Figure 16.5**    Micrographs of the electrospun webs.

were made of PVA, finding an appropriate liquid for the test to be performed is almost impossible because of the solubility of PVA in both organic and inorganic liquids.

As an alternative, image analysis was employed to measure pore structure parameters in electrospun nanofiber webs. PSD curves of the webs, determined using the image analysis method, are shown in Figure 16.6. Pore characteristics of the webs ($O_{50}$, $O_{95}$, $C_u$, number of pores, and porosity) measured by this method are presented in Table 16.2. It is seen that by decreasing the porosity, $O_{50}$ and $O_{95}$ decrease. $C_u$ also decreases with respect to porosity, that is, to say, increasing the uniformity of the pores. Number of pores has an increasing trend with decreasing porosity.

The image analysis method presents valuable and comprehensive information regarding to pore structure parameters in nanofiber webs. This information may be exploited in preparing the webs with needed pore characteristics to use in filtration, biomedical

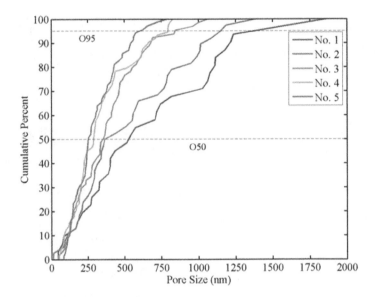

**Figure 16.6**    PSD curves of electrospun webs.

applications, nanoparticle deposition, and other purposes. The advantages of the method
are listed below:

1. The method is capable of measuring pore structure parameters in any nanofiber webs
   with any pore features and it is applicable even when other methods may not be
   employed.
2. It is so fast. It takes less than a second for an image to be analyzed (with a 3 GHz
   processor).
3. The method is direct and simple. Pore characteristics are measured from the area of the
   pores that is defined as the number of pixels of the pores.
4. There is no systematic error in measurement (such as assuming pores to be cylindrical
   in mercury and flow porosimetry and the errors associated with the sieving methods
   which were mentioned). Once the segmentation is successful, the pore sizes will be
   measured accurately. The quality of images affects the segmentation procedure. High-
   quality images reduce the possibility of poor segmentation and enhance the accuracy of
   the results.
5. It gives a complete PSD curve.
6. There is no cost involved in the method and minimal technical equipments are needed
   (SEM for obtaining the micrographs of the samples and a computer for analysis).
7. It has the capability of being used as an on-line quality control technique for large-scale
   production.
8. The results obtained by image analysis are reproducible.
9. It is not a destructive method. A very small amount of sample is required for measurement.

In an attempt to establish the effects of some structural properties on pore characteristics
of electrospun nanofibers, two sets of simulated images with varying properties were
generated. The simulated images reveal the degree to which fiber diameter and density

**Table 16.2**   Pore characteristics of electrospun webs.

| No. | $O_{50}$ Pixel | $O_{50}$ nm | $O_{95}$ Pixel | $O_{95}$ nm | $C_u$ | Pore no. | Porosity |
|-----|-------|------|-------|--------|------|----------|----------|
| 1 | 39.28 | 513.9 | 94.56 | 1237.1 | 8.43 | 31 | 48.64 |
| 2 | 27.87 | 364.7 | 87.66 | 1146.8 | 5.92 | 38 | 34.57 |
| 3 | 26.94 | 352.5 | 64.01 | 837.4 | 3.73 | 64 | 26.71 |
| 4 | 22.09 | 289.0 | 60.75 | 794.8 | 3.68 | 73 | 24.45 |
| 5 | 19.26 | 252.0 | 44.03 | 576.1 | 2.73 | 69 | 15.74 |

**Table 16.3**   Structural characteristics of first set images.

| No. | Angular range | Line density | Line thickness |
|-----|---------------|--------------|----------------|
| 1 | 0–360 | 20 | 5 |
| 2 | 0–360 | 30 | 5 |
| 3 | 0–360 | 40 | 5 |
| 4 | 0–360 | 20 | 10 |
| 5 | 0–360 | 30 | 10 |
| 6 | 0–360 | 40 | 10 |
| 7 | 0–360 | 20 | 20 |
| 8 | 0–360 | 30 | 20 |
| 9 | 0–360 | 40 | 20 |

**Table 16.4**   Structural characteristics of second set images.

| No. | Angular range | Line density | Line thickness Mean | Std |
|-----|---------------|--------------|------|-----|
| 1 | 0–360 | 30 | 15 | 0 |
| 2 | 0–360 | 30 | 15 | 4 |
| 3 | 0–360 | 30 | 15 | 8 |
| 4 | 0–360 | 30 | 15 | 10 |

affect the pore structure parameters. The first set contained images with the same density varying in fiber diameter and images with the same fiber diameter varying in density. Each image had a constant diameter. The second set contained images with the same density and mean fiber diameter while the standard deviation of fiber diameter varied. The details are given in Tables 16.3 and 16.4, and typical images are shown in Figures 16.7 and 16.8.

Pore structure parameters of the simulated webs were measured using image analysis method. Table 16.5 summarizes the pore characteristics of the simulated images in the first set. For the webs with the same density, increasing fiber diameter resulted in a decrease in $O_{95}$, number of pores, and porosity. No particular trends were observed for $O_{50}$ and $C_u$. Figures 16.9 and 16.10 show the PSD curves of the simulated images in the first set. As the web density increases, the effects of fiber diameter are less pronounced since the PSD curves of the webs become closer to each other. For the webs with the same fiber diameter, increasing the density resulted in a decrease in $O_{50}$, $O_{95}$, $C_u$, and porosity whereas number of pores increased with the density.

Table 16.6 summarizes the pore characteristics of the simulated images in the second set. No significant effects for variation of fiber diameter on pore characteristics were

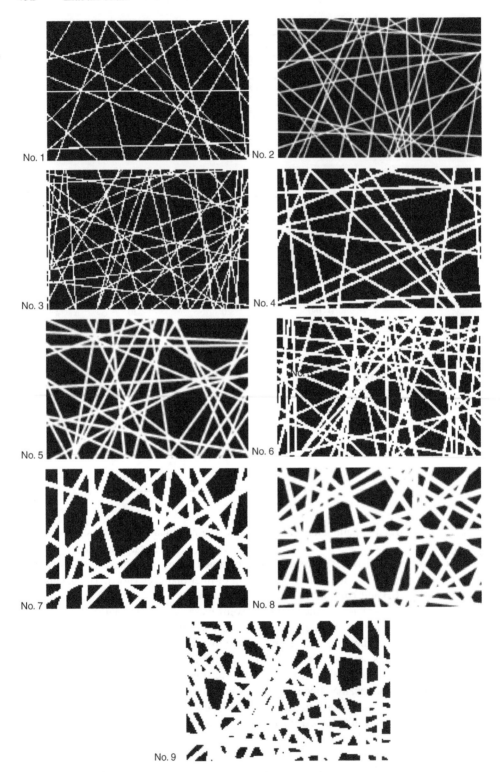

**Figure 16.7**  Simulated images of the first set.

No. 1

No. 2

No. 3

No. 4

**Figure 16.8**   Simulated images of the second set.

**Table 16.5**   Pore characteristics of the first set of simulated images.

| No. | $O_{50}$ | $O_{95}$ | $C_u$ | Pore no. | Porosity |
|---|---|---|---|---|---|
| 1 | 27.18 | 100.13 | 38.38 | 84 | 79.91 |
| 2 | 15.52 | 67.31 | 22.20 | 182 | 71.78 |
| 3 | 13.78 | 52.32 | 18.71 | 308 | 69.89 |
| 4 | 36.65 | 94.31 | 43.71 | 67 | 66.10 |
| 5 | 17.89 | 61.64 | 22.67 | 144 | 53.67 |
| 6 | 12.41 | 51.60 | 16.70 | 245 | 47.87 |
| 7 | 24.49 | 86.90 | 33.11 | 58 | 41.05 |
| 8 | 16.31 | 56.07 | 21.66 | 108 | 32.53 |
| 9 | 13.11 | 45.38 | 17.75 | 126 | 22.01 |

**Table 16.6**   Pore characteristics of the second set of simulated images.

| No. | $O_{50}$ | $O_{95}$ | $C_u$ | Pore no. | Porosity |
|---|---|---|---|---|---|
| 1 | 14.18 | 53.56 | 18.79 | 133 | 35.73 |
| 2 | 13.38 | 61.66 | 20.15 | 136 | 41.89 |
| 3 | 18.14 | 59.35 | 22.07 | 121 | 41.03 |
| 4 | 15.59 | 62.71 | 20.20 | 112 | 37.77 |

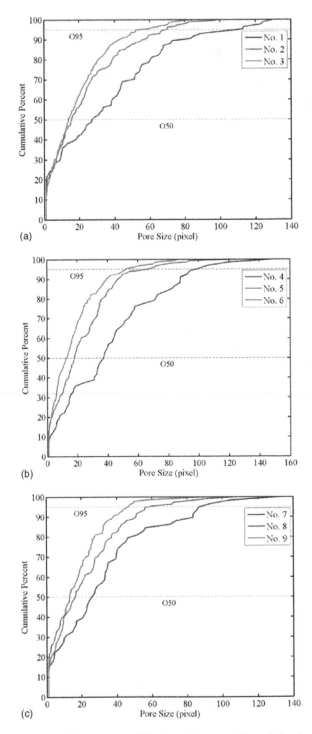

**Figure 16.9**  PSD curves of the first set of simulated images; effect of density, images with the diameter of (a) 5, (b) 10, and (c) 20 pixels.

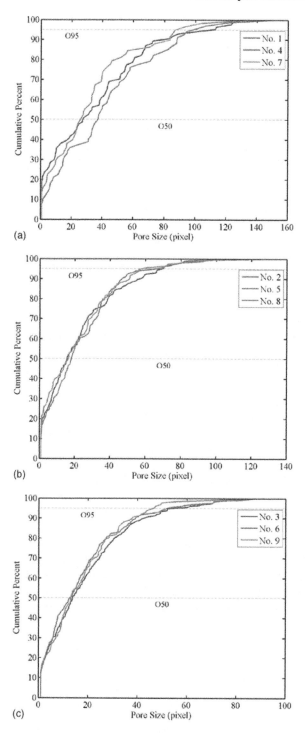

**Figure 16.10**  PSD curves of the first set of simulated images; effect of fiber diameter, images with the density of (a) 20, (b) 30, and (c) 40 lines.

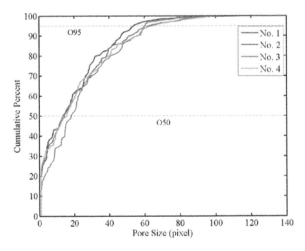

**Figure 16.11**    PSD curves of the second set of simulated images; the effect of fiber diameter variation.

observed, suggesting that average fiber diameter is determining factor not variation of diameter. Figure 16.11 shows the PSD curves of the simulated images in the second set.

## 16.5. Conclusion

The evaluation of electrospun nanofiber pore structure parameters is necessary as it facilitates the improvement of the design process and its eventual applications. Various techniques have been developed to assess pore characteristics in porous materials. However, most of these methods are indirect, have inherent problems, and are not applicable for measuring pore structure parameters of electrospun webs. In this investigation, we have successfully developed an image analysis-based method as a response to this need. The method is simple, comprehensive, and very fast and directly measures the pore structure parameters.

The effects of web density, fiber diameter, and its variation on pore characteristics of the webs were also explored using some simulated images. As fiber diameter increased, $O_{95}$, number of pores, and porosity decreased. No particular trends were observed for $O_{50}$ and $C_u$. Increasing the density resulted in a decrease in $O_{50}$, $O_{95}$, $C_u$, and porosity, whereas number of pores increased with the density. The effects of variation of fiber diameter on pore characteristics were insignificant.

## References

1. A. K. Haghi, M. Akbari, Trends in electrospinning of natural nanofibers, Physica Status Solidi (a), 204, 1830–1834 (2007).
2. D. H. Reneker, I. Chun, Nanometre diameter fibers of polymer, Produced by Electrospinning, Nonotechnology, 7, 216–223 (1996).
3. D. R. Salem, Structure formation in polymeric fibers, in Chapter 6, H. Fong, , D. H. Reneker (Eds.), Electrospinning and the Formation of Nanofibers, Hanser Publications, Cincinnati, OH, USA, 2001.

4. Th. Subbiah, G. S. Bhat, R. W. Tock, S. Parameswaran, S. S. Ramkumar, Electrospinning of nanofibers, Journal of Applied Polymer Science, 96, 557–569 (2005).
5. A. Frenot, I. S. Chronakis, Polymer nanofibers assembled by electrsopinning, Current Opinion in Colloid and Interface Science, 8, 64–75 (2003).
6. Ch. L. Casper, J. S. Stephens, N. G. Tassi, D. B. Chase, J. F. Rabolt, Controlling surface morphology of electrospun polystyrene fibers: effect of humidity and molecular weight in the electrospinning process, Macromolecules, 37, 573–578 (2004).
7. W. Dierickx, Opening size determination of technical textiles used in agricultural applications, Geotextiles and Geomembranes, 17 (4), 231–245 (1999).
8. S. K. Bhatia, J. L. Smith, Geotextile Characterization and pore size distribution: Part II. A review of test methods and results, Geosynthetics International, 3 (2), 155–180 (1996).
9. S. K. Bhatia, J. L. Smith, B. R. Christopher, Geotextile characterization and pore size distribution: Part III. Comparison of methods and application to design, Geosynthetics International, 3 (3), 301–328 (1996).
10. S. T. Ho, D. W. Hutmacher, A comparison of micro CT with other techniques used in the characterization of scaffolds, Biomaterials, 27, 1362–1376 (2006).
11. A. H. Aydilek, T. B. Edil, Evaluation of woven geotextile pore structure parameters using image analysis, Geotechnical Testing Journal, 27 (1), 1–12 (2004).
12. A. H. Aydilek, S. H. Oguz, T. B. Edil, Digital image analysis to determine pore opening size distribution of nonwoven geotextiles, Journal of Computing in Civil Engineering, 280–290 (2002).
13. R. C. Gonzalez, R. E. Woods, Digital Image Processing, Prentice-Hall, New Jersey, Second Edition, 2001.
14. B. Jähne, Digital Image Processing, Springer, UK, 5th Revised and Extended Edition, 2002.
15. H. S. Kim, B. Pourdeyhimi, A note on the effect of fiber diameter, fiber crimp and fiber orientation on pore size in thin webs, International Nonwoven Journal, 15–19 (Winter 2000).
16. M. S. Abdel-Ghani, G. A. Davis, Simulation of nonwoven fiber mats and the application to coalescers, Chemical Engineering Science, 117 (1985).
17. B. Pourdeyhimi, R. Ramanathan, R. Dent, Measuring fiber orientation in nonwovens, part I: simulation, Textile Research Journal, 66 (11), 713–722 (1996).

Printed and bound by CPI Group (UK) Ltd, Croydon, CR0 4YY

23/10/2024

01778246-0008